Tephra Studies

NATO ADVANCED STUDY INSTITUTES SERIES

Proceedings of the Advanced Study Institute Programme, which aims at the dissemination of advanced knowledge and the formation of contacts among scientists from different countries

The series is published by an international board of publishers in conjunction with NATO Scientific Affairs Division

A	Life Sciences	Plenum Publishing Corporation
B	Physics	London and New York
C	Mathematical and Physical Sciences	D. Reidel Publishing Company Dordrecht, Boston and London
D	Behavioural and Social Sciences	Sijthoff & Noordhoff International Publishers
E	Applied Sciences	Alphen aan den Rijn and Germantown U.S.A.

Series C – Mathematical and Physical Sciences

Volume 75 – Tephra Studies

Tephra Studies

Proceedings of the NATO Advanced Study Institute "Tephra Studies as a Tool in Quaternary Research", held in Laugarvatn and Reykjavík, Iceland, June 18-29, 1980

edited by

S. SELF

Department of Geology, Arizona State University, Tempe, U.S.A.

and

R. S. J. SPARKS

Department of Earth Sciences, University of Cambridge, England

D. Reidel Publishing Company

Dordrecht : Holland / Boston : U.S.A. / London : England

Published in cooperation with NATO Scientific Affairs Division

Library of Congress Cataloging in Publication Data

CIP

NATO Advanced Study Institute 'Tephra Studies as a Tool in Quaternary Research' (1980: Laugarvatn, Iceland, and Reykjavík, Iceland).
Tephra studies.

(NATO Advanced Study Institutes Series. Series C, Mathematical and Physical Sciences ; v. 75)
Includes index.
1. Volcanic ash, tuff, etc.-Congresses. 2. Geology, stratigraphic-Quaternary-Congresses. I. Self, S. (Stephen), 1946- . II. Sparks, R. S. J. (Robert Stephen John), 1949- . III. Title. VI. Series.

QE461.N365 1980 552'.23 81-11970
AACR2

ISBN-13: 978-94-009-8539-1 e-ISBN-13: 978-94-009-8537-7
DOI: 10.1007/978-94-009-8537-7

Published by D. Reidel Publishing Company
P.O. Box 17, 3300 AA Dordrecht, Holland

Sold and distributed in the U.S.A. and Canada
by Kluwer Boston Inc.,
190 Old Derby Street, Hingham, MA 02043, U.S.A.

In all other countries, sold and distributed
by Kluwer Academic Publishers Group,
P.O. Box 322, 3300 AH Dordrecht, Holland

D. Reidel Publishing Company is a member of the Kluwer Group

CONTENTS

Sigurdur Thorarinsson, Iceland, June 1980

DEDICATION

Sigurdur Thorarinsson was born on January 8, 1912 in Vopnafjördur, northeastern Iceland, the son of a farmer. After completing high school in Iceland, he went to Copenhagen and then to Stockholm to pursue his university studies in geology, geography and botany. He took his doctorate degree at the University of Stockholm in 1944.

After returning to Iceland, Sigurdur Thorarinsson became Director of the Department of Geology and Geography, Museum of Natural History, Reykjavik in 1947, a position that he held until 1969. He was appointed Professor and Director of the Geographical Institute of Stockholm during the years 1950/51 and 1953. Since 1969, he has been Professor of Geology and Geography at the University of Iceland in Reykjavik.

His main fields of study have been glaciology, geomorphology, and tephrochronology, the branch of geological science that he has pioneered. He has witnessed and documented all eruptions that have taken place in Iceland since 1934. Some of his major publications are <u>Tefrokronologiska studier på Island</u> (1944); <u>Surtsey, The New Island in the North Atlantic</u> (1966); and <u>The Eruptions of Hekla in Historical Times</u> (1967). He has published about 200 papers in various scientific and popular periodicals, and has lectured in most western European countries, as well as on other continents.

Sigurdur Thorarinsson is Doctor *honoris causa* of the University of Iceland and a member of many academies and societies of science, including honorary memberships of the Geological Society of London, Geological Society of America and the International Glaciological Society. He has been awarded the Swedish Vega medal, the Danish Steno and Vitus Bering medals and the Spendiaroff prize of the XXI International Geological Congress. In 1939 Sigurdur Thorarinsson married Inga Backlund; they have two children.

This volume is the result of a proposal by members of the INQUA (International Union for Quaternary Research) Commission on Tephra, made at the 10th International Congress in Birmingham, England, 1977. The Commission resolved to hold a meeting dedicated to tephra studies to be located, if possible in Iceland. A grant from NATO made this a reality. The Advanced Study Institute was held in Laugarvatn, Iceland from June 18-29, 1980, and included 3 field trips in Southern Iceland.

Organizing Commission Members:

Sigurdur Thorarinsson,
Honorary President

R. Stephen J. Sparks,
President

Stephen Self,
Secretary

Co-ordinator in Iceland:

Gudrún Larsen

"TEPHRA STUDIES AS A TOOL IN QUATERNARY RESEARCH"

The Editors and organizers thank all those who participated and have submitted manuscripts. Our thanks are extended to Sue Selkirk (Arizona State University) for designing the emblem for the ASI and Patti Stephens for typing and collating a large part of this volume. The final preparation of the manuscript was greatly aided by a generous grant from the Deans of Liberal Arts and the Graduate College at Arizona State University.

TEPHRA is used herein as a collective term for all airborne pyroclasts, including both air-fall and pyroclastic flow material. This usage complements rather than replaces terms such as ignimbrite, welded tuff, pumice, etc., that are used to designate specific types of tephra produced by distinctive types of eruptions. S. Thorarinsson in The World Bibliography and Index of Quaternary Tephrochronology, Eds. J.A. Westgate and C.M. Gold, University of Alberta, 1974, 528 p.

PREFACE

The fundamental principles of tephrochronology were developed in Iceland through the classic research of Professor Sigurdur Thorarinsson in the 1940's and 1950's. By his studies on the volcanic ash layers (tephra) produced by the historic explosive eruptions of the volcano Hekla, he established that individual tephra layers could be correlated over large areas of Iceland. As the deposition of such layers was essentially instantaneous on a geological time-scale, the tephra layers provided a distinctive and widespread isochronous stratigraphic marker. Since Thorarinsson's pioneering work, the study of tephra has become a powerful and increasingly important research tool in many branches of the geological sciences and related disciplines.

In response to the recent rapid growth and diversification of tephra studies we decided to coordinate a NATO Advanced Studies Institute under the auspices of the INQUA Commission on Tephra. The ASI, entitled "Tephra Studies as a Tool in Quaternary Research," was held from June 18th to June 29th, 1980, at Laugarvatn in Iceland. This book represents the proceedings of that Institute and is the first book dedicated to tephra studies, its many uses and applications. The subject matter combines disciplines as diverse as Quaternary stratigraphy, isotope geochronology, petrology, deep-sea geology, volcanology, volcanic hazards mapping, archaeology and ecology in one volume, and illustrates both present uses and future potential of tephra research.

The book should appeal to those wishing to discover the latest developments in tephra studies and how tephra research might apply to their own problems. It contains examples of large regional studies such as in Japan and North America, where there now exists fairly comprehensive documentation of Quaternary tephrochronology. With the improvement in analytical techniques of recent years, many sophisticated methods for correlation and age dating of tephra layers have been developed and are illustrated. The basic principles underlying the systematic investigation of tephra layers are discussed in several chapters.

S. Self and R. S. J. Sparks (eds.), Tephra Studies, xiii–xiv.

Another important development in tephra research has been in the field of volcanology. Tephrochronology, combined with studies of the distribution, grain size and compositional variations in tephra layers, provides fundamental information on the frequencies, styles and magnitudes of volcanic eruptions from individual volcanic centers. Several contributions show the way such studies enable volcanologists to assess the past behavior of a volcano and allow general predictions for future behavior. The style of an eruption is directly reflected in the individual characteristics of a tephra layer. Igneous petrologists are also now showing increased interest in tephra studies because compositional variations in tephra layers provide clues to the evolution of magma chambers. The book provides examples of these developments in volcanology and petrology.

A great many people contributed to the success of the meeting. We would particularly like to acknowledge the efforts of Gudrún Larsen, Sigurdur Thorarinsson, Elsa Vilmundardóttir, Sheila Tuffnell, and Sarah Brazier for organizational and editorial assistance.

S. Self
Department of Geology
Arizona State University
Tempe
Arizona 85281
USA

R.S.J. Sparks
Department of Earth Sciences
University of Cambridge
Cambridge CB2 3EW
England

October 1980

OPENING ADDRESS

TEPHRA STUDIES AND TEPHROCHRONOLOGY: A HISTORICAL REVIEW WITH SPECIAL REFERENCE TO ICELAND

Sigurdur Thorarinsson

Science Department, University of Iceland,
Reykjavik, Iceland

In the fourteenth century Icelandic chroniclers already discerned between three types of tephra: pumice, sand and ash. From 1625 onwards, detailed accounts of the tephra fall in all major Icelandic eruptions were written. The earliest description of tephra layers in soil sections is from 1638 and somewhat more comprehensive studies of tephra layers in soil were carried out in the middle eighteenth century.

Studies of tephra layers as a tool for various research began in Iceland and other countries around 1930. In Iceland these studies led to a doctoral thesis: Tefrokronologiska studier på Island, published in 1944, where the terms tephra and tephrochronology were introduced. In this thesis its author promoted tephra as an important tool in volcanological, pollenanalytical, glaciological, geomorphological and archaeological research.

In the famous Eddic poem called Völuspá (The Sybil's Prophesy), an inspired vision of the beginning and end of the world, is a description of the ultimate fight of the gods against the giants coming from the east and the fire-demon Surtr, the black one, coming from the south. In this description we find this stanza:

The sun grows dark,
the earth sinks into the sea,
the clear, bright stars
disappear from the sky;
vapour pours out
and fire, life's nourisher,
the high flame
plays on heaven itself.

S. Self and R. S. J. Sparks (eds.), Tephra Studies, 1–12.

Many scholars believe now that Völuspá was composed in Iceland about the time the country was converted to Christianity in A.D. 1000. If this is really so, it is tempting to regard the above stanza as inspired by the sight of a volcanic eruption in Iceland, presumably an explosive one with tephra fall and lightning playing in the eruption column. I have ventured the hypothesis that the eruption which inspired the author was a Katla eruption, which was likely to have occurred about the year 1000 (1, pp. 21-23).

Be this as it may, the first direct mention in written sources of tephra in Iceland refers to a Hekla eruption in 1104 A.D. This eruption was the first of this famous volcano since the beginning of settlement in Iceland, about 870 A.D. The mention is found in an annal, Lögmannsannáll, written in the fourteenth century by a priest, Einar Haflidason, who for a time was administrator of the episcopal see at Hólar in North Iceland. Nothing is said there about the eruption except that it was "The first coming up of fire in Mount Hekla", but referring to the winter 1104/1105, the annal has the entry "Sandfall winter" (2, p. 111). The words quoted are certainly based on an older, written source. In Haflidason's annal--his autographed copy of it still exists--we also find some detail. This was the great Hekla eruption that began near the middle of July in the year 1300. The tephra fall in the initial phase of this eruption is described by Einar Haflidason as follows:

"So much pumice flew unto the shading at Næfurholt that the roofs of the buildings were burnt off. The wind was from the southeast, and it carried northward over the country such dense sand between Vatnsskard and Axarfjardarheidi, along with such great darkness, that no-one inside or outside could tell whether it was night or day, while it rained the sand down on the earth and so covered all the ground with it. On the following day the sand was so blown about that in some places men could hardly find their way. On those two days people in the north did not dare to put to sea on account of the darkness". Recent studies (1, pp. 47-50) confirm what Einar Haflidason says about the spreading of the Hekla 1300 tephra.

Another 14th century annal, Annales Vetustissimi, mentions "Sandfall in the northern country and ashfall and great darkness" (2, p. 52).

From the quoted descriptions it is clear that in the 14th century Icelanders discerned between three types of tephra, viz. Pumice (Icelandic vikr), sand (sandr) amd ash (aska). This distinction becomes more clear by a contemporary description of the tremendous explosive eruption of Öræfajökull in 1362. This description is in an annalistic fragment, probably written in

North Iceland. From it I quote: "Two parishes, those of Hof and Raudilækr (both at the foot of Öræfajökull), were entirely wiped out. On even ground one sank in the sand up to the middle of the leg, and winds swept it into such drifts that buildings were almost obliterated. Ash was carried over the northern country to such a degree that foot-prints became visible on it. As an accompaniment to this, pumice was seen floating off the Northwest coast in such masses that ships could hardly make their way through" (2, p. 226).

Here it is clear what the chronicler means by the three kinds of tephra. Ash is the most fine-grained one, sand is more coarse-grained, and pumice so light that it keeps floating on the sea.

Now we proceed nearly three centuries forward. On September 2, 1625 a violent explosive eruption started in the subglacial Katla volcano and lasted until September 12. Although basaltic, Katla's eruptions are wholly explosive and often produce a great amount of tephra, and so it was this time. Twenty-three farms in the districts east and southeast of the volcano were temporarily deserted. Thorsteinn Magnússon (ca. 1570-1655), keeper of the former monastery Thykkvabæjarklaustur, situated 30 km ESE of Katla, wrote a detailed day by day account of the eruption (3, pp. 200-215), which is the most thorough eyewitness report of a tephra fall written anywhere until then and, on the whole, a very remarkable account of a volcanic eruption (Fig. 1). Of a special interest is his description of the various electrical phenomena accompanying the tephra fall.

The 1625 tephra spread far east and southeast. In a description of Iceland, written in 1647 by Bishop Thorlákur Skúlason in Hólar, he mentions that he happened to be on his way to Denmark, when so much sand fell on the ship, which was then at a distance of about 600 km from the volcano, that the sails turned black and the crew filled jars and vessels with ash, which they gathered together with their hands. Thorlákur Skúlason also writes that the ashfall badly damaged grassland on the Faeroe Islands, and furthermore he mentions that ash fell somewhere in Northern Norway (4, pp. 14-15). A contemporary Icelandic chronicler, Björn Jónsson, mentions that ash fell in Norway (5, p. 223).

That ash from the 1625 eruption really fell in Norway is proved by a letter the Danish scientist and historian Ole Worm wrote to a colleague in Holland, August 29, 1642. In this letter he tells that he got a sample of ash from the 1625 eruption from a sailor who collected it on a ship near Trondheim. There the ash fall caused complete darkness. Worm writes that he also got a sample of "blue earth" that fell with rain in Scania in 1619 (6, p. 408). It is not quite certain that this was volcanic ash, but if so, it was ash from an eruption in Iceland that year, most probably at Grímsvötn.

Sandferdig oc kort Jßlandiske
RELATION,
Om det forferdelige oc gru-
elige Jordskelff / som skedde for Østen
paa Jßland / hoss Tyckebey Kloster / forgangen
Septembris, met Torden oc Liunet / Jldens nedfald aff
Lufften / met stort Mørck / Aske / fyrige oc gloende Stene
oc Brande / Deßligeste it forskreckeligt Vandfald oc Exundation,
aff det Bierg Jøckelen / som varede fra den 2. ind til
den 14. Sept. dag / aldrig tilforne enten
seet eller hørt.

Malachiæ 4. Cap.
See / der kommer en Dag / hand skal brende som en Ovn / da
skulle alle foractere oc ugudelige være Halm / oc den Tilkommen-
des dag skal optende dem / siger den HErre Zebaoth.

Prentet i Kiøbenhaffn / Aar 1627.

Fig. 1. The title page of a booklet, printed in Copenhagen in 1627, which is an extract of Thorsteinn Magnússon's remarkable account of the Katla eruption in 1625 and is the oldest publication about a volcanic eruption in Iceland. Only two copies are now known to exist.

As far as I know, the above reports are the oldest ones existing of long distance transport of tephra.

In his treatise De mirabilibus Islandiae (On the wonders of Iceland), written in April 1638, Gísli Oddsson (1593-1638) in Skálholt writes about tephra layers in Icelandic soils and is the first to do so. He writes as follows:

"In many places, where the ground is not only a cover of stones, one finds in ditches in the soil, ash, which forms a compact and uninterrupted layer in the mould. The thickness of this layer is one to two inches, or even a hand's breadth. This proves indisputably that ash from the fire-spreading mountains has covered the main part of the country. In such profiles there is underneath this ashlayer soil which has the colour of humus and contains remnants of trees, so that it is burnable and can be used as fuel. Then comes another ashlayer, more than a hand's breadth thick, and in its uppermost part tree trunks that are still better fuel. Under a compressed layer of humus soil is the third layer, its thickness a span, which also contains remnants of trees, so little rotten that one can use them to carve small objects. All this proves that volcanic ash has repeatedly damaged this country" (7, p. 69).

Considering that the passage quoted was written nearly three and a half centuries ago, one cannot help being impressed by the Bishop's keen observations and the logical conclusions he draws from them.

Eleven decades later, in 1749, there appeared in print in Copenhagen a booklet in Latin, written by a 22 year old Icelander, Eggert Ólafsson, then baccalaureate at the University of Copenhagen. The title of the booklet is: Enarrationes historicae de natura et constitutione Islandiae formatae et transformatae per eruptiones ignis" (Historical comments on the nature and character of Iceland, which is formed and transformed by volcanic eruptions). In this book Ólafsson describes, among other things, two types of pumice in the soils of Iceland, black and white. "The black one", he writes, "is heavier and more solid, its nature being similar to that of lava, whereas the light one, which is well known abroad and called in Danish pimpe-steen, has completely lost natural weight" (8, p. 22). Ólafsson also writes, that in peat pits he has found light fine-grained pumice two ells beneath the surface and that it has gradually sunk down to this depth. He did not realize that this "sinking" in reality was the thickening of the soil and such thickening is very rapid in Iceland, yet the phenomenon continued to interest him. In his monumental two volume account (9) of his travels through Iceland in the years 1752-57, together with Bjarni Pálsson, later Surgeon-General of Iceland, he writes that during their travels through the districts south of Vatnajökull in the summer of 1756, he and Pálsson had many holes dug in the soil in order to find the ash layers of the Öræfajökull eruption of 1727 and the Katla eruption of 1755. "The last mentioned one we found to have sunk one inch down into the greensward, but the layer of ash and pumice deposited over the entire area in the year 1727, when Öræfajökull was on fire, had in high and dry spots, where the soil was hard and dense, sunk a little more than half a foot, but in lower and marshy places to double that depth" (9, p. 782). Ólafsson and Pálsson also studied the ash layers in many pits dug into the soil in North Iceland, and stated that there black and light ash layers alternated with layers of sand and dust.

Almost certainly Ólafsson and Pálsson collected in Iceland the coarse-grained (pea-size grains) sample of the Katla tephra of 1755, which was brought to the Royal Academy of Science in Copenhagen. This sample, and another fine dust one from the same eruption, collected on a ship south of the Faeroe Islands, were both examined by Christian Gottlieb Kratzenstein, professor physices experimentalis at the University of Copenhagen. As far as I know, this is the very first laboratory study of tephra. Kratzenstein melted the samples, did some other experiments with them and concluded that they were a mixture of coal and iron pyrite and maybe petroleum (10, p. 192). In those days it was a common belief that volcanic eruptions were caused by burning of combustible matter, such as coal, in the earth.

Contemporary descriptions of big eruptions in Iceland in the seventeenth and eighteenth centuries, some of which are very thorough, contain a lot of information on tephra fall, on the extension and thickness of tephra layers and on their effects on habitation, grassland, fishing in lakes and rivers, birdlife, etc. Of a special interest are the description of diseases, especially fluorosis caused by the tephra. The fluorine adheres to the surface of the grains, and fine-grained tephra is thus likely to transport more fluorine than a coarse-grained one. Most dangerous for grazing animals is a layer of fluorine-contaminated tephra, so thin and so fine-grained that the tephra particles stick easily to the vegetation and thus get into the digestive system of the animals. One millimeter's thickness of fluorine-contaminated tephra is sufficient to cause chronic fluorosis.

Early symptoms of fluorine in animals are dental lesions and damage to membranes of the joints. The earliest description in Iceland, and elsewhere, of dental lesions in animals, known to have been due to toxication by fluorine in volcanic ash, is to be found in a contemporary account of the Hekla eruption that commenced on February 13, 1693. The description was written by a farmer and chronicler in West Iceland, Oddur Eiríksson. He writes about the effects of the eruption:

"In the following autumn and winter people noticed that on the teeth of grazing sheep were yellow spots and some black ones; in some animals the teeth were all black; the teeth fell out in some cases, but small, round-pointed teeth came up afresh, like the teeth of a dog or a catfish; where the spots came the tooth turned soft so it could be shaved like wood. In some animals the flesh peeled away from around the front teeth and molars. . . People thought that this was due to the sand-fall from Hekla" (2, pp. 103-104).

A thorough and detailed description of various diseases in animals and people, caused by volcanic ash and volcanic gases, is to be found in the classical account of the Lakagígar eruption of 1783 by Reverend Jón Steingrímsson (1728-1791). In this account he gives the first known description of Pele's hair. The eruption began on June 8, and on that day and the two following, Pele's hair was observed in the inhabited districts SE of the volcano. On June 14, to quote Reverend Jón, there was. . . "calm weather. A great amount of sand fell here (he was living 28 km SSE of the volcano) containing still more hairs than were observed in the sand-fall on June 9. These hairs were bluish-black and glittering, their length and thickness like seal's whiskers. . . they formed a continuous layer on the ground and where they fell on desert sand the wind went under them so that they were coiled to form oblong, hollow rolls" (11, pp. 9-10).

Reverend Jón states that in some places there are 5, in other places 11, volcanic pumice and sand layers in the soil (11, p. 51). He counted the layers in order to find out the number of eruptions that had affected the district where he lived. He also discusses whether or not the tephra can be a fertilizer. His conclusion is that this may sometimes be the case in the first 2-3 years after the deposition on the ground, but that, on the whole, the tephra is a damaging factor on grasslands.

The knowledge of tephra and tephra layers in Iceland did not increase considerably during the nineteenth century, and this holds also for the first three decades of the twentieth. The geographer and geologist Thoroddsen (1855-1921) won great fame as a volcanologist, but his important contribution to volcanology lay mainly in his thorough description of Iceland's volcanic areas and their many types of volcanoes and in his works on volcanic activity in historical time, based on written sources, which at that time existed mainly in unpublished manuscripts. Curiously enough, Thoroddsen never witnessed a volcanic eruption. In his monumental four volume work, Lýsing Islands (Description of Iceland), he has the following to say about tephra layers:

"Ash layers can be observed in peat pits all over the country, although in varying number and of various thickness, depending on the distance to the larger volcanoes. Most of these layers are black basalt ash. Such eruptions, however, have been much more rare" (12, p. 201).

On March 28-29, 1875 Askja erupted. This was a big, explosive rhyolitic eruption and the most fine-grained tephra was carried all the way to the Baltic Sea. The dust-fall in Norway and Sweden aroused scientists' interest and led to the first map ever made of a widespread tephra layer. A clergyman in East Iceland, Sigurdur Gunnarsson at Hallormsstadur, studied the thickness and extent of the layer in Iceland.

As already mentioned, ash layers may prove noxious for grazing animals, even when their thickness is only a millimeter or so. When travelling in East Iceland many years ago I was told about an odd, but presumably reliable, method that was practiced in order to establish the southern limit of the 1875 tephra sector. A man crossed the border-zone on foot. Now and then he picked up a pellet of sheep droppings and took a bite at it. When he no longer felt grains of fine sand between his teeth while doing so, he decided that he was outside the tephra sector.

The historical review I have so far presented deals exclusively with Iceland. I lack sufficient knowledge to relate in a similar way how tephra studies and knowledge about tephra had developed in other actively volcanic countries. Yet I venture to say, that in regard to knowledge of tephra and various effects of tephra fall the Icelanders kept apace with other nations, at least until the 19th century, and were at times somewhat ahead, especially in committing their experience to writing.

The Krakatoa tephra of 1883 increased interest in tephra and its distribution, as did the eruption of Katmai in 1912. The first thorough studies of tephra layers in soil sections are the studies of the Eocene tephra layers in Denmark, which began at the beginning of this century (13, 14), but the late 1920's and the early 1930's was about the time when work on tephra as a tool in other research began, although on a limited scale, in many places. In 1928-29 the Finnish geologist V. Auer led an expedition to Tierra del Fuego. When working on the vegetation history of this island he used three discernible tephra layers in peat profiles as guide horizons (15). His countryman, Th. G. Sahlstein (later Sahama), studied the chemistry and mineralogy of these layers (16). Later a third Finnish geologist, M. Salmi, who took part in an expedition to Patagonia led by V. Auer in 1937-38, made use of pollenanalysis of peat soils in Patagonia to throw light on the postglacial eruption history of the southernmost Andes. Using the pollen diagrams to connect the soil profiles, and combining this with chemical and petrographical studies of the tephra layers, he could discern between four eruption periods and eight eruption sites could be localized. No layer could be absolutely dated (17).

In connection with geological mapping of the Rotorua-Taupo district in New Zealand in the late 1920's, the eight youngest tephra layers found in the soil profiles were measured. One reason for the mapping was that a sheep-disease, "bush-sickness", had been found to be most common in areas where coarse-grained acid tephra layers appeared (18-20; 21; 22). In Japan studies of tephra layers began in Hokkaido in the early 1930's (23, 24).

Turning again to Iceland, in 1934, when I began field work in my homeland in the summer, my knowledge of tephra layers was restricted to what was then common knowledge in the district in the northeast, Vopnafjördur, where I grew up as a farmer's son. Peat was an important fuel at that time and, like others who had worked in peat pits, I knew that besides dark layers of volcanic sand in the peat soils, there were two conspicuous layers called the upper and lower light layers. These layers are now known as the Hekla H3 and H4 tephras, with C^{14} ages of 2800 and 4000 yr B.P. respectively.

In 1932 I had begun to study Quaternary geology at the University of Stockholm. My professor was Lennart von Post, the founder of the pollenanalytical method, which, besides being used for the study of vegetation history, is a dating method. But when, in 1934, I began, not without optimism, to practice von Post's method in Iceland I soon found that its application there presented considerable difficulties. At that time pollen analysis was still based mainly on tree pollen and birch proved to have been the only forestmaking tree in Iceland through the entire postglacial period. Furthermore, the very high mineral content of the organic soil, partly caused by the frequent tephra falls, rendered the palynological work more difficult, but I also realized that identification and correlation of the numerous tephra layers in Icelandic soils would greatly facilitate the planned pollenanalytical studies. A fellow countryman, Hákon Bjarnason, director of forestry, had, when studying the soils of the Icelandic birchwoods in 1931, become interested in the two light tephra layers in North Iceland. In 1935 we began in cooperation to work on measurement and identification of some widespread tephra layers in soil profiles. These joint studies led to the first paper on Icelandic tephra layers, published in 1940 (25). From being, in the beginning, a study to aid pollen analysis, the establishment of a volcanic ash chronology in Iceland soon became an aim in itself.

In 1939 I took part in the excavations of farm ruins in the deserted Thjórsárdalur valley in South Iceland, 20 km NW of Hekla. These excavations were a joint Nordic enterprise (26) and my task was to try to date, with the aid of tephra layers, the desolation of the valley. These studies resulted in the first part of my doctoral thesis: Tefrokronologiska studier på Island, published in 1944 (27). My conclusion regarding the fate of the valley was that it had been abandoned because of tephra fall from the big Hekla eruption in 1300. In this I was proved wrong and not until 1949 was it definitely established--and then with the aid of tephrochronology--that the valley had been abandoned because of tephra fall from the very large initial eruption of Hekla in 1104.

It was in this doctoral thesis that I suggested the terms tephra as a collective term for pyroclasts, and tephrochronology for the dating method based on tephra layers. I mention this, as I have found that textbooks which mention the coinage of these terms refer only to much later papers of mine that were written in English, probably because my thesis was written in Swedish, which few geologists outside the Nordic countries understand. I will, therefore, end this introduction by saying something about that part of my thesis which deals with the terminology and methodology of tephra studies.

Like the term magma, tephra is a Greek word. I found that I needed a collective term for all pyroclasts. I did not like to use ash as a collective term, because I wanted to avoid the inconsistent use of that term, which on the one hand has, especially in compounds such as ash fall, ash cone, etc., been used as a collective term for pyroclasts and, on the other hand, to denote a certain grain-size fraction of ejecta. Furthermore, I did not like long and awkward terms like pyroclastic ejecta. I also wanted a term in linguistic harmony with magma and lava. In classical Greek there are two words for ash: conis and tephra. I chose tephra, partly because it fitted phonetically with lava and magma, and partly because I found that Aristotle had used the word for volcanic ash in his work Meteorologica in an account of an eruption on the island of Hiera, one of the Lipari islands. That account is, as far as I know, the oldest description of a tephra fall and tephra transport in European literature. The term volcano is derived from the Roman name of that island, Vulcano, and the term tephra is thus closely related to classical volcanology. Tephrochronology can be defined as a dating method based on the identification, correlation and dating of tephra layers.

In my thesis, I criticized the then existing confusion in volcanological terminology, especially regarding the pyroclasts. I proposed that the vague classification with regard to grain-size into dust, ashes, sand, lapilli, etc., should be replaced by classification into groups, the respective particle sizes of which are exactly defined in metric fractions in accordance with Atterberg's grain-size scale (28) (silt-tephra with particle diam. 0.002-0.02 mm; fine sand tephra 0.02-0.2 mm, etc.). I also proposed that classification according to degree of porosity should be accurately defined and that information on thickness of tephra layers should be supplemented with data on volume weight and specific gravity, so that the volume of the tephra layers may be calculated as a volume of solid lava of similar composition and thus be mutually commensurable. An attempt was made to use trace elements to discern between tephra layers of similar chemical composition.

One chapter dealt with the spreading and distant transport of the Askja tephra of 1875 in the light of available data on the meteorological conditions at that time. In that connection I listed eruptions in Iceland of which it was known that tephra had been transported to other European islands or to Scandinavia. I maintained that as the rocks in Scandinavia are, on the whole, crystalline and the fine-grained volcanic ash is mainly glass, it ought to be possible to trace ash from Iceland in Scandinavian peat soils when doing pollenanalytical work and thus establish long distance tephrochronological correlations. Two decades later this was proved to be the case (29). I also discussed the possibility of tephrochronological correlations by identifying the tephra in fossil pumice drifts on raised beaches in Scandinavia and elsewhere.

Besides dealing in my thesis with the use of tephrochronology as a dating method and a tool to aid volcanology, archaeology and pollenanalytical studies of vegetation changes, I briefly touched upon the application of tephrochronology in glaciological and geomorphological research. In short, I did the best I could at that time to promote tephra studies--and particularly tephrochronological ones--as an important tool in Quaternary research. It is satisfactory to realize that tephra studies have gradually developed to become such a tool.

REFERENCES

1. Thorarinsson, S.: 1967, The Eruption of Hekla 1947-1948, in: Soc. Sci. Islandica 1, pp. 1-170.
2. Storm, G. (ed.): 1888, Islandske Annaler indtil 1578, Christiania (Oslo).
3. Magnússon, Th.: 1625, Relatio Thorsteins Magnússonar um jöklabrunann fyrir austan 1625 (Account of the subglacial eruption in the East in 1625), Safn til sögu Íslands IV, Copenhagen and Reykjavík 1907-1915, pp. 200-285.

4. Skúlason, Th., Responsio Subitanea (1647), in: Two treatises on Iceland, Bibl. Arnamagnaeana 3, Copenhagen 1943, pp. 3-19.
5. Annaler 1400-1800, I-II: 1822-1932, Hid Íslenzka bókmenntafélag, Reykjavík.
6. Breve fra og til Ole Worm II, Munksgaard, Copenhagen: 1967.
7. Oddson, G.: 1638, De mirabilibus Islandiae, Islandica 10, 1917, Cornell University Library, pp. 31-82.
8. Ólafsson: 1749, Enarrationes historicæ de natura et constitutione Islandiae formatae and transformæ per eruptiones ignis, Pars I, Hafnia (Copenhagen), 148 p.
9. Ólafsson, E.: 1772, Vice-lavmand Eggert Olafsens og Land-Physici Biarne Povelsens Reise igiennem Island, Soroe.
10. Beskrivelse over det, i Island, den 11 Sept. 1755, Paakomne Jordskiaelv, og den derpaa, den 17de Octobr. samme Aar, fulgte Ilds-Udbrydelse af den forbraendte Bierg-Klofte Katlegiaa. Det Kgl. Danske Vidensk. Selsk. Skr. 7, pp. 188-196, 1758.
11. Steingrímsson, J.: 1788, Fullkomid skrif um Sídueld (A thorough description of the Lakagígar eruption), in: Safn til sögu Íslands IV, Copenhagen and Reykjavik 1907-1915, pp. 1-59.
12. Thoroddsen, Th.: 1911, Lýsing Íslands 2, Hid Íslenzka Bókmenntafélag, Copenhagen.
13. Boggild, O.B.: 1903, Medd. Dansk Geol. Foren. Bd. 2, 9, pp. 1-12.
14. Boggild, O.B.: 1918, Den vulkanske Aske i Moleret. Danmarks Geol. Unders. II, 33, 159 p. plus Atlas with 17 plates.
15. Auer, V.: 1932, Acta Geographica 5, 2, Helsinki, 313 p.
16. Sahlstein, Th.G.: 1932, Acta Geographica 5, 1, Helsinki, pp. 1-35.
17. Salmi, M.: 1941, Annales Acad. Sci. Fenn. Ser. A III, Geologica-geographica 2, 115 p.
18. Grange, L.I.: 1927, N.Z. Geol. Surv. Bull. 31, 61 p.
19. Grange, L.I.: 1929, N.Z. Jour. Sci. Tech. 11, pp. 219-118.
20. Grange, L.I.: 1931, N.Z. Jour. Sci. Tech. 12, pp. 228-240.
21. Grange, L.I. and Taylor, N.H.: 1932, N.Z.D.S.I.R. Bull. 32, 62 p.
22. Taylor, N.H.: 1930, N.Z. Jour. Sci. Tech. 12, pp. 1-10.
23. Uragami, K., Naganuma, I. and Togashi, R.: 1933, Bull. Volc. Soc. Japan 1, pp. 81-94.
24. Uragami, S., Yamada, S. and Naganuma, I.: 1933, Bull. Volc. Soc. Japan 1, pp. 44-60.
25. Bjarnason, H. and Thorarinsson, S.: 1940, Geogr. Tidsskr. 43, pp. 5-30.
26. Stenberger, M. (ed.): 1943, Forntida gardar i Island I. Munksgaard, Copenhagen, 332 p.
27. Thorarinsson, S.: 1944, Geogr. Ann. Stockh. 26, pp. 1-217.
28. Atterberg, A.: 1905, Chemische Zeitung 29, pp. 195-198.
29. Persson, C.: 1960, Geol. Foren. Forh. 88, pp. 361-395.

DATING OF TEPHRA

GEOCHRONOLOGY OF QUATERNARY TEPHRA DEPOSITS

C.W. Naeser, N.D. Briggs, J.D. Obradovich and G.A. Izett

U.S. Geological Survey, Federal Center,
Denver, Colorado 80225, U.S.A.

ABSTRACT. Three radiometric methods have been used to date Quaternary tephra: fission-track (F-T), K-Ar, and radiocarbon (C-14). The fission-track and K-Ar methods provide direct ages for the tephra because they date phenocrysts from the source magma that were deposited by the eruption cloud. Radiocarbon dating provides only indirect ages because the carbon used in this method comes from material, such as pre-existing wood, included in the tephra during deposition or collected from underlying or overlying strata.

Contamination of the sample is a problem with all three methods. In both the F-T and K-Ar method the contamination is usually by older, detrital minerals, which are incorporated in the tephra during eruption or deposition. However, primary zircons can be identified by adhering glass, and sanidine can be easily distinguished from detrital microcline or orthoclase. The roots of plants and animal burrows can also provide a conduit for contamination to enter the tephra at any time after deposition. This latter source of contamination can be minimized by careful sampling. Samples used for the C-14 method can be contaminated by both older and younger carbon at any time during their history.

Each method has its own analytical problems and limitations. For all practical purposes, the C-14 method is limited to samples less than 50,000 years old, although newly developed techniques may increase this limit. The fission-track method can routinely date zircon and glass shards that are older than 100,000 years. Glass shards and zircon have been dated from tephra younger than 100,000 years, but the error can be as large as ± 100%.

S. Self and R. S. J. Sparks (eds.), Tephra Studies, 13–47.

If the zircon contains abnormally high U content (>1000 ppm), ages less than 100,000 years with moderate precision are possible. The younger limits of K-Ar depend on the material to be dated. Sanidine can be routinely dated at ages greater than 70,000 years, but the practical younger limit of plagioclase is about 200,000 years. Dating biotite, because of its high surface area and large quantities of absorbed atmospheric argon, is a problem at ages less than a million years. In rare circumstances some minerals with high potassium contents can be reliably dated as young as 30,000 years.

Several examples of Quaternary tephra dated by these methods and the problems encountered will be cited. Examples include: Salmon Springs ash (Washington), Bailey ash (California), Pearlette family ash beds (western U.S.A.), Bishop ash (western U.S.A.), the Mount St. Helens tephras (western U.S.A. and Canada), and the tephra at Vrica (Italy).

1. FISSION-TRACK DATING

By C.W. Naeser and N.D. Briggs

Theory and Methods

This discussion will be limited to the application of fission-track dating to volcanic ash deposits, but most of the procedures discussed are applicable to the dating of the source volcanic rocks as well. During the last decade there have been a number of examples in the literature of fission-track dating of Quaternary tephra deposits (1-7). Those studies have applied the method to glass shards and/or zircon.

The techniques used to date geologic materials with fission tracks have been developed by physicists and geologists over the last 18 years. The early development of the method has been reviewed by Fleischer and others (8) and Naeser (9).

A fission track is the damage zone formed as a fission fragment passes through a solid. Three naturally occurring isotopes spontaneously fission: ^{232}Th, ^{235}U, and ^{238}U. Of these only ^{238}U produces a significant number of fission events. The half-lives for spontaneous fission for ^{232}Th and ^{235}U are so long that for all practical purposes their contribution can be ignored. Yet even the spontaneous fission of ^{238}U is a rare event. More than a million ^{238}U atoms will decay by alpha emission for each fission decay. When an atom such as ^{238}U fissions two new nuclei are created. The original nucleus breaks up into two lighter nuclei (one about 90 atomic mass units and the other about 135 a.m.u.) with the liberation of about 200 MeV of energy. These two highly charged nuclei recoil from each other in opposite directions and disrupt the electron balance of the atoms in the mineral lattice or glass along their path. This disruption causes the positively charged ions in the lattice to repulse each other and force themselves into the crystal structure, forming the track or damage zone. The new track is only a few angstroms wide and is about 10-20 μm in length. The track is longer in lighter minerals and glasses than in the heavier minerals such as zircon.

The track is stable in most insulating minerals at temperatures of 80°C or less (8), but fission tracks in natural glasses have been shown to be affected at much lower temperatures (10, 7). This fading in glass will be discussed in a later section. Assuming a track formed at a time when the temperature of the host phase was below the temperature at which fading begins, the track will be stable.

The track in its natural state can only be observed with an electron microscope. It is possible, by choosing the proper chemical etchant, to enlarge the damage zone so that it can be observed in an optical microscope at intermediate magnifications (x200-500). Some of the common etchants used include: nitric acid (for apatite), hydrofluoric acid (for glass and micas), concentrated basic solution (for sphene), and basic fluxes (for zircon) (9).

The number of tracks present in a crystal or glass is a function of the age and the uranium content. It is necessary to know the uranium content of the mineral or glass that was counted for the fossil tracks. The easiest and best way to determine the uranium content is to make a new set of tracks using the thermal neutron-induced fission of ^{235}U. The fission-track age equation is as follows:

$$A = \frac{1}{\lambda_d} \ln\left[1 + \frac{\rho_s \lambda_d \sigma I \phi}{\rho_i \lambda_f}\right]$$

where: ρ_s = fossil track density from ^{238}U,
ρ_i = neutron-induced track density from ^{235}U,
λ_d = total decay constant for ^{238}U (1.551×10^{-10} yr^{-1}),
ϕ = neutron fluence (neutrons/cm^2)
σ = cross-section for thermal neutron-induced fission of ^{235}U (580×10^{-24} cm^2),
I = atomic ratio $^{235}U/^{238}U$ (7.252×10^{-3}),

λ_f = decay constant for spontaneous fission of U
(6.85 x 10^{-17} yr^{-1}) (ref. 11).
(7.03 x 10^{-17} yr^{-1}) (ref. 12)*, or
(8.42 x 10^{-17} yr^{-1}) (ref. 13)--and
A = age in years.
*value preferred by authors

Zircons and glass have been used to date Quaternary tephra; other phases present in Quaternary tephra have uranium contents which are too low to be useful for dating. There are different methods used to date zircon and glass, because zircon crystals from the same source tend to have different uranium concentrations, while glass shards from a single eruption tend to have similar uranium contents. Figures 1 and 2 illustrate these methods. Zircon requires the use of the external detector method (Figure 1). Zircon crystals can have both intergranular and intragranular uranium inhomogeneity. It is therefore necessary to count the induced tracks produced from the exact same area of a crystal that is counted for the fossil track count. In the external detector method the fossil tracks are counted in the crystal and the induced tracks are counted in a detector that covered the crystal mount during the neutron irradiation. Either a low-uranium-content (<10' ppb) muscovite or a plastic detector can be used.

The population method (14) can be used to date glass (Figure 2). All of the glass shards from a single deposit have similar uranium concentration. It is therefore possible to split the sample into two groups. One group is mounted in epoxy, polished and etched for the fossil-track density determination. The second group is irradiated, and then mounted in epoxy, polished and etched (it is standard practice to etch both groups at the same time). The irradiated subset contains both fossil and induced tracks (pre-irradiation annealing to remove the fossil tracks is not recommended because it could alter etching characteristics and chemistry). The fossil track density (ρ_s) is subtracted from the total track density in the irradiated sample ($\rho s + \rho i$) to arrive at the induced track density (ρi).

Advantages and Disadvantages

One real advantage to fission-track dating of tephra deposits is that the problem of contamination is minimized. In other methods (C-14 and K-Ar) large samples or many crystals must be analyzed. Contamination of a C-14 sample with recent carbon will result in a younger age, and a few grains of older detrital grains in a K-Ar sample can have a significant effect on a K-Ar age. In the case of fission-track dating with

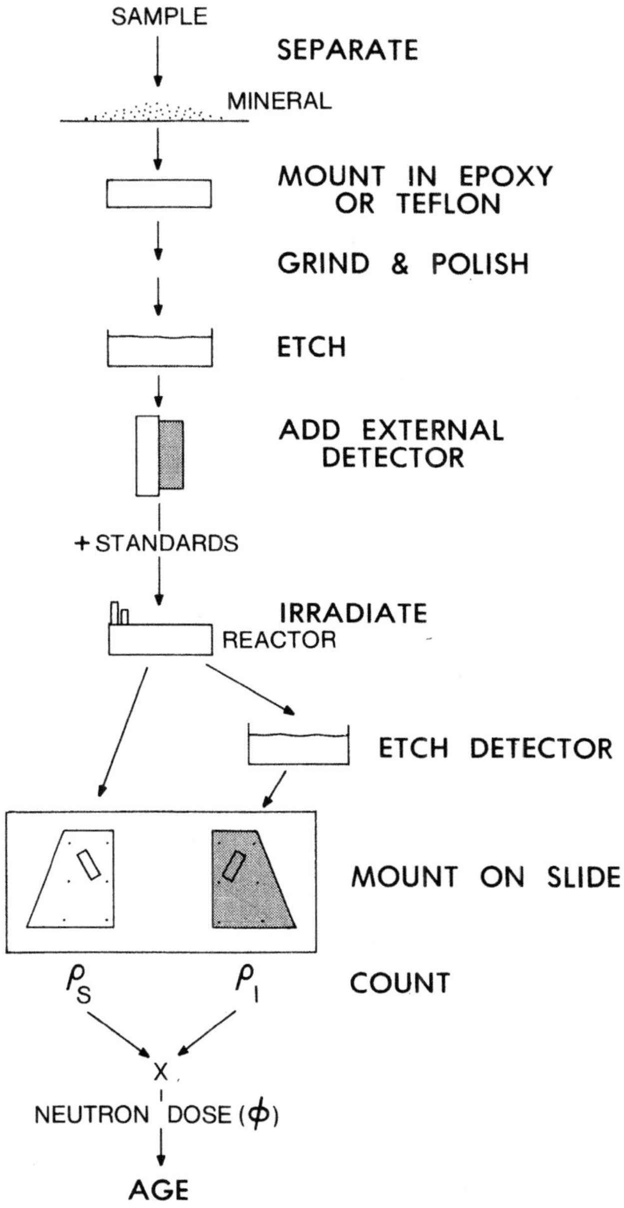

Figure 1. Steps involved in obtaining a fission-track age using the external detector method.

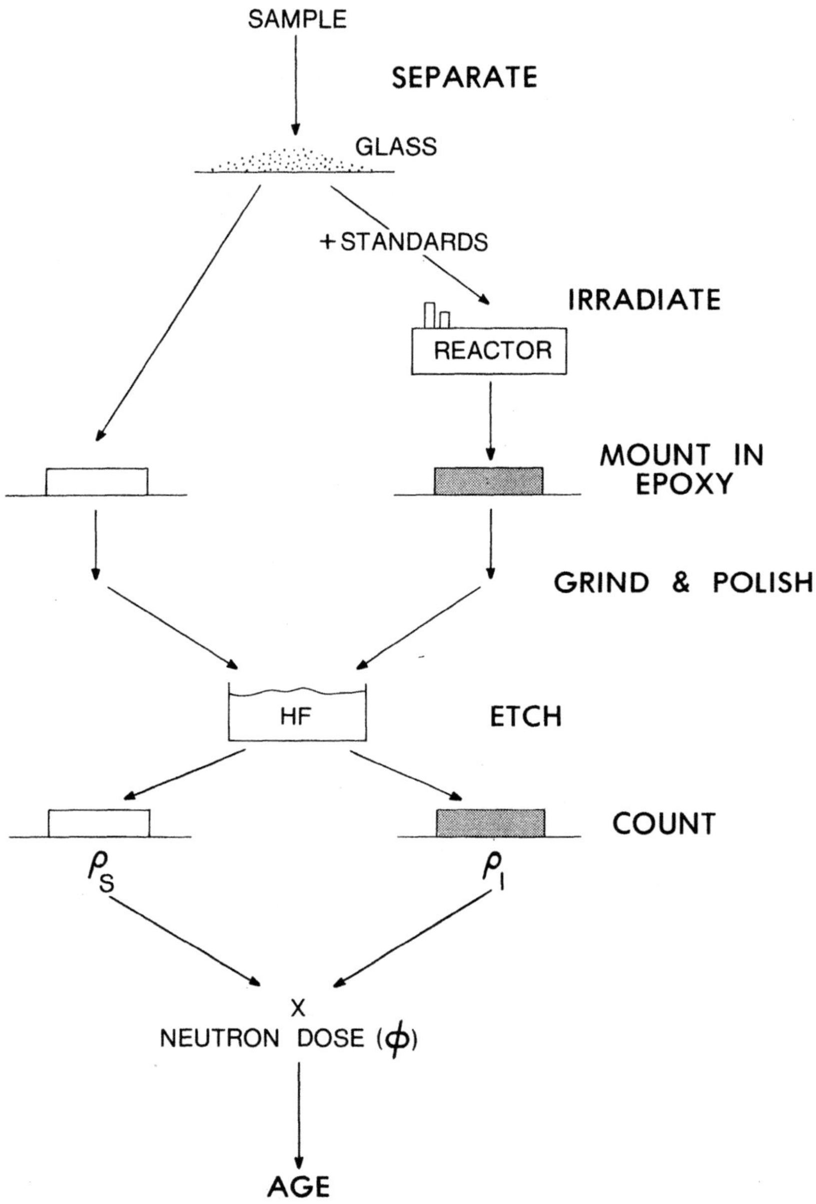

Figure 2. Steps involved in obtaining a fission-track age using the population method.

zircons, single crystals are dated. Older detrital grains show up as contamination. A grain with a Miocene age in a Pleistocene sample is obvious because of its older age. Figure 3 shows a zircon separate that is contaminated. Some of the grains have glass adhering to the sides and are primary; those without glass are detrital. Contamination is usually not a problem with glass, because it would be difficult to contaminate a tephra deposit with significant amounts of older glass.

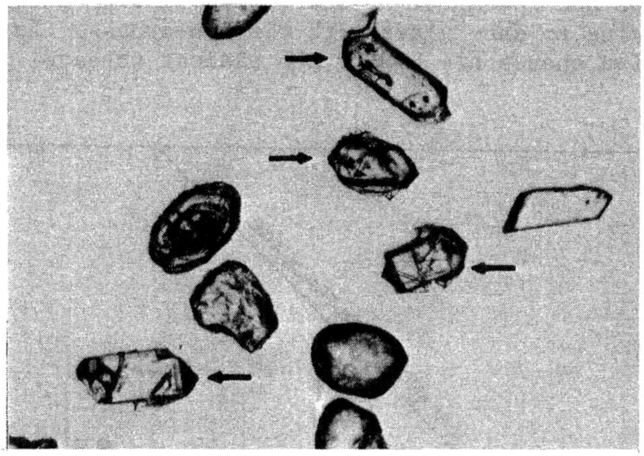

Figure 3. Zircons separated from a tephra deposit. Arrows point to glass mantled primary zircons; the other grains are detrital.

Another advantage of fission-track dating is that less material is needed, and therefore smaller samples need to be collected. Only a few hundred zircon crystals are needed (and only 6-12 are actually counted).

However, there are also several disadvantages to fission-track dating. One is that in very young samples (<100,000 yrs) there are very few tracks present. This lack leads to ages with large analytical uncertainties. Herd and Naeser (15) determined an age on zircon of about 100,000 yr with a 40% standard deviation; in 45 zircons a total of 16 tracks were observed. Briggs and Westgate (1) reported one glass sample in which they did not see any fossil tracks in several thousand shards. Thus for young samples the analytical undertainty is large, but even then the result might answer a geological question.

Another problem is that zircons are not present in all tephra. The presence or absence of zircon is dependent upon several factors: the chemistry of the parent magma and the distance downwind from the eruption vent. Experience has shown that there is a better chance to recover usable zircons from acidic tephra. Basic tephra tends not to have zircon microphenocrysts. If zircons are present but extremely fine grained (<75 μm), they are too small to be dated by fission tracks, as is often the case when the tephra is sampled near the maximum distance from its vent.

The shape and character of shards can also be a disadvantage when trying to date glass with fission-tracks. Large bubble-junction shards are by far the easiest to date.

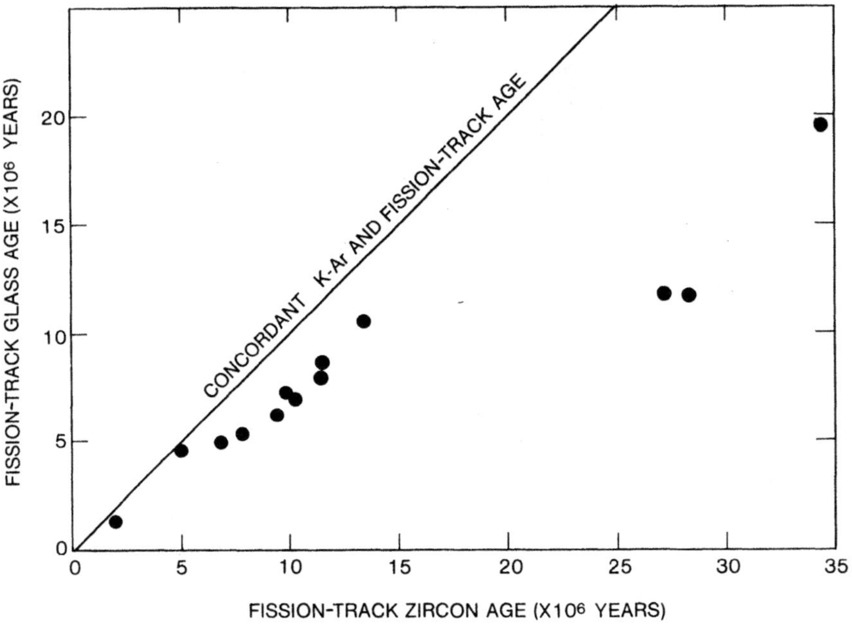

Figure 4. Plot showing the fission-track ages of coexisting glass shards and zircons.

Fine-grained or highly vesicular pumiceous glass is much more difficult to date, because of the small amount of surface area that is available.

Perhaps the greatest disadvantage of fission-track dating of glass is the problem of track fading. A number of studies have documented this phenomenon (16, 17, 10, 7, 18). Track fading is a problem for fission-track dating of all types of glass, but it is especially serious in the case of the hydrated glass shards found in the typical tephra deposit. Seward (7) showed that about 60% of the glass ages on Quaternary tephras in her study were significantly younger than either the stratigraphic age or the fission-track ages of the coexisting zircons. Naeser and others (10), in a study of fourteen tephras from upper Cenozoic (<30 m.y.) deposits of the western United States, found only one glass that has a fission-track age concordant with the coexisting zircon age (Figure 4). All of the remaining samples had ages that were significantly younger than the zircon ages. However, glass fission-track ages reported by Boellstorff (19, 20) are always older than glass ages reported by other workers dating the same tephra (Tables 1, 2, 3). The reason for this discrepancy is unknown at present.

There are two procedures available to correct these lowered ages. It has been shown (18) that the diameter of the etched fossil tracks is smaller than the etched diameter of the induced tracks. Storzer and Wagner (18) developed an empirical curve which can be used to correct for partial track fading. If fading has taken place this diameter reduction is usually visually detectable without actually measuring diameters, but measurements are necessary if a correction to the age is going to be made. Figures 5 and 6 show both the fossil and induced tracks in shards from an ash at Chimney Rock, Nebraska. The tracks in Figure 5 are noticeably smaller than the tracks in Figure 6. The second procedure for correcting glass ages was developed by Storzer and Poupeau (21). This is called the plateau annealing method. In this procedure pairs of glass, fossil and induced, are heated together in a furnace. The usual method is to heat different splits for 1 hour at temperatures between 150^0 and 250^0C. If fading has taken place the induced track density will be reduced before the fossil density is affected. When the amount of fading in the two is equal, they will then fade at the same rate, giving a plateau age. Figure 7 shows the results of plateau annealing for twelve upper Tertiary glasses (both obsidians and tephra shards). These corrections are difficult to apply to Quaternary glasses because the small number of fossil tracks present prevents statistically significant measurements.

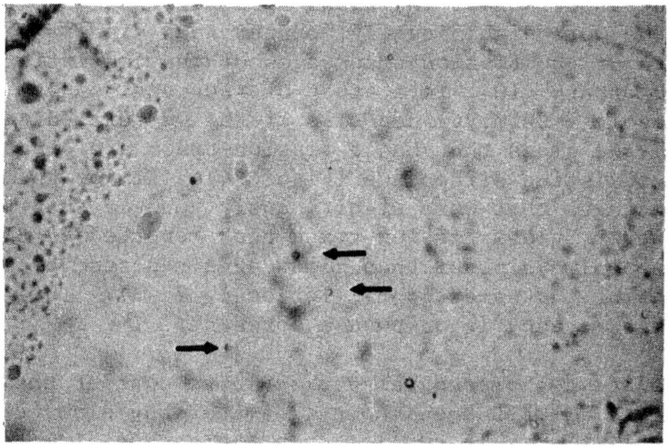

Figure 5. Fossil fission tracks (marked by arrow) in glass shards from Chimney Rock, Nebraska. Average size of track is $\sim 5\mu m$.

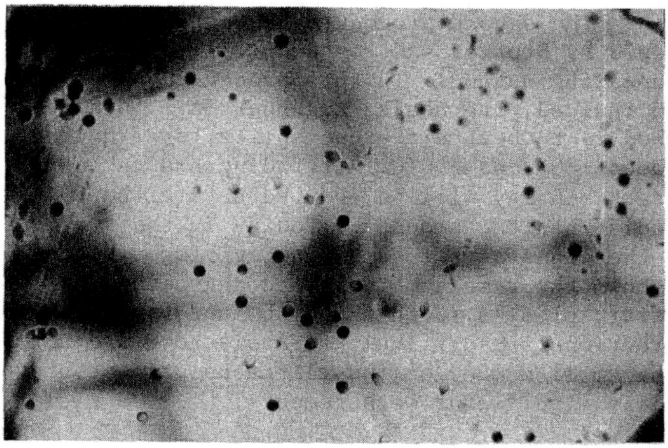

Figure 6. Induced fission tracks in a glass shard from Chimney Rock, Nebraska. Note the larger size of the induced tracks as compared to the fossil tracks in Figure 5. Average size of track is $\sim 10\mu m$.

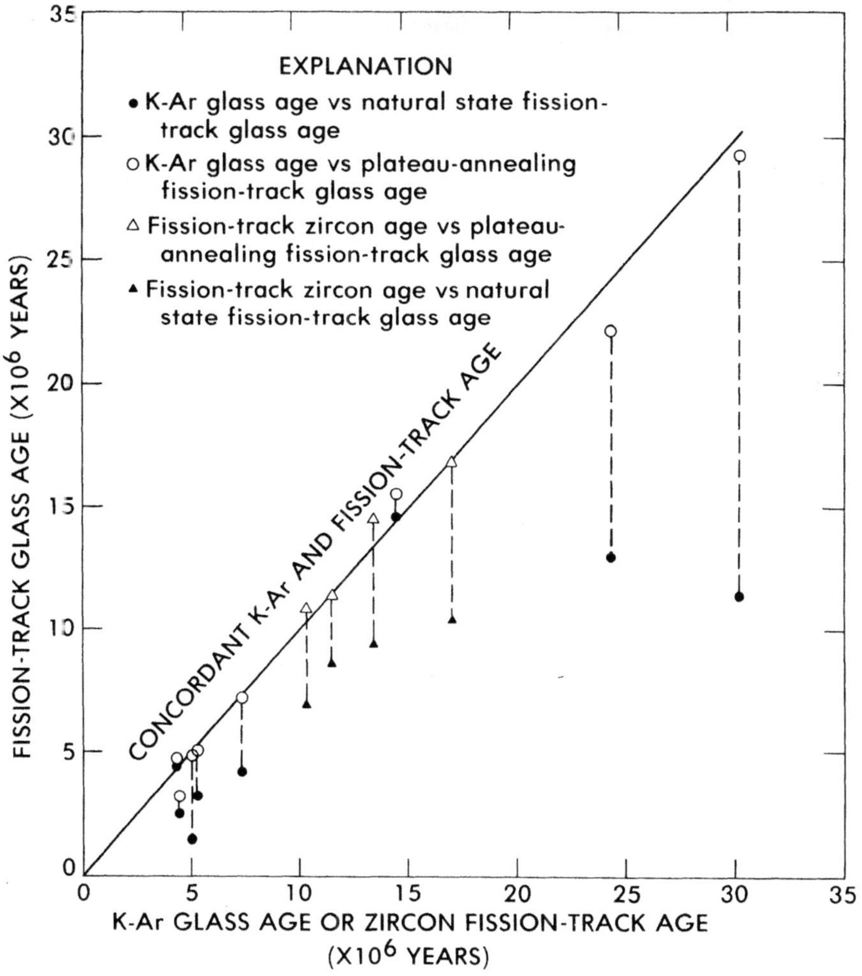

Figure 7. Diagram showing apparent and plateau corrected ages of glasses compared to zircon (F-T) or K-Ar age.

Hydrated glass is particularly susceptible to track fading. Lakatos and Miller (22) found that they could reduce the temperature required for track fading in an obsidian by adding 2.2% H_2O to the glass. The temperature was lowered more than 100°C. This lower temperature means that significant fading can occur at ambient surface temperatures. Thus a fission-track age of a glass must always be considered a minimum age for the deposit. Because of the reality of fading at ambient temperatures, fission-track glass ages should only be used as a guide to the age of the deposit.

Examples of Quaternary Tephra

Pearlette family ash beds. Volcanic ash beds of the Pearlette family occur in Pleistocene deposits of the western United States (23, 24). Prior to about 1970 the Pearlette ash was considered to be a single ash bed to which most mid-continent Quaternary deposits were related. Izett and others (23, 24) recognized that there were minor chemical differences between some of the Pearlette ash localities, and that these ashes could be geochemically correlated to the three major ashflow eruptions that had occurred in the region of Yellowstone National Park, Wyoming. These eruptions have been dated (J.D. Obradovich, 1973, written communication) at 2.02±0.08 m.y., 1.27±0.1 m.y. and 0.616±0.008 m.y. using the K-Ar method. Naeser and others (3) dated zircons from two of the three types and obtained ages of 1.9±0.1 m.y. for the type B ash and 0.6±0.1 m.y. for the type O ash. These ages matched the ages in the source region and confirmed the geochemical evidence of Izett and others (23, 24) that there were three Pearlette ashes rather than just one. Dates obtained for the type B (Borchers) Pearlette ash give an excellent example of the problems caused by track fading in glass (Table 1).

Table 1. Fission-track age of the Borchers ash (Pearlette type B)

Investigator	Glass	Zircon
Naeser and others (3)	1.3±0.17 m.y.	1.94±0.12 m.y.
Seward (7)	1.39±0.08 m.y.	1.93±0.32 m.y.
N.D. Briggs (unpub. data)	1.21±0.07 m.y.	1.9±0.25 m.y.
Boellstorff (19)	1.97±0.25	--

Bishop ash. The Bishop ash, like the Pearlette family of ashes, is a widespread tephra fall in the western United States (2, 24). Izett and others (24) have identified it as the air-fall equivalent of the Bishop Tuff, which originated in the Long Valley Caldera in California about 700,000 years ago. Dalrymple and others (25) reported an average K-Ar age of 0.708±0.015 m.y. (± at the 95 percent confidence interval) on minerals from the Bishop Tuff in its source area. Both glass and zircon from the Bishop ash have been dated by the fission-track method (Table 2).

Table 2. Fission-track ages of zircon and glass from the Bishop ash

Investigator	Glass	Zircon
Izett and Naeser (2)	--	0.74±0.05 m.y.
N.D. Briggs (unpub. data)	0.56±0.05 m.y.	--
Boellstorff (19)	0.82±0.16 m.y.	--

Vrica, Italy, ash. Perhaps the most important tephra to those interested in the Quaternary is the thin (<2 cm), discontinuous layer occurring at Vrica, Italy (about 4 km south of Crotone). This ash layer occurs very close to the Pliocene-Pleistocene boundary. The section of marine mudstone is being considered as the stratotype for the Pliocene-Pleistocene boundary (4). The ash occurs near the middle of unit Y. The boundary between the Pliocene and Pleistocene is also within unit Y. So far only glass from this important tephra has been dated; the ages obtained are listed in Table 3.

Table 3. Ages of glass from the Vrica tephra

Investigator	Fission-track	K-Ar
Selli and others (4)	2.07[1]±0.33 m.y.	2.2±0.2 m.y.
Boellstorff (20)	2.5±0.1 m.y.	

[1]Age corrected for partial track fading; apparent uncorrected age ∿ 1.5 m.y.

Summary

Fission tracks can be used to date Quaternary tephra deposits. The principal problems are track fading in glass, contamination, and low track densities in tephra <100,000 years old. The advantage is that single crystals can be dated, which minimizes contamination problems.

2. RADIOCARBON DATING

By N.D. Briggs

Theory and Methods

The radiocative isotope of carbon, ^{14}C, is continuously produced in the upper atmosphere by the reaction of neutrons produced by cosmic rays with ^{14}N:

$$^{14}N + n \rightarrow {}^{14}C + p$$

Once formed, ^{14}C undergoes radioactive decay back to ^{14}N, by emission of a beta particle. This decay takes place at a constant rate, with a half-life of 5730±40 years (26). When the rate of production equals rate of decay, ^{14}C in the atmosphere reaches equilibrium. The present equilibrium ratio of $^{14}C/^{12}C$ is 1.2×10^{-12}, producing a beta activity of 13.7 disintegrations per minute per gram of total carbon (27). This carbon is taken up by plants, and other organisms in the food chain, and by the oceans as dissolved carbonate. As long as an organism is alive, it maintains the equilibrium activity of ^{14}C through continuous carbon exchange. After death, no new ^{14}C is added and thus the activity begins to decrease at a constant rate. Assuming the ^{14}C activity at the time of death was the same as at present, the amount of time that has elapsed since death can be calculated.

In conventional radiocarbon dating, an age is determined by comparing the beta activity of the sample to that of a modern carbon standard (counting techniques and suitable standards are described in references 28, pp. 44-47, and 29, pp. 29-43). This ratio, when corrected for background radiation (ref. 28, pp. 43-44), is used to calculate the age. By convention, ages are usually calculated using the half-life of 5568±30 years first used by Libby (30), which is about 3% lower than the presently accepted value given above, and are quoted in years before present (B.P.), with "present" taken as 1950. Errors are usually quoted as ±1 standard deviation, calculated from the number of counts recorded for the unknown, standard, and background. One to ten grams or more of carbon are normally required for a date, although with special equipment and long counting times, it is apparently possible to date as little as 10 mg (31).

Limitations and Problems

The effective limit for conventional radiocarbon dating is about 40,000-50,000 years. In samples older than this the beta activity is so low (less than 0.2% of modern activity in samples greater than 50,000 years old) that problems associated with separating sample counts from background radiation and eliminating modern ^{14}C sample contamination are usually impossible to overcome. Lack of suitable calibration for variation in ^{14}C production (see below) is another limiting factor.

Several techniques have been tried in recent years to push back the effective limit of the method, including:

1. Isotope enrichment: Carbon-14 is preferentially concentrated in the carbon sample; this, combined with use of thick shielding to cut down background radiation, increases the detectable activity so that it should be possible to date samples as old as 75,000 years B.P. (see ref. 32 for description of the method; 33). The process requires large samples (60-120 grams), is time consuming (up to 5 weeks for sample preparation), and, most important, does not overcome the major problem of sample contamination. Careful examination of the various separated fractions of the sample may reveal such contamination if it exists.

2. Accelerator (mass separation) techniques: Instead of determining ^{14}C content by counting relatively rare beta decay events, particle accelerators are used as ultrasensitive mass spectrometers to count ^{14}C atoms directly. This makes it possible to date smaller samples (a few milligrams of carbon) and potentially much older samples (the predicted range is 100,000 years B.P.), with greater precision and shorter counting times (measured in minutes rather than days for older samples) (34, 35, 36). Two basic instruments have been tried: the cyclotron (37, 38) and the tandem Van de Graaff accelerator (39, 40, 41). The latter more easily overcomes the major problem of distinguishing ^{14}C from ^{14}N. Samples of "known" age back to about 40,000 years B.P. have been successfully dated (39). The limiting factor at present is the high and variable background in the counters; this may be overcome when instruments built specifically for radiocarbon dating come into use (42, 43, 36). Another drawback for routine use of this technique is the high cost involved in setting up the laboratory.

Care must be used in interpreting radiocarbon ages of tephras, because several factors can produce erroneous results.

1. It is now known that the ^{14}C activity in the atmosphere has not been constant through geologic time (see review in ref. 44). The best calibration of the resulting deviation of radiocarbon years from calendar years over the last 7500 years is based on dating of tree rings, particularly bristlecone pine (see ref. 28, pp. 71-78; and ref. 44). These data show that by 7000 years B.P., radiocarbon dates are about 800 years too

young. In the 9000-32,000-year-B.P. interval, a less precise calibration suggests that the maximum error in radiocarbon dates is probably no more than 2000 years (45).

2. Isotope fractionation and lag time in mixing of ^{14}C within the carbon exchange system also produces anomalous initial ^{14}C activity in some samples (ref. 46, pp. 212-214; and ref. 28, pp. 36-40).

3. Contamination. This is by far the most serious problem because it offers the greatest potential for error and is the most difficult to correct. It may involve incorporation of either older or more modern carbon, and may occur at any time in the sample's history, from original deposition, through burial, to sample collection and laboratory handling (33; 46, pp. 212-215). As a first precaution, care must be used during sample collection and in interpretation of data to note the relationship between the age of the carbon sample and the true age of the tephra. Some tephras are dated from organic carbon in underlying or overlying beds and this will obviously provide only a maximum or minimum age. However, even carbon which appears contemporaneous with the tephra may not be. For example, the innermost rings of a slow-growing tree may record a time of growth much earlier than the time when the tree was actually killed by the tephra eruption, or vegetation charred by forest fire or an older eruption may be incorporated in a younger tephra. Conversely, rootlets or burrowing animals can deposit younger carbon in a tephra. Some of the special sampling and interpretation problems associated with radiocarbon dating of tephras are discussed by Healy and others (42, pp. 7-8).

There can also be contamination within the carbon sample itself, through incorporation of dead carbon (e.g., the "hard water effect") or modern carbon, as through contact with the atmosphere or humic acid. Some materials (for example, bone and shell) are particularly susceptible to this contamination, which can sometimes be removed by pre-treatment before dating (ref. 28, pp. 41-43; ref 29, pp. 3-8; 38).

Introduction of old carbon decreases sample activity, producing an anomalously old age; the magnitude of the error depends only on the percentage of carbon introduced (e.g., 1% dead carbon increases an age by 80 years, regardless of the true sample age; ref. 46, Fig. 8; ref. 28, p. 42). Several methods of checking for this contamination have been suggested 48; R.W. Mathewes and J.A. Westgate, written communication, 1980).

The error resulting from introduction of modern carbon depends both on the amount of contamination and the age of the sample and is particularly severe in old samples that are at or beyond the effective limit of radiocarbon dating (ref. 28, p. 41; and ref. 46, Fig. 9). Addition of only 1% modern carbon gives an apparent age of 38,000 years B.P. for an infinitely old sample. This example emphasizes the caution necessary before accepting old, apparently finite radiocarbon dates.

Examples of Quaternary Tephra

Radiocarbon dating has provided the main chronological control for upper Pleistocene-Holocene tephras younger than about 40,000 years in many parts of the world. For example, Mount St. Helens volcano, Washington, U.S.A., has produced numerous tephras, beginning more than 37,000 years ago and continuing up to the present. The larger, more distinctive of these form marker beds over wide areas of the northwestern United States and western Canada. Numerous radiocarbon dates have helped define the tephras near the source (49) and confirm the identification of the marker beds in areas far removed from the source (50).

Radiocarbon dating has also established the age of the widespread Mazama ash, erupted from Crater Lake, Oregon, about 6600 years ago (51, 52) and the c. 12,000-year-old tephras associated with Glacier Peak, Washington (53).

In New Zealand, radiocarbon dating has helped establish the space-time relationships of the extensive rhyolitic and less widespread andesitic tephras erupted from four major volcanic centers in the central part of the North Island over the last 42,000 years (47, 54, 55, 56).

There are several cases where fission-track dating has demonstrated that tephras are considerably older than indicated by radiocarbon dating. For example, the type locality of the Salmon Springs Glaciation, at Salmon Springs, Washington, comprises two drift sheets separated by peat and volcanic ash (57). The peat grades downward with decreasing organic content into about a meter of silt, which in turn grades into the volcanic ash (D.J. Easterbrook, written communication, April 24, 1980). The peat has been radiocarbon dated at $71,500 ^{+1700}_{-1400}$ by the enrichment method (32). Fission-track data on the ash, and on correlative ash at Auburn, Washington, suggest that the section is much older (Table 4).

Table 4. Comparison of fission-track and radiocarbon ages for Salmon Springs tephra, Washington

Locality	Dating method (Material dated)	Age (yrs B.P.)	Reference
Salmon Springs	C-14 (peat)	71,500 + 1700 - 1400	(89)
	F-T (zircon)	840,000 ± 210,000	*
Auburn	F-T (zircon)	870,000 ± 270,000	*
	F-T (glass)	660,000 ± 70,000	*

*Fission-track dates determined by N.D. Briggs on samples provided by D.J. Easterbrook. Details of the sample localities, F-T data, and stratigraphic significance will be given in a report by D.J. Easterbrook and others (in prep.).

3. PROBLEMS IN APPLYING THE POTASSIUM-ARGON METHOD TO QUATERNARY STUDIES

By J.D. Obradovich and G.A. Izett

The uncertainties associated with K-Ar age determinations can be divided into two general categories: geologic accuracy and instrumental precision and accuracy. The problem of geologic accuracy is far more important that instrumental precision and accuracy and is responsible for many long-standing and outstanding disputes regarding the K-Ar age of specific strata or fossil assemblages. Whereas the problem of instrumental precision and accuracy may typically introduce uncertainties of a few percent, the problem of geologic accuracy can easily introduce uncertainties of a hundred percent or more (the magnitude of error being greatest in the direction of too old ages). Failure to recognize problems relating to geologic accuracy is something that cannot be laid solely at the feet of non-experts in geochronology (geologists, stratigraphers, and paleontologists); recognized experts in geochronology are often equally guilty. One need only to have followed the long-standing controversy regarding the isotopic age of the KBS tuff in Kenya (summarized in ref. 58) to appreciate the complexity of the problem of evaluating the geologic accuracy of K-Ar age determinations.

Recent advances in mass spectrometry have substantially increased the precision with which the radiogenic argon and potassium contents can be measured. Figure 8 depicts the typical precision presently attainable in our Denver potassium-argon laboratory. For radiogenic argon contents greater than 30 percent the associated error (one standard deviation) is 1.5 percent or less; for radiogenic argon contents around 5 percent the error is approximately 10 percent. For radiogenic argon contents less than 5 percent the error rises rapidly to more than 100 percent.

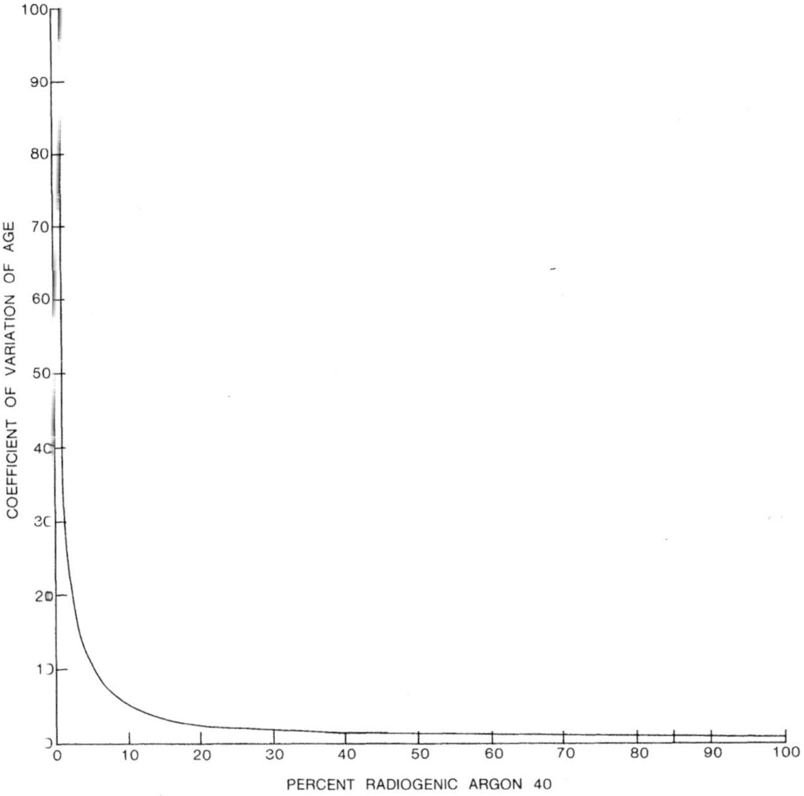

Figure 8. Error in K-Ar ages associated with the content of atmospheric argon in the sample, as measured in U.S. Geological Survey laboratories, Denver.

The ability to date Quaternary tephra by the K-Ar method is primarily a function of the age, the potassium content, and the variable but somewhat distinctive concentrations of atmospheric argon that seem to characterize different minerals found in tephras. If we accept the idea that precise ages (in the analytical sense) can be determined on most potassium-bearing minerals (biotite, sanidine, hornblende, leucite, nepheline, plagioclase, and anorthoclase) found in young tephra, then the problem of geologic accuracy becomes paramount. Basically the problem of evaluating the geologic accuracy of a particular K-Ar age determination may involve one or all of the following common sources of error:

(1) Contamination
(2) Alteration
(3) Excess radiogenic ^{40}Ar
(4) Biostratigraphic

Contamination is an important problem when attempting to date ashes or tuffaceous deposits. It is a simple depositional process for older detritus to become admixed with the primary volcanic material of a tephra during eolian or water transportation and deposition. Even minerals separated from large pumice cobbles found in sedimentary deposits can give erroneous ages. Because many pumice fragments have inter-connecting tubular vesicles, it is not surprising that detrital material can be worked deep into the inner part of pumice fragments. While it is one thing to recognize contaminating material by petrographic examination using refractive oils (59), it is quite another to remove this detritus using standard mineral separation techniques. Two examples follow that stress the importance of detrital contamination of tephras and its influence on K-Ar age determinations. Although the first does not deal with Quaternary tephra it is nonetheless illustrative of the problem.

During eruption of silicic volcanic rocks of the Thirty-nine Mile volcanic field west of Cripple Creek, Colorado, a rhyolitic air-fall tuff was deposited in the Florrissant Lake Beds of early Oligocene age (60). Preparation of a sanidine concentrate of the tuff revealed the abundant presence of microcline derived from the Precambrian Piles Peak Granite ($\sim$1 b.y. old). Viewed under a stereomicroscope with reflected light, thick microcline fragments, because of the opacity and distinctive color, were readily distinguished from the optically clear sanidine. When the microcline fragments were thin, the distinction between the minerals was not as evident, and it was difficult to recognize microcline. An initial effort to hand pick the detrital microcline from the feldspar concentrate reduced the age from 72 m.y. to 43 m.y. Obviously all the detrital microcline was not recognized and removed simply using reflected light. However, this problem was resolved when the minerals were viewed under a stereomicroscope fitted with a polarizer and analyzer using both transmitted and reflected light. With crossed polars or introduction of a first-order red plate the distinctive interference colors and twinning characteristics of microcline were then quite apparent. Once the sample was hand picked under these conditions, an age of 34.9 m.y. was obtained, which is in agreement with the K-Ar ages of underlying and overlying welded rhyolite tuffs (61).

The problem of dealing with a contaminated sanidine sample is a relatively simple one provided the sample is coarse grained enough and older than a few million years. The finer grained and younger (less-than-2 m.y.) samples are difficult to deal with, considering the time necessary to obtain the needed hand-picked feldspar concentrate. There is no satisfactory method for recognizing and removing any of the dark-colored silicate minerals (biotite and hornblende) that might be detrital.

The dating study reported by Yeats and McLaughlin (62) on the Bailey ash, a thin (2-10 cm) lenticular ash found within the lower part of the Pico Formation (then considered upper Pliocene or Wheelerian benthonic foraminiferal stage of ref. 63) of southern California, is one that is often cited as an example where relatively concordant K-Ar ages led to an erroneous age assignment. This in no way is meant to impugn the scientific integrity of Yeats and McLaughlin, but it is difficult not to make use of their work to illustrate a situation where an age was reported that was in error by more than half an order of magnitude. Given the status of the benthonic foraminiferal stages at that time and the view point that the provincial Pliocene of California was as old as 10 m.y. (64), the assignment of an age of 8.9 ± 0.5 m.y. (all ages have been recalculated using the new decay constants and isotopic abundance for ^{40}K (65) for the Bailey ash did not evoke any serious criticism when first reported in 1966 (66). However, Bandy and Ingle (67) considered the results to be anomalous and considered an age of 5 m.y. (68) to be a more reliable estimate for the age of the Pico Formation. The results of Yeats and McLaughlin (62) are summarized in Table 5. Subsequent work by Bandy and Wilcoxon (69) revealed the presence of Globorotalia truncatulinoides near the level of the Bailey ash, implying that this ash could be no older than 1.8 m.y. and probably near to 1.6 m.y., based on the disappearance of Discoaster brouweri just above the ash. Because of this new age implication, we collected samples of the Bailey ash. Petrographic examination of the separated minerals indicated the presence of microcline, obviously a contaminant, along with the sanidine. Electron microprobe studies by the writers indicated the presence of

Table 5. "Bailey ash" age data

K-Ar Ages Reported by Yeats and McLaughlin (62)	
BIOTITE	10.9 ± 2.1 m.y.*
BIOTITE	8.0 ± 1.2 m.y.
BIOTITE (J. Obradovich)	9.5 ± 0.9 m.y.
BIOTITE	18.4 ± 2.0 m.y.
SANIDINE	9.22 ± 1.1 m.y.
SANIDINE + ANORTHOCLASE	24.6 ± 2.0 m.y.
SANIDINE	8.6 ± 0.4 m.y.
SANIDINE	9.8 ± 1.4 m.y.
SANIDINE	8.9 ± 1.3 m.y.
GLASS	< 0.2 m.y.
GLASS	< 0.4 m.y.
GLASS (Yeats, 71)	1.1 ± 0.1 m.y.
Fission Track	
ZIRCON (Izett & others, 70)	1.2 ± 0.2 m.y.
GLASS (Boellstorff & Steineck, 72)	0.79 ± 0.09 m.y.
GLASS (Boellstorff & Steineck, 73)	1.12 ± 0.36 m.y.

*All ages have been recalculated using the new decay constants for ^{40}K (65).

two distinct populations of biotite, which suggests that one biotite population is a contaminant (Figure 9). Clearly this ash is contaminated, despite the statements by Yeats and McLaughlin to the contrary, and their results are to be discounted. Fission-track dating of zircon microphenocrysts of the Bailey ash yielded an age of 1.2 ± 0.2 m.y. (70), in good agreement with the paleontologic constraints set by the presence of G. truncatulinoides but younger than the extinction level of D. brouweri. The specimen of D. brouweri may simply be reworked from older beds and its lack of occurence may not represent an extinction level.

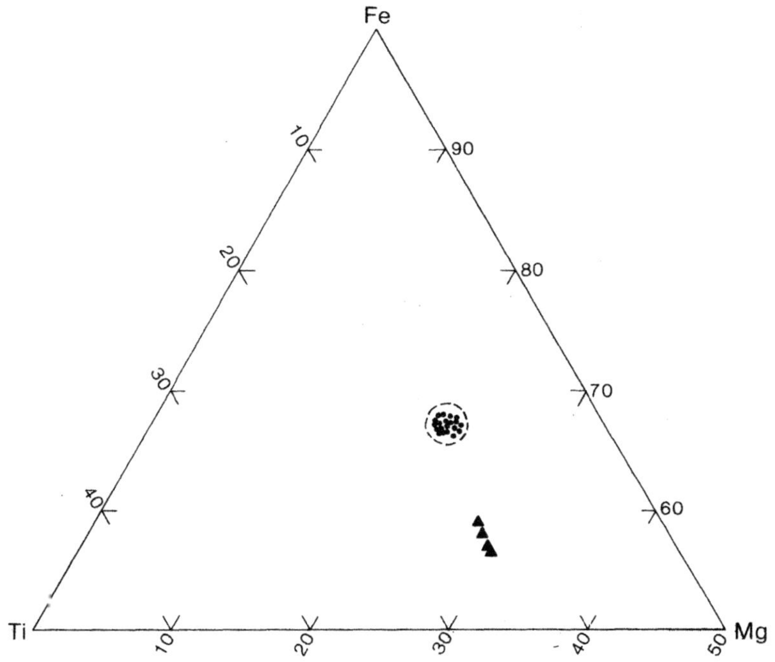

Figure 9. Fe-Ti-Mg content, in percent, of biotite from the Bailey ash bed, as determined by microprobe analysis.

The problem of alteration related to K-Ar ages is primarily concerned with the mineral biotite. The altered outer parts of feldspars can be removed by leaching in HF acid. Loss of potassium from biotite by leaching does not seem to introduce any uncertainty to a K-Ar age until the potassium content falls below 5 percent (Figure 10 and ref. 74). Even then the age may be correct, but it would need to be confirmed by independent results. Below 5 percent the ages are commonly too young, but on occasion too old (74). Alteration of biotite is a commonplace occurrence when the ash is found in a marine environment, particularly so when the ash has been converted to bentonite. However, bentonites of Quaternary age are not common (75).

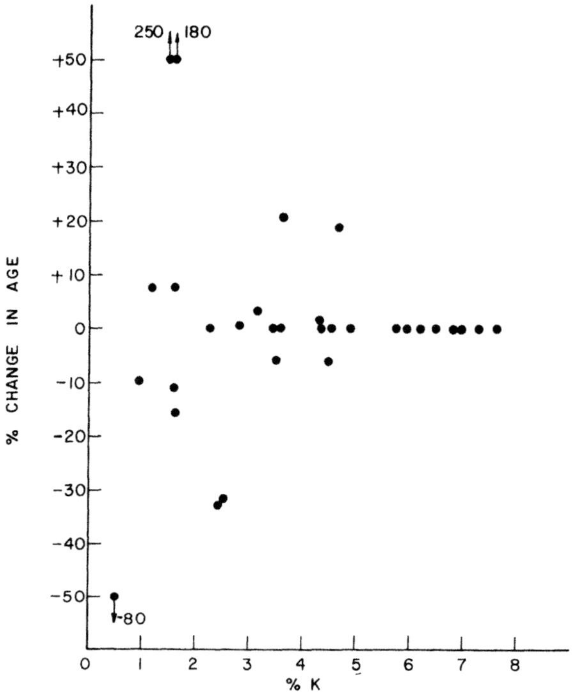

Figure 10. Variations of potassium-argon age with potassium content for altered biotites.

Excess radiogenic argon is rare in primary volcanic minerals associated with tephra deposits. (Basaltic lavas are known to contain excess ^{40}Ar, while obsidian flows apparently do not (J.D. Obradovich, unpub. data, 1980). However, all types of volcanic flows are excluded from this discussion.) However, K-Ar ages are commonly determined on volcanic glass shards, and these are known to give erroneously old ages. Magma prior to its eruption may contain argon with a $^{40}Ar/^{36}Ar$ ratio significantly greater than the $^{40}Ar/^{36}Ar$ ratio of 296, characteristic of the atmosphere. During eruption and rapid chilling of a magma during which glass shards can be formed, excess argon may not be totally expelled. Numerous instances can be cited where "too old" ages have been reported for glass shards, and a few examples are given in Table 6. Nevertheless, there have been K-Ar glass ages reported that are quite close to the age expected (76, 77) based on stratigraphic or biochronologic evidence. But are these K-Ar ages determined on glass shards real or merely artifacts? All glass shards are known to undergo progressive hydration--a process that generally goes on for several million years. During this process the glass takes in potassium as a replacement for sodium. Depending upon the amount of potassium gained (resulting in young ages) and the period of time during which the hydration takes place, any number of different ages could be obtained. Coupled with the possibility of potassium gain is the possibility of argon loss by diffusion both during and after hydration. These latter processes may be reflected in the many "too young" ages reported. If all of the foregoing processes can take place, then just what reliance can be placed upon any age determined on hydrated glass shards? If the age on glass shards agrees with that on another mineral, is the age meaningful or is the agreement strictly fortuitous because of compensating factors? Based upon repeated experiences, no reliance should be given a K-Ar age determined on volcanic glass shards.

While one might question why the problem of biostratigraphic uncertainty is being discussed when we are dealing with isotopic dating methods we must ultimately come to grips with the question "Do we really know what we are dating?". That is just how clear are the geologic relationships between the tephra being dated and the fossil assemblage or zone whose age we wish to know and furthermore just how diagnostic is the paleontologic evidence?

The age of the Lomita Marl Member of the San Pedro Formation (classically regarded as the basal Pleistocene of southern California; ref. 78 and 79) is a worthwhile example of geologic uncertainty, even though we are not dealing with tephra in this instance. Obradovich (80) reported a mean age of 3.2 ± 0.1 m.y. for the Lomita based on K-Ar dating of glauconite from nine different stratigraphic levels at the Lomita quarry. This age was at first accepted by Bandy (81, 82), Ingle (68), and Bandy and Ingle (67), but then later rejected because the age did not agree with the supposed constraints set by planktonic foraminifera (83, 84, 85, 69). The question is, just what is the evidence that requires the age of 3.2 m.y. to be incorrent? First of all it must be pointed out that there are two widely divergent interpretations of the geologic relationships. Woodring and others (78) describe the relationships as follows:

> "At most localities the lower Pleistocene deposits consist chiefly or entirely of the San Pedro sand. At places calcareous beds or silt, or both, of varying thickness are found at the base of the section. The calcareous beds are designated the Lomita marl, and the silt is designated the Timms Point silt. In central San Pedro, where the three units--Lomita marl, Timms Point silt, and the San Pedro sand--are superimposed, they are found in upward sequence in the order named. If each unit represents a distinct

Table 6. Apparent K-Ar ages of hydrated glass shards compared with K-Ar mineral or fission-track zircon ages from certain upper Cenozoic ash beds and tuffs of the western United States [Leaders (--) indicate ages not determined. All ages given in millions of years.]

Formation	K-Ar age (glass)	K-Ar age (mineral)	Fission-track age (zircon)
Pico	< 0.2	--	1.2 ± 0.2
	< 0.4		
	1.1 ± 0.1		
Rio Dell	2.3 ± 0.5	--	1.48 ± 0.56
Tulare (60-100 mesh)	12.3 ± 0.6	--	2.2 ± 0.3
(150-200 mesh)	5.4 ± 0.4	--	
Ogallala	8.4 ± 0.3	--	5.0 ± 0.3
Troublesome	20.3 ± 0.5		13.5 ± 1.1
Gering	40.6 ± 0.4	27.7 ± 0.3 (sanidine)	
		28.5 ± 0.3 (sanidine)	27.2 ± 0.9
		28.9 ± 0.3 (sanidine)	

time interval, discontinuities might be expected to account for the varying thickness of the Lomita marl and the Timms Point silt and for their local absence. With the following exceptions, however, no marked discontinuities are apparent. On Deadman Island, which has been destroyed, the San Pedro sand overlay the Timms Point silt along an abrupt and irregular contact. At Lomita quarry sand doubtfully identified as the San Pedro appears to overlie the Lomita unconformably. At other localities where contacts between the three units were observed the change in lithology is generally gradational. Though continuous exposures along strike are not extensive, the observed relations appear to be most satisfactorily explained by the inference that the lower part of the San Pedro sand grades laterally into the Lomita marl and Timms Point silt and by the further inference that minor discontinuities are found locally between the three units and also within them as shown in Figure 10."

If the above stratigraphic relations are valid, then the following members of the San Pedro Formation--Lomita Marl, Timms Point Silt, and the lower part of the sand member--are time equivalents.

Valentine (86), however, provides a completely different view point. He states: "The Timms Point grades upward into the San Pedro sand in central San Pedro, but elsewhere is unconformable below that unit; it is considered here to be separated from the Lomita marl below by an unconformity. Pods and blebs of marl near the base of the Timms Point, interpreted by Woodring and others as evidence of intergradation between these two units, are believed to represent reworked material. These marly pods contain fossils of Lomita types. Some species recorded from the Timms Point silt, then, may not have lived in the region under Timms Point conditions, but are reworked."

The two interpretations are depicted in Figure 11.

Bandy and others (85) published a photograph of a specimen of Globorotalia truncatulinoides recovered from the Timms Point Silt Member (Plate 4, specimen 2) which indicates that the silt and by inference the Lomita Marl Member are no older than 1.8 m.y. Bandy (83, 84) carried this reasoning even further to show that the Lomita-Timms Point sequence was as young as 700,000 years. Bandy and Casey (87) had earlier defined a major cooling event near the Matuyama and Brunhes paleomagnetic boundary based on the massive influx of left coiling (cold) Globorotalia pachyderma. If the Lomita Marl and Timms Point Silt Members are time equivalents (78), then the extensive geochronologic study on glauconite from the Lomita Marl done by Obradovich (80) is grossly in error, as the age of 3.2 m.y. is too old by a factor of four or more.

If on the other hand, the Lomita Marl and Timms Point Silt Members are separated by an unconformity (86), then what compelling evidence requires the Lomita Marl Member to be as young as 700,000 years? The Lomita Marl Member at the Lomita quarry rests unconformably on the Pliocene Fernando Formation (formerly called Repetto Formation), dated as 3.4 ± 0.3 m.y. (F-T on zircon from an ash interbedded in the Fernando Fm., ref. 88). Glauconite from the Fernando Formation has also been dated as old as 3.9 ± 0.2 m.y. (80). The Timms Point Silt Member (less than 1.8 m.y.) and the sand member have been dated by the amino acid racemization method (89) at 1.5 to 0.7 m.y. (and possibly younger) and 0.45 ± 0.15 m.y., respectively.

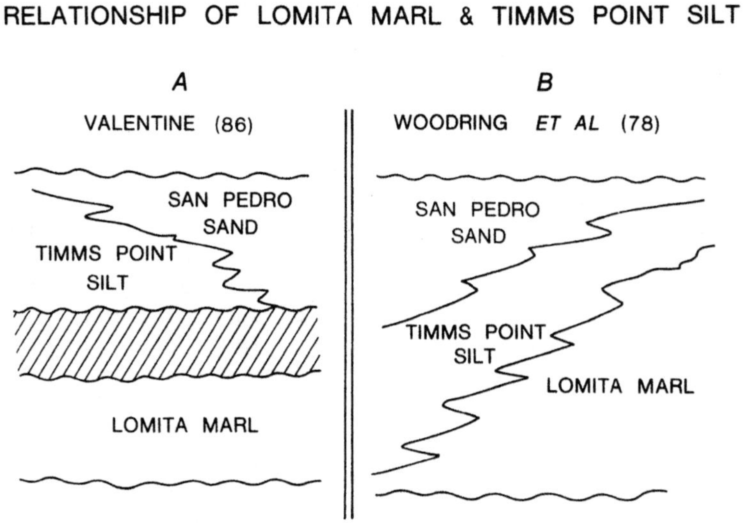

Figure 11. Schematic cross sections for the time interpretation of the Lomita Marl Member.

Jenkins (90), in examining the planktonic foraminifera from the Lomita Marl Member, could not, because of the limited fauna, place it with respect to the Pliocene and Pleistocene boundary of Ericson and others (91) at approximately 1.6 to 1.7 m.y. Based on the time of first appearance of some of the foraminifera, an age as old as 3.0 to 3.5 m.y. cannot be ruled out, while the single specimen of Discoaster sp. (if not reworked) indicates that the Lomita is no younger than 1.6 to 1.7 m.y. (92). Recently Crouch and Poag (93) reported the presence of Cibicorbis hitchcockae from the Lomita. C. hitchcockae is also known from the Gulf Coast area, where it is present in strata as young as late Pliocene (Globorotalia miocenica range zone; G. miocenica becomes extinct at about 2.2 to 2.3 m.y. ago; 94, 95, 92). C. hitchcockae is found in older beds in other parts of the Caribbean.

It should be pointed out that the coiling direction of G. pachyderma is simply indicative of water temperature and not age. If the interpretation of the glacial chronology for the Sierra Nevada is correct (96), then there is evidence of a pre-Pleistocene glacial climate, as the Deadman Pass Till overlies an andesite flow dated at approximately 3.2 m.y. and is overlain by the Two Teats Quartz Latite (97), dated approximately 2.8 m.y. Does the presence of a sinistral population of G. pachyderma relate to this time of alpine glaciation and the movement of colder water into the southern California region?

The presence of Patinopectin caurinus in the Lomita has often been cited as evidence that the Lomita is an age equivalent of the Pico Formation (referred to earlier as 1.2 m.y. old) and the Santa Barbara Formation. In referring back to Woodring and others (78, p. 81), one sees that P. caurinus is reported from two localities. However, at one locality, the fragments are doubtfully assigned to P. caurinus and at the other locality there is only a single small imperfect valve. It should be emphasized that P. caurinus is still living, although not now found in southern California marine waters. How compelling then are age assignments of the Lomita if they are based on questionable identification of this taxa? Should these specimens actually represent P. caurinus then the implied age significance of this form will not be definitely known until the time of its evolutionary first appearance has been established, something which remains to be done. Clearly the isotopic and biochronologic age of the Lomita has not been satisfactorily resolved.

Another example of biostratigraphic uncertainty relative to the dating of young tephra is the problem of the Pearlette ash of the Great Plains region of the United States. It is a classic example of how an assumption led to confusion regarding the evolutionary sequence of vertebrate assemblages related to glacial and interglacial deposits stratigraphically tied to a tephra (23). The assumption, which was seemingly logical, was that the numerous chemically similar Pleistocene ash beds of the Great Plains (Pearlette ash) represented a single ash fall marking an instant of geologic time (98). Hibbard (99) summarized the stratigraphic arrangement of the late Pliocene and Pleistocene deposits and their contained faunas for Meade County, Kansas. In 1958 Hibbard (100) erected and redefined stratigraphic units in southwestern Kansas and assigned the Pearlette ash a late Kansan age. If the Pearlette ash was present the sediments and their contained faunas could be assigned an age based on their position with respect to the ash. The Borchers fauna was assigned to the Yarmouthian Interglacial because it overlay the Pearlette and because the taxa in the fauna had southern affinities and included relicts from the upper Pliocene Rexroad Formation. Figure 12 shows the stratigraphy as summarized by Hibbard and Taylor (101) for Yarmouth and older deposits.

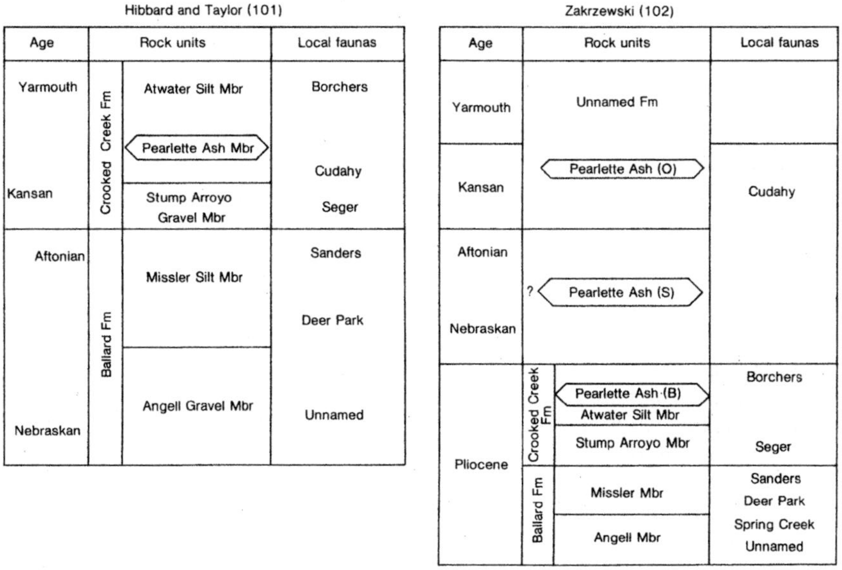

Figure 12. Comparison of the stratigraphy of the lower and middle Pleistocene units of the Great Plains before and after subdivision of the Pearlette family ashes.

When it was determined (23, 24) in the late 1960's that the so-called Pearlette ash consisted of not one but three discrete ash beds of significantly different ages (Pearlette types B, S, and O) derived from the Yellowstone National Park region, the glacial stratigraphy of the Great Plains and the age of some of the vertebrate assemblages underwent drastic revision. A current stratigraphic assessment, Figure 12, is taken from Zakrzewski (102) although modified somewhat. The position of Pearlette type S ash bed shown on Figure 12 relative to the Aftonian or Nebraskan is not known. Ultimately two of the three Pearlette ashes (types B and O) were dated in the Great Plains by the F-T method on zircon microphenocrysts as 1.9 ± 0.1 m.y. and 0.6 ± 0.1 m.y., respectively (3), and in the source region all three ashes (types B, S, and O) were dated by the K-Ar method as 2.02 ± 0.08 m.y., 1.27 ± 0.13 m.y. and 0.616 ± 0.008 m.y., respectively (J.D. Obradovich, unpublished data, 1980).

All of this has been written with the thought in mind that while geochronology can be a routine analytical matter, there are many pitfalls to be avoided. Unless one is made aware of them a great deal of work may simply be accepted without critical judgement. Geochronology can only be as good as the geology, the paleontology, or the samples used.

REFERENCES

1. Briggs, N.D. and Westgate, J.A.: 1978, U.S. Geological Survey, Open-File Report 78-701, pp. 49-52.
2. Izett, G.A. and Naeser, C.W., 1976: Geology 4, pp. 587-590.
3. Naeser, C.W., Izett, G.A. and Wilcox, R.E.: 1973, Geology 1, pp. 187-189.
4. Selli, R., Accorsi, C.A., Bandini, M.M., Bertolani, M.D., Bigazzi, G., Bonadonna, F.P., Borsetti, A.M., Cati, F., Colalongo, M.L., D'Onofrio, S., Landini, W., Menesini, E., Mezzetti, R., Pasini, G., Savelli, C., and Tampieri, R.: 1977, Giornale di Geologia, series 2, 42, pp. 181-204.
5. Seward D.: 1974, Earth Planetary Science Letters 24, pp. 242-248.
6. Seward, D.: 1975, New Zealand Jour. Geol. Geophysics 18, pp. 507-510.
7. Seward, D.: 1979, Geology 7, pp. 479-482.
8. Fleischer, R.L., Price, P.B., and Walker, R.M.: 1975, "Nuclear Tracks in Solids: Principles and Applications" (Univ. of California Press, Berkeley), 605 p.
9. Naeser, C.W.: 1979, "Lectures in Isotope Geology", E. Jager and J.C. Hunziker, Eds. (Springer-Verlag, Berlin), pp. 154-159.
10. Naeser, C.W., Izett, G.A., and Obradovich, J.D.: 1980, U.S. Geological Survey, Bulletin 1489, 31 p.
11. Fleischer, R.L. and Price, P.B.: 1964, Phys. Rev 133, pp. 1363-1364.
12. Roberts, J.H., Gold, R., and Armani, R.J.: 1968, Phys. Rev. 174, pp. 1482-1484.
13. Spadavecchia, A. and Hahn, B.: 1967, Helv. Phys. Acta 40, pp. 1063-1079.
14. Naeser, C.W.: 1967, Geol. Soc. America Bulletin 78, pp. 1523-1526.
15. Hurd, D.G. and Naeser, C.W.: 1974, Geology 2, pp. 603-604.
16. Fleischer, R.L., Price, P.B., and Walker, R.M.: 1965, Jour. Geophysical Research 70, pp. 1497-1502.
17. MacDougall, J.D.: 1976, Earth Planetary Science Letters 30, pp. 19-26.
18. Storzer, D. and Wagner, G.A.: 1969, Earth Planetary Science Letters 5, pp. 463-468.
19. Boellstorff, J.: 1973, Isochron/West 8, pp. 39-43.
20. Boellstorff, J.: 1978, Science 202, pp. 305-307.
21. Storzer, D. and Poupeau, G.: 1973, C. R. Acad. Sc. Paris 276, series D, pp. 137-139.

22. Lakatos, S., and Miller, D.S.: 1972, Earth Planetary Science Letters 14, pp. 128-130.
23. Izett, G.A., Wilcox, R.E., and Borchardt, G.A.: 1972, Quaternary Research 2, pp. 554-578.
24. Izett, G.A., Wilcox, R.E., Powers, H.A., and Desborough, G.A.: 1970, Quaternary Research 1, pp. 121-132.
25. Dalrymple, G.B., Cox, A., Doell, R.R.: 1965, Geol. Soc. America Bulletin 76, pp. 665-674.
26. Godwin, H.: 1962, Nature 195, p. 984.
27. Purser, K.H.: 1978, "Proceedings of the First Conference on Radiocarbon Dating with Accelerators", H.E. Gove, Ed. (Univ. Rochester, New York), pp. 1-32.
28. Aitkin, M.J.: 1974, "Physics and Archeology", 2nd ed., (Clarendon Press, Oxford), 291 p.
29. Ralph, E.K.: 1971, "Dating Techniques for the Archaeologist", H.N. Michael and E.K. Ralph (Massachusetts Inst. Tech. Press, Cambridge, Mass.), pp. 1-48.
30. Libby, W.F.: 1955, "Radioactive Dating" (Univ. Chicago Press, Chicago, Ill.), 175 p.
31. Harbottle, G., Sayre, E.V., and Stoenner, R.W.: 1979, Science 206, pp. 683-685.
32. Stuiver, M., Heusser, C.J., and Yang, I.C.: 1978, Science 200, pp. 16-21.
33. Grootes, P.M.: 1978, Science 10, pp. 11-15.
34. Anon: 1978, Mosaic Nov/Dec, pp. 43-49.
35. Gove, A.E. (Ed.): 1978, "Proceedings of the First Conference on Radiocarbon Dating with Accelerators" (Univ. Rochester, New York), 401 p.
36. Stuiver, M.: 1978, Science 202, pp. 881-883.
37. Muller, R.A.: 1977, Science 196, p. 489.
38. Muller, R.A., Stephenson, E.J., and Mast, T.S.: 1978, Science 201, pp. 347-348.
39. Bennett, C.L., Beukens, R.P., Clover, M.R., Elmore, D., Gove, H.E., Kilius, L., Litherland, A.E., and Purser, K.H.: 1978, Science 201, pp. 345-347.
40. Bennett, C.L., Beukens, R.P., Clover, M.R., Gove, H.E., Liebert, R.B., Litherland, A.E., Purser, K.H., and Sandheim, W.E.: 1977, Science 198, pp. 508-510.
41. Nelson, D.E., Korteling, R.G., and Stott, W.R.: 1977, Science 198, pp. 507-508.
42. Litherland, A.E.: 1978, "Proceedings of the First Conference on Radiocarbon Dating with Accelerators", H.E. Gove, Ed. (Univ. Rochester, New York), pp. 70-113.
43. Muller, R.A.: 1978, "Proceedings of the First Conference on Radiocarbon Dating with Accelerators", H.E. Gove, Ed. (Univ. Rochester, New York), pp. 33-34.
44. Damon, P.E., Lerman, J.C., and Long, A.: 1978, Ann. Review Earth Planetary Sciences 6, pp. 457-494.
45. Stuiver, M.: 1978, Nature 273, pp. 271-273.
46. Olsson, I.U.: 1968, Earth Science Review 4, pp. 203-218.

47. Healy, J., Vucetich, C.G., and Pullar, W.A.: 1964, New Zealand Geological Survey, Bulletin n.s. 73, 88 p.
48. Nambudiri, E.M.V., Teller, J.T., and Last, W.M.: 1980, Geology 8, pp. 123-126.
49. Mullineaux, D.R., Hyde, J.H., and Rubin, M.: 1975, U.S. Geological Survey Jour. of Research 3, pp. 329-335.
50. Westgate, J.A.: 1977, Canadian Jour. Earth Sciences 14, pp. 2593-2600.
51. Fryxell, R.: 1965, Science 147, pp. 1288-1290.
52. Powers, H.A. and Wilcox, R.E.: 1964, Science 144, pp. 1334-1336.
53. Porter, S.C.: 1978, Quaternary Research 10, pp. 30-41.
54. Kohn, B.P. and Topping, W.W.: 1978, Geol. Soc. America Bulletin 89, pp. 1265-1271.
55. Pullar, W.A. and Birrell, K.S.: 1973, New Zealand Soil Survey, Report 1.
56. Vucetich, C.G. and Pullar, W.A.: 1969, New Zealand Jour. Geol. Geophysics 12, pp. 784-837.
57. Crandell, D.R., Mullineaux, D.R., and Waldron, H.H.: 1958, Amer. Jour. Science 256, pp. 384-397.
58. Hay, R.L.: 1980, Nature 284, p. 401.
59. Curtis, G.H.: 1966, "Potassium Argon Dating", compiled by C.A. Schaeffer and J. Zahringer (Springer-Verlag, New York), pp. 151-162.
60. MacGinitie, H.D.: 1953, Carnegie Institute of Washington Publication 599, Contribution to Palentology, 198 p.
61. Epis, R.C. and Chapin, C.E.: 1974, U.S. Geological Survey, Bulletin 1395-C, 21 p.
62. Yeats, R.L. and McLaughlin, W.A.: 1970, Geol. Soc. America, Special Paper 124, pp. 173-206.
63. Natland, Manley: 1953, Pacific Petroleum Geol. News Letter, Pacific Section, Amer. Assoc. Petrol. Geol 7(2), p. 2.
64. Evernden, J.F., Savage, D.E., Curtis, G.H., and James, G.T.: 1964, Amer. Jour. Science 262, pp. 145-198.
65. Steiger, R.H. and Jager, E.: 1977, Earth Planetary Science Letters 36, pp. 359-362.
66. Yeats, R.L., McLaughlin, W.A., and Edwards, G.: 1967, Geol. Soc. America, Special Paper 101, p. 348.
67. Bandy, O.L. and Ingle, Jr., J.C.: 1970, Geol. Soc. America, Special Paper 124, pp. 131-172.
68. Ingle, Jr., J.C.: 1967, Bulletin Amer. Paleontology 52 (236), pp. 217-394.
69. Bandy, O.L. and Wilcoxon, J.A.: 1970, Geol. Soc. America Bulletin 81, pp. 2939-2948.
70. Izett, G.A., Naeser, C.W., and Obradovich, J.D.: 1974, Geol. Soc. America, Abs. with Programs 6, p. 197.
71. Yeats, R.S.: 1965, Amer. Assoc. Petroleum Geologists Bulletin 49, pp. 526-546.
72. Boellstorff, J. and Steineck, P.L.: 1974, Geol. Soc. America, Abs. with Programs 6, pp. 660-661.

73. Boellstroff, J. and Steineck, P.L.: 1975, Earth Planetary Science Letters 27, pp. 143-154.
74. Obradovich, J.D. and Cobban, W.A.: 1975, Geol. Assoc. Canada, Special Paper 13, pp. 31-54.
75. Grim, R.E. and Guven, N.: 1978, "Developments in Sedimentology no. 24" (Elsevier Sci. Pub. Co., New York), 256 p.
76. Dymond, J.: 1969, Earth Planetary Science Letters 6, pp. 9-14.
77. Selli, R.: 1970, Giornale di Geologia 35(1), pp. 51-59.
78. Woodring, W.P., Bramlette, M.N., and Kew, W.S.W.: 1946, U.S. Geological Survey, Prof. Paper 207, 145 p.
79. Woodring, W.P.: 1952, Amer. Jour. Science 250, pp. 401-410.
80. Obradovich, J.D.: 1968, "Means of Correlation of Quaternary Successions", R.B. Morrison and H.E. Wright, Jr., Eds. (Proc. VII Congress Int'l Assoc. Quaternary Research 8), pp. 267-279.
81. Bandy, O.L.: 1967, "Progress in Oceanography 4", M. Sears, Ed. (Pergamon, Oxford), pp. 27-49.
82. Bandy, O.L.: 1968, Paleogeography, Paleoclimatology, Paleoecology 5, pp. 63-75.
83. Bandy, O.L.: 1972, "Proc. Pacific Coast Miocene Biostratigraphic Symposium, 47th Ann. Pacific Section SEPM Convention, March 9-19, Bakersfield, Calif.", E.H. Stinemeyer, Ed, pp. 37-51.
84. Bandy, O.L.: 1972. Paleography, Paleoclimatology, Paleoecology 12, pp. 131-151.
85. Bandy, O.L., Casey, R.E., and Wright, R.C.: 1971, Amer. Geophysical Union, Antarctic Research Series 15, pp. 1-26.
86. Valentine, J.W.: 1961, Univ. Calif. Publications in Geological Sciences 34(7), pp. 309-442.
87. Bandy, O.L. and Casey, R.E.: 1969, Antarctic Journal U.S. 4 (5), pp. 170-171.
88. Obradovich, J.D., Naeser, C.W., and Izett, G.A.: 1978, Stanford Univ. Publications, Geological Sciences 13, pp. 40-41.
89. Wehmiller, J.G., Lajoie, K.R., Kvenvolden, K.H., Peterson, E., Belknap, D.F., Kennedy, G.L., Addicott, W.O., Vedder, J.G., and Wright, R.W.: 1977, U.S. Geological Survey, Open File Report 77-680, 103 p.
90. Jenkins, D.G.: 1974, Contributions, Cushman Foundation for Foraminiferal Research 15, part 1, no. 276, pp. 25-27.
91. Ericson, D.B., Ewing, M., and Wallin, G.: 1963, Science 139, pp. 727-737.
92. Briskin, M. and Berggren, W.A.: 1975, Micropaleontology, Special Publication 1, pp. 167-198.
93. Crouch, R.W. and Poag, C.W.: 1979, Jour. Foraminiferal Research 9(2), pp. 85-105.
94. Berggren, W.A., Phillips, J.D., Bertels, A., and Wall, D.: 1967, Nature 216, pp. 253-254.
95. Berggren, W.A. and Van Couvering, J.A.: 1974, Paleogeography, Paleoclimatology, Paleoecology 16, pp. 1-216.
96. Curry, R.R.: 1966, Science 154, pp. 770-771.

97. Sheridan, M.F.: 1971, Friends of the Pleistocene, Rocky Mt. Sec., Guidebook 16, pp. 1-60.
98. Swineford, A.: 1949, Jour. Geology 57, pp. 307-311.
99. Hibbard, C.W.: 1944, Geol. Soc. America Bulletin 55, pp. 707-754.
100. Hibbard, C.W.: 1958, Amer. Jour. Science 256, pp. 54-59.
101. Hibbard, C.W. and Taylor, D.W.: 1960, Contr. Univ. Michigan Museum Paleontology 16, pp. 1-223.
102. Zakrzewski, R.J.: 1975, Papers on Paleontology, no. 12, in "The Claude W. Hibbard Memorial Volumes, no. 4, Studies on Paleontology and Stratigraphy", J.A. Dorr, Jr. and N.E. Friedland (Museum of Paleontology, Univ. Michigan, Ann Arbor, Michigan), pp. 121-128.

APPROXIMATE DATING OF TEPHRA

Virginia Steen-McIntyre

Department of Anthropology
Colorado State University
Fort Collins, Colorado 80523 USA

ABSTRACT. Tephra studies would benefit greatly from the development of simple, inexpensive, approximate-dating methods that would allow us to estimate the age of a tephra sample in the field office. Colleagues outside the discipline, whose main interest in tephra layers lie in the dates they represent, would benefit also. Two methods that look promising are tephra-hydration dating, and etching of heavy mineral phenocrysts. Basic research into the weathering of tephra will be needed before either method can be used with confidence, but the potential is there. The paper discusses these methods, the problems associated with them, the uses to which they have been put, and the data that have been collected.

1. INTRODUCTION

Tephrochronology has "come of age" in North America, as people in industry, business, and branches of the scientific community other than geology discover the value of tephra layers as time-marker horizons for Quaternary sediments. To such, questions of source vent, chemical composition, or phenocryst content are academic. What they want to know--and quickly--is the age.

To help meet this special need, we must develop approximate dating methods that can be applied in the field office or even under more primitive conditions. The methods should be simple in that they require little in the way of time or special equipment, inexpensive, and roughly accurate. To be of most benefit, they should utilize only components available at the collecting

S. Self and R. S. J. Sparks (eds.), Tephra Studies, 49–64.

site or from the sample bag. In other words, regional studies should not be a necessary prerequisite for the dating of the sample.

Approximate dating methods that meet all the requirements listed above are still in the future. However, two approaches look like promising candidates: tephra-hydration dating and etching of heavy mineral phenocrysts by intrastratal solution.

2. TEPHRA-HYDRATION DATING

The tephra-hydration dating method is similar to obsidian-hydration dating (1,2), in that the hydration of volcanic glass plays an important role in both. In the former, however, the volcanic glass occurs as pumiceous shards of fine-sand size rather than as dense, large fragments of obsidian. The method is described in detail elsewhere (3,4,5 pp. 119-124).

Hydration of glassy tephra occurs as water molecules enter the volcanic glass from surfaces of shards and pumice fragments during weathering. The water source can be either liquid or vapor. As it diffuses slowly into the interior of the glass, it raises the refractive index (n) approximately 0.01 (6, p. 1075 and Table 1,7 pp. 53-58). In glasses of rhyolitic to dacitic composition, hydration is succeeded by superhydration, in which enclosed vesicles slowly fill with excess water through the process of diffusion (8).

The time required for water to fully penetrate volcanic glass and fill the vesicles depends on hydration rate and the specific surface of the fragment. Specific surface is the ratio of total surface area of a fragment to its volume and, for material coarser than silt (62 m), it depends in large part on the vesicularity of the glass and whether the vesicles are connected with the particle surface. For pumiceous tephra from a single blast of an eruption, coarse bombs may hydrate as thoroughly as volcanic ash, provided the samples were collected from a similar environment. Hydration rate is defined as the distance water moves through glass per unit time. It depends greatly upon glass composition and post-depositional environment, especially in regard to soil temperature and chemistry of the groundwater (1,2,8). The presence of crystallites may affect hydration rate (8), and also mode of deposition. Dated samples I have examined suggest that pumiceous dacite collected in a temperate climate hydrates in approximately 15,000 years, whereas fragments of dense glass take somewhat longer. Samples collected in the tropics hydrate at a much faster rate; those from arctic regions much slower, although Friedman and Long (2) feel that hydration will not be affected by water changing to ice and will proceed at surficial temperatures somewhat below $0^{o}C$. Superhydration in cold climates may take more than ten million years to completely fill vesicles with water (8).

2.1 Procedure

To apply the tephra-hydration dating method, one first notes the volume of hydrated glass in individual, naturally-fragmented shards of fine-sand size (extent of hydration), then estimates the average volume of water in selected glass vesicles of 100 different fragments (extent of superhydration). The measurements are then compared with those for similar samples of dated tephra in order to obtain an approximate age for the undated sample. Other factors equal, the greater the volume of hydrated glass and the more water in the vesicles, the older the sample.

To estimate extent of hydration, a sprinkle of loose grains of fine-sand size (-100+300 mesh) is mounted in a medium slightly higher than that of the hydrated glass. For a semi-permanent mount, I find that Preservaslide ($\underline{n} \sim 1.52$) works well. In such a mount, the Becke line moves outward from the nonhydrated core to the hydrated rind, and from the rind to the mounting medium when the focus is raised. The hydrated rind will often show strain birefringence.

Another approach would be to mount the grains in a high-dispersion immersion oil that approximately matches the refractive index of the nonhydrated glass (about 0.01 lower than $\underline{n}$ of the hydrated glass), and view them with a microscope equipped with focal masking. The combination of a white-light illumination source, high-dispersion index oil (9), and the focal masking methods of illumination (10) produces brilliant rims of color at the apparent boundary between the hydrated and nonhydrated glass, and at the oil-shard interface. Using the central focal masking method, for example, which produces limpid colors on a black background, the nonhydrated glass cores are stained violet or blue, and the hydrated glass rinds are outlined in pink or orange. With either of the above methods, the apparent thickness of the hydration rinds can be measured directly with a calibrated micrometer ocular and a magnification of 400-500x.

To determine extent of superhydration, one estimates visually the volume of water in spindle-shaped vesicles 10 to 50 µm long that are enclosed within 100 separate glass shards. Fragments either can be mounted in oil for this examination or in a synthetic resin like Preservaslide, which has a low $\underline{n}$ and hardens at room temperature. Conventional mounting media such as Canada balsam ($\underline{n}$. 1.53-1.55) and Lakeside #70 ($\underline{n}$ 1.54) are not recommended. The relatively high $\underline{n}$ causes the shards to appear in strong negative relief which masks internal features such as vesicles. Also, they require high heat to liquify; a condition that may alter the hydration patterns of the glass shards.

A slide containing the glass fragments is placed in a mechanical stage and the shards are scanned with the microscope, using a previously calibrated lens system consisting of a x45 or x50 dry objective and a micrometer ocular or cross-hair ocular with magnification in the range x8 to x10. When a shard is found that contains one or more enclosed vesicles of the specified size and shape, the vesicles are examined carefully, and the average water volume, determined by visual estimation, is tallied in one of seven columns ($\leq$0.1 (none), $\leq$1, $\leq$5, $\leq$10, $\leq$33, $\leq$67, $\leq$100% water). When the vesicles of 100 shards have been examined in this manner, the total marks per column are tallied and plotted as a cumulative curve on semi-log paper. The curve represents extent of superhydration for the sample.

2.2 Examples

In Tables 1 through 4, examples are given of hydration data for relatively young, glassy tephra samples collected in the State of Washington, central El Salvador, and the Mediterranean area. Additional data for Glacier Peak tephra samples and for the younger St. Helens eruptions are given elsewhere in this volume (11, Table 3). Hydration data for tephra layers from the Laguna de Fuquene cores, Colombia, can be found in (12).

TABLE 1. HYDRATION DATA FOR SAMPLES OF PUMICEOUS GLASS FROM COARSE-GRAINED GLACIER PEAK TEPHRA, 12,000 YEARS OLD, COLLECTED ALONG AN EAST-WEST TRANSECT, CENTRAL WASHINGTON STATE

[A]SAMPLE NO.	SITE INFORMATION				HYDRATION DATA							
(FRAGMENT SIZE, MM)	ELEV. (M)	MEAN ANNUAL T., °C. (MAX.)	MEAN ANNUAL PRECIP. (CM)	VEGETATION ZONE	EXTENT HYDRATION: APPARENT RIND THICKNESS IN μM (ACTUAL VALUES)	EXTENT SUPERHYDRATION: VOL. % H_2O IN VESICLES OF 100 SHARDS $\leq$0.1	$\leq$1	$\leq$5	$\leq$10	$\leq$33	$\leq$67	$\leq$100 %
GP-1 (15-66)	1,830	--	>152	ABIES AMABILIS - TSUGA MERTENSIANA	5,5 (6,5,5,6)	97	3					
GP-2 (20-55)	1,160	6 (11)	>89	ABIES LASIOCARPA - PICEA ENGELMANNI	5+ (5+,5+,5+,5+)	96	4					
GP-3 (5-35)	730	--	~127	TSUGA HETEROPHYLLA - THUJA PLICATA	5,5 (5+,6,5,6)	95	5					
GP-4 (1-11)	460	10 (16)	>28	PSEUDOTSUGA MENZIESII	[B]7 (7,-,-,-)	89	11					
[C]GP-5 (0.5-2)	340	10 (16)	>25	ARTEMISIA TRIDENTATA - AGROPYRON SPICATUM	7 (7,7,7,-)	66	30	4				

[A] SEE STEEN (7) FOR MORE DETAILED INFORMATION.

[B] I COULD FIND ONLY ONE RIND THAT COULD BE MEASURED: THE REST WERE TOO THICK. THEORETICALLY, OTHER RINDS COULD BE MEASURED IF THE TOP OF THE SHARDS WERE FIRST GROUND AWAY.

[C] SAMPLE COLLECTED FROM SOUTH-FACING SLOPE. GP-4 COLLECTED FROM WEST-FACING SLOPE.

Samples of coarse-grained Glacier Peak tephra are examined in Table 1. The samples were collected along a transect that extended from a cool, wet site (GP-1) situated in a stunted stand of western white pine (*Pinus monticola*) and subalpine fir (*Abies lasiocarpa*) eastward to a warm, dry, treeless site (GP-5) with sage (*Artemesia tridentata*) and cheat grass (*Bromus tectorum*) as the prominent vegetation. As one moves eastward along the transect, the samples become progressively more hydrated. This so, even though annual precipitation decreases. Friedman and colleagues (1,2) find that elevated temperatures cause dense obsidian to hydrate more rapidly, and the same appears to be true for pumiceous dacite. Particle size may also play a role. Even the aspect of the collection site may influence hydration rate, at least for samples collected at or near the ground surface. Note that sample GP-5, collected from a bare quarry on a south-facing slope has significantly more water in the vesicles than sample GP-4, collected in a shaded wood on the east side of a valley, even though the present mean air temperature for the two sites is very similar.

In Table 2, hydration data are given for glassy fragments collected from a soil profile at site GP-2 (Table 1). Samples from the A2 and B2 soil horizons consist of naturally fragmented shards of fine-sand size. The "B3" sample is of crushed pumice fragments 20-55 mm in length, collected from the B2 horizon. The hydration pattern is very similar for the coarse and fine fragments collected from the B2 horizon. This would suggest that, at least at site GP-2, rate of hydration is not influenced by particle size for fragments as coarse as fine sand. Fragments silt size and smaller may pose another problem (12). Tephra fragments from the A2 horizon are much less hydrated than the fragments from the underlying B2 horizon. In this particular case, the lack of significant hydration is not a function of soil profile development, but rather of layered parent material. The volcanic glass shards from the A2 horizon, which in morphology

TABLE 2. HYDRATION DATA FOR GLASSY TEPHRA FRAGMENTS FROM A SOIL PROFILE IN THE *ABIES LASIOCARPA* - *PICEA ENGELMANII* VEGETATION ZONE, GLACIER PEAK AREA, CASCADE RANGE, WASHINGTON

[A] SAMPLE NO. (FRAGMENT SIZE)	SITE INFORMATION: ELEVATION (M)	MEAN ANNUAL T., °C (MAX.)	MEAN ANNUAL PRECIP. (CM)	HYDRATION DATA: EXTENT HYDRATION — APPARENT RIND THICKNESS IN μM (ACTUAL VALUES)	EXTENT SUPERHYDRATION — VOL. % H_2O IN VESICLES OF 100 SHARDS: ≤0.1	≤1	≤5	≤10	≤33	≤67	≤100 %
GP-2A2 (FINE SAND)	1,160	6 (11)	>89	2 (2,1,3,2)	97	3					
GP-2B2 (FINE SAND)				5+ (5,6,5,5)	98	2					
GP-2B3 (20-55 MM)				5+ (5+,5+,5+,5+)	96	4					

[A] SEE STEEN 7), STEEN AND FRYXELL (13) FOR MORE DETAILED INFORMATION.

and refractive index are quite different from the Glacier Peak material, are from a younger ashfall.

TABLE 3. HYDRATION DATA FOR GLASSY FRAGMENTS OF TBJ TEPHRA, 1700 YEARS OLD, CAMBIO ARCHAEOLOGIC SITE EL SALVADOR, CENTRAL AMERICA

A. SAMPLE NO.	TYPE OF DEPOSIT	SIZE OF FRAGMENTS SAMPLED	HYDRATION DATA							
			EXTENT HYDRATION	EXTENT SUPERHYDRATION						
			APPARENT RIND THICKNESS IN μM (ACTUAL VALUES)	VOL. % H_2O IN VESICLES OF 100 SHARDS						
				≤0.1	≤1	≤5	≤10	≤33	≤67	≤100 %
BASE										
1-FINE	AIRFALL	FINE SAND	6- (6,6,6,5)	81	18	1				
1-COARSE		5-25 MM	5 (5+,6+,5,3)	80	20	-				
2		FINE SAND	6- (6,6,5+,6)	84	14	2				
3			6 (6,6,6,6)	93	6	-	-	-	1	
4			6+ (7,6,6,6)	83	16	-	-	-	1	
5	ASHFLOW		8+ (10,9,7.7)	81	13	6				
6	AIRFALL		6,5 (9,6,6,5)	94	6					
TOP										

A. SEE HART AND STEEN-McINTYRE (14) FOR MORE DETAILED INFORMATION.

Table 3 displays data for a vertical series of samples taken from a wall in the Cambio excavation, central El Salvador. Like Glacier Peak tephra, the tbj tephra has a chemical composition somewhere in the dacite range. Note, however, how much more rapidly dacite hydrates in a tropical climate. The six tbj samples are essentially the same age; petrographic studies (14) show they also are of essentially the same composition. They have not hydrated at the same rate, however. Sample 5, which we believe from toher evidence came from the distal end of an ashflow, has hydrated at a significantly higher rate compared to the other samples. Both it and sample 6, thought to represent the airfall portion of the ignimbrite eruption, exhibit a relatively wide range in rind-thickness measurements even for the small number of rinds measured. Apparently, something other than weathering environment and composition affected the hydration rate of these samples, and that "something" probably involves conditions that existed at or near the source vent at the time of the eruption.

The Franchthi ash and deep-sea ash V10-58 are actually two samples of the same tephra horizon (15). In Table 4, hydration data are compared for these samples. The pattern of hydration is very similar for both, confirming the fact that the amount of water does not play a significant part in hydration rate provided vapor pressure is sufficient to saturate the glass surface (2, p. 352). The data also imply that hydration rate has been very similar for the two samples, even though one was collected from a cave, the other from the sea bottom.

TABLE 4. HYDRATION DATA FOR GLASSY FRAGMENTS OF FRANCHTHI TEPHRA AND DEEP-SEA ASH V10-58, [A]27,000 YEARS (?), MEDITERRANEAN AREA

[B]SAMPLE	DEPOSITIONAL ENVIRONMENT	HYDRATION DATA: EXTENT HYDRATION — APPARENT RIND THICKNESS IN μM (ACTUAL VALUES)	EXTENT SUPERHYDRATION — VOL. % H_2O IN VESICLES OF 100 SHARDS: ≤0.1	≤1	≤5	≤10	≤33	≤67	≤100 %
FRANCHTHI ASH	CAVE	5+ (5,5,6,5)	27	51	19	-	3		
[C]V10-58 (650 CM)	SEA BOTTOM	5+ (5,6,5,5)	35	39	20	5	1		

[A]SEE VITALIANO, THIS VOLUME, FOR MORE DETAILED INFORMATION.

[B]SAMPLE OF FRANCHTHI ASH SUPPLIED BY W.R. FARRAND, UNIVERSITY OF MICHIGAN; SAMPLE OF V10-58 TEPHRA BY F. McCOY, LAMONT-DOHERTY GEOLOGICAL OBSERVATORY. THE TWO SAMPLES ARE PART OF THE SAME TEPHRA FALL.

[C]SAMPLE IS FROM THE LOWER, COARSE-GRAINED UNIT.

To conclude our examples of hydration data, we will consider an older set of samples--a suite of dated tephra from vents in Yellowstone National Park. Figure 1 gives superhydration curves for eight dated samples of Yellowstone tephra. They were provided by G.M. Richmond and R.E. Wilcox, U.S. Geological Survey.

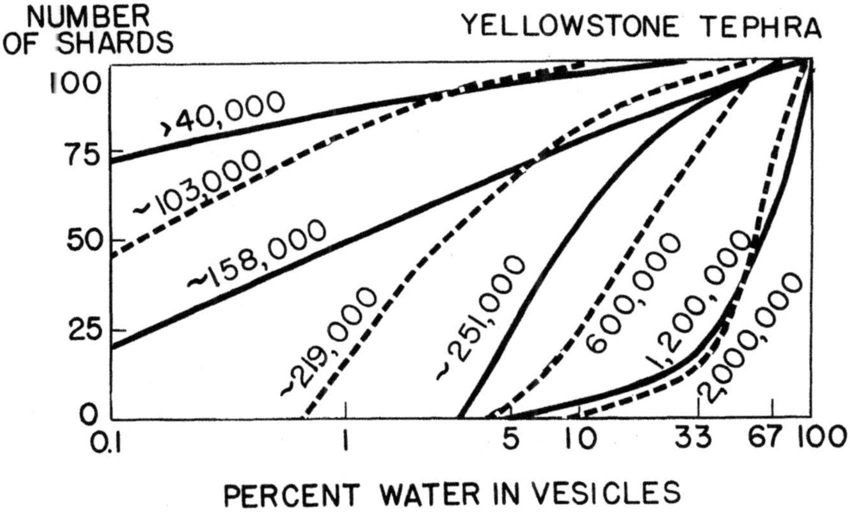

Figure 1. Superhydration curves for a suite of dated samples of Yellowstone tephra. Each curve represents one sample. Curves 40,000 to 251,000 years are unnamed. Older curves are: 600,000 years, Pearlette type-O; 1,200,000 years, Pearlette type-S, 2,000,000 years, Pearlette type-B.

Samples that range in age from 40,000 to 251,000 years were collected in or near the park in a cool, moist environment. The older samples were collected from farther away. A good spread in the curves can be seen, at least for the younger set of samples. These not only have a similar chemical composition, but were collected from a similar weathering environment. The spread is not as good for the older samples. Part of this may be due to a difference in depositional environment, but I believe most of the blame must be placed on the crude method that I use to measure the water of superhydration. It is a simple task to decide by visual examination whether a vesicle is empty or has a trace of water; it is impossible to decide if the vesicle is 65 or 70 per cent filled.

Figure 2 shows the range in extent of superhydration for four samples of dated Pearlette type-O tephra, 600,000 years old. Mean temperature at the Onion Creek site is high; that at Bear Canyon relatively low. Once again, temperature appears to play an important part in the rate of superhydration of pumiceous glass, although almost certainly there are other factors involved.

The data in Tables 1-3 demonstrate how hydration rate and extent are affected by: environmental factors and, possibly,

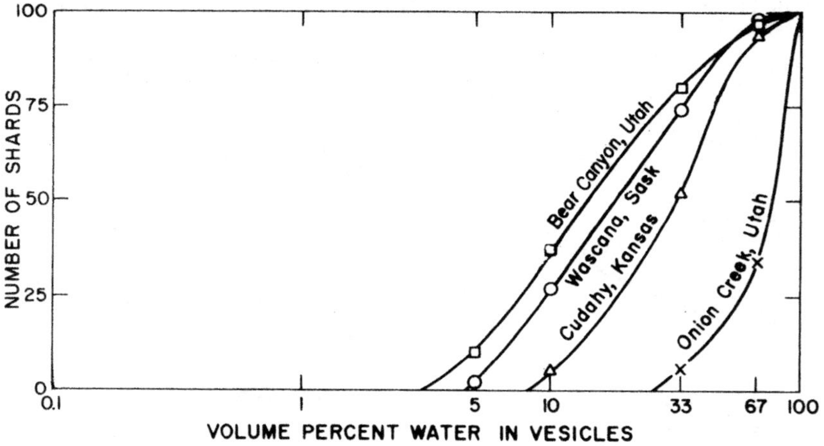

Figure 2. Superhydration curves for four dated samples of Pearlette type-O tephra, 600,000 years old (4). The difference in extent of superhydration is due primarily to environmental factors that have existed at the sites.

particle size (Table 1); time (Table 2); and mode of emplacement (Table 3). Table 4 shows that hydration rate is independent of the amount of water present. From Figure 1, it appears that a suite of samples from the same magma body will show progressive superhydration with age, provided the samples are collected in the same type of environment. Figure 2 shows the range of superhydration that can occur when samples are collected from different environments. In cases where the glass is incompletely hydrated, extent of hydration, as evidenced by apparent thickness of the hydration rinds, shows a more consistent change than extent of superhydration; a result, at least in part, of the measuring methods employed.

3. ETCHING OF HEAVY-MINERAL PHENOCRYSTS

Far less developed than the tephra-hydration dating method, but just as promising as an approximate dating technique is that which estimates age by noting the extent of etching of heavy-mineral phenocrysts in the tephra sample. The following information is taken from "A Manual for Tephrochronology" (5, pp. 124-128).

Intrastratal solution is that which takes place within a sedimentary bed after deposition (16, p. 674). It produces in certain heavy minerals a characteristic etched appearance that I have referred to as "picket-fence" structure and that has also been called "hacksaw" or "cockscomb" structure. Not all minerals in a sample are equally affected by intrastratal solution; they seem instead to be selectively etched, with the phenocrysts of basic magmas being most readily attacked (17).

Several workers have made effective use of the selective solution of heavy minerals to obtain relative ages for deposits in North America (18-24). Others point out that intrastratal solution depends not only on time, but also on grain size of the sedimentary unit (25), weathering environment (23,26) (Hay, written communication 1975), and Eh, pH, and hydraulic gradient of the groundwater (24).

Heavy-mineral phenocrysts from within pumiceous tephra clasts are ideal specimens to observe for signs of etching. Because they are contained within larger fragments, there is no danger of contamination from older, more highly etched minerals of the same species. In addition, the glassy envelope that surrounds the crystal protects it from the immediate effects of weathering (26). This is an especially valuable consideration for tephra samples collected in the tropics, where chemical weathering is both rapid and severe.

The extent of etching of phenocrysts seems to depend on mineral species, time, climate, and on the chemical composition of the minerals and glass. On the Island of Hawaii, phenocrysts from within clasts of basaltic Pahala tephra (10,000 -17,000 years) have begun to alter, while those from younger beds have not (26). On St. Vincent in the West Indies, andesitic tephra deposits approximately 30,000 years old show decomposed lapilli fragments and slightly etched hypersthene crystals (27) (Hay, written communication, 1975). In temperate Central Mexico, etching of hypersthene phenocrysts from dacitic tephra deposits is rare and incipient in units 22,000-24,000 years old, and entirely lacking in younger units (P.W. Lambert, personal communication, 1973). For dacitic tephra fragments approximately 250,000 years old, collected in the same general area in Mexico, hypersthene is severely etched, augite is slightly etched, and the hornblende still looks fresh.

3.1 Procedure

The method I use to measure extent of mineral etching is very crude indeed. Fine-grained samples are washed in an ultrasonic cleaning bath to remove clay and silt, and the residue is rinsed in acetone, dried, and sieved. Coarse fragments are first gently crushed, then treated in like manner. Fragments of fine-sand size are mounted on a microscope slide and the heavy minerals are examined for signs of etching. The mineral species is noted, then whether the crystals appear fresh, slightly etched (tips only), strongly etched (tips and sides), or very strongly etched (almost completely eaten away).

3.2 Examples

Within recent years I have twice used the etching of heavy-mineral phenocrysts as an approximate dating technique to help solve archaeological problems. The first case involved the Classic Maya and Their affect on the environment in which they lived and worked; the second, the antiquity of Man in the New World.

Deevy and his associates have recently published a major paper describing paleoecological studies of lake-core sediments from the Peten, northern Guatemala (28). Two of the lakes, Yaxha and Sacnab, were surrounded by a Mayan urban center, especially during Classis and Post-classic time. The authors make special mention of "Maya clay", thick deposits of silty montmorillonitic clay they believed to be the insoluble residue of the limestone country rock, washed into the lakes as a result of human activity. The sediment was deposited rapidly. It is silica-rich and, in case of Late Preclassic and Early Classic Yaxha material, it is notably deficient in phosphorous. Such data would also characterize fine-grained dacitic ash.

At the time the paper was published, Hart and I had been preparing a manuscript on a major tephra eruption, the tbj eruption, that occurred in El Salvador in Protoclassic (very Early Classic) time (14). How far the tephra cloud had extended we did not know, but we felt quite certain it had covered the Peten (11,14). Could the "Maya clay" actually be, at least in part, fine-grained tbj tephra? Deevy supplied us with samples of his core sediments to test this hypothesis. When the clay was removed, a few silt-size grains remained--primarily devitrified shards of volcanic glass along with some relatively fresh shards on which hydration rinds could be measured. Also present were fresh, glass-mantled phenocrysts, including olive-green amphibole and hypersthene, common constituents of the tbj tephra. Most assuredly, these fragments could not have been the insoluble residue of a Tertiary limestone. Work on the problem continues.

The second case concerned the dating of Hueyatlaco, an Early Man site in Central Mexico (11,29,30). Several lines of geologic evidence, including uranium-series dates on butchered bone found associated with bifacial tools, suggested an age for the site of 250,000-300,000 years, an age that few archaeologists are willing to consider. The sedimentary section at the site contains four tephra units; one unit occurs in the zone that has produced the bifacial tools and three are found topographically (and stratigraphically) higher. Two of the upper units have been dated by the fission-track method, and the dates obtained agree with the bone dates. The tephra unit associated with deposits in which the bifacial tools were found contains no zircon and could not be dated by this means. Fortunately, this lowest tephra layer, a coarse ash, contains phenocrysts of hypersthene. When the ash fragments were crushed and the fine-sand fraction viewed under the microscope, the hypersthene crystals were found to be strongly etched, even more so that hypersthene from one of the upper, dated tephra units. The possibility seems remote that this tephra unit could be younger than the dated tephra horizon several metres above it, or that the series of tool-producing deposits, of which it is an integral part, could be younger than 250,000 years.

4. PROBLEMS AND POSSIBLE SOLUTIONS

It is obvious from the above that, while the approximate dating methods discussed in this paper hold promise, much work remains to be done before they can be applied confidently to undated tephra samples. The problems to be solved are to be found both in the laboratory and in the field.

4.1 Laboratory

To begin with, it will be necessary to develop more accurate methods of measuring data. In the case of hydration extent, this could be accomplished by first grinding away the tops of partially-hydrated shards to expose true rind thickness, then measuring carefully with a x100 oil immersion lens and a split-image (Filar) micrometer (1). Pierce (30) finds he is able to measure rinds on obsidian fragments to ±0.3 μm using this method. Friedman and Long claim an accuracy of ±0.1 μm (2).

Obtaining an accurate measure of water of superhydration will be more difficult, perhaps next to impossible without a well equipped laboratory. Glass continues to take up water after the hydration front has passed through (Friedman, oral communication 1979). This may account for the slight but steady increase in $\underline{n}$ with progressively finer fractions of volcanic glass from the same sample (12). Measuring techniques that use differential thermal analysis or a combination of oven and continuous weighing balance should be tried, but with caution. Pumice fragments that are superhydrated tend to puff and pop if heated too rapidly--a potential danger to expensive equipment.

And how does one describe "degree of shagginess" of mineral crystals beyond the terms already suggested? Locke (23) has developed a quantitative method that measures the maximum depth of etching on 100 grains from one sample, then uses the data to compute the mean maximum etching depth (MMED) for the sample. This method, of course, can only be used on grains that still show a trace of the original outline.

Finally, there is the composition of the glass itself. Hydration rate depends in large part on chemical composition. Increased SiO_2 increases hydration rate, but increased CaO and MgO decreases it. Al_2O_3, FeO, Na_2O and K_2O content appear to have little effect, one way or the other (2). To study the effect of glass composition on hydration rate, it will be best to first limit investigation to a series of dated tephra samples that originated from the same parent magma; the St. Helens suite, for example, or the Yellowstone suite. Within such a group of samples, one could then concentrate on layers that contain more than one type of glass--the St. Helens Y_n would be a good choice (11, Table 3). It goes without saying that the samples should be collected from a similar weathering environment.

4.2 Field

This brings us to the second area beset by problems facing those interested in using the approximate dating methods described in this paper. The field. How can one possibly assess all the environmental and other factors that can influence hydration rate of glassy tephra? Etching of heavy-mineral phenocrysts? Where does one collect samples so that the effect of these factors can be minimized?

To help answer the first question, we may borrow a mode of thinking used by soil scientists when they attempt to isolate the various factors that interact to form a soil. Jenny (32) recognized five independent variables which describe the soil system and placed them in a fundamental "equation" of soil forming factors:

$$s \propto f(cl, o, r, p, t \ldots);$$

i.e., the characteristics of any given soil, including its chemical and physical properties, are dependent upon climate (cl), organisms (o), relief (r), parent material (p), time (t), and perhaps on other factors not yet recognized.

We may treat hydration rate and extent of volcanic glass in like manner. Drawing upon what little we know:

$$HR \propto f(p, tp, ch? \ldots);$$

or in other words, as far as we can tell, hydration rate (HR) is a function of parent material (p), especially chemical composition of the glass fragments and possibly particle size; temperature (tp), both short-term elevated temperatures that may be experienced during transport and long-term temperature variations at the site of deposition; and probably the chemistry of fluids circulating through the deposit (ch). Hydration extent (HE) is a function of hydration rate, specific surface (ss), and time (t).

$$HE \propto f(HR, ss, t \ldots);$$

Mineral etching may be treated in much the same way:

$$E \propto f(p, cl, ch, r, t \ldots);$$

where extent of etching (E) is dependent on parent material (p), especially as related to the size of the fragments, mineral species, chemical composition of the mineral and glassy envelope, and specific surface of the surrounding glass; climate (cl), especially temperature and effective moisture; chemistry of the groundwater (ch); relief (r), a measure of hydraulic gradient for near-surface samples; and time (t). Undoubtedly, in all three "equations" there are other factors that we have yet to recognize.

Of the factors listed above, two--time and parent material--can be regarded as being relatively constant for coarse fragments (bombs to coarse ash) from a single tephra blast. Specific surface of the particles will vary with size and with conditions that prevailed at the vent during the eruption. The other factors; temperature, chemistry of circulating fluids, effective precipitation, and relief are dependent in most part on local conditions and vary from site to site.

To minimize the effect of temperature on hydration rate of a sample, samples should be collected from sites that experience little diurnal or seasonal temperature variation. Deep-sea cores look like a promising choice if such a choice exists, or caves and lava tubes. For surface sites, samples should be collected at a depth of at least 0.5 m (2). Pierce (30) favors a minimum depth of 1.5 m, and prefers 3 m. In the Northern Hemisphere, sites on the north side of hills may prove to be a good selection. Such sites experience little direct sunlight, if any. Be especially wary of tephra associated with charcoal, samples collected from a geothermal area, and ashflow deposits.

It is a difficult task to minimize the effects of environmental factors on the rates of glass hydration and mineral etching. One usually must take what the site of interest provides and make the best of it. Be cognizant, however, of the possible influences of these factors. The effect of temperature has already been discussed. In addition, note the dominant vegetation, soil profile development, and drainage at a collecting site. The chemicals released by decomposing pine needles, for example, are very different from those found under rotted oak leaves. Volcanic glass taken from a podzolic A2 horizon may differ significantly from the same glass taken from the underlying B or C horizons because of the intense leaching that takes place in the A2. If a choice exists, it is best always to sample B or C soil horizons, be they buried or active, instead of A horizons. Tephra fragments collected from a well-drained site may be more weathered than those collected in a bog--or less! It will depend upon the chemistry of the groundwater.

5. CONCLUSION

Clearly, many problems await those who wish to use weathering phenomena to develop approximate dating methods for samples of glassy tephra. Some progress has been made in recognizing factors other than time that affect the hydration extent of volcanic glass and the etching of heavy-mineral phenocrysts, but much still needs to be learned. The reward, however, seems well worth the effort that is required to do the necessary basic research.

REFERENCES

1. Friedman, I. and Smith, R.L.: 1960, Am. Antiquity 25, pp. 476-522.
2. Friedman, I. and Long, W.: 1976, Science 191, pp. 347-352.
3. Steen-McIntyre, V.: 1973, Internat. Union Quat. Res. (INQUA), Ninth Congress, Abs., pp. 342-343.
4. Steen-McIntyre, V.: 1975, in Quaternary Studies, R.P. Suggate and M.M. Cresswell, eds., Royal Soc. New Zealand, Wellington, pp. 271-278.
5. Steen-McIntyre, V.: 1977, A manual for tephrochronology, published by the author, Idaho Springs, Colorado 80452, 157 p.
6. Ross, C.S. and Smith, R.L.: 1955, Am. Mineralogist 40, pp. 1071-1089.
7. Steen, V.C.: 1965, unpublished M.S. thesis, Geology, Washington State Univ., Pullman, 147 p.
8. Roedder, E. and Smith, R.L.: 1965, Geol. Soc. America Sp. Paper 82, p. 164.
9. Wilcox, R.E.: 1964, Am. Mineralogist 49, pp. 683-688.
10. Wilcox, R.E.: 1962, in Proc. Internat. Microscopy Symposium, 1960, Chicago, W.C. McCrone, ed., McCrone Associates, pp. 160-165.
11. Steen-McIntyre, V.: this volume, Tephrochronology and its application to problems in New-World archaeology.
12. Riezebos, P.A.: 1978, Quat. Res. 10, pp. 401-424.
13. Steen, V. and Fryxell, R.: 1965, Science 150, pp. 878-880.
14. Hart, W.J., Jr. and Steen-McIntyre, V.: in press, in Volcanism and the prehistory of the Zapotitan Valley, P.D. Sheets, ed.
15. Vitaliano, C.J.: this volume, Tephra deposits of the Franchti Cave, Kheilada, Peleponesus, Greece.
16. Pettijohn, F.J.: 1957, Sedimentary rocks, 2nd ed., Harper and Brothers, New York, 718 p.
17. Goldich, S.S.: 1938, Jour. Geology 46, pp. 17-58.
18. Birkeland, P.W.: 1964, Jour. Geol. 72, pp. 810-825.
19. Birkeland, P.W.: 1974, Pedology, weathering and geomorphological research, Oxford Univ. Press, New York, 285 p.
20. Bradley, W.C.: 1957, Geol. Soc. America Bull. 68, pp. 421-444.
21. Bradley, W.C. and Addicott, W.O.: 1968, Geol. Soc. America Bull. 79, pp. 1203-1210.

22. Shroba, R.R.: 1976, Geol. Soc. America Abs. with Programs 8, p. 629.
23. Locke, W.W., III: 1979, Quat. Res. 11, pp. 197-212.
24. Walker, T.R.: 1967, Geol. Soc. America Bull. 78, pp. 353-368.
25. Blatt, H. and Sutherland, B.: 1969, Jour. Sed. Petrology 39, pp. 591-600.
26. Hay, R.L. and Jones, B.F.: 1972, Geol. Soc. America Bull. 83, pp. 317-332.
27. Hay, R.L.: 1959, Jour. Geology 67, pp. 540-562.
28. Deevy, E.S., et al.: 1979, Science 206, pp. 298-306.
29. Steen-McIntyre, V., Fryxell, R. and Malde, H.E.: 1973, Geol. Soc. America Abstracts with Programs 5, p. 820.
30. Steen-McIntyre, V. Fryxell, R. and Malde, H.E.: in preparation.
31. Pierce, K., Obradovich, J.D. and Friedman, I.: 1976, Geol. Soc. America 87, pp. 703-710.
32. Jenny, H.: 1941, Factors of soil formation, McGraw-Hill Book Company, Inc., New York, 281 p.

THE INTERRELATIONSHIP BETWEEN MAGNETOSTRATIGRAPHY AND TEPHROCHRONOLOGY

Kenneth L. Verosub

Department of Geology
University of California - Davis
Davis, California 95616 USA

ABSTRACT. Magnetostratigraphy and tephrochronology represent complementary techniques for the determination of the ages of sedimentary sequences. Each method is subject to its own limitations. The magnetic polarity zonation of a sequence may not be sufficiently characteristic to correlate it to the magnetic polarity time scale. The presence of a single characteristic tephra layer does not provide detailed information about the time span represented by the entire sequence. The combined use of magnetostratigraphy and tephrochronology can often overcome these limitations as in the case of a single, dated tephra layer which serves to remove any ambiguity from the correlation of the magnetic polarity zonation. For sediments deposited entirely within a single polarity interval, geomagnetic excursions may eventually serve as useful stratigraphic markers; however at the present time problems associated with the existence and nature of geomagnetic excursions preclude their use as marker horizons.

Tephrochronology and magnetostratigraphy have both evolved relatively recently as new techniques which can be used for the dating of late Cenozoic sedimentary deposits. The two techniques are complementary and, when used together, can often provide considerably more information about the temporal context of a sedimentary deposit than either technique can provide by itself. This paper discusses some of the ways in which the paleomagnetic study of sediments can be combined with the analysis of one or more intercalated tephra layers.

The technique of magnetostratigraphy is based on the recovery of a record of the intervals of normal and reversed

S. Self and R. S. J. Sparks (eds.), Tephra Studies, 65–72.

polarity of the earth's magnetic field which occurred during the accumulation of a sedimentary sequence. A primary requirement for application of the technique is that the sedimentary sequence span sufficient time as to encompass at least one and preferably several polarity changes. During the late Cenozoic, periods of constant polarity have been as short as 20,000 years and as long as 730,000 years (Fig. 1). Thus, in general, a sedimentary sequence must span at least several hundred thousand years in order that the magnetostratigraphic technique can be applied.

The first step in applying magnetostratigraphy to a sedimentary sequence is to develop a magnetic polarity zonation for the sequence. This represents the record of normal and reversed polarities which occurred during the deposition of the sequence and is obtained by analysis of the directions of magnetization of paleomagnetic samples. Typically, three samples are taken from each sampling horizon, and the vertical spacing between horizons is a fixed interval which depends upon the thickness of the sequence and the approximate sedimentation rate inferred from the depositional environment. For sediments which accumulate rapidly, such as estuarine, lacustrine and terrestrial deposits, an interval of 1 to 10 meters is usually sufficient to ensure that even the shortest polarity interval is recorded by at least three to five successive horizons.

One of the advantages of the magnetostratigraphic technique is that it is only necessary to determine the polarity of a given horizon rather than to determine the actual original direction of magnetization of that horizon. Thus in many cases, sediments which contain some secondary magnetization, or overprint, can still be studied through magnetostratigraphy. The reason for this is that the secondary magnetization is often a viscous remanence which, in the present earth's magnetic field would represent a normal component of magnetization superimposed on the original normal or reversed component. By examing the final direction of magnetization of a sample after a single magnetic cleaning, or even better, by examining the changes in the direction of magnetization during a step-wise magnetic cleaning, one can usually determine unambiguously the polarity of the original component of the magnetization. Thus a sample with a normal primary component and a normal overprint will have a normal direction which is not affected by magnetic cleaning. On the other hand a sample with a reversed primary component and a normal overprint will usually exhibit either an actual reversed direction or else a normal direction which clearly moves toward a reversed direction during magnetic cleaning.

Different workers have used different criteria for determining the polarity zonation of a sedimentary sequence (4 - 7). All of the different methods have one point in common however,

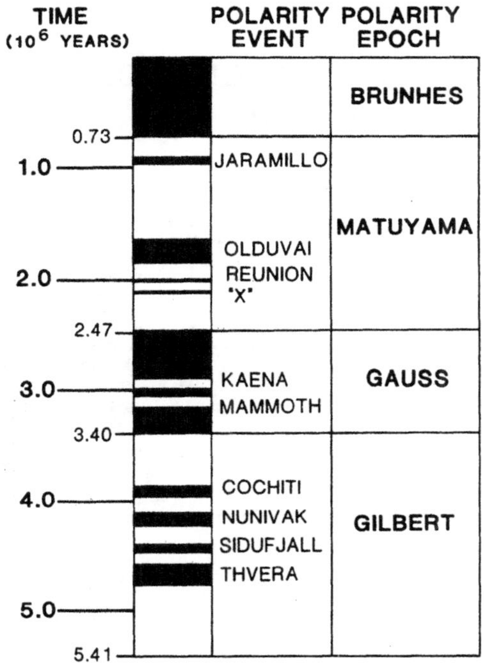

Fig. 1. The magnetic polarity time scale for the last 5.4 million years. From (1) as modified by (2, 3). Solid area is normal polarity, open area is reversed polarity.

namely that they discriminate in favor of samples which show any tendancy towards a reversed polarity. The justification for this is that since the secondary component is very likely to be of normal polarity, any indication of a reversed direction in the measured magnetization probably represents a reversed original component. In our work (8), we have established a hierarchy of criteria to determine first whether a given sample has normal, reversed, or intermediate polarity, then whether a given horizon has normal, reversed, or intermediate polarity, and finally whether a given interval of sediment has a normal, reversed, or intermediate polarity. Although these criteria are rigorously applied, they usually serve to confirm the conclusions drawn by visual inspection of data (Fig. 2).

The procedures outlined above determine the magnetic polarity zonation of the sedimentary sequence. In order to use this magnetic polarity zonation to date the sedimentary sequence, one must correlate the zones of the sequence to the known magnetic polarity time scale. The magnetic polarity time scale

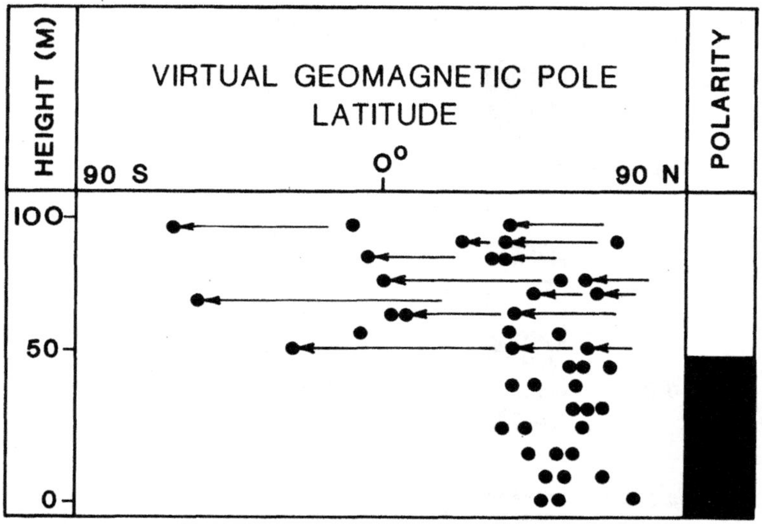

Fig. 2. Magnetic polarity zonation of a portion of a sedimentary sequence. Circles denote the latitude of the virtual geomagnetic pole corresponding to the final paleomagnetic direction, arrows indicate change in latitude as a result of magnetic cleaning. There are three samples per sampling horizon.

represents the best determination of the behavior of the magnetic field through time (2). The basis for the magnetic polarity time scale is ultimately the record of marine magnetic anomalies on the seafloor (9). However, the pattern of the last 6.5 million years has been most accurately determined and dated radiometrically through the study of sequences of Icelandic lavas (1). The problem in making a correlation is that the only known feature which distinguishes an interval of one polarity from another of the same polarity is the duration of the polarity interval. A correlation of the magnetic polarity zonation to the magnetic polarity time scale on the basis of the pattern of normal and reversed polarities, that is, on the apparent relative duration of each polarity interval, involves the very explicit assumption of a constant sedimentation rate. In a deep-sea environment where the nature of the sediment and the rate of deposition may be quite uniform, such an assumption is probably valid. However in a near-shore or terrestrial sequence there may be important changes in the sedimentary facies as well as climatically-controlled changes in sedimentation within the same facies. Under these circumstances the assumption that an entire sequence has been deposited at a constant rate of sedimentation is usually unacceptable. In fact, one of the prime advantages of a magnetostratigraphic study of a sedimentary sequence is that if the correlation to the magnetic polarity time scale can be made without any assumptions concerning rates of sedimentation, then the actual rate of sedimentation in each polarity interval can be accurately determined.

The problem then is to find another means of correlating the magnetic polarity zonation to the magnetic polarity time scale. It is here that the complementary nature of magnetostratigraphy and tephrochronology is most apparent. As the simplest situation we can consider a sedimentary sequence containing a prominent tephra layer which can itself be dated or correlated to a dated tephra layer elsewhere. By itself this dated layer provides some useful information about a particular point within the sedimentary sequence, but it provides little or no information about the total time span during which the sediment was deposited. The dated tephra does, however, represent the single datum needed to determine the correlation of the magnetic polarity zonation to the magnetic polarity time scale. By fixing one date within one particular polarity zone, the tephra layer may reduce the choice from several possible correlations to a single, unique one. Once the correlation has been established, the age of each polarity boundary within the sedimentary sequence can be determined by reference to the magnetic polarity time scale, thereby providing many dated horizons distributed throughout the sequence (Fig. 3).

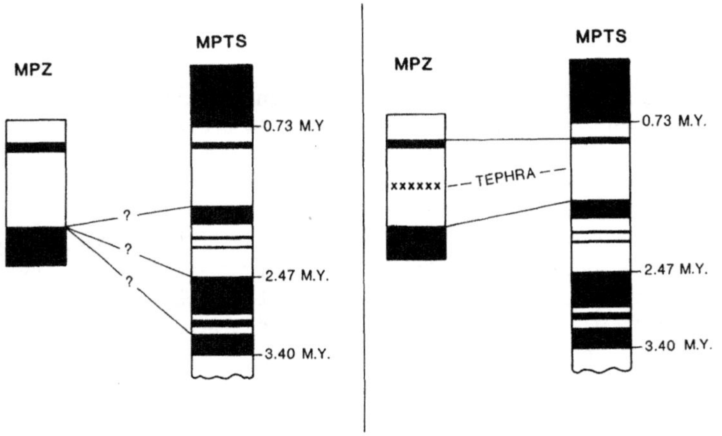

Fig. 3. Correlation of a magnetic polarity zonation (MPZ) to the magnetic polarity time scale (MPTS). The presence of a dated tephra layer removes the ambiguity in the correlation.

This particular approach to the combined use of magnetostratigraphy and tephrochronology has recently been employed in the study of hominid-bearing strata of Plio-Pleistocene age in Ethiopia (10), Kenya (6, 11) and Java (12). It has also been used to calibrate several important late Cenozoic vertebrate faunal assemblages. Recent studies of Plio-Pleistocene vertebrate-bearing terrestrial sediments include sites in North America (4, 13), Asia (14) and Africa (15).

In actual practice, the joint use of magnetostratigraphy and tephrochronology can be more varied and complicated than the simple approach described above. In many cases some degree of biostratigraphic age-control already exists for a sedimentary sequence. The faunal data then serves as an additional constraint on the magnetostratigraphic correlation. For example, the biostratigraphy of the sediments of the Ridge Basin of California (8), was sufficient to determine uniquely the correlation of the magnetic polarity zonation to the magnetic polarity time scale. In this case, the magnetostratigraphy and biostratigraphy can be used to date a prominent tephra layer within the studied portion of the sedimentary sequence, and to place approximate ages on two other prominent tephra layers within an unstudied, stratigraphically higher portion of the section. These age determinations will be useful when we attempt to correlate these tephra layers to others in southern and central California. In this context, it should be pointed out that the magnetic polarity of sediments with which a tephra layer is intercalated represents one of the parameters which characterizes that tephra. Thus two tephra layers at different sites which may have similar petrologic and geochemical properties must be considered distinct if the polarities of their associated sediments are different.

The foregoing disucssion is concerned with sedimentary sequences which contain a record of at least one, and preferably several, intervals of normal and reversed polarity of the earth's magnetic field. This requirement is usually met for sequences which span relatively long portions of the Quaternary or which include both Pleistocene and Pliocene sediments. For sequences which accumulate entirely within a single polarity interval, the magnetostratigraphic approach described above is, in general, not very fruitful. The situation is particularly unfortunate for relatively young sequences since the most recent well-documented polarity transition is that between the present Brunhes normal interval and the preceding Matuyama reversed interval, dated at 730,000 years ago (3). Although this date precludes use of magnetostratigraphy for many recent deposits of considerable importance, another feature of the geomagnetic field, namely geomagnetic excursions, may eventually provide a means of extending the magnetostratigraphic technique, on a local scale, to these young sequences.

Geomagnetic excursions are short-term, large-scale fluctuations of the geomagnetic field which would manifest themselves as a series of intermediate or reversed directions in a sedimentary sequence of Brunhes age. Many geomagnetic excursions have been reported in the literature, however at the present time their precise nature, and in many cases their very existence, is uncertain (16). This uncertainty arises for two reasons. The first is that in many cases it is difficult to determine whether a short series of unusual paleomagnetic directions represents an actual geomagnetic excursion or rather reflects some mechanical or chemical disturbance of the sediment which has affected the fidelity of the paleomagnetic recorder (17). This question is particularly important when the proposed geomagnetic excursion occurs at a lithologic boundary which by its very presence attests to a major disturbance of the depositional regime. The second problem arises from our inability to find consistent records of the same geomagnetic excursion in nearby, contemporaneous sedimentary sequences (18).

An accurate assessment of the present state-of-the-art concerning geomagnetic excursions is that although they probably do exist, there is no general agreement on when they have occurred, how long they have lasted, or whether they have been global or regional in their effect. Until these questions are fully answered, the use of presumed geomagnetic excursions as magnetostratigraphic horizons for tephra studies is probably premature. In fact, the principal interrelationship in this area at the present time is that a single widespread identifiable tephra layer may be the key to our understanding of geomagnetic excursions. Thus if such a layer is stratigraphically near a proposed geomagnetic excursion at one site, other sites at which the tephra is present could be sampled for paleomagnetic study, and the tephra layer could be used to provide an unambiguous marker horizon for the study of the temporal and spatial variations which characterize the excursion. Perhaps only in this way will the actual nature of geomagnetic excursions be determined.

REFERENCES

1. McDougall, I., Saemundsson, K., Johannesson, H., Watkins, N.D. and Kristjansson, L.: 1977, Bull. Geol. Soc. Amer. 88, pp. 1-15.
2. Ness, G., Levi, S. and Couch, R.: 1980 (in press), Rev. Geophys. Space Phys. 18.

3. Mankinen, E.A. and Dalrymple, G.B.: 1979, J. Geophys. Res. 84, pp. 615-626.
4. Opdyke, N.D., Lindsay, E.H., Johnson, N.M. and Downs, T.: 1977, Quat. Res. 7, pp. 316-329.
5. MacFadden, B.J.: 1977, Amer. J. Sci. 277, pp. 769-800.
6. Hillhouse, J.W., Ndombi, J.W.M., Cox, A. and Brock, A.: 1977, Nature 265, pp. 411-415.
7. Kodama, K. and Cox, A.: 1979, Geophys, Res. Lett. 6, pp. 253-256.
8. Ensley, R.A. and Verosub, K.L.: 1979, Geol. Soc. Amer. Absts. Prog. 7, p. 421.
9. Hiertzler, J.R., Dickson, G.O., Herron, E.M., Pitman, W.C. and LePichon, X.: 1968, J. Geophys. Res. 73, pp. 2119-2136.
10. Aronson, J.L., Schmitt, T.J., Walter, R.C., Taieb, M., Tiercelin, J.J., Johanson, D.C., Naeser, C.W. and Nairn, A.E.M.: 1977, Nature 267, pp. 323-327.
11. Brock, A. and Issac, G.L.P.: 1974, Nature 247, pp. 344-348.
12. Ninkovich, D. and Burckle, L.H.: 1978, Nature 275, pp. 306-308.
13. Johnson, N.M., Opdyke, N.D. and Lindsay, E.H.: 1975, Bull. Geol. Soc. Amer. 86, pp. 5-12.
14. Keller, H.M., Tahirkheli, R.A.K., Mirza, M.A., Johnson, G.D., Johnson, N.M. and Opdyke, N.D.: 1977, Earth Planet. Sci. Lett. 36, pp. 187-201.
15. Brown, F.H., Shuey, R.T. and Croes, M.K.: 1978, Geophys. J. R. Astr. Soc. 54, pp. 519-538.
16. Verosub, K.L. and Banerjee, S.K.: 1977, Rev. Geophys. Space Phys. 15, pp. 145-155.
17. Verosub, K.L.: 1975, Science 190, pp. 48-50.
18. Verosub, K.L.: 1977, Earth Planet. Sci. Lett. 36, pp. 219-230.

CORRELATION TECHNIQUES IN TEPHRA STUDIES

John A. Westgate and Michael P. Gorton

Department of Geology, University of Toronto,
Toronto, Ontario, Canada, M5S 1A1

ABSTRACT. Distinctive tephra layers constitute important time-parallel markers, which if widespread, offer the potential for reliable correlation over long distances. Confident correlations require a multiple criterion approach to tephra characterisation and equivalence of samples should only be considered firmly established if their stratigraphic, palaeontologic, palaeomagnetic, and radiometric age relations are compatible, and the physico-chemical properties of their glass shards and phenocrysts agree. The strong susceptibility of tephra to reworking further argues for use of several stratigraphic controls in order to safeguard against gross errors.

Grain-discrete methods of analysis have the advantage of being sensitive to contamination effects, and under most circumstances, are to be preferred over those methods that require use of bulk separates. The major element geochemistry of glass and FeTi oxides, as determined by use of an electron microprobe, in conjunction with the minor and trace element geochemistry of glass, as determined by XRF and neutron activation techniques, provide especially useful data for the identification and correlation of tephra beds.

1. INTRODUCTION

A wide range of studies are currently focused on tephra. For example, insights into the character, magnitude, explosivity, and duration of past volcanic eruptions are being sought through analysis of the physical properties of tephra, especially granulometry, shard morphology, and size and shape of the dispersal

S. Self and R. S. J. Sparks (eds.), Tephra Studies, 73–94.

fan (1-5). Secondly, mineralogical and geochemical gradients in tephra piles monitor compositional stratification of the parent magma body and so constitute a valuable source of information pertinent to differentiation mechanisms (6-8). Thirdly, distinctive and widespread tephra layers have much stratigraphic value for they represent time-parallel markers that permit reliable correlation over long distances (9-11). Thus, tephra has been recognised in deep-sea sediments at distances as great as 3000 km from its source (12), and even on the continents, where preservation potential is not as good, discrete beds discontinuously

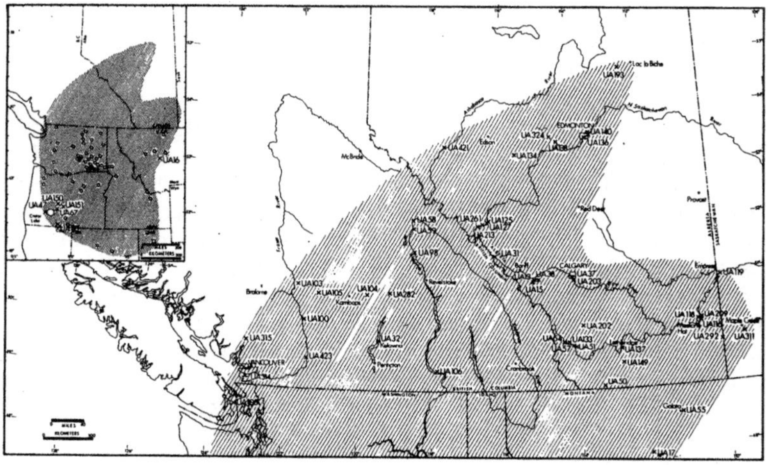

Fig. 1. Distribution of Mazama tephra in western North America. The source vent is located at Crater Lake (inset map), which formed after the collapse of Mount Mazama about 6700 years ago. The volume of tephra blasted from the top of the volcano during its climactic outburst was approximately 40 km^3 (13). The most distant locality, where tephra occurs as a discrete bed, is Lac la Biche, Alberta, about 1550 km northeast of Crater Lake, Oregon. Crosses and circles represent sample localities of Westgate and other workers, respectively.

persist to distances exceeding 1500 km (Figs. 1 and 2, Table 1). This paper is concerned with the latter aspect of tephra studies, a field commonly referred to as "tephrochronology" (14).

The stratigraphic value of a tephra layer is realised only when its distinguishing criteria have been recognised and its

Table 1. Data on tephra layers mentioned in text

Stratigraphic unit	Source	Ferromagnesian phenocrysts	Distribution from vent of occurrence ; maximum distance from vent	Age (k.y. BP)	References
White River Ash	Mt. Bona, Alaska;1				
eastern layer		Hb,Opx	E; 950 km	1.25	[17]
northern layer		Hb,Opx	N; 500 km	1.85	[17]
Bridge River tephra	Plinth–Meager Mt., British Columbia;2	Cpx,Opx,Hb,Bi	ENE; 550 km	2.35	[18]
Mount St. Helens, set Y tephra	Mount St. Helens, Washington;4				
Yn layer		Cu,Hb	NE; 1000 km	3.4	[18,19]
older, unnamed layer		Cu,Hb	NE; 1150 km	4.3	[18,19]
Mazama tephra	Crater Lake, Oregon;5	Hb,Opx,Cpx	E,N, and NE; 1550 km	6.7	[20,21]
Glacier Peak tephra	Glacier Peak Washington;3				
layer B		Opx,Hb	SE; 950 km	11.2	[22,23]
layer G		Opx,Hb	E; 800 km	~13.0	[22,23]
Old Crow tephra	Alaska Peninsula?	Opx,Cpx,Hb,Bi	Old Crow Flats, Yukon;?	50 – 100	[24]
Pearlette "O" tephra	Yellowstone Natl.Park;6	Ch,Hb,Cpx	SE,E,NE; 1200 km	600.0	[25,26]
Bishop Tuff	Long Valley, Calif.,;7	Bi,Hb	E; 1850 km	700.0	[25,27]
Wellsch Valley tephra	Bend, Oregon?;8	Hb,Opx,Cpx,Bi	Stewart Valley,Sask.; 1250 km?	700.0	[28]
Salmon Springs tephra	Unknown vent in Cascades	Hb,Opx,Bi	Southern Puget Lowland, Washington;?	850.0	[29,30]
Pearlette "S" tephra	Yellowstone Natl.Park;6	Ch;Hb	SE; 1200 km	1200.00	[25,31]

Notes: Locations of source vents are shown on Fig. 2, where each is identified by the number shown here in the "source" column.
Hb = hornblende, Opx = orthopyroxene, Cpx = clinopyroxene, Bi = biotite, Cu = cummingtonite, Ch = chevkinite

age determined. Efforts have been made to find a rapid, single-parameter method for tephra identification (15,16) but most workers now acknowledge that reliable correlation requires a multiple criterion approach to tephra characterisation. Equivalence of samples should only be considered firmly established if: (1) their stratigraphic, palaeontologic, palaeomagnetic, and radiometric age relations are compatible, (2) properties of the glass shards and phenocrysts agree, and (3) the combination of these characteristics is distinctive from that of other tephra beds in the area (32). The last requirement presupposes a comprehensive knowledge of the geologic record, which is still unattained for many areas, so that although it is generally agreed that tephra beds provide one of the most useful and precise means of time-correlation, one must still accept Weller's comment that "very rarely or never is the evidence absolutely conclusive" (33).

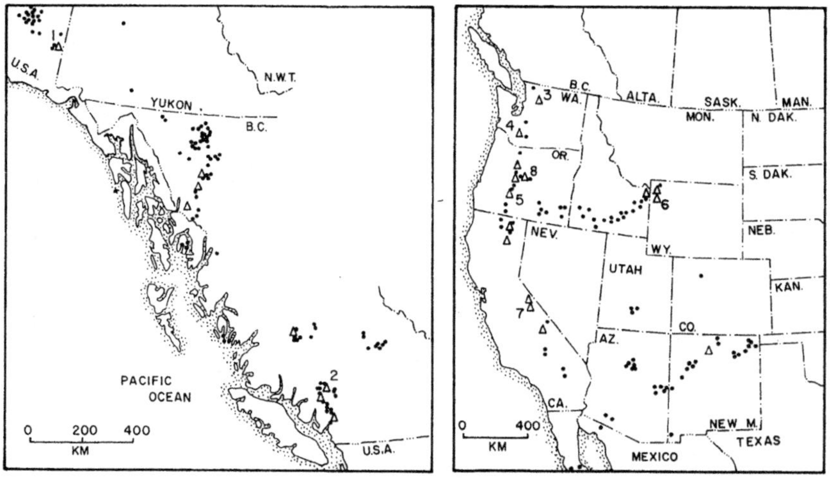

Fig. 2. Distribution of Quaternary vents in western North America based on maps by Wilcox (9) and Souther (34). Open triangles represent vents that have erupted large volumes of silica-rich tephra; the numbers are identified in Table 1. Filled circles represent vents that have furnished silica-poor tephra; the resultant tephra layers are restricted in extent because they are associated with relatively weak eruptions.

We discuss in this paper the relative merits of tephra characterisation techniques for correlation purposes, illustrated to a large extent by our own data; we comment briefly on appropriate age determination methods, and conclude with a discussion on the problem of reworking.

2. CHARACTERISATION OF TEPHRA: USEFUL METHODS FOR CORRELATION

Attributes of tephra beds that are readily discernible in the field, such as colouration, degree of weathering, lithic content, granulometric parameters, sedimentary structures, distribution, thickness, and stratigraphic context, may provide sufficient control for identification and correlation in regions close to source vents (35,36) but detailed petrographic and petrochemical studies are usually required to identify distal tephra units embedded in non-volcanic sedimentary sequences (25,37). In the latter context, desirable parameters are those that do not change over the fall-out zone--namely, the physico-chemical properties of the glass shards and primary phenocrysts. Bulk tephra analyses do not satisfy this condition because the tephra has been subjected to sedimentary fractionation. However, such information is of local value, provides data pertinent to location of the source vent (38), and, in some cases, has been used successfully to correlate tephra beds over large distances (39,40).

The value of mineral assemblage as a correlation parameter is governed by the extent to which sedimentary fractionation has taken place during atmospheric transport over distances in which the tephra is recognisable as a discrete bed. We have found that all mineral species in a given tephra bed persist to the most distant part of the documented fall-out zone; for example, this is true of Mazama, Mount St. Helens set Y beds, and Glacier Peak tephra (Table 1). Mineral assemblage, therefore, should form part of the correlation matrix; in fact, it may provide a strong clue as to the identity of the source vent. Chevkinite in the Pearlette tephra and cummingtonite in Mount St. Helens tephra are pertinent examples. Mineral assemblage and other petrographic characteristics are commonly the same in co-magmatic units derived from closely-spaced eruptions.

Felsic tephra beds are commonly distinguished from one another on the basis of the properties of their volcanic glass. The popularity of glass in this respect lies in its abundance, ease of concentration, and homogeneous composition or narrow compositional range within a single eruptive unit. However, susceptibility to alteration limits or prevents its use for older beds and in those areas where chemical weathering is vigorous. Some workers have stressed in their studies the physical properties of glass shards, documenting morphology, transparency, vesicularity, microphenocryst content, hydration, and refractive index (41), whereas others have concentrated on chemical characteristics.

Important controls on the morphology of glass shards are the properties of the magma and the eruptive mechanism (3), so that shard morphology is commonly an excellent diagnostic character of a particular tephra. The three tephra units illustrated in Figure 3 can be readily distinguished by their glass shards. The extremely thin, clear bubble-walls of shards in the Old Crow tephra and their remarkable preservation (Fig. 3 c and d) are sufficiently rare in tephra layers to serve as excellent criteria for identification and correlation. Indeed, petrographic idiosyncrasies are one of the most sensitive means of identifying a tephra bed. The inclusion-riddled ilmenite of

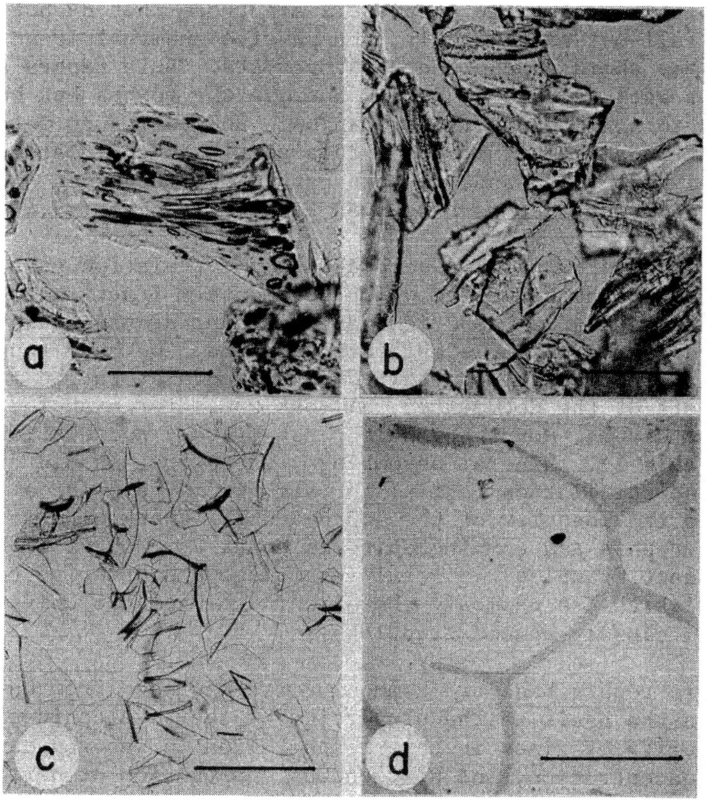

Fig. 3. The contrast in glass shard morphology of three separate tephra beds. (1) Bridge River tephra: the shards are chunky and possess lineated, lensoid vesicles; in transmitted light; scale bar is 0.1 mm long. (b) Salmon Springs tephra: the shards are chunky but free of vesicles; in transmitted light; scale bar is 0.1 mm. (c) and (d) Old Crow tephra: the shards are fragments of very thin bubble-walls and contain no microlites or vesicles. (c) is an oil mount in transmitted light; scale bar is 0.5 mm. (d) is a polished section in reflected light; scale bar is 0.05 mm. Further details on these eruptive units is given in Table 1.

the tephra in the Salmon Springs Drift at Sumner, Washington (Fig. 4, Table 1) further illustrates this point.

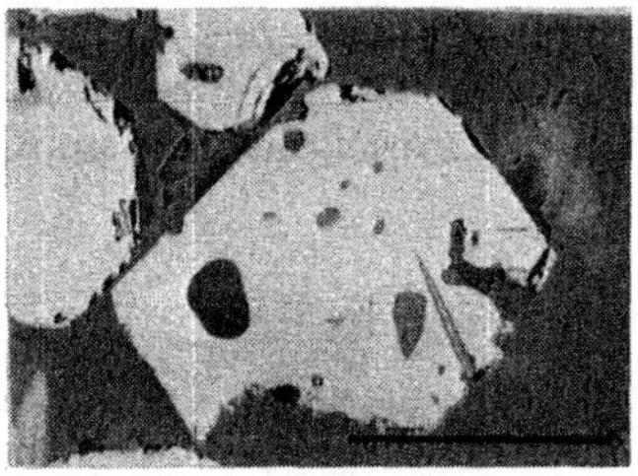

Fig. 4. Inclusion-riddled ilmenite of the Salmon Springs tephra. Scale bar is 0.1 mm long.

Much attention has been given to use of the range and modal value of the refractive index of volcanic glass, as measured with refractive index oils, but these efforts have met with mixed success. Silicic volcanic glasses do not vary greatly in their major element composition so that differences in their refractive indices are small, and, in some cases, insufficient to permit distinction between separate eruptive units (42,43). Further difficulties arise from differential hydration, which increases the dispersion of such data (9). Correlations based on refractive index histograms (41) defined by several hundred measurements are obviously more securely based but the time-consuming nature of the method has discouraged its use.

The major element composition of volcanic glass is most reliably determined by the electron microprobe technique, which permits distinction of minor variation in the chemistry of glasses (16,37). This is a grain-discrete method and so the problems associated with analysis of bulk separates are avoided. Specifically, bulk glass analyses suffer from the inevitable presence of inclusions and microlites, the possible occurrence of foreign particles and weathering products on the surface of shards and pumice fragments, and more rarely, the occurrence of detrital glass reworked from older tephra (Fig. 5). Similarly, grain-discrete methods of analysis are to be preferred over the use of bulk separates in the determination of the major element composition of phenocrysts because they permit detection of post-depositional changes, contamination effects, inhomogeneities, and the presence of more than one indigenous phase of a given mineral. For example, in seeking possible correlatives of a thin tephra layer in South Island, New Zealand, Kohn (44) found that bulk titanomagnetite analyses could not be used because of the presence of detrital titanomagnetite derived from the enclosing loess. However, electron microprobe analyses of a few homogeneous titanomagnetite grains soon suggested a correlative. Similarly, bulk titanomagnetite analyses of the Pearlette "O" layer (Table 1) would have given meaningless information because of the presence of five distinct, primary species (Fig. 6) (45). Hildreth (46) has noted that tephra beds related to central-vent volcanoes commonly possess multiple titanomagnetite populations,

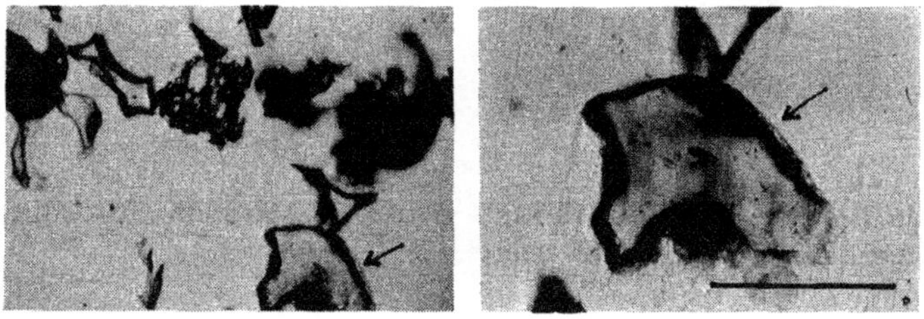

Fig. 5. The glass contaminant (arrowed) in the Wellsch Valley tephra (Table 1) is noticeably darker than the indigeneous glass. The smaller spots on the former are fossil fission-tracks. Note that the f-t density on this grain is much larger than that on the primary shards. Scale bar is 0.1 mm.

and, in our experience, multiple populations of hornblendes are not uncommon (22). However, sample size may well be the critical factor in determining the most appropriate analytical technique; small samples can be readily analysed with a microprobe but preclude use of most bulk methods. Thus, many distal tephra deposits do not contain a sufficiently large phenocryst crop to allow the use of bulk analytical techniques.

Many of the widespread tephra layers of Quaternary age in western North America can be distinguished on the basis of the major element composition of their glass, as determined by the microprobe technique (Table 2) (16,22,25,37,47). Iron, Ca, and K are particularly useful for diagnostic purposes (Fig. 7), although, in the case of late Quaternary tephra beds in the Pacific Northwest and southwestern Canada, Ti, Al, and K are the most important elements for separation, according to a discriminant function analysis (22). All tephra units are not discernible in this way, as is shown by data on the two members of Mount St. Helens set Y tephra (Table 2). In this case, separation can be made on the basis of trace element geochemistry of the glass, as determined by neutron activation analysis (Table 5).

Table 2. Average glass composition of some widespread tephra layers in western North America, as determined by the microprobe technique, weight per cent, without water, recalculated to 100 per cent.

	Bridge River tephra	Mt. St. Helens set Y tephra		Mazama tephra	Glacier Peak tephra		Pearlette tephra	
		layer Yn	unnamed older layer		layer B	layer G	layer O	layer S
SiO_2	74.35 ± 0.73	74.94 ± 0.19	75.04 ± 0.36	72.58 ± 0.31	76.82 ± 0.25	77.12 ± 0.20	76.75 ± 0.26	76.79 ± 0.31
TiO_2	0.36 ± 0.04	0.17 ± 0.02	0.16 ± 0.01	0.48 ± 0.03	0.23 ± 0.01	0.21 ± 0.02	0.11 ± 0.02	0.10 ± 0.02
Al_2O_3	13.72 ± 0.27	14.48 ± 0.15	14.34 ± 0.20	14.40 ± 0.16	12.89 ± 0.21	12.73 ± 0.12	12.22 ± 0.07	12.21 ± 0.16
FeO*	1.62 ± 0.18	1.41 ± 0.05	1.38 ± 0.04	2.11 ± 0.08	1.20 ± 0.10	1.06 ± 0.03	1.51 ± 0.12	1.42 ± 0.01
MnO	0.04 ± 0.01	0.03 ± 0.00	0.02 ± 0.00	0.04 ± 0.00	0.03 ± 0.01	0.03 ± 0.01	n.d.	n.d.
MgO	0.44 ± 0.09	0.49 ± 0.08	0.52 ± 0.07	0.56 ± 0.08	0.32 ± 0.06	0.32 ± 0.04	0.04 ± 0.02	0.05 ± 0.01
CaO	1.40 ± 0.18	1.85 ± 0.06	1.85 ± 0.06	1.66 ± 0.10	1.38 ± 0.06	1.19 ± 0.04	0.54 ± 0.01	0.55 ± 0.02
Na_2O	4.61 ± 0.15	4.44 ± 0.21	4.37 ± 0.27	5.18 ± 0.20	3.94 ± 0.05	3.87 ± 0.09	3.64 ± 0.15	3.33 ± 0.27
K_2O	3.16 ± 0.14	2.03 ± 0.08	2.09 ± 0.05	2.70 ± 0.11	2.99 ± 0.09	3.32 ± 0.06	5.02 ± 0.10	5.42 ± 0.16
Cl	0.12 ± 0.03	0.11 ± 0.05	0.11 ± 0.01	0.18 ± 0.02	0.17 ± 0.02	0.15 ± 0.01	0.16 ± 0.02	0.12 ± 0.02
P_2O_5	0.04 ± 0.01	0.04 ± 0.00	0.04 ± 0.00	0.06 ± 0.02	0.04 ± 0.00	0.04 ± 0.01	n.d.	n.d.
D.I.	89.09	85.36	85.44	86.44	89.44	90.70	94.70	94.73
n	37	14	11	98	17	14	13	3

Notes: All determinations were made by J. A. Westgate using an A.R.L. "EMX" microprobe at 15kV, 200μA emission current, 0.1μA beam current, ~0.01μA sample current. See [37] for information on standards and correction procedures. Location of source vents and additional data on these units are given in Figs. 1 and 2 and Table 1. Glass shards in Mount St. Helens tephra have a calc-alkaline dacitic composition of the K-poor series; Mazama glass has a calc-alkaline rhyolitic composition of the K-poor series; all other units are calc-alkaline rhyolites of the average series [48].

* Total iron as FeO. D.I. = differentiation index. n = number of analyses; each is based on at least 20 grains. One standard deviation is given. n.d. = not determined.

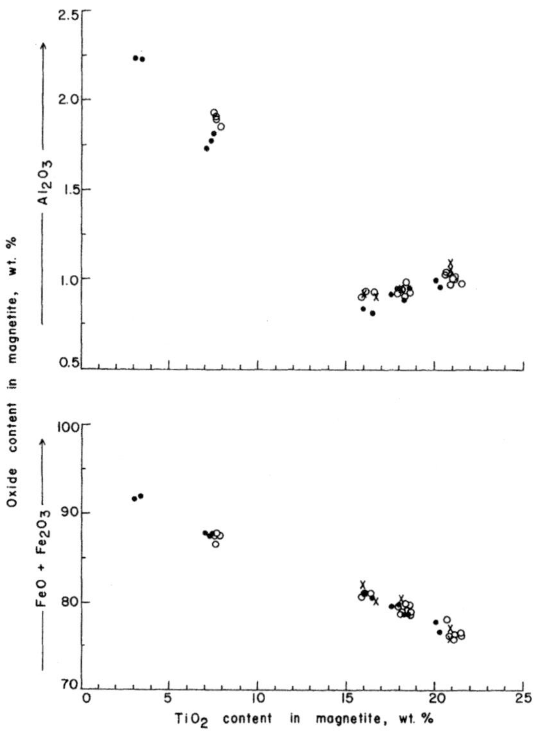

Fig. 6. Variation diagram of oxide content in titanomagnetite versus TiO_2 content for the Pearlette "O" tephra. Filled circle, Wascana Creek Ash in southern Saskatchewan; open circle, localities in the United States; cross, Hartford Ash of Boellstorff (45).

Iron-titanium oxides are the most used minerals in chemical characterisation studies. They are very sensitive to the environmental conditions of initial crystallisation and bulk rock chemistry (49) and so exhibit a considerable range in compositions and modal abundances in igneous rocks. Other factors that have encouraged their use include ubiquitous occurrence in volcanic rocks, relative stability during weathering, and ease of extraction by magnetic methods. Late Quaternary tephra layers in the Pacific Northwest can be successfully differentiated in this way (Table 3, Fig. 8), the ilmenites being more diagnostic than the titanomagnetites. However, as in the case of glass, the major element geochemistry of the FeTi oxides does not effectively separate the two members of Mount St. Helens set Y tephra. A particularly strong case for equivalence of samples can be argued if each contains the same assemblage of different titanomagnetite phases (Fig. 6) (45).

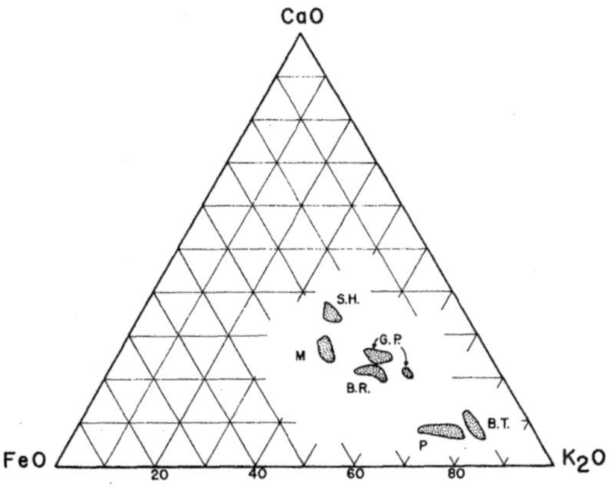

Fig. 7. Relative abundance of FeO, CaO, and K_2O in glass of some widespread Quaternary tephra layers in western North America. All determinations were done on an electron microprobe. Compositional range is based on 43 samples in the case of Mazama (M) tephra, 21 for Bridge River (B.R.) tephra, 15 for Mount St. Helens (S.H.) set Y tephra, 16 for Pearlette (P) tephra, and 18 samples for the Bishop Tuff (B.T.). The Bishop Tuff data and some of that for Pearlette tephra come from Izett et al. (25), the remainder is the work of Westgate (18). Table 1 contains further details on these eruptive units.

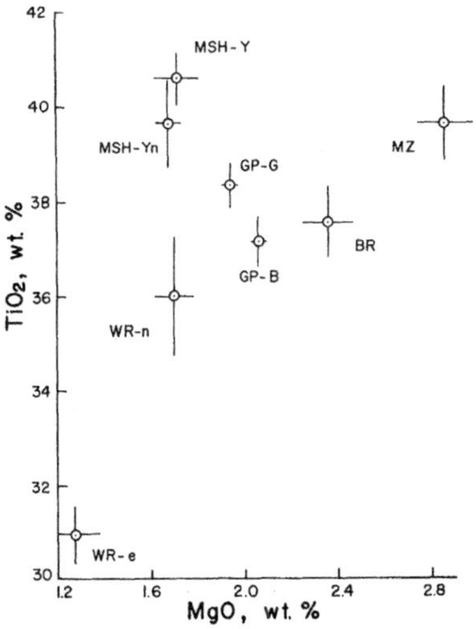

Fig. 8. TiO_2 versus MgO plot for ilmenites of some widespread late Quaternary tephra layers in western North America. Rays represent one standard deviation from the mean. See Table 3 for details of the microprobe analyses and an explanation of the abbreviations.

Table 3. Average composition of FeTi oxides in some widespread tephra layers in western North America as determined by the microprobe technique, weight per cent.

	White River Ash		Bridge River tephra	Mount St. Helens set Y tephra		Mazama tephra	Glacier Peak tephra	
	eastern layer	northern layer		layer Yn	unnamed older layer		layer B	layer G
Magnetite:								
TiO_2	5.84 ± 0.12	6.14 ± 0.24	7.61 ± 0.27	5.96 ± 0.02	5.92 ± 0.26	8.52 ± 0.44	5.96 ± 0.24	6.12 ± 0.02
FeO	35.05 ± 0.57	35.33 ± 0.52	35.49 ± 0.51	34.91 ± 0.54	35.35 ± 0.23	35.75 ± 0.49	34.57 ± 0.34	34.76 ± 0.17
Fe_2O_3	54.07 ± 0.76	53.66 ± 0.77	52.34 ± 0.70	54.58 ± 1.08	52.95 ± 0.95	48.95 ± 1.25	55.09 ± 0.46	54.97 ± 0.04
MgO	1.34 ± 0.08	1.25 ± 0.12	1.81 ± 0.12	1.16 ± 0.23	0.99 ± 0.02	2.13 ± 0.18	1.25 ± 0.21	1.30 ± 0.01
MnO	0.30 ± 0.01	0.33 ± 0.02	0.40 ± 0.12	0.32 ± 0.01	0.29 ± 0.03	0.43 ± 0.07	0.33 ± 0.02	0.39 ± 0.02
Al_2O_3	2.96 ± 0.07	2.93 ± 0.13	2.02 ± 0.12	2.64 ± 0.64	3.27 ± 0.06	2.51 ± 0.16	2.27 ± 0.40	2.26 ± 0.06
SiO_2	0.35 ± 0.35	0.27 ± 0.25	0.18 ± 0.18	0.07 ± 0.07	0.27 ± 0.03	0.37 ± 0.20	0.01 ± 0.01	0.06 ± 0.04
Total	99.91	99.91	99.85	99.64	99.04	98.66	99.48	99.86
Ulvösp.*	15.68 ± 0.69	16.91 ± 0.61	22.02 ± 1.05	17.08 ± 0.40	17.77 ± 0.69	25.40 ± 1.49	16.96 ± 0.76	17.55 ± 0.23
Ilmenite:								
TiO_2	30.98 ± 0.61	35.98 ± 1.28	37.55 ± 0.75	39.64 ± 0.95	40.55 ± 0.55	39.66 ± 0.79	37.18 ± 0.53	38.32 ± 0.49
FeO	25.63 ± 0.53	29.25 ± 0.99	29.37 ± 0.85	32.28 ± 0.88	33.19 ± 0.40	30.33 ± 0.53	29.41 ± 0.49	30.72 ± 0.37
Fe_2O_3	40.89 ± 1.13	32.12 ± 2.18	29.92 ± 0.90	25.09 ± 1.40	23.46 ± 1.29	25.14 ± 1.23	30.25 ± 0.94	29.05 ± 0.94
MgO	1.27 ± 0.13	1.70 ± 0.08	2.36 ± 0.11	1.68 ± 0.05	1.71 ± 0.09	2.86 ± 0.12	2.06 ± 0.02	1.94 ± 0.01
MnO	0.16 ± 0.02	0.28 ± 0.05	0.36 ± 0.11	0.39 ± 0.03	0.37 ± 0.02	0.39 ± 0.03	0.35 ± 0.02	0.38 ± 0.00
Al_2O_3	0.54 ± 0.02	0.46 ± 0.06	0.33 ± 0.03	0.36 ± 0.05	0.35 ± 0.01	0.36 ± 0.02	0.30 ± 0.03	0.30 ± 0.00
SiO_2	0.17 ± 0.17	0.17 ± 0.17	0.14 ± 0.13	0.03 ± 0.03	0.12 ± 0.03	0.14 ± 0.09	0.01 ± 0.01	0.07 ± 0.07
Total	99.64	99.96	100.03	99.47	99.75	98.88	99.56	100.78
Hem.**	42.36 ± 1.08	31.34 ± 3.20	28.77 ± 1.06	24.50 ± 1.41	22.78 ± 1.24	24.42 ± 1.23	29.25 ± 0.89	27.80 ± 0.82
T(°C)	~1065	~915	~945	~815	~815	~915	~875	~875
n	15	11	15	4	5	24	3	2

Notes: Operating conditions, standards, and correction procedures as specified in Table 2. Fe_2O_3 contents were determined by assuming stochiometry. Temperature values deduced by extrapolation of data of Buddington and Lindsley [50]. *Mol. % ulvospinel in cubic oxide. **Mol. % hematite in rhombohedral oxide. n = number of analyses; each is based on at least 10 grains. One standard deviation is given.

Minor and trace element compositions of glasses offer one of the best means of discriminating between tephra layers for they can differ between one another by a large factor. Bulk analytical techniques must be used in this case because concentrations of these elements are below the useful detection limit of the microprobe. Small, highly-charged ions are more stable under weathering conditions than large ions with a single charge, such as Na, K, Rb, Cs, Ba, and Cl, and so are potentially of greater value for correlation purposes; they include P, Sc, Y, Zr, Hf, and the rare earth elements. The analysis of these elements presents special problems, primarily because of their low concentrations. Analytical techniques most commonly used in determining their concentration include X-ray fluorescence, atomic absorption, and neutron activation.

The average trace element concentrations in glass separates of some widespread tephra layers in western North America, as

Table 4. Average trace element concentrations in some widespread tephra layers in western North America as determined by instrumental neutron activation on glass separates.

	Old Crow tephra 1	Mazama tephra 2	Bridge R. tephra 3	Mount St. Helens "Y" 4	Bishop Tuff 5	Pearlette "O" tephra 6
La	23.6	20.1	18.5	13.7	16.6	82.5
Ce	50.5	42.9	35.9	28.3	39.9	156.
Nd	25.	21.	15.	11.	15.	56.
Sm	5.3	4.4	2.8	2.6	2.4	13.2
Eu	.80	.85	.51	.72	.12	.45
Tb	.90	.64	.36	.30	.68	2.1
Ho	1.5	1.13	.77	.37	.51	3.6
Er	3.3	2.5	1.6	.77	3.0	8.2
Yb	3.2	2.2	1.5	.85	2.9	8.0
Lu	.51	.35	.26	.11	.35	1.0
U	3.7	2.3	2.0	1.5	6.8	6.1
Th	9.0	5.5	5.4	3.2	20.	28
Rb	87.	51.	44.	37.	156.	172.
Cr	7.	6.	6.	7.	1.5	3.
Hf	6.3	5.8	–	3.1	3.5	8.1
Sb	5.2	4.1	–	4.1	3.9	4.2
Sc	6.2	5.8	4.3	3.3	2.9	1.3
Ta	.74	.53	.83	.51	2.1	4.0

Notes: 250 mg. samples were irradiated in a slowpoke reactor for 16 hours, counted after 6 days for Ho, Nd and Sm on an LEPD detector, after 7 days for U, Lu and La on a coaxial detector, and after 35 days for the remaining elements. Method calibrated against BCR1, AGV1 and an internal standard UTB1.

See Table 5 for estimates of precision.

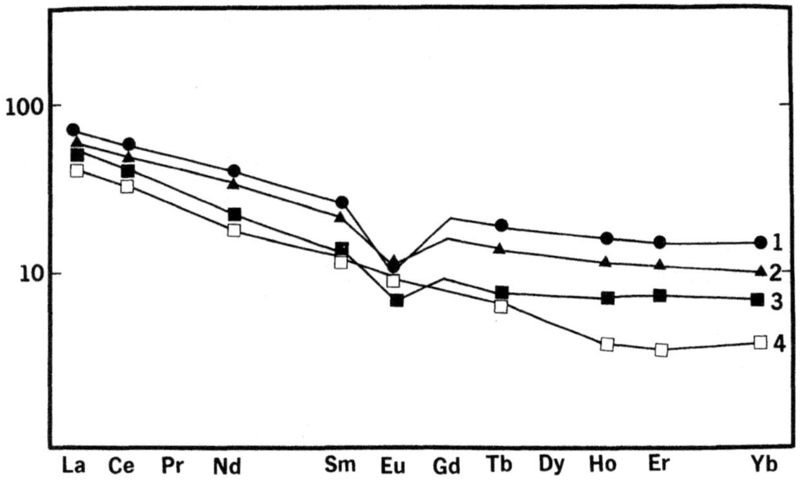

Fig. 9. Chrondrite normalised rare earth concentrations for selected tephra layers: 1. Old Crow tephra, 2. Mazama tephra, 3. Bridge River tephra, 4. Mount St. Helens set Y tephra.

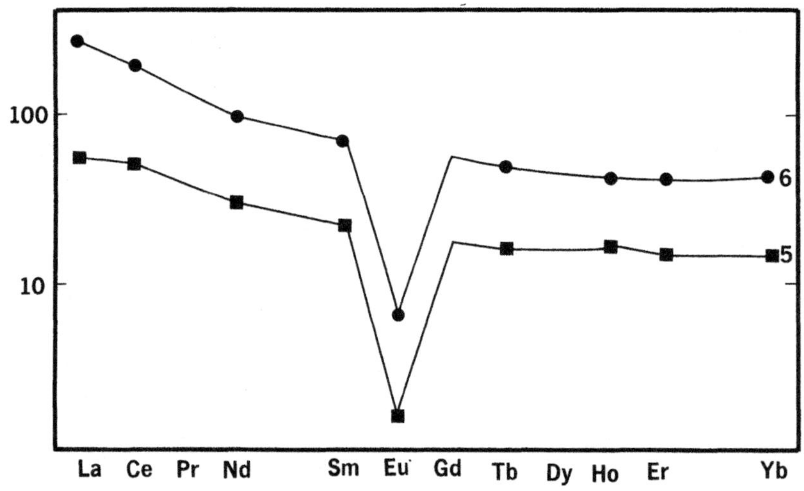

Fig. 10. Chrondrite normalised rare earth patterns for selected tephra layers: 5. Bishop Tuff, 6. Pearlette "O" tephra.

determined by neutron activation, are listed in Table 4. Some elements are readily determined but do not show a useful range in concentration; Sb and Sc are cases in point. In contrast, the rare earth elements, particularly the heavy rare earths, may vary in concentration by a factor of 10 (Figs. 9 and 10). Our data show that the latter group of elements differs markedly from one volcanic centre to another but that within a given eruptive centre differences are much more subtle. Table 5 shows small differences in trace element concentrations between two members of the set Y tephra of Mount St. Helens. These differences are insignificant for most elements. However, separation of these two beds, which differ in age by approximately a thousand years (Table 1), can be done by their light rare earth elements, which systematically differ by about 20%--a value significantly greater than the ∿5% precision and greater than the range observed in glass separates of the Old Crow tephra, samples of which come from a wide geographic area.

Table 5. Reproducibility of trace element determinations by instrumental neutron activation illustrated by reference to the Old Crow tephra. Subtle differences visible in the Mount St. Helens Set Y tephras.

	Old Crow tephra				Mount St. Helens Set Y tephra		
	Alaska UA739	Alaska UT114	Yukon UT1	2 sigma	Yn UA560	unnamed Y UA320	2 sigma
La	24.1	23.0	23.6	± .4	12.3	15.1	± .3
Ce	53.	50.	49.	± 1.	26.	30.	± 1.
Nd	24.	27.	25.	± 3.	10.	12.	± 2.
Sm	5.2	5.5	5.1	± .03	2.4	2.7	± .02
Eu	.80	.80	.80	± .11	.74	.69	± .10
Tb	.93	.93	.85	± .07	.23	.37	± .08
Ho	1.2	1.3	1.2	± .13	.30	.28	± .08
Er	3.4	3.5	3.1	± .30	.72	.82	± .26
Yb	3.2	3.1	3.3	± .25	.87	.83	± .25
Lu	.57	.38	.59	± .04	.08	.13	± .03
U	3.9	3.5	3.7	± .3	1.6	1.3	± .2
Th	8.9	8.9	9.2	± .17	3.0	3.3	± .13
Rb	85.	84.	91.	± 6.	33.	41.	± 5.
Cr	8.	6.	7.	± 1.	6.	8.	± 1.
Hf	6.4	6.2	6.4	± .23	3.2	2.9	± .20
Sb	5.4	4.8	5.3	± .06	3.6	4.5	± .06
Sc	6.3	6.5	5.9	± .06	3.9	2.6	± .05
Ta	.83	.65	.75	± .15	.47	.55	± .15

Notes: 2 sigma limits based on counting statistics alone. Actual precision may be considerably worse than this due to such factors as counting geometry and sample homogeneity.

Rare earth patterns have diagnostic value. Mount St. Helens set Y tephra shows a steep pattern with no Eu anomaly. On the other hand, Mazama tephra has a less steep pattern with an obvious

Eu anomaly (Fig. 9), which becomes even more accentuated in the Bishop Tuff and the Pearlette "O" tephra (Fig. 10).

Selected trace elements, determined by the XRF method on glass separates and bulk samples of three different tephra beds, are shown in Table 6. Again, it can be seen that each tephra is quite distinctive, despite the significant difference between bulk and glass samples. The most useful elements are Sr, Y, and Zr.

Table 6: XRF trace element determinations in glass separates (G) and bulk samples (B) of selected tephras.

	Old Crow tephra				Salmon Springs tephra				Fort Selkirk tephra			
	UT1		UT50		UT52		UT88		UT81		UT82	
	B	G	B	G	B	G	B	G	B	G	B	G
Rb	116.	76.	85.	117.	69.	79.	54.	71.	54.	58.	54.	69.
Sr	145.	187.	164.	139.	148.	126.	184.	127.	292.	274.	286.	278.
Y	30.9	30.3	30.1	29.8	11.7	9.3	13.9	11.8	12.7	7.8	9.3	6.3
Zr	270.	290.	239.	263.	132.	105.	155.	110.	151.	103.	106.	106.
U	5.	3.	4.	4.	4.	4.	4.	4.	3.	3.	3.	2.
Th	8.	7.	10.	8.	10.	10.	11.	11.	7.	4.	7.	3.

Notes: Trace elements determined in duplicate on pressed powder pellets using a Molybdenum X-ray tube and calculating absorption coefficients from background measurements.
At these concentrations precision is about 2% for Rb, Sr, Y and Zr, and about 25% for U and Th.
The Fort Selkirk tephra occurs close to the confluence of the Yukon and Pelly Rivers in the Yukon Territory.

Table 7 is a rough guide to the applicability of the several analytical methods to geochemically important groups of elements. The XRF technique cannot match the high sensitivity of atomic absorption and neutron activation, although it generally offers

Table 7. Comparison of the suitability of analytical techniques for the geochemically important groups of elements.

	Probe	XRF	AA	NAA
Major elements	1	1	2	3
Transition/heavy metals	2–3	2	1	2
Large cations	3	1	3	3
Rare earth elements	3	2–3	3	1
High at. no. metals	3	2–3	3	1

Notes: 1 Excellent, 2 Satisfactory, 3. Unsatisfactory

high precision. Thus, it is best restricted to determination of those elements present in relatively high concentrations, such as the major elements--microprobe and conventional whole rock analysis are likewise appropriate in this respect--and the larger cations such as Rb, Sr, Ba, Y, and Zr. Interferences are usually minor at these concentrations. Neutron activation offers high but variable sensitivity for many elements of high atomic number and consequently is most useful for the rare earth elements. Interferences are a serious problem but can usually be corrected. Atomic absorption is most useful for the first transition and heavy metals, such as Ag, Cd, Hg, Pb, etc. However, these elements are usually present in low concentrations in tephra and interferences are difficult to eliminate.

It should be noted that the quality of analyses obtained by use of these various methods is often limited more by the availability of well determined standards than by considerations of detection limit or analytical precision. Furthermore, attempts to correlate data produced in different laboratories is complicated by use of different standards and slightly different analytical procedures. There is a clear need for agreement on the glass and mineral standards that should be used in tephrochronology--at least at the regional level. Analyses should be accompanied by the identity and determined compositions of the standards used in order that independent assessment can be made on their accuracy, but such information is commonly lacking.

The need for careful methodological descriptions is best illustrated by a case history related to the Wellsch Valley tephra (Table 1). The fission-track age of glass from this tephra was provisionally determined as 18 m.y. by John Boellstorff of the University of Nebraska (personal communication, 1980)--an age considerably older than our estimate of 0.7 m.y. (28). However, this discrepancy was readily explained by the different techniques used. We mounted the glass shards in epoxy, then polished and etched them. The spontaneous fission-track density was determined by counting tracks on all shards that traversed the field of view of the microscope. We noted two populations: abundant, clear shards with a very low spontaneous track density, and scarce, larger dark shards with a very high track density (Fig. 5). The latter group is obviously a contaminant. Boellstorff, however, etched a bulk sample of glass shards and determined the track density of the surviving shards. In this way the contaminant glass was concentrated, explaining the much older date (Table 8). This example further illustrates the desirability of using grain-discrete methods.

Because of the large amount of geochemical data that may be generated and the expectation that they should fall into distinctive groups, rigorous statistical methods of assessment should be used. Nonetheless, many workers simply use graphical methods to aid in the problem of identification and correlation of tephra samples (e.g. Figs. 6-8). Such methods are appropriate for the assessment of rare earth elements, however, because of the strong geochemical coherence of this group. If a sufficiently large number of analyses have been made, the standard deviation can be used to assess the significance of the difference in elemental concentrations between samples (see Tables 2 and 3). Borchardt et al. (51) devised a similarity coefficient that allows all analyzed variables for a pair of samples to be compared. Therefore this coefficient, which is 1 for identical analyses, can be used to evaluate the likelihood of equivalence of tephra samples. A disadvantage for the rare earth elements is that it does not convey information on their pattern, which is perhaps more useful than the absolute concentrations. Sarna-Wojcicki et al. (52) claim that the similarity coefficient is also useful for determining provenance. Cluster analysis is another statistical technique that has been successfully used to evaluate chemical data for correlation purposes (53,54).

Table 8. Etch time versus spontaneous fission track density for glass shards of the Wellsch Valley tephra

Etch time	Spontaneous track density, tr cm^{-2}	Shards surviving etch
2 min	217,300	trace
1 min	142,280	trace
0.5 min	1,132*	abundant
0.5 min	181,083†	abundant

Data by John Boellstorff. * Smaller glass shards with low track density.
† Larger glass shards with high track density.

The stratigraphic value of a tephra layer is greatly enhanced if its age and palaeomagnetic properties are known. The K-Ar, fission-track, ^{40}Ar-^{39}Ar, and ionium dating techniques, along with the qualitative hydration method (55), can be applied to Quaternary tephra, and tephra beds in deep-sea sediments can be dated by the oxygen-isotope and biostratigraphic record (56). However, all these methods are too imprecise to be used alone for identification and correlation purposes. The ^{14}C age of associated organic matter is likewise only suggestive of the identity of a tephra layer. Palaeomagnetic characteristics facilitate correlation but in themselves do not date tephra beds. The magnetic polarity, however, does provide a means of assessing the reliability of radiometric age data in that the chronology of polarity reversals for the Quaternary Period is reasonably well-established (57). Caution is necessary at present for our understanding of the polarity sequence is still evolving (58). The pattern of secular variation of the earth's magnetic field, as recorded in the tephra-bearing sediments, is another useful signature for correlation on a local and regional scale (22,59).

3. THE PROBLEM OF REWORKING

A variety of mixing processes act on tephra once it is deposited unless it is rapidly buried by younger sediment. Disturbances by soil-forming processes, creep, frost activity, bioturbation, and reworking by wind and running water are important in subaerial regions whereas bioturbation and resedimentation by slumping and associated turbidity flows occur in subaqueous depositional environments. Identification and correlation of tephra layers is obviously hindered by this reworking, which can be detected by: (1) presence of non-volcanic sediment, (2) separation of phenocrysts from the vitric component, (3) abnormal thickness or grain-size, given the distance from source, and (4) diffuse boundaries to the bed. Proximal tephra beds may escape full reworking because of their thickness but thin distal beds can be lost as discrete units by these mixing processes.

A most serious hazard is relocation into a stratigraphic position much younger than the intrinsic age of the tephra. Thus, the stratigraphic position in deep-sea sediments of a tephra-rich horizon derived immediately from ablating icebergs does not equate with the time that that material was erupted. Such situations would be expected in sediments of the North Atlantic and Southern Ocean. Similarly, tephra-bearing sediments in thrust slices would overlie younger sediments. In the western Canadian plains the possibility that such a condition exists must be carefully evaluated at each tephra site for sediments over large tracts of these plains have been deformed by glaciotectonic processes (45). These examples show that careful examination for reworking and consideration of other local stratigraphic controls are necessary in order to safeguard against gross errors.

ACKNOWLEDGEMENTS

Work in Quaternary tephrochronology at the University of Toronto has been supported by grants from the Natural Sciences and Engineering Research Council of Canada and the Department of Energy, Mines, and Resources. We thank John Boellstorff, University of Nebraska, for information on his fission-track studies of the Wellsch Valley tephra.

REFERENCES

1. Walker, G.P.L.: 1971, J. Geol. 79, pp. 696-714.
2. Walker, G.P.L.: 1973, Geol. Rund. 62, pp. 431-446.
3. Heiken, G.: 1972, Geol. Soc. Amer. Bull. 83, pp. 1961-1988.
4. Huang, T.C. and Watkins, N.D.: 1976, Science 193, p. 576-579.
5. Ledbetter, M.T. and Sparks, R.S.J.: 1979, Geology 7, pp. 240-244.
6. Lipman, P.W.: 1971, J. Geol. 79, pp. 438-456.
7. Hildreth, W.: 1976, Geol. Soc. Amer., Abst. with Programs 8, p. 918.
8. Shaw, H.R., Smith, R.L. and Hildreth, W.: 1976, Geol. Soc. Amer., Abst. with Programs 8, p. 1102.
9. Wilcox, R.E.: 1965, in: The Quaternary of the United States", ed. Wright, H.E., Jr. and Frey, D.G., Princeton Univ. Press, pp. 807-816.
10. Vucetich, C.G. and Pullar, W.A.: 1969, New Zealand J. Geol. and Geophys. 12, pp. 784-837.
11. Machida, H.: 1976, Geog. Repts. Tokyo Metropolitan Univ. 11, pp. 109-132.
12. Huang, T.C., Watkins, N.D., Shaw, D.M. and Kennett, J.P.: 1973, Earth and Planet. Sci. Letters 20, pp. 119-124.
13. Williams, H. and Goles, G.: 1968, in: Andesite Conference Guidebook, ed. Dole, H.M., Dept. Geol. and Min. Industries, Oregon, Bull. 62, pp. 37-41.
14. Thorarinsson, S.: 1974, in: World Bibliography and Index of Quaternary Tephrochronology, ed. Westgate, J.A. and Gold, C.M., Univ. of Alberta, Edmonton, pp. xvii-xviii.
15. Czamanske, G.K. and Porter, S.C.: 1965, Science 150, pp. 1022-1025.
16. Smith, D.G.W. and Westgate, J.A.: 1969, Earth and Planet. Sci. Letters 5, pp. 313-319.
17. Lerbekmo, J.F., Westgate, J.A., Smith, D.G.W. and Denton, G.H.: 1975, Roy. Soc. New Zealand, Bull. 13, pp. 203-209.
18. Westgate, J.A.: 1977, Can. J. Earth Sci. 14, pp. 2593-2600.
19. Mullineaux, D.R.: 1974, U.S. Geol. Survey, Bull 1326, 83 p.
20. Powers, H.A. and Wilcox, R.E.: 1964, Science 144, pp. 1334-1336.
21. Fryxell, R.: 1965, Science 147, pp. 1288-1290.
22. Westgate, J.A. and Evans, M.E.: 1978, Can. J. Earth Sci. 15, pp. 1554-1567.
23. Porter, S.C.: 1978, Quat. Res. 10, pp. 30-41.
24. Briggs, N.D. and Westgate, J.A.: 1978, U.S. Geol. Survey, Open-File Report 78-701, pp. 49-52.
25. Izett, G.A., Wilcox, R.E., Powers, H.A. and Desborough, G.A.: 1970, Quat. Res. 1, pp. 121-132.
26. Naeser, C.W., Izett, G.A. and Wilcox, R.E.: 1973, Geology 1, pp. 187-189.

27. Izett, G.A. and Naeser, C.W.: 1976, Geology 4, pp. 587-590.
28. Westgate, J.A., Briggs, N.D., Stalker, A. Macs. and Churcher, C.S.: 1978, Geol. Soc. Amer., Abst. with Programs 10, pp. 514-515.
29. Crandell, D.R., Mullineaux, D.R. and Waldron, H.H.: 1958, Am. J. Sci. 256, pp. 384-397.
30. Easterbrook, D.J. and Briggs, N.D.: 1979, Geol. Soc. Amer., Abst. with Programs 11, pp. 76-77.
31. Boellstorff, J.D.: 1976, Proc. 24th Annual Meeting of the Midwestern Friends of the Pleistocene, Kansas Geol. Survey, pp. 37-71.
32. Wilcox, R.E. and Izett, G.A.: 1973, Geol. Soc. Amer., Abst. with Programs 5, p. 863.
33. Weller, J.M.: 1960, in: Stratigraphic principles and practice, Harper and Row, New York, 725 p.
34. Souther, J.G.: 1976, Geosci. Canada 3, pp. 14-20.
35. Topping, W.W.: 1973, New Zealand J. Geol. and Geophys. 16, pp. 397-423.
36. Larsen, G. and Thorarinsson, S.: 1978, Jokull 27, pp. 28-46.
37. Westgate, J.A. and Fulton, R.J.: 1975, Can. J. Earth Sci. 12, pp. 489-502.
38. Lerbekmo, J.F. and Campbell, F.A.: 1969, Can. J. Earth Sci. 6, pp. 109-116.
39. Bowles, F.A., Jack, R.N. and Carmichael, I.S.E.: 1973, Geol. Soc. Amer. Bull. 84 pp. 2371-2388.
40. Hahn, G.A., Rose, W.I., Jr. and Meyers, T.: 1979, Geol. Soc. Amer., Spec. Paper 180, pp. 101-112.
41. Steen-McIntyre, V.: 1977, in: A manual for tephrochronology, published by the author, Idaho Springs, Colorado, 167 p.
42. Ninkovich, D.: 1968, Earth and Planet. Sci. Letters 4, pp. 89-102.
43. Kohn, B.P.: 1970, Lithos 3, pp. 361-368.
44. Kohn, B.P.: 1979, Quat. Res. 11, pp. 78-92.
45. Westgate, J.A., Christiansen, E.A. and Boellstorff, J.D.: 1977, Can. J. Earth Sci. 14, pp. 357-374.
46. Hildreth, W.: 1979, Geol. Soc. Amer., Spec. Paper 180, pp. 43-75.
47. Westgate, J.A., Smith, D.G.W. and Tomlinson, M.: 1970, in: Early man and environments in northwestern North America, ed. Smith, R.A. and Smith, J.W., Univ. of Calgary Archaeological Assoc., The Students' Press, Calgary, pp. 13-34.
48. Irvine, T.N. and Baragar, W.R.A.: 1971, Can. J. Earth Sci. 8, pp. 523-548.
49. Haggerty, S.E.: 1979, Can. J. Earth Sci. 16, pp. 1281-1293.
50. Buddington, A.F. and Lindsley, D.H.: 1964, J. of Petrology 5, pp. 310-357.
51. Borchardt, G.A., Aruscavage, P.J. and Millard, H.T., Jr.: 1972, J. Sed. Petrol. 42, pp. 301-306.
52. Sarna-Wojcicki, A.M., Bowman, H.W. and Russell, P.C.: 1979, U.S. Geol. Survey, Prof. Paper 1147, 15 p.

53. Howorth, R. and Rankin, P.C.: 1975, Chem. Geol. 15, pp. 239-250.
54. Sarna-Wojcicki, A.M.: 1976, U.S. Geol. Survey, Prof. Paper 972, 30 p.
55. Steen-McIntyre, V.: 1975, Roy. Soc. New Zealand, Bull. 13, pp. 271-278.
56. Ninkovich, D., Shackleton, N.J., Abdel-Monem, A.A. Obradovich, J.D. and Izett, G.A.: 1978, Nature 276, pp. 574-577.
57. Mankinen, E.A. and Dalrymple, G.B.: 1979, J. Geophys. Res. 84, pp. 615-626.
58. Mankinen, E.A., Donnelly, J.M. and Gromme, C.S.: 1978, Geology 6, pp. 653-656.
59. Creer, K.M., Gross, D.L. and Lineback, J.A.: 1976, Geol. Soc. Amer. Bull 87, pp. 531-540.

TEPHROCHRONOLOGY BY MICROPROBE GLASS ANALYSIS

Gudrún Larsen

Nordic Volcanological Institute
University of Iceland, Reykjavik

ABSTRACT. Basaltic tephra layers form a considerable part of the postglacial tephra deposits in Iceland. They can be related to the volcanic systems that produced them by characteristics based on major and minor element chemistry. Microprobe analysis of volcanic glass shards have made it possible to relate even very thin (1-2 mm) basaltic tephra layers in distal areas outside the volcanic zones to their source.

1. INTRODUCTION

Most eruptions that occur in Iceland are dominantly effusive, basaltic fissure eruptions. In areas covered by ice, at sea and where ground water table is high, these eruptions are partly or wholly explosive due to interaction between water and magma. Widespread basaltic tephra layers are therefore common in Iceland. Silicic tephra layers form the framework of tephrochronology in Iceland (1,2) but basaltic tephras add considerable detail to the tephrochronological record. In soil sections in southern Iceland covering the last 11 centuries, between 30 and 80% of all tephra layers are basaltic.

2. CORRELATION PROBLEMS

Tephrochronological studies covering the period 900 AD to present in Iceland have so far mostly been based on field studies, combined with studies of old Icelandic literature, where descriptions of volcanic eruptions, their source and the year of eruption are recorded (1,2,3,4).

S. Self and R. S. J. Sparks (eds.), Tephra Studies, 95–102.

In proximal areas, thickness and grain size variations of individual tephra layers are often sufficient criteria to relate the tephras to their source and to establish a dispersal pattern. In distal areas, however, chemical analysis may be the only way to relate a tephra layer to its source. There the thickness of the layers may be less than 1 cm and the samples that can be collected may be either too small for bulk chemical analysis, contaminated by soil or differentiated during transport and deposition.

The electron microprobe has greatly facilitated the use of chemical composition of tephras as a tool in tephrochronology. Volcanic glass and minerals can be analysed separately and the grains need only be a few microns in diameter. This method has proved especially promising for detailed chronological work on basaltic tephra layers. The relevant advantages are: (a) very small samples can be analysed, (b) rapidity of the method, and (c) favourable cost/efficiency ratio.

Basaltic tephra has a distribution pattern different from that of silicic tephra. The latter is generally erupted during a relatively short plinian phase forming a tephra layer with a distinct dispersal axis (2,5). Explosive basaltic eruptions produce moderate amounts of tephra for days or weeks, sometimes forming thin dispersed deposits, even near source (4). Correlating such deposits by field methods is both time consuming and unrewarding. A large number of soil profiles can, however, be efficiently studied by the microprobe technique in a relatively short time.

Each tephra layer, however insignificant, represents an eruption. If a tephra layer is omitted, information on an eruption can be lost, as it may not be recorded anywhere else. Detailed tephrochronology may be most useful to scientific studies such as archaeology in areas far outside the volcanic zones. In such cases the microprobe may often be the only way to relate thin basaltic tephra layers to their source.

3. CHEMICAL CHARACTERIZATION OF BASALTIC TEPHRA

In Iceland chemical "fingerprinting" of basaltic tephra is based on the fact that postglacial volcanic activity occurs along distinct volcanic fissure systems and on central volcanoes, collectively termed volcanic systems (6,7,8). Each system has chemical characteristics which in many cases can be used to identify its products.

The situation is especially favorable on the Eastern Volcanic Zone (Figure 1), where the volcanic systems are tholeiitic in the

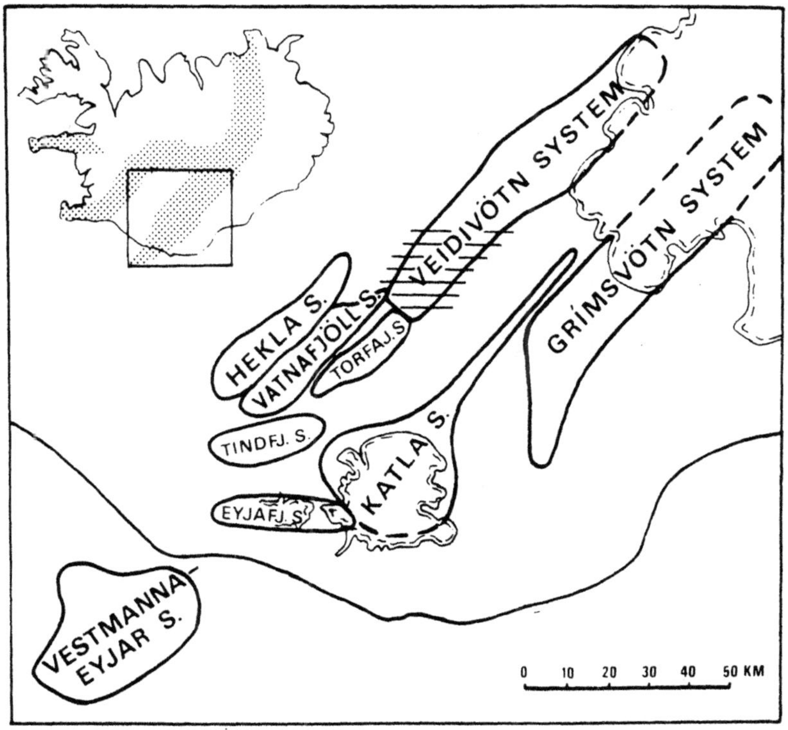

Fig. 1. The nine volcanic systems on the Eastern Volcanic Zone, after Jakobsson (7), and the areas covered by glaciers or affected by a high ground water table (shown by horizontal shading). Three have produced most of the basaltic tephra during the last 11 centuries: the Katla, the Grímsvötn and the Veidvötn systems. Those with names in small print have produced negligible amounts of basalts in postglacial times.

north but increasingly alkaline towards the south (7,8). This is important for tephra correlation work, since most of the postglacial tephra deposits in Iceland, both basaltic and silicic, originate from the Eastern Volcanic Zone. The compositional range of postglacial basalts in this region is well documented. During the timespan in question each volcanic system has produced basalts of relatively homogenous composition. The products of each system are chemically different from those of other systems, regardless of petrographic nomenclature (7,8).

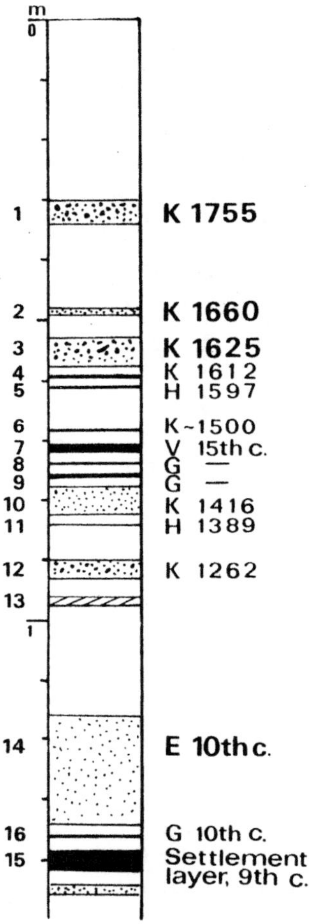

Fig. 2. Uppermost 1.5 m of a soil section from the Katla area, Southern Iceland. Origin and date of eruption of four layers (bold print) was known from previous studies (3,4). Eleven of the remaining layers have been related to their source by chemical characteristics (Figure 3) and given an absolute or approximate date of eruption (small print). G; eruptions within the ice covered part of the Grímsvötn system; H; eruptions of Hekla proper (intermediate to acid); K; eruptions within the ice covered Katla caldera; E; Eldgjá eruption (Katla system); V; Veidivötn eruption (Veidivötn system), Settlement layer: Vatnaöldur eruption (Veidivötn system).

The use of chemical characteristics of basaltic tephra for correlation purposes is demonstrated by a soil section from the Katla area in southern Iceland. The uppermost 1.5 m of the soil section (Figure 2), contains some 16 tephra layers erupted during the last 11 centuries. All the tephra layers except no. 13 are dark colored. The origin of four layers, no. 1, 2, 3, 14 is known through field studies. Most of the remaining tephra layers are thin and difficult to trace in the field. The origin of some of them was anticipated but awaited confirmation (3,4). Five to ten shards of glass from samples of each of the tephra layers in the above section were analysed on the microprobe.

The minor elements TiO_2 and K_2O have a larger compositional variation in basalts than most major elements and the difference between volcanic systems is pronounced on such a plot. Some other element ratio may resolve boundary cases and therefore a complete analysis of each tephra grain is performed. Figure 3 is a plot of K_2O against TiO_2 showing the average value and the range of these elements in postglacial basalts from six volcanic systems on the Eastern Volcanic Zone, based on bulk analyses from Jakobsson (7). Microprobe glass analyses of tephra are compared to this data. Each number represents a point analysis of a single glass shard and refers to a tephra layer in the soil section on Figure 2.

A clear separation between volcanic systems results from the plot in Figure 3. The fit is adequate given the fact that glass analyses are being compared to bulk analyses. The layers 4, 6, 10 and 12 originate within the ice covered Katla caldera; 8, 9, 16 and 7, 15 come from the Grímsvötn and Veidivötn systems, respectively. Three tephra layers were of silicic composition and are not plotted. Since all the tephra layers were deposited during historical times, the year of the eruption could be deduced from written sources in five cases and an approximate date could be assigned in six cases, see Figure 2.

The products of each volcanic system have a limited compositional range which is, however, considerably larger than that of individual eruptions from the same system (7,9). This is partly illustrated on Figure 3 by the analyses from layers 12 and 14 which form small groups within the field occupied by the Katla system. Another example is provided by tephra layers collected from soil sections 30 km apart in order to distinguish between and correlate individual layers using this difference in glass composition (Figure 4). The difference between the respective layers is small but distinct, although the number of grains so far analysed from each sample is too low to test conclusively the homogeneity of these layers.

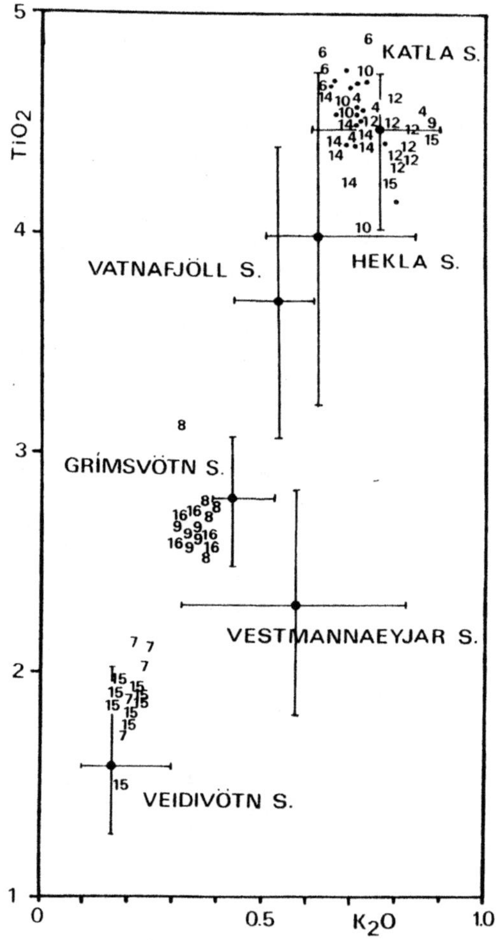

Fig. 3. Plot of K_2O against TiO_2 (wt%) showing the average value and range (filled circles and bars) of these elements in postglacial basalts from six volcanic systems on the Eastern Volcanic Zone (7) and corresponding values for basaltic tephra layers (numbers) from the soil section shown on Figure 2. Tephra layers 1-3 are represented by dots.

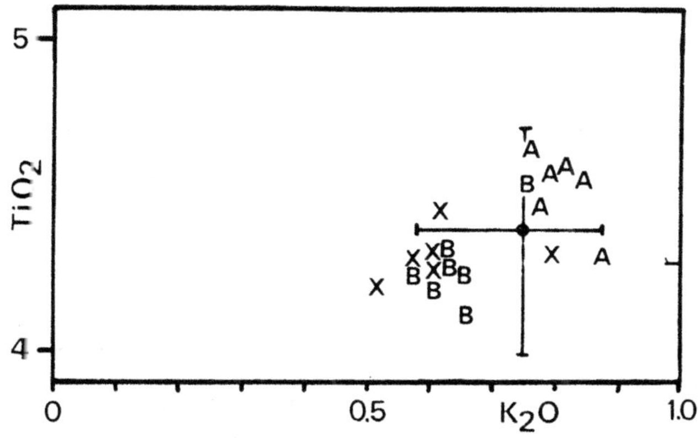

Fig. 4. Two Katla layers, A and B, from a known horizon in a soil section in the Katla area compared to Katla layer X in a soil section 30 km away, thought to correspond to one of them. Apparently X and B are the same layer.

Two tephra layers have been so tested for homogeneity; those from the 1918 eruption of Katla and the 9th century eruption from the basaltic part of Vatnaöldur fissure. The Katla tephra was found to be homogenous. A slight change in glass composition was observed along the eruptive fissure of the latter and minor amounts of glass with different chemical composition were found, corresponding to the glass of an underlying palagonite formation (9).

It should be emphasized that microprobe analyses of glass are directly comparable only with maximum precision. This is done by coating and analyzing the samples that are being compared in the same batch under the same conditions. The difference in chemical composition between eruptions from the same volcanic system are so small that they are easily obscured unless the above-mentioned conditions are fulfilled.

4. SUMMARY

Microprobe glass analysis of basaltic tephra layers have proved to be useful for correlation purposes in Iceland. Two levels of detail can be defined:

1) Basaltic tephra layers can be related to individual volcanic systems by their chemical characteristics, provided the composition of the system is well known.

2) A limited compositional range within each system offers the possibility of distinguishing between individual tephra layers from the same system by major and minor element chemistry alone. More data is, however, needed to firmly establish this and to what extent it needs to be supplemented by trace elements and mineral chemistry.

REFERENCES

1. Thorarinsson, S.: 1944, Geogr. Ann. Stockh. 26, pp. 1-217.
2. Thorarinsson, S.: 1967, The Eruption of Hekla 1947-48, I, Soc. Sci. Islandica, pp. 1-170.
3. Thorarinsson, S.: 1959, Acta Nat. Isl. 22, pp. 1-100.
4. Larsen, G.: 1978, B. Sc. hon. Thesis, University of Iceland, pp. 1-59 (mimeographed).
5. Larsen, G. and Thorarinsson, S.: 1977, Jökull 27, pp. 28-46.
6. Saemundsson, K.: 1978, Geol. Journ. Spec. Iss. 10, pp. 415-432.
7. Jakobsson, S.P.: 1979, Acta Nat. Isl. 26, pp. 1-103.
8. Óskarsson, N., et al.: 1979, Nord. Volc. Inst. and Science Inst. Report 79 05/79 16, University of Iceland, pp. 1-104.
9. Larsen, G.: in preparation.

X-RAY FLUORESCENCE ANALYSIS AS A RAPID METHOD OF IDENTIFYING TEPHRAS DISCOVERED IN ARCHAEOLOGICAL SITES

A.B. Cormie, D.E. Nelson, and D.J. Huntley

Archaeology and Physics Departments, Simon Fraser University, Burnaby, B.C., Canada V5A 1S6

ABSTRACT. Volcanic ash layers from three major Holocene eruptions are found throughout southern British Columbia and are useful stratigraphic markers. Here we show that energy dispersive X-ray fluorescence analysis and alpha counting are useful techniques for the rapid identification of these ashes.

1. INTRODUCTION

Ashes from three Holocene eruptions (Mazama, 6,600 y.B.P.; Mt. St. Helens Yn, 3,400 y.B.P.; and Bridge River, 2,350 y.B.P.), are distributed throughout southern British Columbia, Canada. These tephras are commonly found in local archaeological sites and once identified can provide archaeologists with excellent time-stratigraphic markers. We have concentrated on developing rapid methods by which samples from these tephra layers can be routinely identified.

To date, one of the most widely applied methods for tephra identification is microprobe analysis of the major element concentrations in the glass or mineral grains (eg. 1,2). Neutron activation and X-ray fluorescence analyses of glass separates have also been used (eg. 3,4). These studies have shown that the major and trace element concentrations in glass or selected mineral separates are usually sufficiently homogeneous for ashes from a single eruption, yet sufficiently different for ashes from different eruptions so that tephras may be chemically finger-printed.

In this study, we have investigated the use of energy dispersive X-ray fluorescence (XES) analysis and alpha counting as

S. Self and R. S. J. Sparks (eds.), Tephra Studies, 103–107.

methods for identifying tephras found in British Columbia. A major objective was to find a satisfactory technique which required a minimum of sample preparation and we hoped that these methods would offer several advantages in this regard.

2. ENERGY DISPERSIVE X-RAY FLUORESCENCE ANALYSIS

With XES one can simultaneously analyze a wide range of elements including the major elements K to Fe and the trace elements Rb to Nb in typical analysis times of 5 to 10 minutes per sample. Measurement of a large number of analytes gives a greater possibility that characteristic concentrations will be found.

In our preliminary work we wished to determine whether XES analysis of the glass separates would allow us to distinguish between the three tephras mentioned earlier. A few samples of known material were selected from each source and ground (if necessary), sieved, and pretreated with 20% HCl and 5% NaOCl to remove organic stains, metal oxides and carbonates. (We have verified that these treatments do not significantly alter the concentrations of the elements of interest in pulverized obsidian or tephra glass.) The glass portion was then isolated from the 62-210 μm size fraction using bromoform/acetone mixtures and a Franz magnetic separator. The samples were analyzed using a silver secondary target to scan the elements K to Nb. Relative concentrations were determined for each element by calculating the ratio of the peak areas to the Compton scatter peak. We preferred to simplify identification procedures by using relative rather than absolute concentrations. We found that with purified specimens, several elements would allow us to distinguish between the ash groups. These tests further suggested that the high Zr concentrations in specimens of Mazama and low K concentrations in specimens of Mt. St. Helens Yn would allow us to distinguish between the tephras even with less pure samples.

We compared the effects of different sample treatments on the pure glass, whole ash and various size ranges of whole ashes and found that the <62 μm size range of whole ash was chemically similar to the glass separates. Apparently, the glass is concentrated in the fine fraction of sieved ashes and this portion should therefore allow us to identify tephras with only simple pretreatments. Acid and peroxide treatments are effective for removing carbonates, organics and some clays, and can therefore be used to remove post-depositional K-containing contaminants. These treatments, however, have little effect on the concentration of Zr. We therefore decided to: a) directly analyze the <62 μm portion to determine if Mazama tephra could be distinguished by its Zr concentration without pretreatment, and b) analyze Mt. St. Helens Yn and Bridge River ashes for differences in K concentration following pretreatment with 20% HCl and 30% H_2O_2. The results appear in figures 1 and 2.

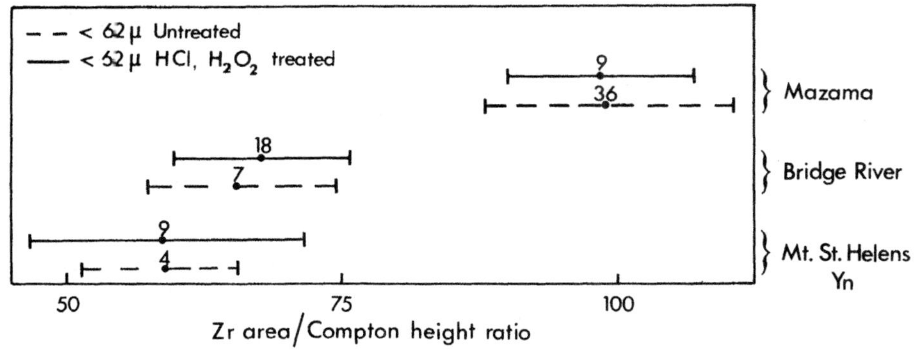

Figure 1. Zr area/Compton peak height ratios for the three ashes measured by XES. The number of samples analyzed is shown above each point and the error bars are ± 2σ in all three figures.

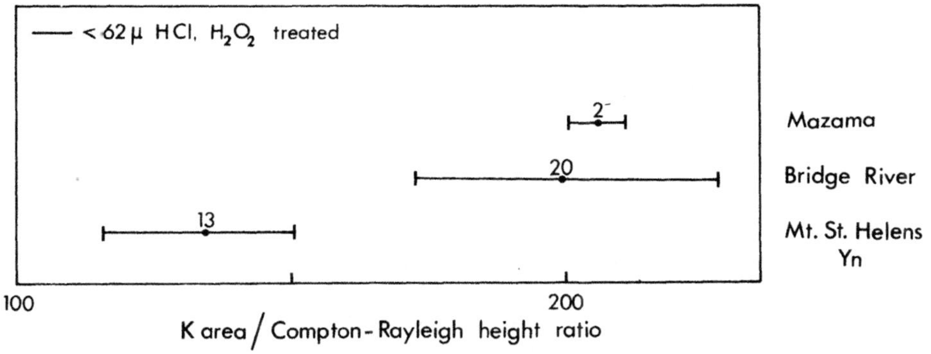

Figure 2. K area/Compton-Rayleigh peak height ratios measured by XES.

The results given in figures 1 and 2 include a number of standard samples as well as previously unknown samples from land, lake or bog deposits. Standard samples are those which have previously been identified by other researchers, or have been collected near the volcanic origin of a tephra. The results in figure 1 indicate that untreated samples of sieved Mazama ash can be distinguished from the other two ash groups on the basis of Zr. Additional treatments do not improve our ability to distinguish between the ash groups. In figure 2 we show that treated samples of Mt. St. Helens and Bridge River ashes can be distinguished on the basis of their K concentrations. No other analytes were useful for distinguishing between tephra samples of intermediate purity.

3. ALPHA COUNTING

Alpha counting can be used to determine the uranium and thorium content of a sample as the alpha activity provides an indirect measurement of concentrations of these two trace elements, (cf. 5). Because the alpha counter is a relatively simple piece of equipment, it seemed feasible that if tephras could be distinguished on the basis of their U and Th contents, alpha counting could be developed for identifying tephras in the field. For these analyses, both purified glass samples and fine grained samples treated with HCl and NaOCl were selected from each source. The alpha count rates were determined as described by Huntley and Wintle, (6). Samples were counted for about four days each, background counts (∿.05/ks) were subtracted, and count rates per unit area were calculated.

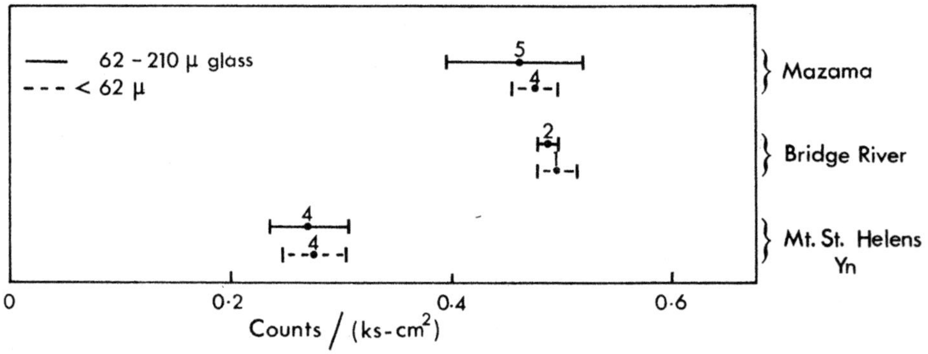

Figure 3. Alpha particle count rates.

The results, (figure 3), indicate that we can use the alpha count rates to distinguish between Mt. St. Helens Yn and the other two tephra groups. We cannot, however, distinguish between Bridge River and Mazama by this method. Most of the uncertainty is due to sample inhomogeneity rather than counting statistics. A reduction in the counting times to ∿1 day will therefore not affect our ability to distinguish Yn from the other ash groups. The similarity between the glass and sieved whole ash samples indicates that alpha counting is effective with limited amounts of sample preparation and alpha counting could potentially be used in the field to distinguish between some ash groups.

4. APPLICATIONS

With XES analysis we were able to identify rapidly a number of both geological and archaeological samples collected in the Okanagan region of British Columbia. For example, at the Gore Cree

Site (EeQw 48, cf. 7), skeletal remains were discovered 0.5 to 1 m below the bottom ash layer in a deposit which contains two ash layers, (8). We identified the bottom layer as Mazama ash (6,600 y.B.P.), and the top layer as Mt. St. Helens Yn ash (3,400 y.B.P.). By calculating an average rate of deposition for the site we can estimate that the age of the human remains is about 8,400 y.B.P.. A radiocarbon age of 8,200 y.B.P. obtained on the bone indicates that this age estimate is reasonably accurate. At the Moulton Creek Site (EdQx5), we identified Mt. St. Helens Yn as the tephra which separates two cultural components, (9). We have finger-printed ashes from other sites in the Okanagan region of British Columbia. Most of these sites have not yet been studied in detail but identification of tephras will not only facilitate the interpretation of site stratigraphy but will also provide regional archaeologists with a means of making inter-site comparisons of cultural material.

REFERENCES

1. Smith, D.G.W., Westgate, J.A., 1969. Earth Planet. Sci. Lett. 5: 313.
2. Westgate, J.A., 1977. Can. J. Earth Sci. 14: 2593.
3. Borchardt, G.A., Harward, M.E., Schmitt, R.A, 1971. Quaternary Res. 1: 247.
4. Smith, R.P., Nash, W.P., 1976. J. Sediment. Petrol. 46: 930.
5. Cherry, R.D., 1963. Geochim. Cosmochim. Acta 27: 183.
6. Huntley, D.J., Wintle, A.G., 1980. Can. J. Earth Sci. (in press).
7. Borden, C.E., 1952. A Uniform Site Designation Scheme for Canada; Anthrop. in British Columbia 3: 44.
8. Elmore, D., Hanson, W., Knox, L., 1979. Gore Creek Site: Excavation of a Burial (unpublished report for Archaeol. Sites Advisory Board of B.C.; Vict., B.C., Can.).
9. Eldridge, M., 1974. Report Submitted to the Archaeol. Sites Advisory Board of B.C.; Vict., B.C., Can..

REGIONAL STUDIES

THE APPLICATION OF TEPHROCHRONOLOGY IN ICELAND

Sigurdur Thorarinsson

Science Department, University of Iceland
Reykjavik, Iceland

ABSTRACT. The situation in Iceland for establishing a tephrochronological time scale is very good. This is due to the great number of volcanic eruptions in post glacial time, a great range of variation in chemistry between tephra layers, a rapid thickening of the soil, separating layers with small age difference, and the existence of detailed accounts of many historical eruptions.

In Iceland, tephrochronology has been applied to studies of the eruption history of the most active volcanoes. It has also been utilised in archaeological research and palynology, studies of fluvial erosion, wind erosion, and in cryopedological and glaciological research. A tephrochronology correlation between Iceland, the European continent and Greenland has been established.

1. INTRODUCTION

Few countries offer better possibilities than Iceland for studies of various geological and geomorphological processes. The endogenic and exogenic agencies which shape the surface of our planet operate there more effectively and clearly than in most other countries. Iceland is one of the most active volcanic areas in the world as well as one of the most active glacial regions. Wind erosion, marine abrasion, frost weathering and solifluction also operate vigorously. Not without reason, Iceland has been called a giant geological and geomorphological--one might also add geophysical--laboratory. However, processes can hardly be studied quantitatively in a laboratory without some measure of time, and for any detailed, quantitative story

S. Self and R. S. J. Sparks (eds.), Tephra Studies, 109–134.

of the geological and geomorphological processes in Iceland, it is highly desirable that an exact and, if possible, absolute time-scale should be established.

The dating methods most successfully applied in post-glacial geology before the Second World War were G. De Geer's varve-chronology, dating back to 1884 and called by him geochronology, A.E. Douglas' tree-ring method or dendrochronology, which goes back to 1901; and L. von Post's pollenanalysis or palynology, worked out during the 1910's. These three methods were all developed in areas where the possibilities of putting them into practice were very favourable. In Middle Sweden, where De Geer and von Post worked, the sections of clay suitable for varve measuring are abundant. This part of Sweden is a border zone between conifer forests and hardwood forests, so that the climatic changes are clearly reflected by the tree pollen spectrum in the peat bogs. Douglas' tree-ring method was first applied in areas where the climatic conditions determining the growth of trees are rather simple. The growth pattern in the southwestern United States is mainly influenced by a single climatic factor, the precipitation. These three methods, however, present considerable difficulties in their application to Iceland. The varve series are in places disturbed by glacier bursts (jökulhlaups), caused by subglacial volcanic activity. Because of the isolation of the country and the harsh climate, birch has been the only forest tree, besides some scattered mountain ashes, through the entire post-glacial time. Also the very high mineral content of the peat soil, partly caused by the frequent tephra falls, render palynological work more difficult.

2. THE SUITABILITY OF ICELAND TO TEPHROCHRONOLOGICAL STUDIES

Nature has arranged it so that the very factors which complicate the application of the above mentioned dating methods, provide unusually good conditions for the establishment of another dating method, tephrochronology. With the reservation, that in Iceland it is difficult to say what is a single volcano and a single eruption, it may be said that about 200 eruption sites have been active there in post-glacial time, many of them, however, erupting only once, and between 30 and 40 have erupted since Nordic settlement began in the country, 11 centuries ago (Fig. 1). Every fifth year or so an eruption begins, or is going on, in the country.

In Iceland extensive tephra layers are more numerous than one would expect in a mainly basalt producing area. This is due to the fact that some of the most active volcanoes, such as Katla and Grímsvötn, are blanketed by ice and their eruptions are phreatomagmatic and wholly explosive, or almost so. Black basaltic

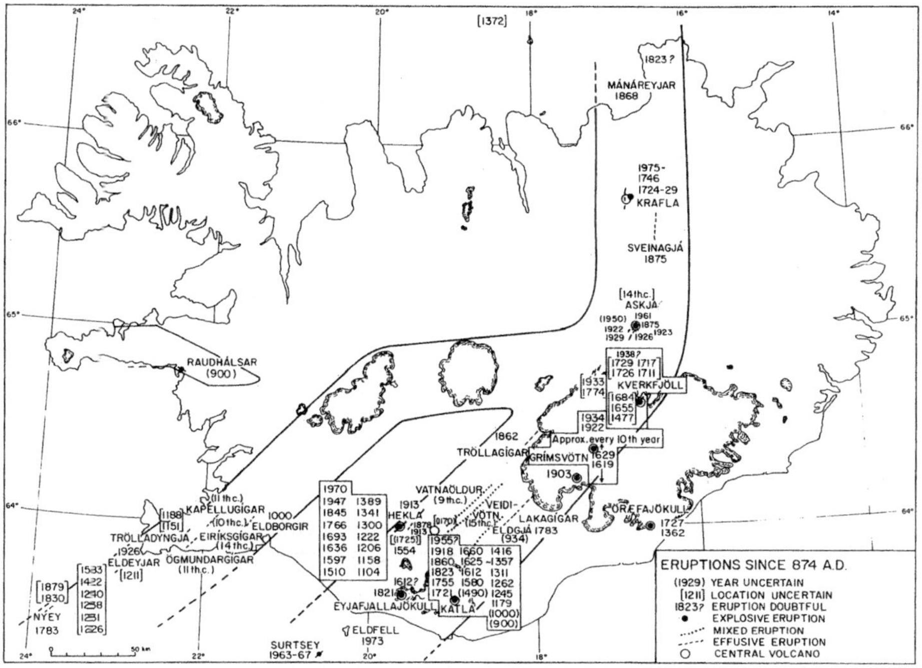

Fig. 1. Volcanic activity in Iceland in historical time (1).

tephra layers alternate in soil sections with brownish intermediate and light acidic layers. The last mentioned ones stand out most clearly in soil sections and are, therefore, easiest to follow from one place to another. There are 12 widespread acidic tephra layers in the post-glacial soils.

The establishment of tephrochronology in Iceland by field studies is facilitated by the country being one of the few highly active volcanic areas in the world where formation of both peat soil and loessial soil is unusually rapid. In the harsh climate the disintegration of organic material is slow and wind erosion has greatly increased the rate of thickening, especially since the arrival of man. As a consequence, tephra layers of only a few years difference in age can be clearly separated in soil sections. It has proved possible to discern in the field between 150 tephra layers in a single post-glacial loessial soil section. Furthermore, Iceland is a country where satisfactorily reliable accounts of volcanic eruptions and tephra falls are more plentiful and go farther back in time than in most other volcanic areas. The oldest eruptions mentioned in Icelandic sources occurred during the settlement time, 870-930 A.D. The oldest layer accurately dated through written sources is the Hekla layer of the 1104 eruption.

The identification and correlation of tephra layers in the laboratory has, in the last decades, been greatly facilitated by use of the microprobe.

3. TEPHROCHRONOLOGY AND THE C-14 METHOD

The C-14 method has, of course, been used in Iceland as elsewhere. To some extent it has replaced tephrochronological dating, but in Iceland and other volcanic areas, it is far from being able to replace it entirely. As far back in time as tephra layers can be dated exactly by written records, they can, in some cases, be used for a more exact dating than is possible with C-14. An identified tephra layer, whether absolutely dated or not, makes an exact time correlation possible within the entire area of its dispersal, which is often very wide. In Iceland, and maybe also in other active volcanic areas, errors in the C-14 datings have occurred which have not yet been explained, but could possibly be due to volcanic CO_2, which could, occasionally, change the $^{12}C/^{14}$ ratio. It is important that C-14 dating and tephrochronology can supplement and aid each other in active volcanic areas. Tephra layers embedded in organic soils, or that themselves enclose organic remains, can be dated by the C-14 method. Such a layer then carries this date with it wherever it is found, even in pure mineral soils where C-14 dating cannot be applied.

4. APPLICATIONS OF TEPHROCHRONOLOGY IN ICELAND

In Iceland tephrochronology has been applied to:

Studies of the eruption history and eruption mechanism of volcanoes.
Archaeological studies, especially dating of farm ruins.
Pollenanalytical studies of vegetation changes.
Studies of fluvial erosion.
Studies of wind erosion.
Studies of periglacial phenomena, especially frost crack polygons.
Dating of ice cores from glaciers.
Dating of glacier oscillations.
Establishing tephrochronological teleconnections between Iceland and other countries.

Examples of each of these applications will be discussed.

4.1 Studies on Icelandic volcanoes eruption history

These studies have had a threefold aim:

a) To supplement knowledge from written sources regarding volcanic activity during the 11 centuries of Nordic settlement in Iceland.
b) To gain knowledge of volcanic activity in prehistoric post-glacial time.
c) To build up a time-scale based on the tephra layers.

Detailed studies have so far been concentrated on the eruption history of the most active volcanoes in historical time, and as such they have necessarily been combined with a critical evaluation of the written sources, especially the older ones. These studies have so far resulted in monographs on Hekla, Öræfajokull and Grímsvotn (2-4). Work on Katla is now going on (5-7), greatly facilitated by the use of the electron microprobe analytical technique.

The studies of the Hekla tephra soon revealed its cyclic activity with five big acidic, explosive initial eruptions during the last 7000 years or so. Two of these layers, H_3 and H_4 (Fig. 2), C-14 age about 2800 years and 4000 years respectively, are the most widespread tephra layers in the post-glacial soils of Iceland, covering 80,000 and 78,000 km^2 on land respectively (8). In common with most Hekla tephras, the SiO_2-content decreases upward in the layers.

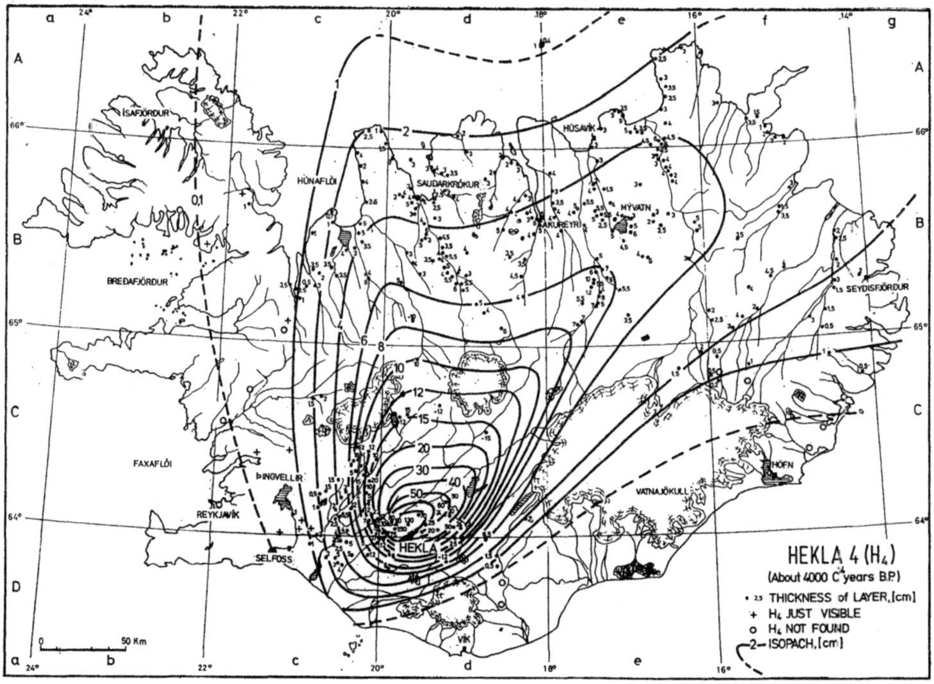

Fig. 2. Isopach map of the tephra layer H_4 (6).

According to written sources, the first Hekla eruption after the beginning of Nordic settlement, about 870 A.D., was the eruption in 1104 A.D. The written sources have nothing to say about this eruption except the five words, "The first eruption of Hekla". But tephra studies have revealed that this eruption was one of the five acidic, explosive initial eruptions.

From written sources it has been concluded that there were 18 historical Hekla eruptions until this century. Tephrochronological studies have reduced that number to 13 and since then, Hekla has added three more eruptions, in 1947, 1970 and 1980. All historical tephra layers have been identified (Fig. 3).

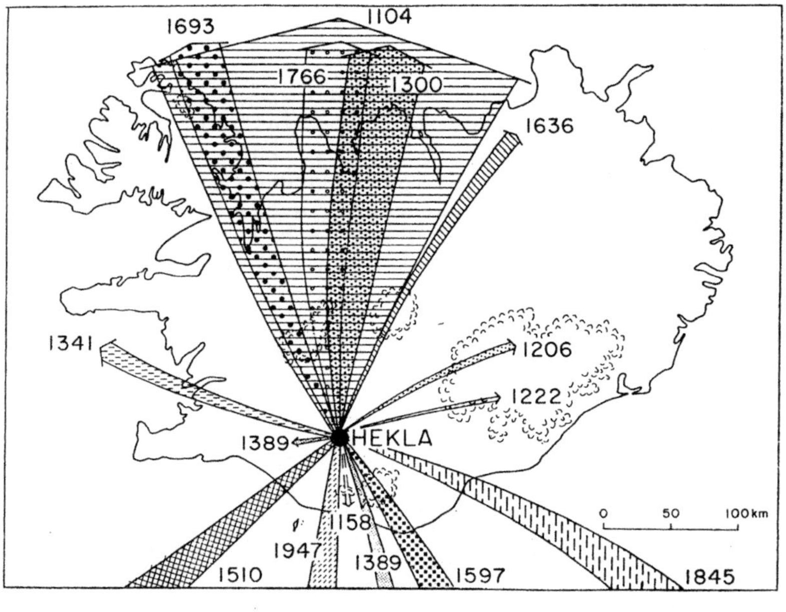

Fig. 3. Map showing in which direction tephra was spread during the initial phase of each of Hekla's 15 eruptions in historical time. The width of each arrow indicates roughly the relative size of the estimated volume of the layer.

Through the studies of Katla tephra, it has already been established that, contrary to former opinion, Katla has erupted more often than Hekla in historical time. Seventeen historical eruptions are now known with certainty. Three big eruptions, one occurring about the year 1000, one in 1357 ± 3 years, and one about 1490 have been detected by tephrochronological studies.

From an isopach map of the $\sim$1357 Katla layer (Fig. 4), the conclusion can be drawn that the tephra fall must have laid waste, for a year or more, the main part of the rural district Mýrdalur, south of Mýrdalsjokull. This disaster is mentioned in a few words by contemporary chroniclers, but they connected the tephra fall in Mýrdalur with a volcano, Trölladyngja, which cannot have wrought havoc upon Mýrdalur. Therefore, no historians seem to have taken the mentioning of a disastrous tephra fall in Mýrdalur seriously, but the tephra layer is conclusive evidence of the disaster (7,5).

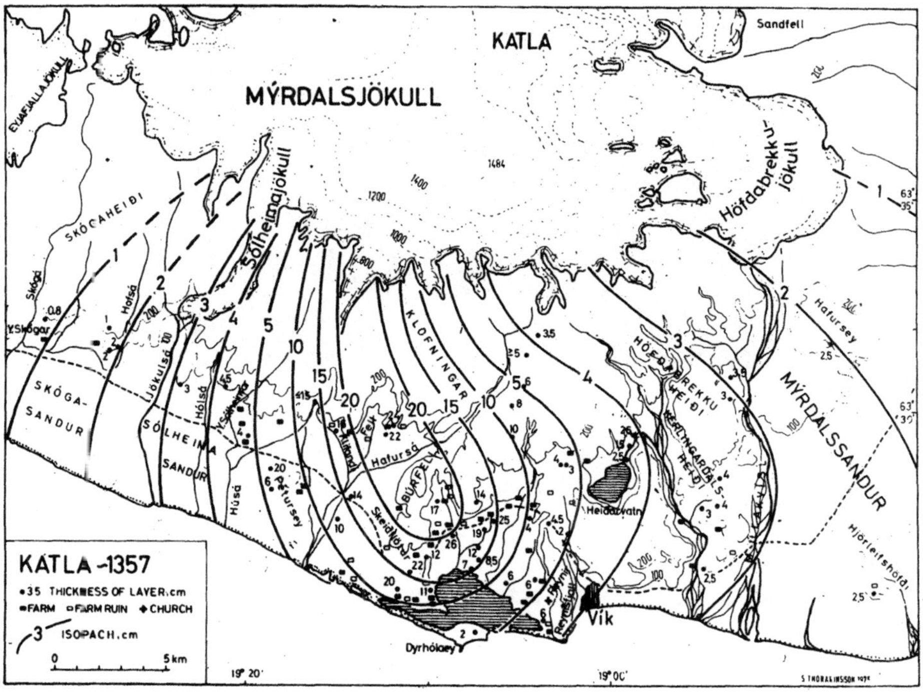

Fig. 4. Isopach map of the $\sim$1357 Katla layer.

Tephra layers have proved very useful for interpreting the eruption history of the Mývatn area (9,10). To mention an example; the famous tuffring Hverfjall was thought to be older than the last glaciation (6), but it was easy to prove that it was a few hundred years younger than H_3 (Fig. 5), and consequently ca. 2500 years old. Recently E. Vilmunardóttir has made successful use of Hekla tephras in mapping Tungnarhraun (12).

Fig. 5. The tephra layers H_3 and H_4 in humus soil beneath stratified Hverfjall tuff 2 km north of Hverfjall. Photo: S. Thorarinsson.

5. TEPHROCHRONOLOGY AND ARCHAEOLOGY

As already mentioned, tephrochronological and C-14 dating are of mutual use. The same applied to tephrochronology and archaeological dating (13). From an archaeological point of view, tephra is also of aid as being one of the best preservations agencies possible (when forming layers thick enough). I need only mention Pompeii and Akrotiri.

The most important contribution of tephrochronology to archaeology in Iceland has been the dating of farm ruins. In my opening address I mention the 1939 excavations of farm ruins in the deserted Thjórsárdalur valley, NW of Hekla (14). The final and decision regarding the age of the farm ruins in the inner part of the valley was that the farms were abandoned in 1104 A.D. because of tephra fall from Hekla. One of these ruins, at Stöng, was extremely well preserved, embedded as it was in the light Hekla pumice.

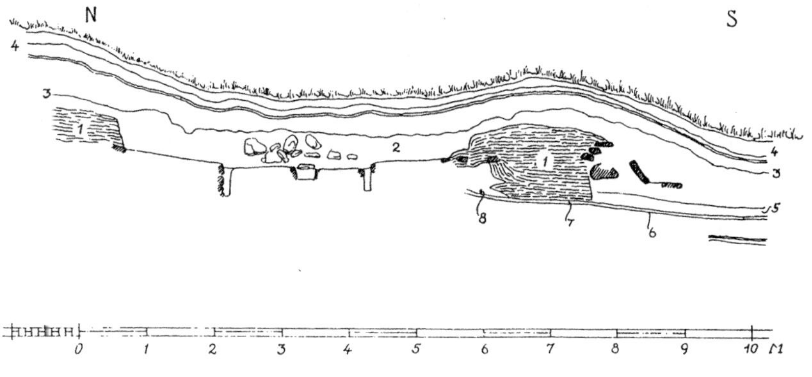

Fig. 6. Section through the main hall of the farm ruin Skallakot in Thjórsárdalur (14). Explanation in text.

In the most southwestern part of the valley, which was not seriously hit by the 1104-eruption, a farm ruin was excavated at a place called Skallakot. Figure 6 shows a section across the main hall of that farm as it illustrates well how convenient and time saving dated tephra layers can be in determining the approximate age of farm ruins. The uppermost tephra layer in the section is Katla 1918, then follows a series of three layers, designated with 4 in the figure. They are H 1766, K 1721 and H 1693. A light layer 3, which here was not more than 1 cm thick, proved to be the same layer, H 1104, which buried the farms in the inner part of the valley. If this layer had been identified in this area before the excavation, it would, obviously, have been sufficient to dig or drill a hole down to the floor of the hall to ascertain that this farm was abandoned long before 1104. In the turf wall in the section were lenses of another light layer, which is therefore older than the wall. Outside the farm, this layer, designated 6, was found almost immediately under a thin layer of charcoal formed when the first settlers cleared the land around the farm with fire. This light layer was deposited about 900 A.D. and is now called the Landnam layer. This layer is found all over south and southwest Iceland and is a very important time marker, not least for studies of the changes in vegetation that were caused by the arrival of man and his livestock in Iceland which before was devoid of grazing mammals. Dated tephra layers have also greatly facilitated the palynological work that has been carried out on this subject in Iceland (15,16).

Beautifully preserved farm ruins abandoned because of the Öræfajökull eruption in 1362, the Katla eruption about 1357 and the Katla eruption about 1490, have been excavated. Together with the 1104 farm ruins and older ones, they elucidate the gradual development of the Icelandic turf and stone built farmhouses from settlement type longhouses of simple design via more differentiated longhouses to a type pointing towards the so-called passage farmhouse of the last three centuries (17).

In the last decades use has been made of dated tephra layers for a survey of the approximate age of farm ruins in Northeast Iceland (18). These studies have revealed that before the middle of the eleventh century the frontier of the Icelandic rural settlements had already reached its most advanced position toward the barren interior, but was soon withdrawn because the border zone proved very vulnerable to the destructive influence of man and his grazing livestock.

In 1978, a week was spent looking for farm ruins in the Hrafnkelsdalur valley in eastern Iceland. In this valley the famous Icelandic sagas, Hrafnkelssaga, took place. According to the saga the valley was rather densely populated in the latter half of the 10th ventury. Some literary historians have maintained that this inland valley, the floor of which is 400 m above sea level, could never have been densely populated, but it not only came out that there are many farm ruins in the valley, but also the tephra layers show that these ruins date back to the 10th century. Thus tephrochronology supports the saga in this respect.

6. THE DEVELOPMENT OF JÖKULSÁRGLJÚFUR AND HVÍTÁRGLJÚFUR

The two most impressive canyons in Iceland are Jökulsárgljúfur (the Jökulsá Canyon) on the river Jökulsá a Fjöllum, in northern Iceland, and Hvítárgljúfur (the Hvítá Canyon) on the river Hvita, in southern Iceland. At the head of Hvítárgljúfur is Iceland's best known waterfall, Gullfoss. Impressive waterfalls are also found where Jökulsá plunges into its canyon. One of these, the Dettifoss, 44 m high, is Iceland's mightiest. Tephrochronology has thrown light on the age and development of both those canyons.

What is collectively called Jökulsárgljúfur is a 30 km long stretch of the Jökulsá river valley. This stretch is morphologically divided into three parts of approximately equal length. The middle part is mainly a glacially eroded valley, but the southernmost and northernmost parts are real canyons, clearly tectonically controlled. The canyon walls, up to 140 m in height, consist of basalt layers, more or less jointed, alternating with poorly cemented sedimentary layers.

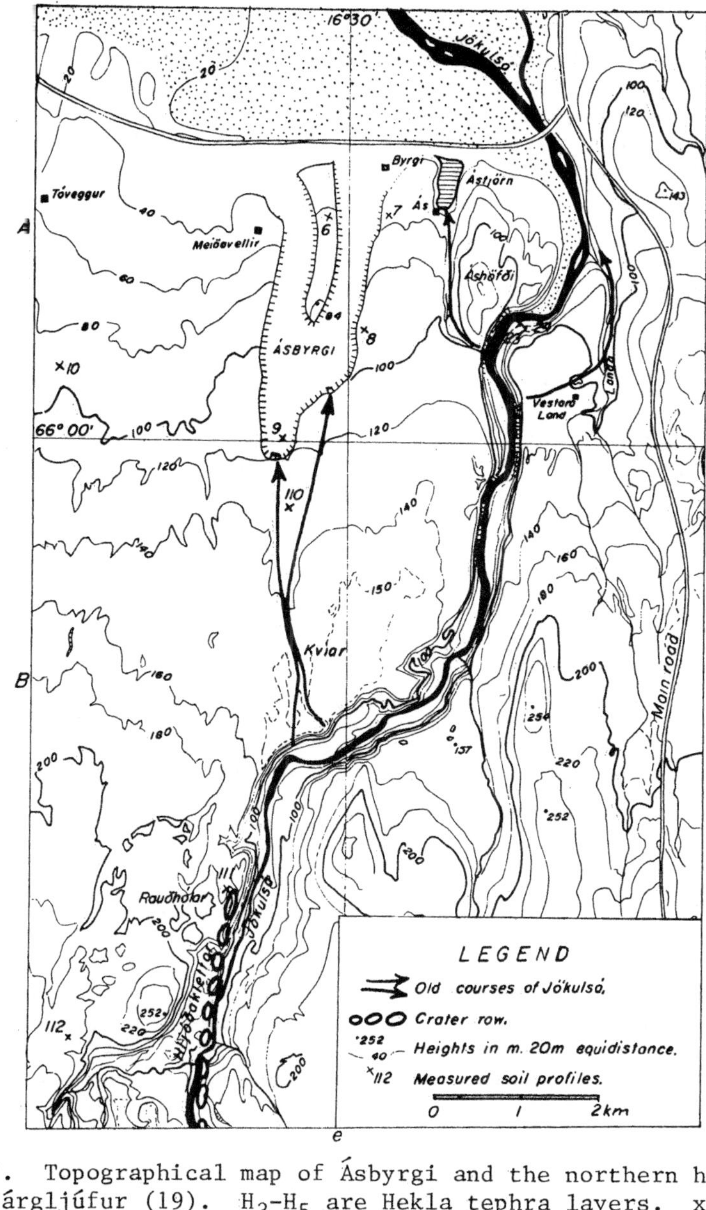

Fig. 7. Topographical map of Ásbyrgi and the northern half of Jökulsárgljúfur (19). H_3-H_5 are Hekla tephra layers. x = location of soil profiles shown on next figure.

Figure 7 is a map of the northern canyon and the northern part of the glacially eroded valley. Both east and west of the canyon are dead falls and dry canyons. By far the biggest and most impressive of these canyons in the 2.5 km long Ásbyrgi--

the shelter of the gods--west of Jökulsá. In school geography it was stated that Ásbyrgi is a beautiful and peculiar horseshoe shaped tectonic depression; only half of this statement is true. Ásbyrgi is indubitably both beautiful and peculiar, but it is neither horseshoe shaped nor a tectonic depression. It is an erosional feature, a real canyon eroded into the flanks of an interglacial shield volcano and, like other inactive canyons in the area, a former riverbed of Jökulsá.

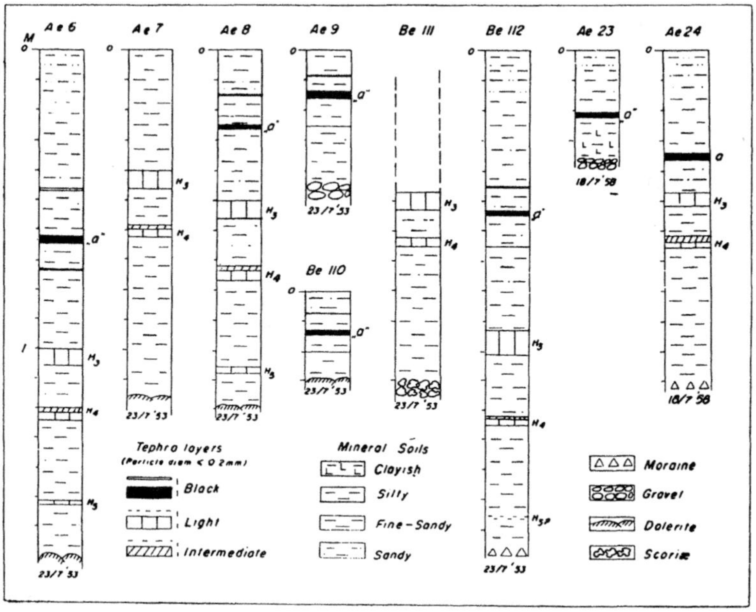

Fig. 8. Soil profiles in the Ásbyrgi area (19).

Before further research was carried out it seemed natural to suppose--and I still think this is the case (cf. also 20)--that Ásbyrgi was eroded during the withdrawal of the inland ice of the last glaciation form northern Iceland. As the area draining to Jökulsá from the west is a volcanically active belt, one can assume that great jökulhlaups have played a decisive role in the formation of the canyon. However, the results of a study of tephra layers in Ásbyrgi and surrounding areas (Fig. 8) were rather surprising. On the rim of the Ásbyrgi walls and all around the old riverbeds one finds complete post-glacial soil profiles, containing all the light Hekla layers. However, on the floor of Ásbyrgi and in the other inactive river beds one finds only a thin layer of soil and no light Hekla layers. In the glaciated part of Jökulsárgljúfur the light layers are also missing below a niveau, corresponding to a water level so high that the water could have flowed to Ásbyrgi, viz. the height of the col between the old and new riverbeds. From this I concluded that Jökulsá flooded Ásbyrgi for the last time after H_3 had been deposited 2800 C-14 years ago, and the northern canyon of Jökulsá was less than 2800 C-14 years old (19). Later, H. Tomasson and G. Sigbjarnarson, working along the southern part of the Jökulsá, found convincing evidence of an enormous jökulhlaup in the river after the deposition of H_3 (21, cf. also 20). They maintained that the canyons of Jökulsá, Ásbyrgi included, were almost exclusively eroded by this single flood, which was big and rapid enough to cause so-called cavitation. I agree that the canyons through which the river now flows, were to a rather great extent, eroded by that flood. There are parallels on the Columbia River basalt plateau where a much bigger canyon, Grand Coulée, was eroded by tremendous "jökulhlaups" from ice dammed lakes.

Th. Einarsson has used tephra layers to elucidate the stepwise, backward erosion of Hvítárgljúfur (unpublished work). In the southernmost part of the canyon all the light Hekla tephra layers are found, but on the terrace just south of Gullfoss the oldest layer is H 1693.

7. TEPHROCHRONOLOGICAL STUDIES OF WIND EROSION

I have discovered that one of the things that surprises most foreign geologists when they travel through Iceland is the great role played by wind erosion on this oceanic island with its humid, cold temperate climate. It is estimated that since the beginning of Nordic settlement 11 centuries ago, at least 20,000 km^2, or nearly 50% of the area then vegetation covered, has been deprived of its vegetation and loessial soil cover. Birch woods are supposed to have covered at least 20,000 km^2 when the settlement began, now only 1,200 km^2 remain.

Opinions have long differed as to the main cause of the destruction of the birch forests and the soil erosion. Some students of the problem have largely blamed volcanic activity and deterioration of the climate. Others have maintained that man and his grazing livestock are the main cause, mindful of the fact that, before the arrival of man, there was no grazing mammal in Iceland.

With the aid of dated tephra layers it has been possible to measure rather exactly the rate of thickening of the soil cover in many places. This rate is an indicator of the soil erosion going on elsewhere. The wind-borne material is partly deposited in the areas still vegetation-covered and bound there by the vegetation. Figure 9 shows to the left a soil profile resting on an 8,000 year old lava flow, 15 km SW of Hekla.

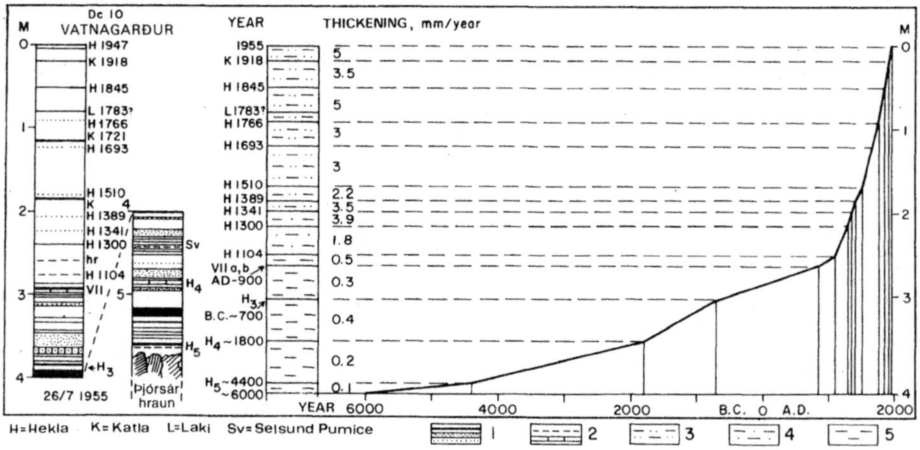

Fig. 9. Soil profile and diagram showing tephra layers, rate of soil thickening and variations in the coarseness of the loessial soil at Vatnagardur, 15 km W of Hekla. The profile to the right is the same as the left one except that the tephra layers have been extracted. 1: black tephra, 2: light tephra, 3: sandy soil, 4: fine-sandy soil, 5: silty soil (17).

Having extracted the thickness of the tephra layers in the soil as they have little to do with the thickening of the loess soil that is caused by wind erosion elsewhere, we get the profile to the right, where the tephra layers are only time markers with no thickness of their own. From that profile we can then construct the diagram to the right, showing the soil thickening versus time. The diagram shows that up to a certain point the rate of thickening is relatively constant, although increasing somewhat, with the climatic deterioration that began about 2,500 years ago. According to this diagram, the great increase comes between the deposition of the Landnam layer about 900 A.D. and H 1104, and that increase can neither be caused by increased volcanism nor by deteriorating climate. Its main cause is the arrival of man and his livestock (22).

Since this diagram was drawn 20 years ago, similar studies have been carried out in other areas and they show on the whole a similar trend, although, naturally enough, with local variations. Figure 10 (23) shows a diagram from Skagafjördur in North Iceland constructed in a similar way, although with a logarithmic vertical scale, and based on the light Hekla tephras. The dot-dash line shows the average rate of thickening in postglacial time.

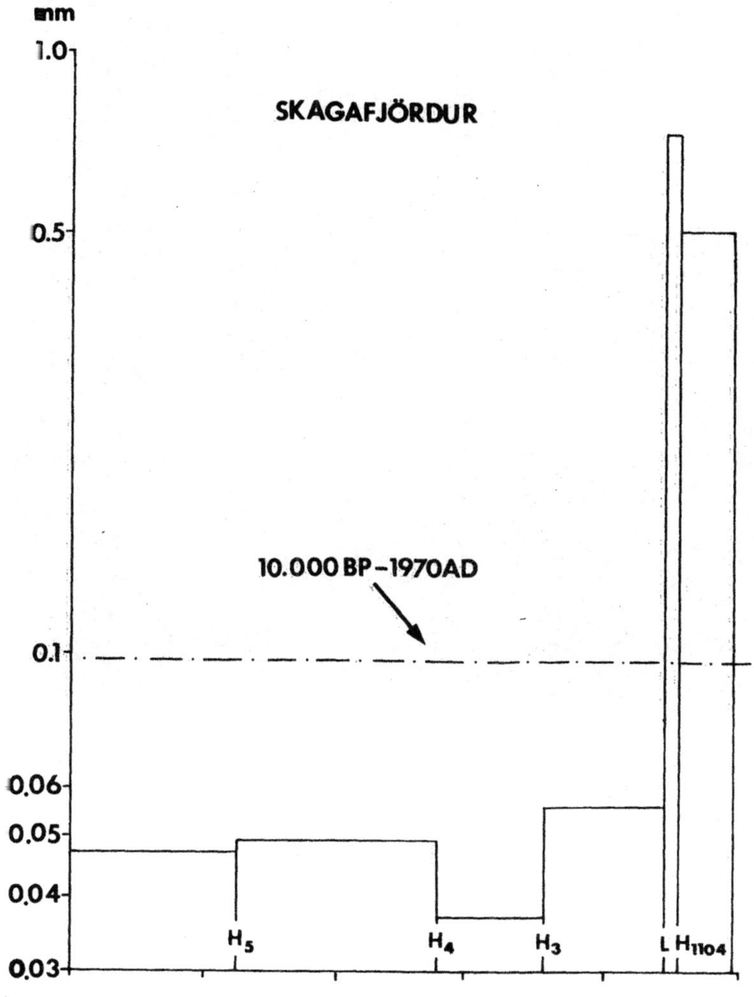

Fig. 10. The rate of soil thickening in Skagafjördur, N. Iceland (23).

The studies of the rate of soil thickening have been supplemented by studies of variations in grain-size of the soil with time, carried out in a similar way (23,24). One indicator of increased soil erosion is an increasing mineral content of the peat soils. As shown on Figure 11, taken on the outskirts of Reykjavík, the increase changes the colour of the peat and this change begins just above the Landnam layer.

Fig. 11. Profile in a bog near Reykjavík showing the change in colour of the peat soil soon after the deposition of the Landnam layer (LL), because of increased eolian sedimentation (22). H and K are Hekla and Katla tephras, respectively.

8. LARGE-SIZE POLYGONS IN ICELAND

Among the phenomena that can favourably be studied in Iceland are various frost phenomena, such as many types of patterned ground; thufur, palsas, polygons etc. and their dependence on climatic parameters. As the climate has changed a lot, not only in post-glacial time but also during Iceland's history, it is important for the study of these frost phenomena to bring the morphogenetic events into their absolute chronological context, and here the tephrochronology is a useful tool. I will exemplify this with a short mention of a study of large-size polygons in Iceland carried out in 1970 by a team of Swedish, Icelandic and American scientists (25).

Large-scale polygons were first noticed in Iceland by Th. Thoroddsen in 1889 at Helliskvísl, NE of Hekla, and at Veidivötn (26). In the highland between Hofsjökull and Tungnafellsjökull they were observed from an airplane in 1954 (27). Their approximate distribution as known at present is shown in Figure 12. The average common diameter of these polygons in between 20 and 40 m. They are found both in loessial soil cover and in subsoil where the ground is bare.

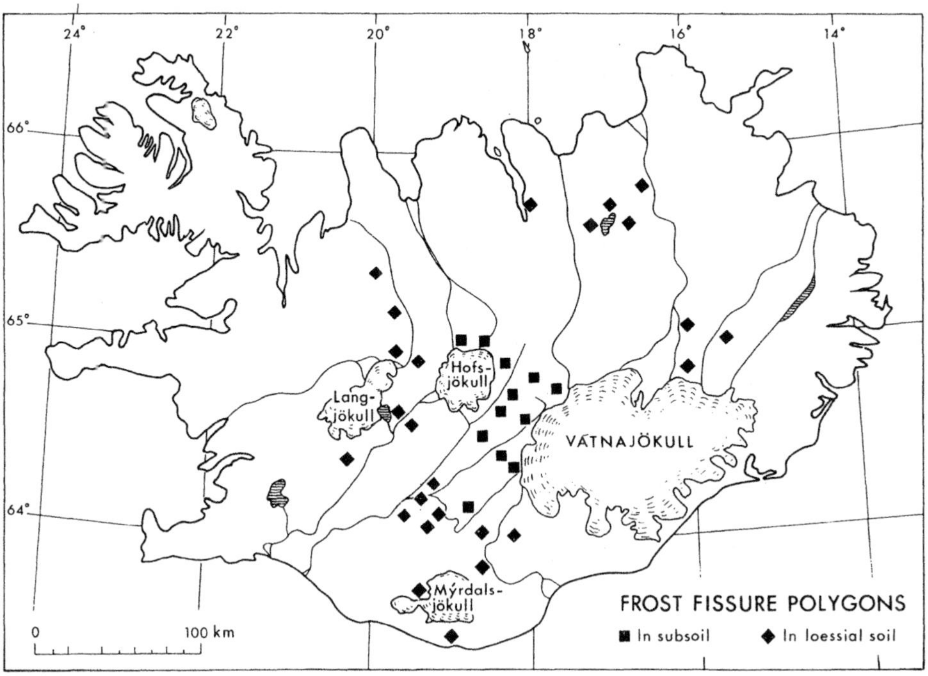

Fig. 12. The occurrence of large-size polygons in Iceland (25).

The subsoil polygons are limited to areas at, or above the 600 m level, while those in loessial soil are found down to 250 m, or even somewhat lower.

The large-size polygons in Iceland are frost crack polygons and the fissures are not the real ice wedge type which need an actual permafrost regime for their formation. The first stage is the formation of a frost crack. Figure 13 is a section across a polygon furrow on a lava plan 16 km N of Hekla and 300 m above sea level. In this section there are 8 well-defined and well-dated tephra layers, the oldest of which is from the Hekla eruption of 1104. All layers down to Hekla 1300 have been disturbed and that layer is slightly affected also. The layer of

1636 has the largest filling and reflects break-up and crack formation of a greater magnitude than that of the overlying layers. The crack and filling has occurred in a surface position, which means that about 1636 there were conditions for formation of frost fissures on the verge of being of the ice wedge type. There may have been an inactive period before the crack higher up was formed. The top layer is split by a crack from the winter 1969/70.

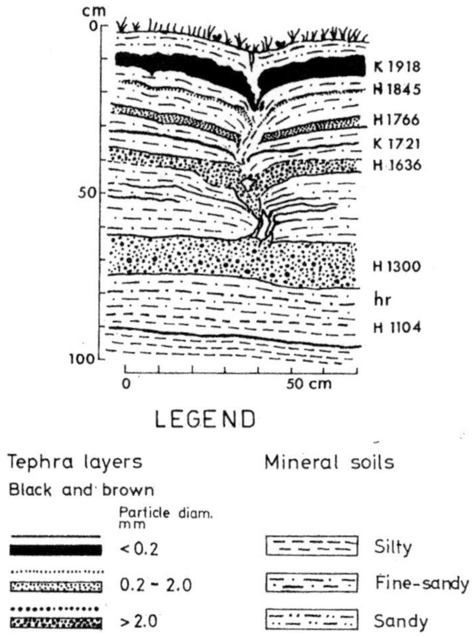

Fig. 13. Section below a polygon furrow in Ferjufit, S. Iceland (25).

Summing up the results of the studies of large-scale polygons in Iceland, we find that at altitudes between 250 and 500 m their time of origin is correlative with the so-called "Little Ice Age", namely the period from about 1550 to about 1920, and during this period real ice wedges may have formed occasionally. At levels below 500 m altitude, frost-crack polygons rarely, or even never, formed during the warm period between the earlier 1920's and early 1960's, but since then, the frequency of frost cracks with a tendency to polygon formation has noticeably increased.

9. TEPHRA LAYERS AND GLACIOLOGY

A tephra layer as a time marker in glaciers was first used by the Swedish-Icelandic Vatnajökull expeditions 1936-37-38. The primary object of these expeditions was to gain knowledge of the regime of Vatnajökull, i.e. the gain through solid precipitation and the loss through melting and evaporation. In March-April 1934 an eruption in Grímsvötn spread a thin, black tephra layer over most of Vatnajökull. This layer was an accurately dated key horizon by which the accumulation since its deposition could be determined.

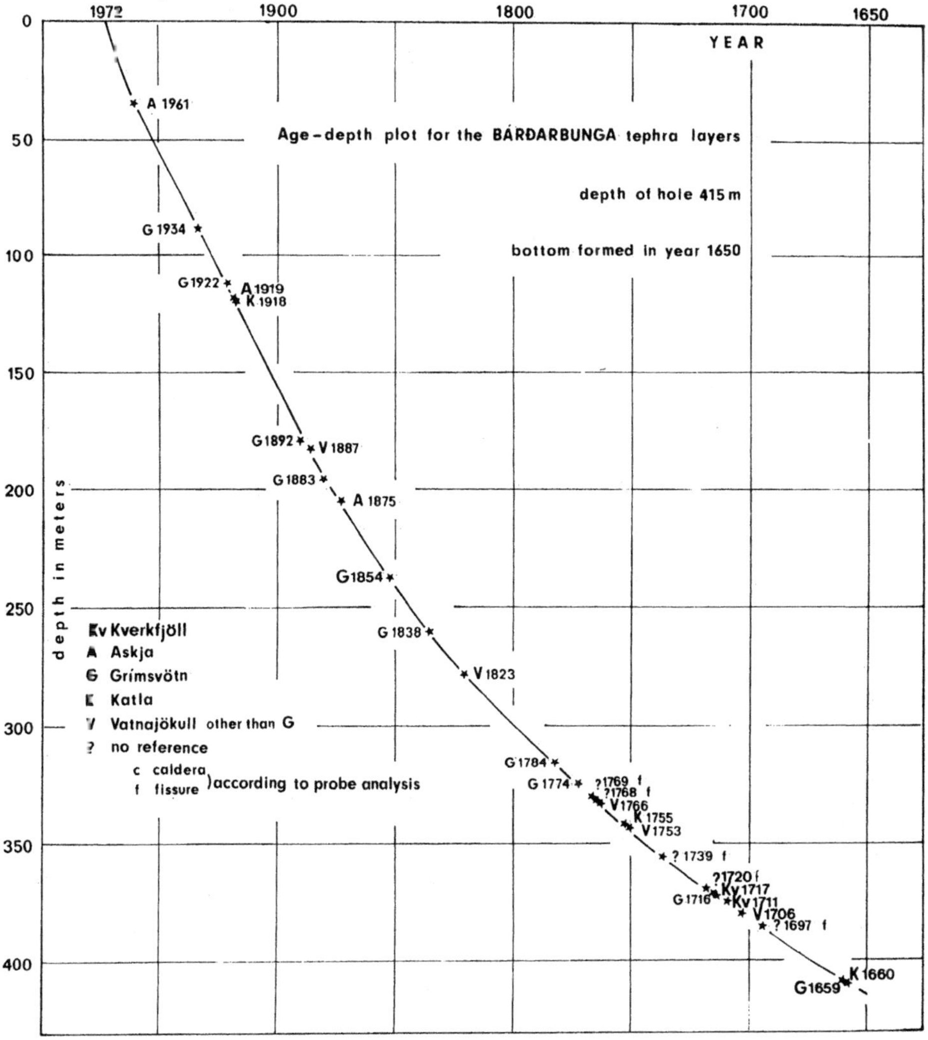

Fig. 14. Age-depth plot for the Bárdarbunga tephra layers, after Steinthórsson (1977)(28).

In 1972 a 415 m long ice core was recovered through drilling on Bárdarbunga in northwestern Vatnajökull, about 1800 m above sea level. Tephrochronological work on this core was carried out by S. Steinthorsson (28). It was established that the core represented the years from 1972 to 1650 (Figure 14). The core was found to contain 30 tephra layers, 22 of which could be attributed to known eruptions. There is no doubt that drilling to the bottom of the glacier, which is there about 800 m thick, would greatly add to our knowledge of the volcanic activity in Iceland during the last one or two thousand years.

10. DATING OF GLACIER OSCILLATIONS

Dating of glacier oscillations will here be exemplified by tephrochronological studies of the oscillations of Hagafellsjökull eystri, a southern outlet of Lanjökull, Iceland's second largest

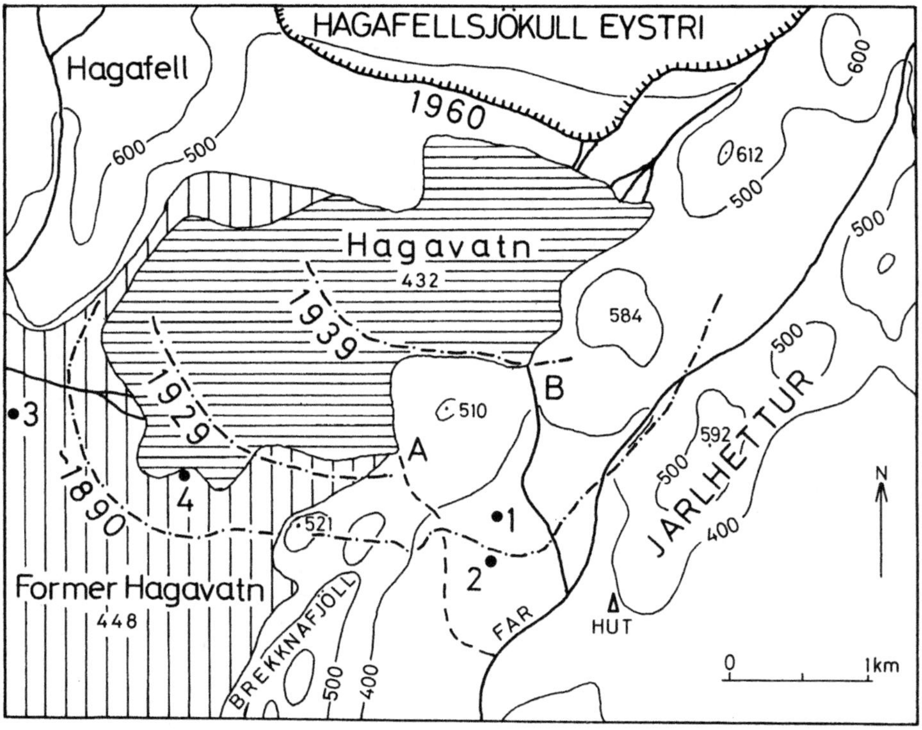

Fig. 15. Lake Hagavatn and the glacier Hagafellsjökull eystri in 1960. Thin broken line shows the course of the river Far between 1929 and 1939. Thick broken line marks previous limits of the glacier (29), see text.

ice sheet (953 km^3 according to satellite images). In front of the glacier is Lake Hagavatn, caused by the closing of a NE-SW valley by a shield volcano, Lambahraun, which was built up after the deposition of H_4 but before H_3 was deposited. Its approximate age is thus about 3500 years.

On Figure 15 the outermost broken line shows the frontal position of the glacier about 1890. The glacier was also near that position about 1840 and 1920. At that time the level of the lake, dammed up in front of the glacier, was ca. 448 m above sea level. In early 1939 it had receded so much that in the night of August 9 it broke through pass A on Figure 15. The lake level was lowered 6 m to 442 m, with a catastrophic flood as a result. A great flood occurred again a decade later, on August 13-15, when the glacier had receded so that the lake found its present outlet through the lowest pass in the moberg ridge. Its level was again lowered, this time nearly 10 m, to its present 432 m level.

We have been able to follow in some detail the recession of the glacier (30). The question is, when did it advance to the positions we have discussed? Section 2 on Figure 16 is through a loessic soil patch, immediately outside the glacier front of about 1840 and 1920. The tephra layers in this soil are all undisturbed and prove definitely that during the last 5 or 6 millenia the glacier never reached a more advanced position than it had about 1840, 1890 and 1920. In reality, this means that this position was the most advanced during most of the post-glacial period.

During the summer of 1950 a student expedition from Durham University in England studied the varved lake sediments in those parts of the Hagavatn basin that had been exposed by the lowering of the lake level in 1929 and 1939. One of the participants, R. Green, searched especially in the sediments for tephra layers that had been identified in the above mentioned profile 2 (31). Just west of the present lake he found the typical tephra layer succession, H 1766, Katla 1721 and H 1693. The 1693 Hekla tephra was underlain by only 10 to 12 annual glacial varves. This was proof that the glacier had not advanced to its 1939 position and blocked pass B until, at most, two decades or so before 1693. In varved lake sediments thrust up from the bottom of the present lake, Green found the Landnam layer, deposited about 900 A.D., which proved that the glacier cannot, at that time, have had a position as advanced as in 1939.

In 1965 loessial soil was found under moraine cover just inside the outermost terminal moraines (point 1 on Figure 15). The glacier had advanced over this spot without eroding more than the uppermost part of the loessial soil cover. The youngest,

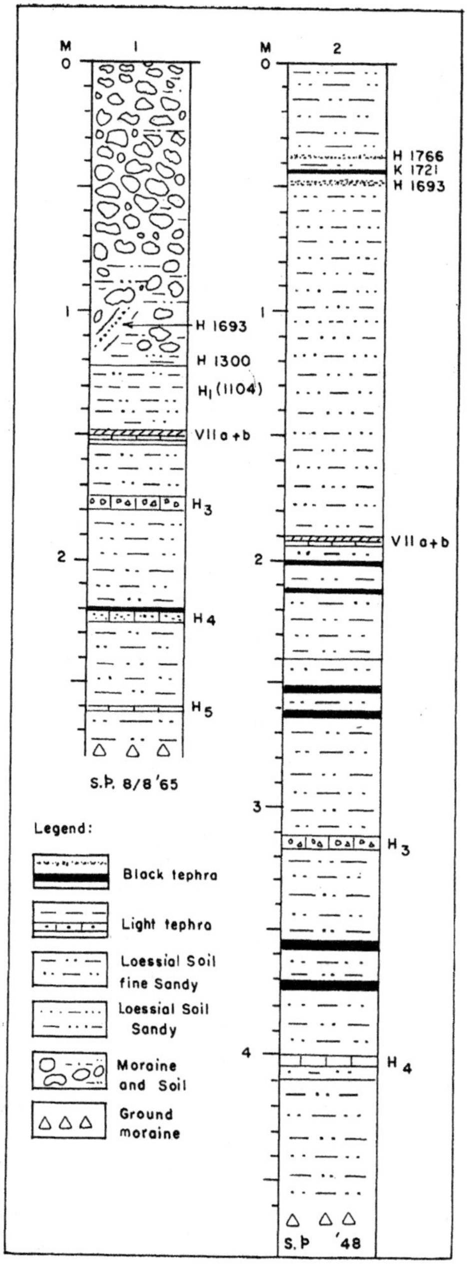

Fig. 16. Two soil profiles south of Hagavatn. Their location is shown on Fig. 15 (29).

undisturbed tephra layer in the soil profile was Hekla 1300 (profile 1, Figure 16), but in the overlying moraine was embedded a 4-5 cm thick lens of the loessial soil and in its middle was the coarse grained Hekla tephra H 1693.. The rate of thickening of the soil between H 1693 and 1766 in the nearby profile 2 did not exceed 0.7 mm/year. The thickness of the soil covering H 1693, when it was overrun, was at least 2-3 cm and probably a lot more, as it is likely that some soil was removed by the advancing glacier. We have here conclusive proof that the Hagafellsjökull eystri did not reach its maximum post-glacial extension before 1730. Most likely it reached it much later. This is in good agreement with the tephrochronological datings of terminal moraines of other glaciers in Iceland, which have shown that the flat, lobe-shaped outlets of Vatnajökull reached their maximum post-glacial extension during the latter half of the cold period between 1150 and 1890, whereas the valley glaciers of Öræfajökull and other steep Alpine type glaciers reached that maximum before the settlement of Iceland, probably during the first centuries of the rapid climatic deterioration that began with the Nordic Iron Age ca. 500 B.C. (32).

11. LONG DISTANCE TEPHROCHRONOLOGICAL CORRELATIONS

Icelandic eruptions have occasionally spread tephra to other islands in the Northern Atlantic or to the European continent. The eruptions known, through written sources, to have done so are Katla 1625, 1660 and 1755; Hekla 1693, 1845; Lakagígar 1783, and possibly an eruption in Vatnajökull 1619. Askja tephra was sampled in many places in Scandinavia in 1875 and Hekla tephra in Finland in 1947. The Swedish scientist Chr. Persson established a tephrochronological connection between Iceland and Scandinavia by counting and identifying volcanic glass grains in peat bogs. Persson measured the concentration of tephra grains in the soil, determined refractive index and trace elements, dated the organic material associated with the horizons of tephra concentrations by C-14 and succeeded in identifying 5 acidic eruptions: Askja 1875, Öræfajökull 1362, Hekla 1104 and the prehistoric H_3 and H_4 (33,34). Icelandic tephra has been detected in a similar way on the Faeroe Islands (35,36). It is almost certain that tephra from Iceland can be found and identified in Scotland, and on the Orkney and Shetland Islands. Such connections might prove useful for the correlation of climate and vegetation changes and might also be of some help in archaeological research.

Another possibility for spreading Icelandic tephra to other countries is by sea. Drifted pumice is found at various levels on raised beaches in Spitsbergen, Norway and Denmark (37,38). In the British Isles, pumice has been found mainly on archaeological sites. Occasional finds of pumice are reported

There may, of course, be many sources for the drifted tephra on the coasts of NW Europe and Spitsbergen, e.g. Jan Mayen, the Azores, the West Indies--but Iceland is certainly one of them. We must also not forget submarine eruptions in shallow waters near Iceland. The exact sources of most of the drifted pumice have not yet been determined, but the electron microprobe has greatly improved the possibilities of doing so. Drifted pumice offers, no doubt, possibilities for tephrochronological correlations between Iceland and countries around the northernmost Atlantic and adjacent parts of the Arctic.

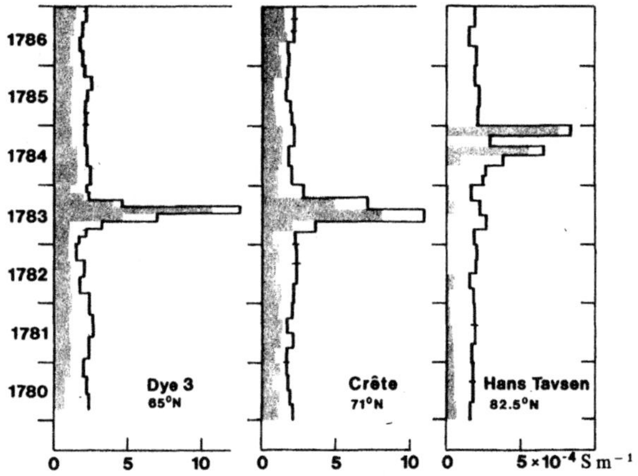

Fig. 17. The Lakagígar eruption revealed in three ice-cores from Greenland by peak values of specific conductivity in melted samples ($5m^{-1}$). The Hans Tavsen ice-core is not dated absolutely. The shaded areas show the contribution of hydrogen ions not induced by CO_2 (39).

An exciting third possibility for correlation teleconnection is through the aerosol fall-out from volcanic eruptions that is preserved in ice cores. Important research work of W. Dansgaard and his collaborators on cores form the Greenland ice that has revealed these fascinating possibilities. Measuring the $^{18}O/^{16}O$ ratios in the ice has enabled datings of the Greenland ice cores that go back more than 100,000 years in time. For at least the last millenium the dating is exact almost to the year as annual layers can be counted. C.U. Hammer has shown convincingly that ice layers of high acidity, due to the fallout of sulphuric acid aerosols from large eruptions in the past, can be detected by measuring the electric conductivity of the ice (39). One of the first eruptions so detected with absolute certainty was the Lakagígar eruption of 1783 (Figure 17). Among other Icelandic eruptions detected in the ice cores is Hekla 1104 and, at my request, Hammer looked for an eruption of similar magnitude as Lakagígar between 850 and 950 A.D. He found that such an eruption had taken place in 934 ± 2 A.D. There is hardly any doubt that this was the eruption that formed the great fissure chasm of Eldgjá.

Tephrochronological connections have, through the work on the Greenland ice cores, become global, and enable absolute and exact dating reaching back in time far beyond written records for many volcanic countries.

REFERENCES

1. Thorarinsson, S. and Saemundsson, K.: 1979, Jökull 29, pp. 27-30.
2. Thorarinsson, S.: 1967, I. Soc. Sci. Islandica, 170 p. + 13 pl.
3. Thorarinsson, S.: 1958, Acta Nat. Isl. II, 2, pp. 1-100.
4. Thorarinsson, S.: 1974, Bókaútgáfa Menningarsjóds, Reykavik, 254 p.
5. Einarsson, E.H., Larsen, G. and Thorarinsson, S.: 1980, Acta Nat. Islandica 28, Mus. Nat. Hist. Reykjavik.
6. Larsen, G.: 1978, Gjoskulog i nagrenni Kotlu/Tephra layers in the neighbourhood of Katla/. B.Sc. hon. thesis, Faculty of Science and Engineering, Univ. of Iceland (Mimeographed).
7. Thorarinsson, S.: 1975, Árbók Ferdafélags Íslands 1975, pp. 125-149.
8. Larsen, G. and Thorarinsson, S.: 1978, Jökull 27, pp. 28-46.
9. Thorarinsson, S.: 1951, Geograf. Annaler 33, pp. 1-80.
10. Thorarinsson, S.: 1952, Nátturufrædingurinn 22, pp. 113-129 and 145-172.
11. Einarsson, Tr.: 1948, Nátturufrædingurinn 18, pp. 113-121.
12. Vilmundardóttir, E.: 1977, Tungnarhraun, National Energy Authority, OS-ROD7702, Reykjavik, 156 p.
13. Eldjarn, K.: 1961, The Fourth Viking Congress, Aberdeen Univ. Studies 149, pp. 10-19.
14. Stenberger, M. Ed.: 1943, Forntida gårdar i Island, Ejnar Munksgaard, Copenhagen, 334 p.
15. Einarsson, Th.: 1956, Sonderveroeffentlichungen des Geol. Inst. der Univ. Koln. V. 6, 52 p.
16. Einarsson, Th.: 1963, in: North Atlantic Biota and Their History, Pergamon Press, London, pp. 355-365.
17. Thorarinsson, S.: 1970, in: Scientific Methods in Medieval Archaeology, Berger, R. (ed.), Univ. of California Press, pp. 295-328.
18. Thorarinsson, S.: 1976, Árbók Hins ísl. fornleifafél., pp. 5-38.
19. Thorarinsson, S.: 1960, Peterm. Geogr. Mitt. 104, pp. 154-162.

20. Eliasson, S.: 1974, Nátturufrædingurinn 44, pp. 52-70.
21. Tomasson, H.: 1973, Nátturufrædingurinn 43, pp. 12-34.
22. Thorarinsson, S.: 1962, Revue Géomorphol. Dynamique 13, pp. 107-134.
23. Güdbergsson, G.: 1975, Jour. Agr. Res. Icel. 7, 1-2, pp. 20-45.
24. Sigbjarnarson, G.: 1969, Nátturufrædingurinn 39, pp. 68-118.
25. Friedman, J.D., Johanson, C.E., Óskarsson, N., Svensson, H., Thorarinsson, S. and Williams, R.S.: 1971, Geogr. Ann. Stockh. 53, Ser. A, pp. 115-145.
26. Thoroddsen, Th.: Andvari 16, pp. 46-115.
27. Thorarinsson, S.: 1954, Jökull 4, pp. 38-39.
28. Steinthórsson, S.: 1977, Jökull 27, pp. 2-27.
29. Thorarinsson, S.: 1966, Jökull 16, pp. 207-210.
30. Sigbjarnarson, G.: 1967, Jökull 17, pp. 263-278.
31. Green, R.: 1952, Jökull 2, pp. 10-16.
32. Thorarinsson, S.: 1964, Jökull 14, pp. 67-75.
33. Persson, Chr.: 1966, Geol. Foren. Stockh. Forh. 88, pp. 361-394.
34. Persson, Chr.: 1967, Geol. Foren. Stockh. Forh. 89, pp. 181-197.
35. Persson, Chr.: 1968, Geol. Foren. Stockh. Forh. 90, pp. 241-266.
36. Waagstein, R. and Johansen, J.: 1968, Medd. Dansk Geol. Forening 18, pp. 257-264.
37. Binns, R.E.: 1967, Tromso Mus. A. Scientia 24, pp. 1-63.
38. Blake, W., Jr.: 1961, in: Geology of the Arctic, Univ. of Tromso Press, pp. 132-145.
39. Hammer, C.U.: 1977, Nature 270 (5637), pp. 482-486.

USE OF TEPHROCHRONOLOGY IN THE QUATERNARY GEOLOGY OF THE UNITED STATES

Stephen C. Porter

Department of Geological Sciences and Quaternary Research Center, University of Washington, Seattle, Washington 98195

ABSTRACT. The most extensive volcanic ash units of western United States that are important for tephrochronologic studies include the Pearlette-type ash layers (0.6, 1.2, and 1.9 x 10^6 years old), the Bishop ash (0.7 x 10^6 years old), Glacier Peak tephra (ca. 12,500-11,250 years old), Mazama ash (6700 years old), several layers from Mt. St. Helens (3400, 450, and 180 years old), two lobes of White River ash in eastern Alaska and Yukon Territory (1890 and 1250 years old), and the Katmai tephra of southern Alaska (A.D. 1912). These and other tephra layers of western United States have been used to assess the age of Pleistocene and Holocene glacial, pluvial, and alluvial records, to estimate the relative magnitude and character of eruptive events, and to infer prevailing wind direction(s) at times of major eruptions. Pollen influx and pollen types have been employed to determine the duration and season(s) of deposition of certain ancient ash layers. Potential applications include use of tephra layers to evaluate rates of geologic processes, to correlate marine and terrestrial stratigraphic sequences, to determine rates of reforestation at the end of the last glaciation, and to assess the extent to which volcanic ash in the atmosphere can contribute to climatic change.

INTRODUCTION

Reconstruction of Quaternary environmental history commonly involves stratigraphic studies of geologic deposits, analysis of landforms, and finding a means of placing morphologic and stratigraphic units in a temporal framework. In areas near of downwind from tephra-producing volcanoes where pyroclastic layers

S. Self and R. S. J. Sparks (eds.), Tephra Studies, 135–160.

form a part of the stratigraphic record, ash layers constitute an extremely useful basis for obtaining limiting ages for deposits with which they are associated and can provide important information regarding the frequency and magnitude of eruptive events and climatic conditions during eruptions. Although application of tephrochronologic methods in Quaternary geologic studies is a rather recent development in the United States, significant advances have been made during the last decade in determining the character, extent, and age of major tephra units that have the greatest potential for regional correlation and dating. Tephrochronology has been employed in a variety of investigations involved with dating of glacier fluctuations, establishment of alluvial and lacustrine chronologies, determination of paleowind directions, and paleoecologic reconstructions. In the following summary, characteristics of some major Quaternary tephra units are discussed, and examples are given of their application to the solution of geologic problems in the United States.

DISTRIBUTION OF TEPHRA-PRODUCING VOLCANOES

Tephra-producing volcanoes of the United States are located within and along the eastern margin of the north Pacific Ocean basin in the Hawaiian Islands, southern Alaska, and western conterminous United States (Fig. 1). Although tephra deposits of Quaternary age are found over vast areas in the latter two regions, they often are thin and discontinuous. They form a more-or-less continuous mantle only along and near major eruptive centers such as the Aleutian arc and the Cascade Range.

Hawaiian Islands

Tephra deposits are associated with each of the major Hawaiian volcanoes, but layers of Quaternary age are found primarily on the main island of Hawaii (Fig. 2) (1). Tephra is associated with all five of the massive volcanoes of the island, and it is especially widespread on Mauna Kea and along the crests of Kohala and Hualalai, each of which has passed from the tholeiitic phase into the alkalic phase of its eruptive history. Although Kohala apparently last erupted about 60,000 years ago, Mauna Kea remained an active local tephra producer until about 3600 years ago (2). The lower windward (northeast) slopes of Mauna Kea and parts of the lower southeast flank of Mauna Loa are mantled with weathered late Quaternary tephra referred to collectively as the Pahala ash, the source(s) and age of which are largely unknown.

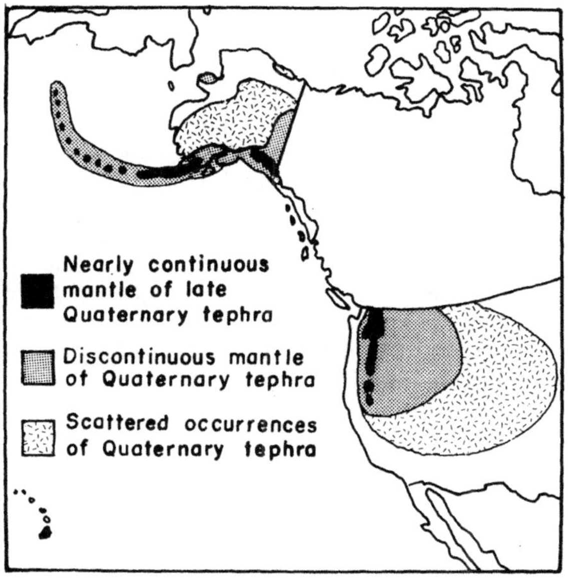

Figure 1. Map showing distribution of tephra deposits in Alaska, Hawaii, and western conterminous United States.

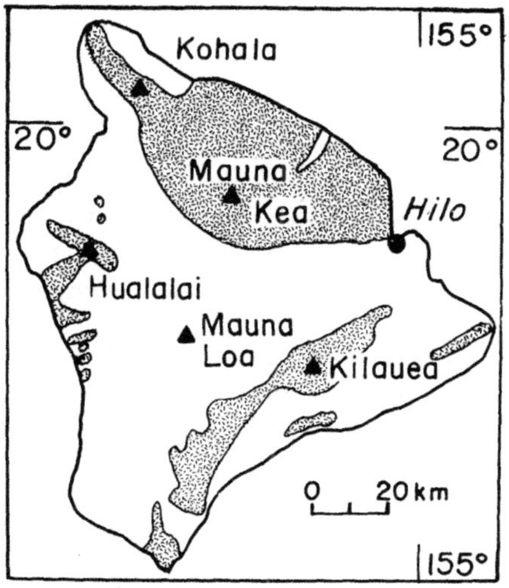

Figure 2. Distribution of major surface and near-surface tephra deposits associated with the five major shield volcanoes of the island of Hawaii.

Southern Alaska

Most of Alaska's Quaternary volcanoes are found along the Aleutian Island arc and its landward extension west of Cook Inlet (Fig. 1). Other important local eruptive centers lie to the east and southeast in the Wrangell Mountains and at Mt. Edgecumbe. Tephra is widespread along the Aleutian Chain (3) and in south-central Alaska (4). Discontinuous deposits have been found in central Alaska and locally on the Seward Peninsula. Although a few scattered exposures are known from the southern Brooks Range, little if any tephra apparently is to be found on the Arctic Slope. Much of the tephra derived from volcanoes of the Aleutian arc lies within sediments of the adjacent North Pacific Ocean and southern Bering Sea (5).

Western conterminous United States

Major late Quaternary tephra layers of western United States have been produced mainly by volcanoes of the Cascade Range which extends from northern California into southern Canada (Fig. 1) (6). However, many minor eruptive centers both in the Cascades and scattered throughout the Cordilleran region have generated tephra deposits of local extent. Major explosive eruptions of middle to early Pleistocene age in the Yellowstone Park region and in east-central California spread tephra over extensive areas of western United States as far as the Great Plains states, but the tephra layers are found only as scattered occurrences, and the full extent of their distribution is poorly known.

MAJOR TEPHRA EVENTS

Tephra layers that have proven especially applicable to Quaternary geologic studies include several early and middle Pleistocene units that constitute regional stratigraphic marker beds useful in regional correlation and dating, and a series of late Pleistocene and Holocene units that help define important events during the final phases of the last (Wisconsin) glaciation and during postglacial time in northwestern North America.

Pearlette-type tephra layers

Fine-grained volcanic ash of middle- to early-Pleistocene age found at scattered places throughout the Rocky Mountain region and Great Plains of western United States has generally been referred to as the "Pearlette Ash." Recent studies indicate that instead of one layer, a family of ash layers is present and that the individual units differ in petrographic and chemical characteristics and in age (7, 8). The oldest layer, type B,

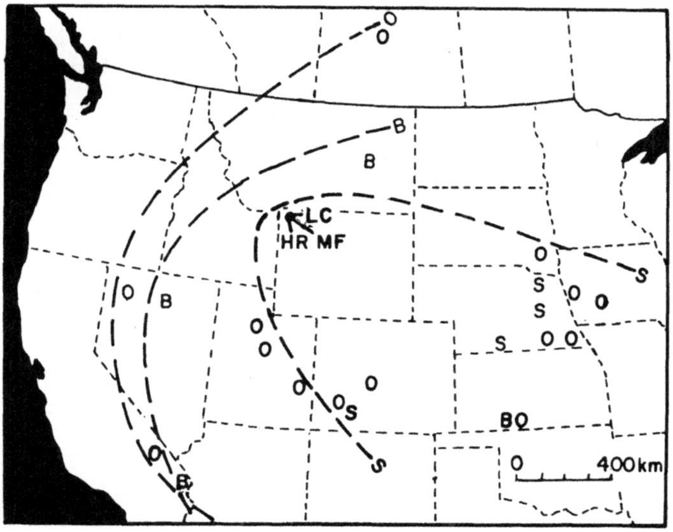

Figure 3. Map of western United States showing localities where Pearlette-type ash layers (B = 1.9 million yr; S = 1.2 million years; O = 0.6 million years) have been reported, the inferred distribution of layers S and O, and the petrographically equivalent tuffs in the source area at Yellowstone Park (HR = Huckleberry Ridge Tuff; MF = Mesa Falls Tuff; LC = Lava Creek Tuff).

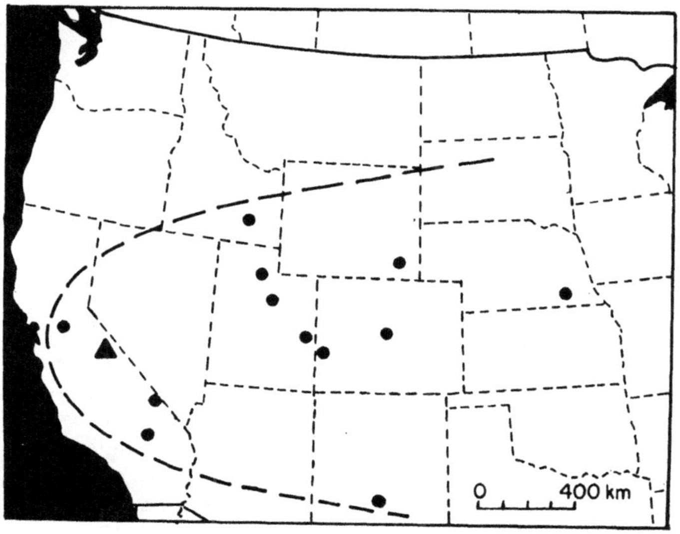

Figure 4. Map of western United States showing localities where Bishop tephra has been reported, and the inferred minimum distribution of the ash layer with respect to its source area in eastern California.

has a fission-track age of about 1.9 ± 0.1 million years and is known mainly from Kansas (Fig. 3). Its source is believed to be the petrographically similar Huckleberry Ridge Tuff of the Yellowstone Park area which has a K/Ar age of 1.96 ± 0.05 million years (8). Type S ash is distributed southeast of Yellowstone Park with reported outcrops in Colorado and Nebraska (Fig. 3). It is believed to be the downwind correlative of the Mesa Falls Tuff of the Yellowstone region which has a K/Ar age of 1.2 ± 0.04 million years. The youngest and most widespread ash, designated type O, is known from occurrences as far apart as southeastern California, southern Saskatchewan, and Iowa (Fig. 3). K/Ar dates of zircons from the ash and from the Lava Creek Tuff, its presumed correlative at Yellowstone, are close to 0.6 million years, while a fission-track age of 0.6 ± 0.1 million years was obtained from the ash in Kansas (8).

Bishop tephra

An ash layer found at scattered localities from California to Nebraska has been correlated on the basis of petrography, chemistry, and radiometric age with the Bishop Tuff, a large middle-Pleistocene ash flow of eastern California (7). Known ash distribution suggests that the tephra apparently forms an east-trending plume extending from southeastern California across the southwestern United States to the western Great Plains (Fig. 4). Fission-track dates and normal remnant magnetism indicate that the ash is approximately 700,000 years old (9, 10).

Glacier Peak tephra layers

Glacier Peak volcano in the North Cascade Range of Washington was the source of weveral major tephra eruptions near the end of the Pleistocene. At least nine layers were produced by the eruptions, but only two, layers G and B, are of broad regional extent (Fig. 5) (11). Layer G, the older layer, extends east across Washington into Idaho and Montana, and has also been identified in Alberta and Saskatchewan. It is not closely dated, but available radiocarbon dates suggest that it may be between about 12,750 and 12,000 years old. Layer B extends southeast from the volcano across northern Idaho, southern Montana, and into northwestern Wyoming. At Lost Trail Pass bog in Montana it consists of two distinct units separated by organic sediments representing a 10-25 year eruptive hiatus (12). There the ash couplet is bracketed by radiocarbon dates and has a mean age of 11,250 years. Both layers G and B consist of pumice and lithic fragments within about 100 km fo the source, but they pass downwind into fine, silty ash.

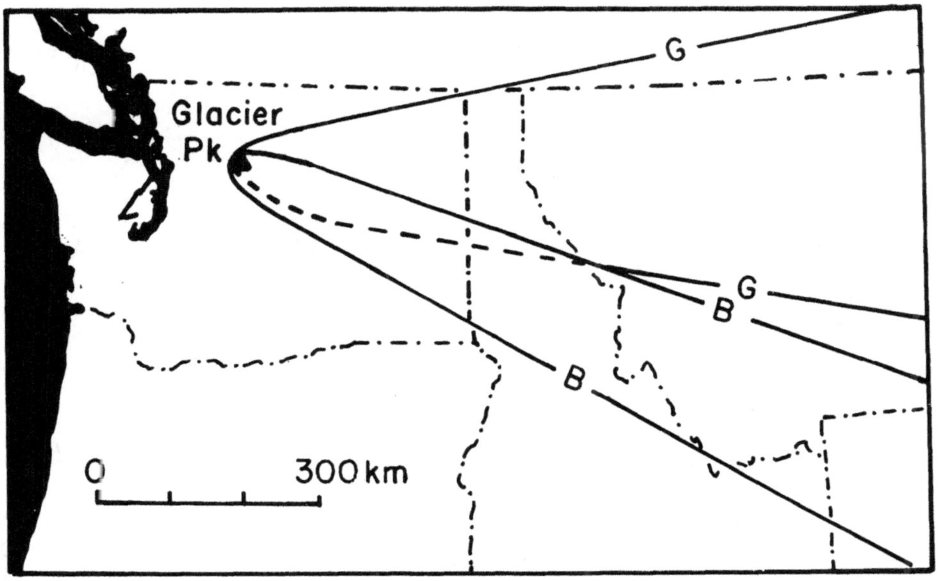

Figure 5. Map of northwestern United States showing inferred distribution of tephra layers G and B from Glacier Peak volcano.

Mazama tephra

The most widespread Holocene tephra layer in the United States originated in southern Oregon at Mt. Mazama, a major stratovolcano estimated to have been about 3700 m high prior to a series of catastrophic eruptions that spread large amounts of tephra over most of northwestern United States (Fig. 6) (13). Although generally regarded as a single layer, recent studies indicate that there are several layers comprising a Mazama tephra set (14). A single layer at Lost Trail Pass bog has been bracketed by radiocarbon dates of 6720 ± 120 and 6700 ± 100 yr, while at Wildcat Lake in southeastern Washington two Mazama layers are bracketed by dates of 6940 ± 120 and 5389 ± 130 yr, and an inter-ash organic unit has an age of 6750 ± 90 yr (12). The Mazama tephra is ubiquitous in the Oregon and Washington Cascades, and it forms a thick mantle (>10 cm) downwind from the vent at distances up to 200 km. Mazama tephra also has been found in marine sediments in the northeast Pacific Ocean off the coasts of Washington and Oregon (15). Between 30 and 38 km^3 of ash-fall deposits and an additional 25-33 km^3 of ash-flow deposits are estimated to have been produced during the Mazama eruptions (16).

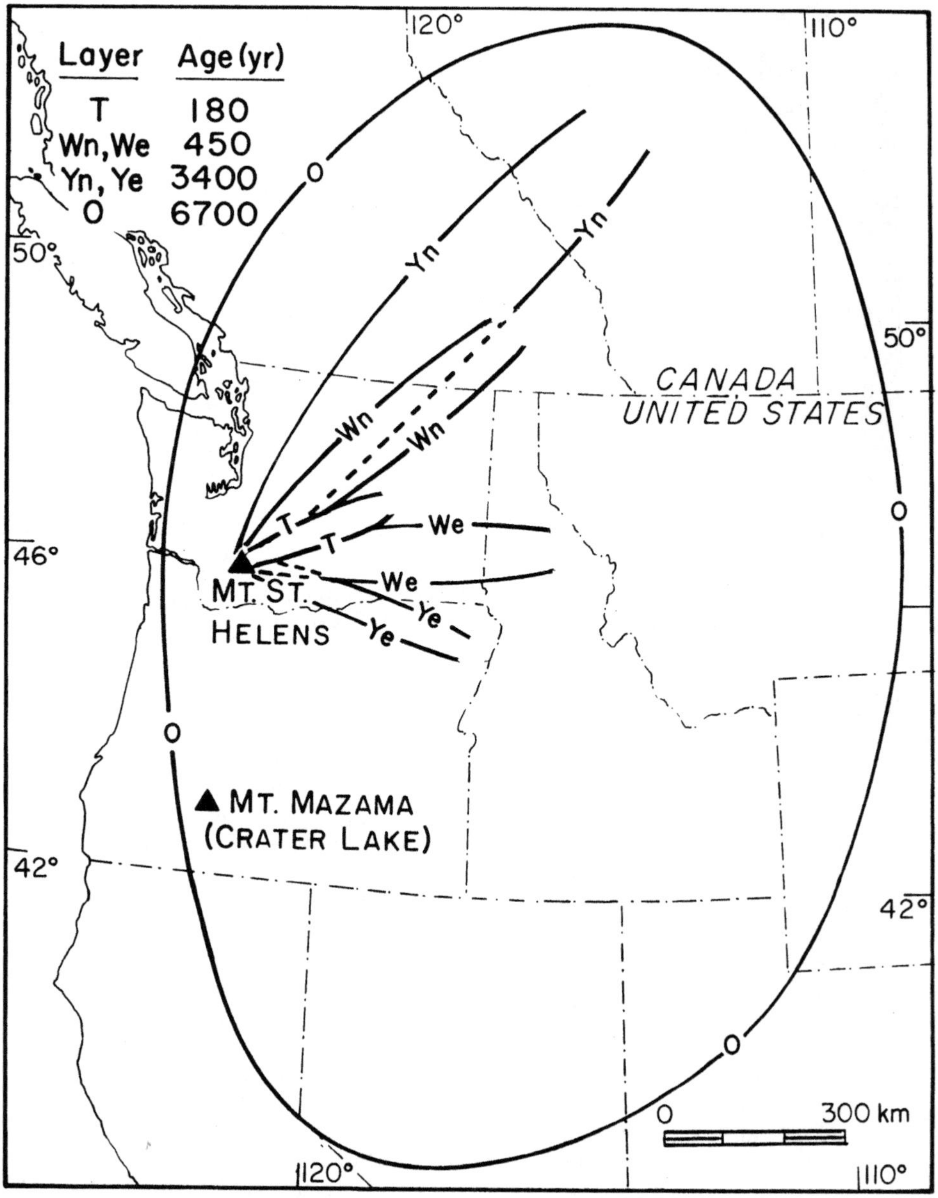

Figure 6. Map of western United States showing distribution of widespread Holocene tephra layers from Mt. Mazama and Mt. St. Helens.

Mt. St. Helens tephra layers

Mt. St. Helens in southwestern Washington has produced several tephra layers of regional extent. Among the most important are layers Yn and Ye, the plumes of which trend northeast and east, respectively, and were erupted about 3400 ^{14}C years ago; layers We and Wn, both of which were erupted close to 450 ^{14}C years ago are distributed east and northeast of the volcano (Fig. 6) (17). Layer T, erupted within a year or two of A.D. 1800, forms a northeast-trending plume that has been traced nearly 200 km from the volcano and may extend as far as northern Idaho and Montana (18).

Renewed activity began suddenly in late March, 1980 and continued intermittently until May 18 when a major explosive eruption blasted away the summit of the mountain, created a huge north-facing crater, and led to dispersal of fine ash across eastern Washington, Idaho and Montana. A second large eruption a week later spread ash from northwest Oregon to the northwestern Olympic Peninsula.

White River tephra

Two prominent lobes of tephra extending north and east from the St. Elias Mountains near the Alaska-Yukon boundary have been referred to as the White River Ash (19). The formation covers at least 300,000 km^2, mainly in Canada (Fig. 7). Its source is believed to lie on the northeast flank of Mt. Bona in an area now covered by glaciers. The older northern lobe dates to about 1890 ^{14}C years ago, whereas the larger eastern lobe was erupted about 1250 ^{14}C years ago.

Katmai tephra

The last major tephra eruption in Alaska occurred in June, 1912 when major volcanic explosions in the vicinity of Mt. Katmai on the Alaska Peninsula caused tephra ≥ 3 cm thick to accumulate over an area of at least 60,000 km^2 extending from upper Bristol Bay to beyond Kodiak Island (Fig. 7) (20, 21, 22). This eruption, one of the greatest of historic times, produced an estimated dust-veil index of about 500 and resulted in a weakening of the direct solar beam by about 25% (23). At least 28 km^3 of tephra is believed to have been produced during the eruption.

APPLICATION OF TEPHRA STUDIES TO GEOLOGICAL PROBLEMS

The recognition and characterization of widespread tephra layers in Quaternary deposits of western United States has

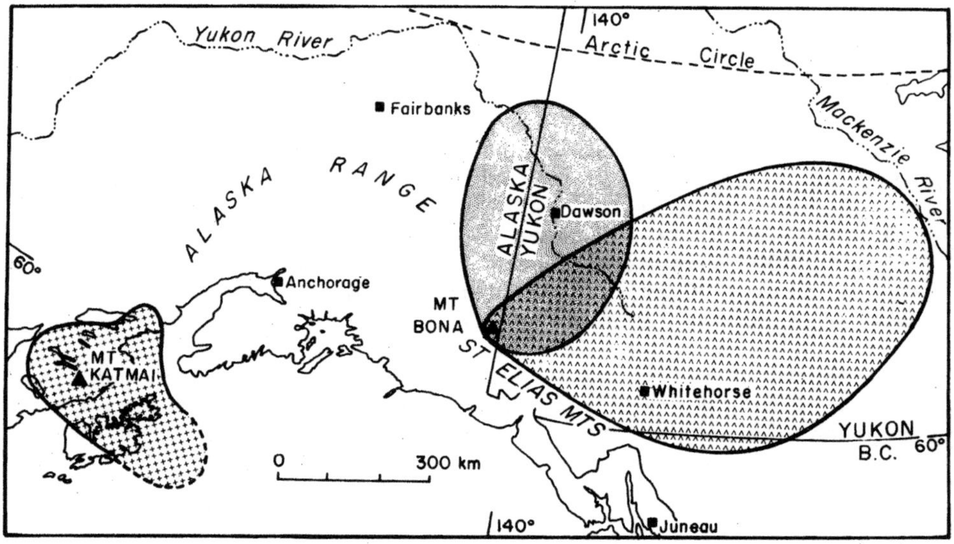

Figure 7. Map of south-central Alaska showing distribution of tephra from the 1912 eruption of Mount Katmai and of the two lobes of White River Ash.

prompted geologists to use these convenient time-stratigraphic marker beds to help solve a variety of geologic problems. The examples presented here do not constitute an exhaustive compilation, but are merely representative cases that illustrate some of the approaches that have been taken.

Mid-continent Pleistocene chronology

Tephra layers have played a key role in obtaining ages for some of the classic glacial-stratigraphic units of the mid-continent region with which local stratigraphic successions elsewhere in the United States are often correlated. Although radiocarbon dates have helped determine the timing of Wisconsin-age glacial events within and beyond the limit of the last glaciation, the absolute age of older glacial drifts and inter-glacial weathering profiles could only be given relative ages until radiometric dating of several widespread tephra layers of early and middle Pleistocene age (8) helped provide a basis for assigning dates to early Pleistocene drifts.

The type-S Pearlette ash (1.2 million years) has been found in Nebraska in silts of reported late Nebraskan age that lie below Kansan till (24) (Table 1). Type-B ash (1.9 million years) in Meade County, Kansas, directly underlies the Borchers fauna that may be earliest Nebraskan in age or older (25). The type-O ash (0.6 million years) has been reported overlying till of possible Kansan age in Nebraska, South Dakota, and Iowa (24, 26), and in Kansas it directly overlies fossil rodent remains (Cudahy fauna) of late Kansan age (25). However, type-O ash and Hartford ash (=type O?) (0.7 million years) have been reported in the classic Aftonian interglacial sequence in Iowa (27). The dated tephra units help demonstrate that the standard four-fold glacial sequence of central North America probably encompasses at least a million years, and have helped point out probable errors in interregional correlation (27). The Kansan, Illinoian, and Wisconsin glaciations encompass an interval during which 7 or 8 glacial-interglacial cycles are represented in the succession of marine oxygen-isotope stages (28), pointing to the probable incompleteness of the continental glacial-stratigraphic record.

Table 1. Stratigraphic relationships of tephra layers of western United States to standard Northern American Pleistocene time-stratigraphic units.

Series	Stage	Tephra layer	Approximate age (yr)
		Glacier Peak	11,250-12,500
	Wisconsinan		
	Sangamonian		
	Illinoian		
Pleistocene	Yarmouthian		
		Pearlette O	600,000
		Bishop	710,000
	Kansan		
	Aftonian		
	Nebraskan		
		Pearlette S	1,200,000
		Pearlette B	1,900,000
Pliocene			

Bishop Tuff and the age of the Sherwin till

The age of glacial deposits antedating the last glaciation is poorly known in the mountains of western United States, for such deposits lie beyond the conventional range of radiocarbon dating and few are directly related to lavas that can be dated by the K/Ar method. The Sherwin till, which crops out along the east side of the Sierra Nevada in California, is an important glacial rock-stratigraphic unit that was tentatively inferred to by a correlative of the Kansan drift in midwestern United States

(29). The type Sherwin till apparently is the same as a nearby till that lies beneath a layer of bedded ash and pumice, dated as 710,000 K/Ar years old (31), which is capped by Bishop Tuff (30). Weathering of the uppermost part of the till suggests that the till may be more than a few tens of thousands of years, but not as much as 100,000 years, older than the tephra, and probably dates to about 750,000 years ago (30). Such an age supports Eliot Blackwelder's inferred early Pleistocene age for the drift (29) and makes the Sherwin till one of the few ancient tills of western United States for which a reasonably close limiting date is available.

Deglaciation of the North Cascade Range

Moraines at or close to the outer limit of glaciers during the last (Fraser) glaciation in the eastern North Cascade Range are believed to have been built close to 14,000 years ago (11, 32). Through a broad region lying east and southeast of Glacier Peak volcano, pumiceous tephra of several late-glacial eruptions overlies the youngest drift. The oldest layer (G), which may date to between 12,750 and 12,000 years ago, has been found near valley heads and in some cirque basins as a primary airfall deposit, indicating that by the time of the initial eruption deglaciation was at an advanced stage and termini of some major valley glaciers had receded 40 km or more from their outermost moraines (11, 33). Moraines of a late-glacial readvance at Stevens Pass that lie within the fallout pattern of a younger layer (M) which may date to about 12,000 years ago apparently lack a cover of that tephra, but are mantled with Mazama tephra (layer O) as well as younger tephra layers from Mt. St. Helens (Fig. 8). These relationships indicate that the moraines were built after about 12,000 years ago but before 6700 years ago. Wood overlying till that is traceable to end moraines at Snoqualmie Pass which are inferred to correlate with the Stevens Pass moraines is 11,050 ± 50 years old. If the moraines are indeed correlative, then the age of the advance is bracketed between about 12,000 and 11,000 years B.P. (11).

Holocene glacier fluctuations in the Cascade Range

The widespread distribution of Holocene tephra layers in the Cascade Range of Washington makes it possible to bracket the ages of many recent moraines and to correlate moraines throughout a reasonably broad area. Moraines of glaciers on and near Mount Rainier in the southern Cascades can be most closely bracketed, for more than 20 Holocene tephra layers are recognized there (34). The major layers have been used to subdivide, correlate, and obtain limiting ages for two major episodes of glacier expansion which were named Burroughs Mountain and Garda (35). Tephra layer O from Mt. Mazama (6700 years old) is not found on

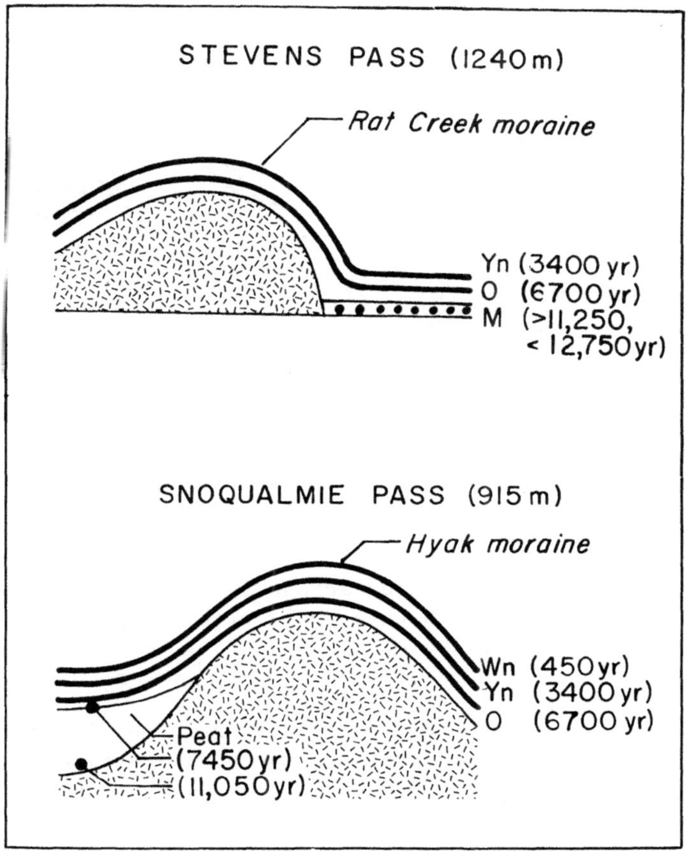

Figure 8. Diagrammatic section through late-glacial end moraines in the North Cascade Range showing relationship of tephra layers to glacial deposits.

the surface of Neoglacial deposits, whereas late Pleistocene moraines and moraines of possible early Holocene age commonly are mantled with this unit. Moraines also lacking a cover of layer Yn from Mt. St. Helens (ca. 3400 years old) are inferred to represent either middle or late Neoglacial advances, and those in the southern Cascades that lack a mantle of layer Wn (450 years old) are inferred to have been deposited since the beginning of the Little Ice Age (Fig. 9).

Fluctuations in the level of Lake Bonneville

Fluctuations of lake level in the Bonneville Basin of western United States are recorded in subaerial stratigraphic exposures and in a 307-m core obtained in Utah near the margin

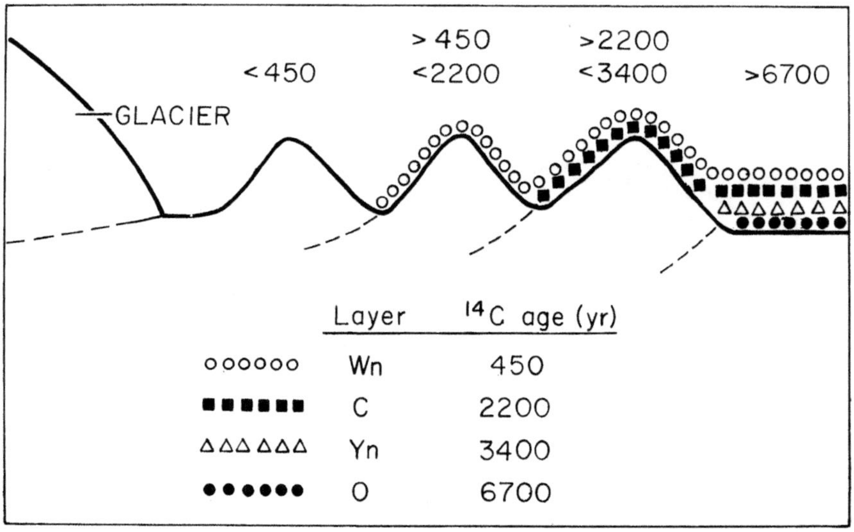

Figure 9. Diagrammatic section across sequence of Holocene moraines in the southern Cascade Range of Washington showing inferred moraine ages based on tephra stratigraphy.

of Great Salt Lake. The core reveals changing environmental conditions in alternating cycles of lacustrine sedimentation and soil development extending back into the Pliocene (37). Dating control is provided partly by paleomagnetic measurements, but also by the presence of two ash layers in the core that have been correlated on the basis of mineralogy with the type-O Pearlette ash and the Bishop ash. The former tephra unit has also been identified in subaerial exposures of lake sediments in the Bonneville Basin (38). The presence of these dated ash units has made it possible not only to date parts of the older lacustrine record, but to correlate this sequence with the stratigraphic succession of the Great Plains which in turn can be linked to the standard North American glacial succession. A correlation is also possible with the deep-sea record and with the central European loess record which show a broadly similar sequence of alternating climatic episodes over approximately the last 800,000 years (37).

Age of the last Scabland flood from Glacial Lake Missoula

The impoundment of Glacial Lake Missoula in northwestern Montana by a lobe of the Cordilleran Ice Sheet resulted in some of the greatest floods in the Quaternary geologic record. The flood waters, released when the ice dam failed, spilled across the lava fields of the Columbia Plateau creating the famed Channeled Scabland and its remarkable assemblage of large-scale flood features (39). It has generally been assumed that the last flood occurred shortly after the culmination of the last glaciation which was thought to date to about 20,000 years ago on the basis of interregional correlations. However, the discovery of three associated ash layers in fine-grained deposits of the last flood, and their correlation with Set-S tephra from Mt. St. Helens, has provided one means of evaluating the age of the flooding. From associated radiocarbon dates, Set-S tephra at Mt. St. Helens was inferred to have been erupted about 13,000 years ago (36). This date has been used as a close minimum age for the last major flood episode, thereby implying that the failure of the ice dam occurred near the end of the last glaciation, and possibly following an advance that culminated about

Table 2. Tephra layers from volcanoes in the Pacific Northwest that have been useful in local and regional tephrochronologic studies (Data from references 11, 12, 17, 34, and 36).

Layer	Source volcano	Approximate age (^{14}C yr)
X	Mt. Rainier	150[a]
T	Mt. St. Helens	180[a]
Set W	Mt. St. Helens	450[a]
B	Mt. St. Helens	1100-2500
C	Mt. Rainier	2200
Set P	Mt. St. Helens	2500-3000
Set Y	Mt. St. Helens	3000-4000
Yn	Mt. St. Helens	3400
B	Mt. Rainier	4500
H	Mt. Rainier	4700
F	Mt. Rainier	5000
S	Mt. Rainier	5200
N	Mt. Rainier	5500
D	Mt. Rainier	6000
L	Mt. Rainier	6400
A	Mt. Rainier	6500
O	Mt. Mazama	6700
R	Mt. Rainier	8750
J	Mt. St. Helens	>8,000 <12,000
B	Glacier Peak	11,250
Set T	Glacier Peak	>11,250, <12,750
M	Glacier Peak	"
F	Glacier Peak	"
C	Glacier Peak	"
N	Glacier Peak	>11,250, <12,750
G	Glacier Peak	"
S	Mt. St. Helens	>12,100, <13,650

[a] Age estimated from tree-ring counts

14,000 years ago at approximately the time when the Puget Lobe achieved its maximum expansion west of the Cascade Range.

Holocene alluviation in the eastern Cascade Range

Postglacial alluvial fans in the east-central Cascades of Washington are currently stabilized by vegetation and entrenched by streams. Exposures along stream banks in several fans display Mazama ash interstratified with fan alluvium (40, 41). The fan surfaces often are discontinuously mantled with a thin layer of Wn tephra from Mt. St. Helens; in at least one fan tephra layer Yn is found in the uppermost fan sediments (Fig. 10). These relationships indicate that widespread alluviation occurred during the early and middle Holocene, a period corresponding to the Hypsithermal interval, but that by the time of the Yn ashfall, aggradation was near an end. The subsequent trenching of fans marked a change in stream regimen that coincided with a shift to generally cooler and moister conditions which favored expansion of alpine glaciers (Burroughs Mountain advance) along the crest of the Cascades.

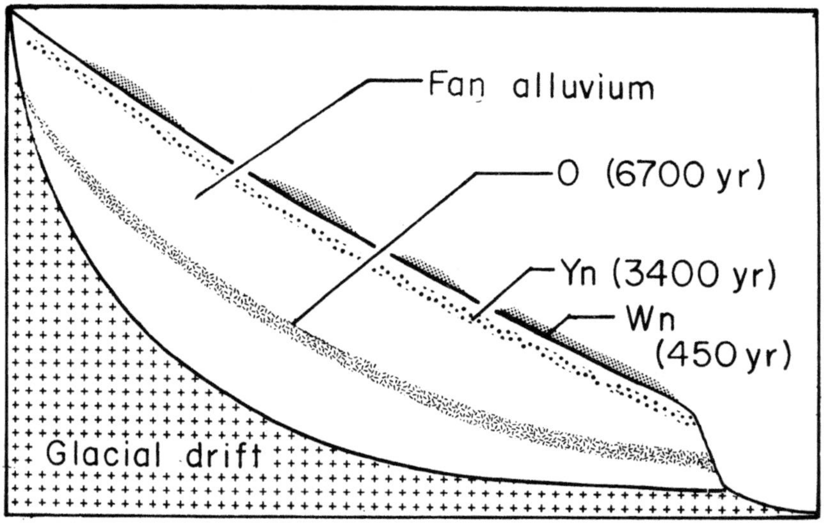

Figure 10. Diagrammatic section through postglacial alluvial fan in eastern North Cascade Range showing stratigraphic position of tephra layers.

Sequence, relative magnitude, and recurrence interval of eruptive events

Stratigraphic relationships of successive tephra layers to

one another and to associated lava flows can provide information about the sequence of eruptive events. For example, cinder layers on the south rift zone of Mauna Kea volcano in Hawaii are interstratified with lave flows that issued from the flanks of cinder cones built during tephra-producing eruptions about 4500 ^{14}C years ago (2). The stratigraphy reveals that initial eruptions occurred at an altitude of about 2740 m, and that most vents opened successively in the downslope direction (Fig. 11). Each eruption began with explosive ejection of cinders and bombs that built up a cone around the vent and distributed tephra downwind as an elongate plume. After explosive release of volatiles, magma welled to the surface as lava that generally buried part of the newly erupted cinders. Where the lava erupted from the upslope side of a cone, the cone commonly was engulfed or partially buried by the lava flow. In only one case does a lava flow have no associated tephra layer. The sequential downslope eruptive activity revealed by the stratigraphy implies progressive movement of magma down the rift zone, a phenomenon also observed during recent eruptions of nearby Kilauea volcano.

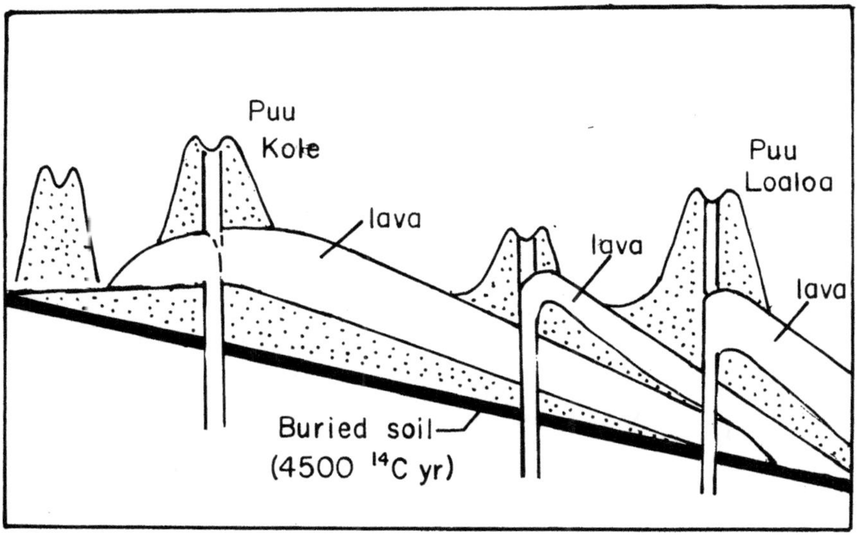

Figure 11. Section along south rift zone of Mauna Kea volcano, Hawaii, showing relationship of successive tephra cones and layers to one another, to associated lava flows, and to a widespread buried soil (2).

A comparison of the relative magnitude of ancient tephra-producing eruptions can be obtained from analysis of downwind grain-size distribution and thickness variations. The latter parameter also can be used to obtain minimum estimates of total volume. Isopach data for selected major tephra layers in the Pacific Northwest have been used to plot tephra thickness as a function of distance (Fig. 12). The relative position of the curves provides an approximate measure of relative volume or magnitude of the eruptions, although the position is determined in part by stratospheric wind velocities during fallout. The layers plotted are of great extent, having been identified at distances of from 500 to as much as 1000 km or more from the source volcanoes, and both isopleth data and areal distribution patterns support the relative order based on thickness data.

Detailed stratigraphic studies of tephra-producing volcanoes may make it possible to reconstruct the eruptive history over a long interval of time. If dating is sufficiently reliable, then recurrence interval of eruptive events can also be estimated. Studies of Mt. St. Helens, for example, have shown that small, intermediate, and large eruptions are represented by the A.D. 1842 tephra layer (ca. 0.01 km^3), layer T (ca. 0.1 km^3), and

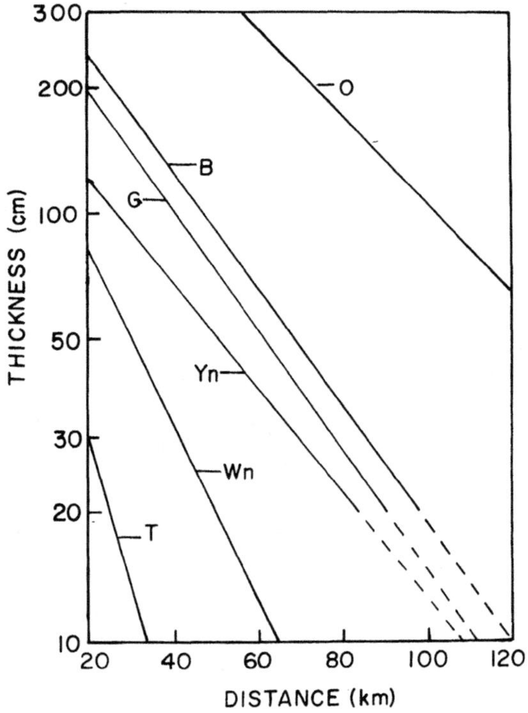

Figure 12. Variation in ash thickness as a function of distance from source vent for six major tephra layers in the Pacific Northwest. Layer O = Mount Mazama; layers G and B = Glacier Peak; layers Yn, Wn, and T = Mount St. Helens.

layer Yn (ca. 1 km^3). Based on the Holocene stratigraphic record of the volcano, it has been estimated that eruptions of these magnitudes may occur as frequently as once every 100, 500, and 3000-4000 years, respectively (42).

Climatic effects of tephra eruptions

Major eruptions of tephra may alter climate in several ways. The much-discussed effect of airborne volcanic dust and aerosols on reducing the incident radiation at the earth's surface was documented in the case of the Mt. Agung eruption of 1963, following which a decrease in insolation was noted (43). It took some 7 years before the atmosphere returned to normal levels. Historic records of volcanic eruptions have been used to estimate the dust veil index which is a measure of the content of volcanic dust in the atmosphere (23). Although extensive records of eruptions exist covering the last 100 years, the number of reported eruptions decreases back in time as records become fewer and less reliable. However, the large number of eruptions (ca. 3750) documented for the past century and a half have made possible realistic appraisal of variations in the hemispheric dust veil as a function of time (44).

Another important effect has recently been pointed out by J. R. Bray (10) who estimated changes in albedo of the ash-covered land surface and of the Northern Hemisphere following deposition of thick and extensive ash layers of early and middle Pleistocene age in western United States. For example, following eruption of the type-O Pearlette ash, which covered an estimated 2.76 million km^2, surface albedo of the ash-covered area increased from 21.5% to 58.7%, resulting in an increase of Northern Hemisphere albedo by 0.41%. For the Bishop ash, the increase was estimated as 0.31%, whereas for both the type-B and type-S Pearlette ashes the changes were 0.13%. From such values, estimates can also be derived for the mean temperature decline resulting from the albedo changes. For the type-O and Bishop ashes, the estimated mean Northern Hemisphere temperature decrease was 0.49° and $0.38^{\circ}C$, respectively. Such changes, if they persisted for many years, could have a detectable influence on global ice masses both directly and through positive feedbacks.

Paleowind directions

Asymmetrical distribution of a tephra deposit with respect to its source vent can result from a directed angle of ejection or from the transport of tephra away from the vent by wind. In the case of very small eruptions, it may be difficult to differentiate between these mechanisms, but for major eruptions the deposition of an elongate tephra plume extending tens or even hundreds of kilometers from the source implies wind transport.

Such plumes can provide information about the prevailing wind at the time of an eruption (45) or may permit inferences to be made about changing wind conditions during eruption intervals.

The Cascade Range of Washington provides numerous examples of tephra layers that can give evidence of former wind directions during geologically recent eruptive episodes, for the plume orientations are known for many of the dated tephra deposits (Fig. 13). Tephra plumes related to late Pleistocene eruptions of Glacier Peak trend between south and east indicating prevailing westerly to northwesterly winds. Holocene layers

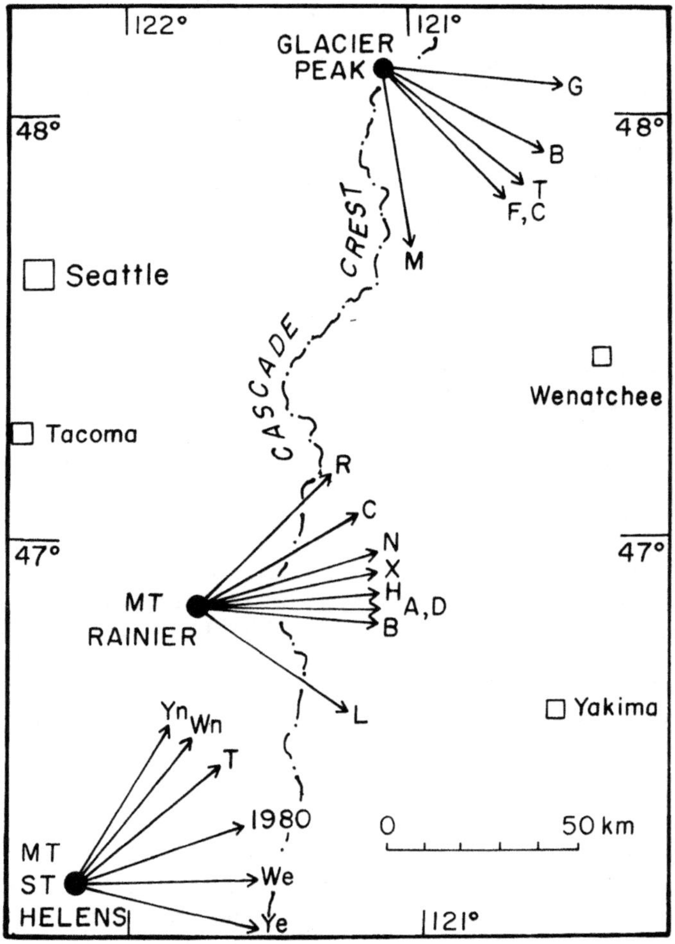

Figure 13. Orientations of axes of tephra lobes from volcanoes in the Washington Cascade Range. See Table 1 for ages of layers.

erupted from Mt. Rainier are spread through a quadrant between southeast and northwest, with most trending approximately east, again indicating westerly winds. Late Holocene ash layers from Mt. St. Helens trend predominantly northeast, although two units related to the Y and W eruptions were spread to the east. The pronounced easterly orientation of axes of these various tephra layers suggests that a westerly wind regime has characterized this region since at least late-glacial times. The northeasterly and southeasterly components shown, respectively, but the Mt. St. Helens and Glacier Peak layers may reflect seasonal contrasts, for southwesterly winds are dominant during the late autumn, winter, and spring, whereas northerly winds are more typical of the summer and early fall. Alternatively, the southeasterly distribution of the Glacier Peak layers, which fell when the shrunken Cordilleran Ice Sheet still lay over part of northern Washington, could reflect the influence of the ice sheet on prevailing wind patterns. The Glacier Peak layers may represent a complex set deposited over a relatively short span of time; if so, they indicate shifting wind conditions. Comparable shifts in wind direction are indicated in the case of the Set W and Set Y layers from Mt. St. Helens which indicate wind shifts of 70° to 80° between major depositional events.

An extensive deposit of dark-gray cinders that mantles much of the lower south rift zone of Mauna Kea volcano on the island of Hawaii was erupted about 4500 ^{14}C years ago from numerous vents that are now marked by cinder cones of various sizes (2). The tephra plumes, which extend 5 km or more from the cones, have been delineated by constructing isopach and isopleth maps. The axes of the plumes trend consistently west to southwest implying persistent easterly to northeasterly trade winds. Older undated tephra plumes of early Holocene to late Pleistocene age also trend predominantly west or southwest implying that trade-wind circulation similar to that of today probably characterized these latitudes during the late Pleistocene as well. This inference is supported by a westerly asymmetry of young cinder cones on the volcano (46) and by west-dipping foreset strata of stabilized sand dunes that are believed, on stratigraphic grounds, to date to the last glacial age (47).

Anomalous timberline of Mt. St. Helens

During the past several thousand years timberlines in mountain regions have fluctuated in altitude in response to changing climates, and in most instances the changes amounted to between 100 and 200 m (49). Such fluctuations also probably occurred on forested volcanoes of the Pacific Northwest that experienced little or no eruptive activity during the late Holocene. However, on active volcanoes present timberlines may reflect the ecologic impact of recent eruptions, rather than climatic change. At

Mt. St. Helens, for example, prior to the 1980 eruption timberline on the northeast slope lay at about 1340 m, or some 640 m below the timberline of adjacent Cascade peaks (50, 51). This anomalously low timberline is believed to have resulted from the major pumice eruptions of ca. 450 and 180 years ago that killed the natural forest on the upper flanks of the volcano. Timberline trees were young and showed little evidence of climatic stunting, and many of them were species found in low- and middle-altitude forests of the Pacific Northwest. The steady advance of timberline, which apparently had been going on since the eruption of layer T pumice, was abruptly terminated on the north side of the volcano during the May 1980 eruption, the blast effects of which obliterated the forest on, and for many kilometers beyond, the north slope of the volcano, thereby once again artifically depressing timberline far below its equilibrium level.

Season and rate of deposition of Cascade ash layers

Ancient ash layers generally cannot be closely dated because of inherent limitations of standard dating techniques. Radiocarbon dates of late Pleistocene tephra units generally can bracket the age of a layer only to within the nearest hundred to several hundred years, due both to laboratory counting errors and to variations in the past rate of radiocarbon production. Consequently, it generally is not possible to determine either the season of deposition or the period of time during which fallout occurs. It recently has been shown, however, that studies of the fossil pollen content of ash layers can provide such information (12, 48). The pollen types present in an ash sample may reflect the season of ash fall, and the number of pollen grains within an ash layer will reflect the duration of accumulation. The latter is a reflection of pollen influx, which is the rate at which pollen is deposited per unit area per unit time.

At Lost Trail Pass Bog in southwestern Montana two ash layers having physical and chemical properties similar to Glacier Peak layer B tephra are bracketed by radiocarbon dates of 11,300 ± 230 (above) and 11,200 ± 250 (below) years (12). Based on calculated pollen influx, each layer is estimated to have been deposited in less than a year. Inter-ash organic sediments represent about 10 to 15 years of deposition. The season of ash deposition is uncertain, but the upper layer may have been deposited in early spring, and very likely before late spring, based on contained pollen types. The lower layer contains little pollen, so rapid deposition is inferred. Mazama ash in the same bog is bracketed by radiocarbon dates of 6270 ± 120 and 6700 ± 100 years (12). An initial 4.9 cm of ash fell in the autumn, and another 1 cm of ash before the next winter.

A final 1 cm of ash accumulated during the following spring or summer. During the next 80 years the ash was reworked and redeposited. Such studies provide important clues about the character and temporal spacing of eruptive episodes, and may help in differentiating original airfall tephra from tephra that has subsequently been redeposited.

POTENTIAL APPLICATIONS

The great number and widespread distribution of Quaternary tephra layers in western United States makes it likely that tephrochronology will play an important role in future paleoenvironmental studies. Several investigations that make use of tephra layers are underway or are being planned.

The multiple Holocene tephra layers in the Cascade Range that originated at Mt. St. Helens, Mt. Rainier, and Mt. Mazama, offer the potential of quantifying depositional and erosional processes over a rather broad area. The tephra layers are often encountered in alluvial and colluvial deposits, and have been seen in dune sands and loess deposits as well. In addition to being useful for dating Neoglacial moraines, they can be used in chronologic investigations of rock glaciers, protalus ramparts, and patterned ground which are common but as yet little-studies features near the crest of the Cascade Range. They may also be of use in evaluating the age and recurrence interval of rockfall and landslide events within the mountains and on the Columbia Plateau.

There are more than 700 glaciers in the North Cascade Range of Washington, and many more in the southern Washington, Oregon, and California Cascades that lie within the fallout zones of certain late Holocene tephra layers. Recognition of tephra layers in these glaciers analysis of ice cores has not yet been attempted, but investigations of large glaciers in which the ice has a reasonably long residence time could lead to identification of recent small-scale eruptive events that have been overlooked in conventional stratigraphic field studies. In coming years, the 1980 tephra layer from Mt. St. Helens can be monitored in glaciers on Mt. Rainier, Mt. Adams, and Goat Rocks that lie within the ash fallout zone to see how thin ash layers are incorporated in and move through Cascade glaciers. Analogous investigations on snowfields can aid in dating stratigraphic levels of perennial snowbanks and in assessing the influence of ashfalls on ablation of snow.

Lakes and bogs in the Puget Lowland of western Washington contain a stratigraphic record of late Quaternary eruptive events in the form of numerous tephra layers. Study and characterization of these layers is now underway using samples from long piston cores obtained from several lakes along a north-south transect. The transect extends from beyond the Fraser (=Wisconsin) limit of the Puget Lobe of the Cordilleran Ice Sheet northward toward the Canadian border. Pollen and plant-macrofossil studies of organic-rich sediments from the cores will be used to determine times of arrival of various arboreal and nonarboreal species, using the tephra layers as regional time-stratigraphic marker horizons. This study should provide information on the time of arrival of species following deglaciation and on rates of reforestation. Similar studies may be possible in the Cascade Range along altitudinal, rather than longitudinal, transects.

Finally, identification of ash layers from marine-sediment cores off the coast of western United States and Alaska may permit direct correlation between marine and terrestrial strata. The potential for such correlations is especially good in the Gulf of Alaska and the Bering Sea where tephra layers from Aleutian volcanoes are likely to be widespread and numerous. If tephra layers predating the last glaciation can be identified in both terrestrial and marine sections, it may be possible to obtain approximate ages for presently undated pre-Wisconsin glacial and interglacial events on land by relating tephra layers to the oxygen-isotope stratigraphy of the marine cores.

REFERENCES

1. Wentworth, C.,: 1938, Hawaiian Volcano Obs. Third Spec. Report, 183 p.
2. Porter, S.C.: 1973, Geol. Soc. America Bull. 84, pp. 1923-1940.
3. Black, R.F.: 1974, Quaternary Res. 4, pp. 264-281.
4. Péwé, T.L.: 1975, U.S. Geol. Survey Prof. Paper 835, 145 p.
5. Scheidegger, K.F., Corliss, J.B., Jezek, P.A. and Ninkovich, D.: 1980, Jour. Volcanology and Geothermal Res. 7, pp. 107-137.
6. Wilcox, R.E.: 1965, in: Wright, H.E., Jr. and Frey, D.G. (eds.), The Quaternary of the United States, Princeton Univ. Press, pp. 807-816.
7. Izett, G.A., Wilcox, R.E., Powers, H.A. and Desborough, G.A.: 1970, Quaternary Res. 1, pp. 121-132.
8. Naeser, C.W., Izett, G.A. and Wilcox, R.E.: 1973, Geology 1, pp. 187-189.
9. Izett, G.A., Wilcox, R.E. and Borchardt, G.A.: 1972, Quaternary Res. 2, pp. 554-578.
10. Bray, J.R.: 1979, Quaternary Res. 12, pp. 204-211.
11. Porter, S.C.: 1978, Quaternary Res. 10, pp. 30-41.

12. Blinman, E., Mehringer, P.J., Jr. and Sheppard, J.C.: 1979, in: Sheets, P.D. and Grayson, D.K. (eds.), Volcanic activity and human ecology, New York, Academic Press, pp. 393-425.
13. Williams, H.: 1942, Carnegie Inst. Washington Publ. 540, 162 p.
14. Mullineaux, D.R. and Wilcox, R.E.: 1979, Am. Geophys. Union Trans. 61, p. 66.
15. Nelson, C.H., Kulm, L.D., Carlson, P.R. and Duncan, J.R.: 1968, Science 161, pp. 47-49.
16. Williams, H. and Goles, G.: 1968, in: Dole, H.M. (ed.), Andesite Conference Guidebook, Oregon Dept. Geol. and Mineral Industries Bull. 62, pp. 37-41.
17. Mullineaux, D.R., Hyde, J.H. and Rubin, M.: 1975, Jour. Res. U.S. Geol. Survey 3, pp. 329-335.
18. Okazaki, R., Smith, H.W., Gilkeson, R.A. and Franklin, J.: 1972, Northwest Sci. 46, pp. 77-89.
19. Lerbekmo, J.F., Westgate, J.A., Smith, D.G.W. and Denton, G.H.: 1975, in: Suggate, R.P. and Cresswell, M.M. (eds.), Quaternary Studies, Roy. Soc. N. Z. Bull. 13, pp. 203-209.
20. Dumond, D.E.: 1979, in: Sheets, P.D. and Grayson, D.K. (eds.), Volcanic activity and human ecology, New York, Academic Press, pp. 373-392.
21. Curtis, G.H.: 1968, in: Coats, R.R., Hay, R.L. and Anderson, C.A. (eds.), Studies in volcanology, Geol. Soc. America Mem. 116, pp. 153-210.
22. Martin, G.A.: 1913, Nat. Geogr. Mag. 25, pp. 131-181.
23. Lamb, H.H.: 1970, Phil. Trans. Royal Soc. A266, pp. 425-533.
24. Boellstorff, J.D.: 1972, Geol. Soc. America Abstr. with Programs 4, p. 274.
25. Naeser, C.W., Izett, G.A. and Wilcox, R.E.: 1971, Geol. Soc. America Abstr. with Programs 3, p. 657.
26. Izett, G.A., Wilcox, R.E., Obradovich, J.D. and Reynolds, R.L.: 1971, Geol. Soc. America Abstr. with Programs 3, p. 610.
27. Boellstorff, J.: 1978, Science 202, pp. 305-307.
28. Shackleton, N.J. and Opdyke, N.D.: 1973, Quaternary Res. 3, pp. 39-55.
29. Blackwelder, E.: 1931, Geol. Soc. America Bull. 42, pp. 865-922.
30. Sharp, R.P.: 1968, Geol. Soc. America Bull. 79, pp. 351-364.
31. Dalrymple, G.B., Cox, A. and Doell, R.R.: 1965, Geol. Soc. America Bull. 76, pp. 665-674.
32. Porter, S.C.: 1976, Geol. Soc. America Bull. 87, pp. 61-75.
33. Beget, J.E.: 1980, Geol. Soc. America Abstr. with Programs 12, p. 96.
34. Mullineaux, D.R.: 1974, U.S. Geol. Survey Bull. 1326, 83 p.
35. Crandell, D.R. and Miller, R.D.: 1965, U.S. Geol. Survey Prof. Paper 501-D, pp. D110-D114.
36. Mullineaux, D.R., Wilcox, R.E., Ebaugh, W.F., Fryxell, R. and Rubin, M.L.: 1978, Quaternary Res. 10, pp. 171-180.

37. Eardley, A.J., Shuey, R.T., Gvosdetsky, V., Nash, W.P., Picard, M.D., Grey, D.C. and Kukla, G.J.: 1973, Geol. Soc. America Bull. 84, pp. 211-216.
38. Morrison, R.B.: 1965, U.S. Geol. Survey Prof. Paper 525-C, pp. C110-C119.
39. Bretz, J.H.: 1969, Jour. Geology 77, pp. 505-543.
40. Hopkins, K.D.: 1966, Unpub. M.S. Thesis, Univ. of Washington.
41. Pavish, M.: 1973, Unpub. M.S. Thesis, Univ. of Washington.
42. Crandell, D.R., Mullineaux, D.R. and Miller, C.D.: 1979, in: Sheets, P.D. and Grayson, D.K. (eds.), Volcanic activity and human ecology, New York, Academic Press, pp. 195-219.
43. Mendonca, B.G., Hanson, K.J. and DeLuisi, J.J.: 1978, Science 202, pp. 513-515.
44. Hirschboek, K.K.: 1980, Paleogeography, paleoclimatology, paleoecology 29, pp. 223-241.
45. Eaton, G.P.: 1963, Jour. Geophys. Res. 68, pp. 521-528.
46. Porter, S.C.: 1972, Geol. Soc. America Bull. 83, pp. 3607-3612.
47. Porter, S.C.: 1979, Quaternary Res. 12, pp. 161-187.
48. Mehringer, P.J., Jr., Blinman, E. and Petersen, K.L.: 1977, Science 198, pp. 257-261.
49. LaMarche, V.C., Jr.: 1973, Quaternary Res. 3, pp. 632-660.
50. Lawrence, D.B.: 1938, Mazama 20, pp. 49-54.
51. Lawrence, D.B.: 1954, Mazama 36, pp. 41-44.

TEPHROCHRONOLOGY AND QUATERNARY STUDIES IN JAPAN

Hiroshi Machida

Department of Geography, Tokyo Metropolitan University, Tokyo, Japan

ABSTRACT. This paper gives an overview of the development of tephrochronology in Japan and discusses several problems regarding the Japanese Quaternary. Tephra from Quaternary stratovolcanoes in Japan is mainly of andesitic composition with subordinate amounts of basalt and rhyolite. In contrast, the eruptions causing the formation of calderas are of rhyodacitic magma and were characterised by extraordinarily violent explosions producing widespread tephras, all of which are important for interregional correlations.

Many eruptions in the Holocene had long been dated with the aid of stratigraphic relations with archaeological sites. Since 1970 recognition of widespread marker tephras has compelled us to revise some former views of archaeological chronology, both in Neolithic and Paleolithic times.

Evidence of marked climatic oscillations and sea level changes is reported from the upper Osaka Group, a sequence of cyclic sediments of lacustrine and marine origin deposited during the last 1 m.y. A stratigraphic framework was established in terms of tephrostratigraphy, paleomagnetic chronology, fission-track and K-Ar ages, and fossil ranges. The middle Pleistocene to Holocene changes of sea level are discussed in detail from the South Kanto district. From the dated tephras and the marine-terrestrial sequences, eight major interglacial episodes during the last 700,000 years have been identified.

The Pleistocene glaciation of Japan, confined to the higher parts of mountain ranges at the northern part of Japan Alps, was studied using four marker tephras interstratified with glacial

S. Self and R. S. J. Sparks (eds.), Tephra Studies, 161–191.

deposits. They indicate that at least three episodes of ice cap glaciation culminated about 20,000 yBP, 50-55,000 yBP and before 100,000 yBP. Valley filling around the Japan Alps might have taken place simultaneously during the glacial culminations.

1. INTRODUCTION

In the Japanese Islands more than 200 volcanoes have been active in the middle-late Quaternary. They display a zonal arrangement along the islands, with a sharp eastern boundary called the volcanic front, paralleling the offshore trenches and troughs. They can be divided into two groups; the eastern Japanese volcanic group and the western Japanese volcanic group. About sixty Quaternary volcanoes continued their activity into historic time. Accordingly, a more or less continuous mantle of tephra exists over some 40% of the four main islands, and the widespread occurrence of isolated patches indicates that most parts of Japan were previously covered by tephra.

The tephra layers not only directly govern the nature of soil and topography of Japan, but are also an important time-marker in interpreting several Quaternary records. Consequently, tephrochronology has been one of the most flourishing fields of study in the Japanese Quaternary records.

In this paper the author first reviews the historical development of tephrochronology in various districts of Japan, and then discusses several problems regarding the Japanese Quaternary from the viewpoint of tephra studies: 1. Growth history of volcanoes interpreted from tephra studies, 2. Archaeological applications, 3. Fluctuations of climate and induced sea level changes, and 4. Glacial chronology.

2. GENERAL REVIEW OF TEPHROCHRONOLOGY

The utilisation of tephra as key marker beds has long been practised in local stratigraphic studies. Apart from such traditional geologic studies, one of the pioneer uses of tephrochronology was a pedological study in Hokkaido, northern Japan, where fresh tephra-fall deposits of Holocene age are widely distributed. Studies there, started in the 1930's under the stimulus of such practical problems as improvement of volcanic ash soil. Classification, distribution, soil texture, age and other aspects of tephras were actively studied (1).

In the Kanto Plain, around Tokyo and Yokohama, where Pleistocene weathered tephra formations are well developed, Suzuki (2) was the first to describe the superficial deposits on uplands as airfall volcanogenic deposits. Volcanological studies on these on these tephra commenced in the 1930's (3) and Haradas' detailed modern pedological study in the early 1940's (4), discriminated some marker tephra deposits and their distribution. Soon after World War II, an archaeological discovery of paleolithic remains in the Pleistocene tephra layers in northern Kanto (5) gave a stimulus to geologists, archaeologists, pedologists and geographers, prompting interdisciplinary group studies. These studies effectively provided scientists with a fresh and attractive tool, the development of tephrochronology.

During the period from the 1950's to the 1960's when tephrochronology was actively studied in Japan, research was conducted by various methods owing to differences in the nature of tephra, and the age and environments of their accumulation. These uniquely regional studies appear to be strongly retained today.

2.1 Local characteristics of tephrochronology

Hokkaido. The Hokkaido area is abundant in Holocene tephra ejected from volcanoes such as Usu, Tarumai, Komagatake, Tokachi, Meakan and Mashu, and the tephra beds are well preserved within peaty layers and alluvial deposits. It should perhaps be emphasised that until the 1950's, the purpose of tephra studies in this district was not to investigate the volcanological history but to provide a better understanding of soil parent materials. However, detailed identification of tephra units and mapping of their aerial extent has confirmed their source volcanoes, as well as providing a basis for the future study of volcanic mechanisms. Tephra studies in Hokkaido have recently been compiled in conjunction with volcanologic and stratigraphic studies (6). Also one special characteristic is the enormous quantity of flow deposits in the late Pleistocene and Holocene related to the collapse of calderas. Volcanological studies on these aspects have also been carried out (7,8).

A situation similar to Hokkaido is found in southern Kyushu, where pedological studies of Holocene tephras were first made by Kanno (9), and chronological studies were later initiated. At the present time, studies on airfall tephras of the Holocene (10) and on large scale pyroclastic flow deposits (11,12,13) are being actively conducted.

Southern Kanto. In this district airfall Pleistocene tephras are abundant and thick, and were investigated mainly in connection with terrace features. The method of classifying and describing tephra groups into several unit formations (Tachikawa loam, Musashino loam, etc.), based on stratigraphic relationships between the tephra and terrace features, as well as on the episode forming the paleosol or erosional breaks, was indeed unique. This style of study is considered to have originated for the following reasons: a) The late Pleistocene tephra

Kanto constituted a massive thick brown soil having few marker beds that are easily identified in outcrop. This is a specific characteristic resulting from the fact that a number of scoria layers, originating mainly from Mt. Fuji, closely resembled one another, and the entire tephra column merges into an apparently homogeneous unit. b) That marine and fluvial terraces were developed extensively in this district at the same time as the deposition of tephra formations. The same tephra sequence is usually found superimposed on the same terrace surface.

This method of study has some shortcomings; however, the role of tephra layers as markers to correlate and classify terraces is particularly emphasised, while consideration of the significance of the tephra as volcanic products is somewhat lacking. On the other hand, these studies have greatly contributed to the promotion of sympathetic studies on Quaternary research involving a number of aspects such as the development of geomorphological features, changes in sea level, crustal movement and archaeological chronology, to mention a few. The standard stratigraphy of Japan for the middle-late Quaternary was established in this district. Furthermore, the method of investigation in southern Kanto strongly influenced the study of late Pleistocene tephras in other areas of Japan such as the Chubu, Chugoku and Tohoku districts.

Kinki. In the Kinki district (around Osaka-Kyoto), within the Plio-Pleistocene Osaka group and its correlatives, cyclic sediments of lacustrine and marine origin are well developed. Among these a number of tephra layers are intercalated. Because of the necessity of correlating formations with each other, and of establishing a stratigraphic framework, detailed descriptions of petrographic and chemical characteristics of tephras were made (14-16). Compared with studies in other districts, these studies are significantly stratigraphic and strongly related to paleogeomagnetic and paleontologic studies. A similar trend is seen in the stratigraphical studies of the middle-lower Pleistocene in various districts of Japan.

3. RECENT DEVELOPMENTS IN TEPHROCHRONOLOGY

In the 1960's, studies concentrating on systematisation of tephrochronology were progressing in such fields as volcanology, stratigraphy, pedology and geomorphology. For instance Nakamura, et al. (17) disclosed the relationship between eruptive phenomena and tephra layers. They stressed that the deposits from one cycle of an eruption should form a standard as a stratigraphic unit. Also, Kobayashi (18) introduced various methods of characterisation of tephras, and emphasised that the development of those characterisation methods should provide the basis for tephrochronology. Several trends of recent studies are summarised below.

3.1 Characterisation of tephra

The characterisation of marker tephras for identification is made from various viewpoints. The distinctive feature of a bed reflects complicated processes and factors: the character of the magma and the physical condition at the time of eruption, as well as the conditions of transportation and accumulation, and the environment of weathering. Table 1 reviews the various characteristics which are used to identify tephra layers.

In Japan by the 1960's emphasis had been placed on field characteristics, together with the properties of heavy mineral assemblages, because a great number of tephra sheets are commonly observed on land and quick characterisation is needed at the outcrop. Until recently, this tendency might have resulted in the tephrochronology of this country being more or less confined to regional studies, due to the general slowness in recognising the most extensive tephras.

Only recently have detailed descriptions been given of not only specific features at the outcrop, but also of the petrographic, chemical and thermomagnetic character of specified minerals (19-22,16). Now, refractive indices of volcanic glass and/or phenocrysts are determined, together with crystal assemblages and micrographic features of glass shards. Crystals are the first and most important criteria for characterising most of the middle-late Quaternary tephras (23). The chemical composition of glass and crystals is used for discriminating specific tephras which have a similar index of refraction and other

Table 1. Various methods of characterisation of tephra.

Stratigraphic method (Field work)		Stratigraphic position (including ages determined by various methods)
		Morphology (thickness, grain size, color, grading, degree of weathering, etc)
Petrographic and chemical method (laboratory work)	Petrographic level	Texture of essential materials
		Mineral assemblage (phenocryst-glass ratio, light mineral-heavy mineral ratio, etc)
		Bulk chemistry
	Mineralogical level	Morphology of specific mineral and glass (crystal habit, shape and color of glass)
		Refractive index of specific minerals and glass
		Thermomagnetic properties
		Chemical composition of specific minerals and glass

Table 2. Representative widespread tephras during the last 100,000 years in Japan.

Tephra & Source	Age(10^3YBP) & Dating method	Mode of emplacement	Bulk volume (km^3)	Reference
Towada-a* (To-a,Towada)	0.9-1; C & A	pfa,pfl	>10-15	Oike(1972) Machida, et al.(1980)
Akahoya*& Koya* (Ah,Kikai)	6.3; C	afa,pfa,pfl	>150	Machida & Arai(1978)
Mashu-g~f* (Ma-f,Mashu)	6.5-7.2; C	pfa,pfl	6-30	Katsui(1955;1963)
Mikata & Oki (Hakutosan?)	10; C	afa	?	Arai & Machida(1980)
Hachinohe* (HP,Towada)	10-13; C	pfa,pfl	>50	Oike, et al.(1959;1964) Tohoku Region Q.R.G,(1969)
Aira-Tn*& Ito* (AT,Aira)	21-22; C	afa,pfl	300	Machida & Arai(1976)
Toya	14-25?; C	pfa,pfl	20	Ishikawa, et al.(1969)
Shikotsu* (Spfa-1,Spfl)	32; C	pfa,pfl	130	Katsui(1958;1959)
Kanuma* (KP,Akagi)	32; C & F	pfa	10	Harada(1943)
Kuju* (KjP-1,Kuju)	30-35; C & S	pfa,pfl	10?	Ono(1963)
Akan	>31.5; C	pfa,pfl	100	Ishikawa, et al.(1969)
Kutcharo	>32; C	pfa,pfl	114	Katsui(1962) Ishikawa, et al.(1969)
Ofudo (Towada)	>33; C	pfa,pfl	50?	Tohoku Region Q.R.G.(1969)
Daisen-Kurayoshi* (DKP,Daisen)	45-47; S	pfa,pfl	>20	Machida & Arai(1979)
Tokyo* (TP,Hakone)	49; F	pfa,pfl	>20	Machida & Moriyama(1968)
Aso-4* (Aso)	50?; C & S	pfl	>300	Ono(1970) Watanabe(1978)
Ata	50?; C & S	pfa,pfl	>40-50	Aramaki & Ui(1966)
Obaradai (OP,Hakone)	66; F	pfa,pfl	10	Machida & Suzuki(1971)
Ontake Pm-I* (Pm-I)	70-90; F	pfa	13	Kobayashi, et al.(1967)

Tephras, the distribution of which is shown in Fig. 1, are marked with an asterisk.
Dating method: C, C-14; A, Archeology; F, Fission-track; S, Stratigraphy.
Mode of emplacement: afa, ash-fall deposit; pfa, pumice-fall deposit; pfl, pumice-flow deposit.

properties. For example, the two middle Pleistocene widespread marker-tephras, called Sakura and Azuki in the Kinki district, have recently been identified in the marine sequences of southern Kanto by minor element composition of glass (by INAA), as well as refractive index (23). By such increases in available information and data it has become possible, not only to obtain correlation between tephras in remote districts, but also to approach various volcanological problems.

3.2 Late Quaternary widespread marker-tephras in Japan

As studies advanced in the description of specific features of tephra, a number of large-volume tephra layers spread over extensive areas were discolsed. Table 2 shows a list of widespread marker-tephras within the last 100,000 years. "Widespread" is here loosely defined as those occurring in a region exceeding 50,000 km^2 in area. Their distribution is shown in Figure 1. Such widespread tephras can be classified as follows, according to difference in their modes of ejection, transportation and sedimentation.

Type a: Airfall deposits of fine-grained materials which occupied the upper part of an eruption column at the same time as the ejection of a huge pyroclastic flow. These are represented by the Akahoya and Aira-Tn ash and are related to the activity of huge caldera volcanoes rather than stratovolcanoes. Their specific features are:

1) very abundant fine-grained volcanic glass fragments (especially characteristic are thin and platy, bubble-walled glass shards caused by extraordinarily delayed vesiculation; 2) quantitatively, they are often equal to or in excess of the volume of pyroclastic flow deposits ejected during the same cycle of eruption. This type of tephra should be the dominant class of widespread tephra. The Akahoya and Aira-Tn ashes (24,10) cover most of the Japanese islands as well as the floor of the northwest Pacific and the Japan Sea, forming important markers in the upper Quaternary sequences. These discoveries further encouraged the identification and correlation of two other older tephras of this type, Sakura and Azuki, important markers in the early-middle Pleistocene of Japan.

Type b: Airfall tephra produced by plinian eruptions. A number of pumice layers of this type are found in Japan. Those defined as type b in Table 2 are the largest of this type. The eruptions were often followed by the ejection of pyroclastic flow deposits.

Type c: Large-scale pyroclastic flow deposits. The Ito pyroclastic flow of the Aira Caldera and the Aso-4 pyroclastic flow from Aso Caldera are representative, and are distributed extensively around large calderas (Figure 1). In many cases they are spread in a concentric pattern around the source, while airfall tephras are distributed in a lobate shape on the lee side of volcanoes (Figure 1).

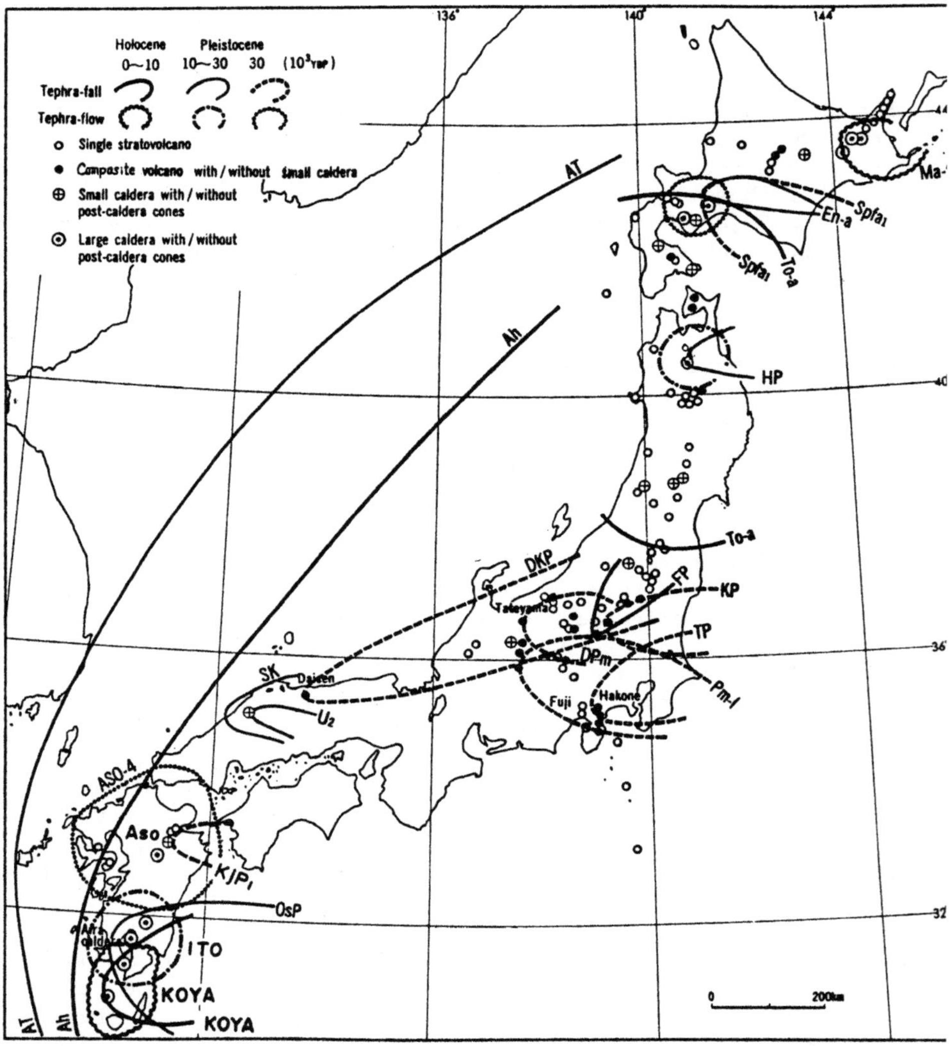

Fig. 1. Distribution of the principal widespread tephras and various types of volcanoes of the late Quaternary in Japan.

3.3 Dating of tephra

The number of tephras whose ages have been determined by several radiometric methods has increased. Ages of the majority of the tephras ejected after about 30,000 years ago have been determined by the radiocarbon method. The accuracy of radiocarbon ages can be checked by such well-identified widespread tephras as Akahoya ash and Aira-Tn ash. However, the ages of a number of tephra sheets remain undetermined, especially for tephras older than 30,000 years. For these, standard ages were given by applying the fission-track method on zircon and/or obsidian fragments from the tephra (25-29), and the ionium method on zircon, apatite and volcanic glass (30,31). In addition, close examination of the stratigraphic relation of dated tephras is often utilised for estimating ages and also revising earlier dates, e.g. Daisen-Kurayoshi pumice (32). The increase in the number of dated marker-tephras has greatly contributed to an understanding of rates of landform development, sea level changes and crustal movements.

3.4 Tephrochronological applications to the Quaternary record

These are of two types: (a) the use of tephra as a time-marker to solve historical problems, and (b) to unravel problems in related fields by the nature of tephra.

As examples of case (a), such studies aid in compiling a history of sea level change and crustal deformation by correlating strata or geomorphic surfaces. They also help in thorough investigation of the history of climate or vegetation supplemented by pollen analysis. In these fields positive achievements were obtained in a number of regional studies. Those in the Osaka area especially contributed to the establishment of the standard sequence for early-middle Pleistocene (33), and those in southern Kanto for middle Pleistocene-Holocene (34). Knowledge of deep-sea tephras around Japan is not yet very abundant (e.g. 35,36). Recently, in Japan Sea cores several land-connected markers, the Akahoya ash, Aira-Tn ash and Aso-4 ash, were identified over vast areas (37). All are significant time planes within deep-sea sediments.

Studies classified as type (b) can be further divided into those which deal with the primary nature of tephras as the main objective, and those dealing with the secondary nature of tephras which were added after accumulation. The former are volcanological studies concerned with growth history of volcanoes and such problems as the behavior and eruption types of magma. Several volcanoes were studied from the viewpoint of tephrochronology: Asama (38), Izu-Oshima (39), Fuji (40), Mashu, Shikotsu (41,7), Sakurajima (42). Zoning of magma at the time of eruption was discussed for the KmP-1 tephra of Hakone Volcano on the basis of petrographic and chemical variations of the tephra column (22). The latter are pedological studies on the origin and age of volcanic ash soils, and the genesis of clay minerals (e.g. 9). In recent years, deformation of tephra layers, e.g. involution caused by periglacial process, has been studied in northern Japan, and environmental changes between the time of tephra accumulation and periods thereafter are being discussed (43). Such features can be observed at two or three specified horizons of the tephra sequence in central-northern Japan relating to the last glaciation.

4. GROWTH HISTORY OF VOLCANOES IN RELATION TO TEPHRAS

This section examines the relationship between tephras and the development of volcanoes by discussing Fuji, Hakone, and Aira-Sakurajima.

4.1 Fuji Volcano

Mt. Fuji is the largest single stratovolcano in Japan, the major portion of which consists of alternating strata or basaltic lavas and coarse-grained pyroclastic materials. At the eastern foot of the volcano airfall tephra layers with maximum thicknesses of more than 110 m are deposited. Each bed of scoriaceous tephra has a thickness of less than several decimetres, and almost 1,000 sheets are layered without being interrupted by thick buried soil or any significant surfaces of unconformity, implying frequent eruptions over long periods. The beginning of those eruptions can be traced to about 80,000 years ago, from the fact that the silicic pumiceous tephra, called the Ontake Pm-I, ejected from the Ontake Volcano, with the fission-track age of about 80,000 years (Table 2), is found in the lowest portion of the Fuji tephra layers. Other dated marker tephras, such as the Obaradai Pumice and the Tokyo Pumice (from Hakone), are found in the lower to middle tephra layers. From about 70,000 years until 10,000 years ago, Mt. Fuji repeatedly erupted at average intervals of some 100 years, and grew to a large stratovolcano. About 10,000 years ago, activity changed to the effusion of enormous amounts of lava, with little tephra. Activities then ceased temporarily during 3,000 to 4,000 years. About 5,000 years ago, a mixed type of eruption started again, forming the present summit and a number of parasitic cones (40,44). The total volume of tephra ejected during the last 80,000 years amounts to about 250 km^3, but the volume of tephra ejected in any one cycle of eruption was comparatively small, not exceeding 1 km^3.

4.2 Hakone Volcano

The growth history of Hakone Volcano was studied by Kuno (45), in terms of lithostratigraphy and the petrologic nature the volcanic products. This volcano has a complicated history, i.e. first, a large stratovolcano ("Older Somma") consisting of basalt and andesite formed, followed by a caldera depression of about 10 km in diameter. Inside the caldera the "Younger Somma" was formed by felsic andesite-dacite magma. Later, another new caldera was produced by the eruption of huge amounts of tephra with pyroclastic flow deposits, inside which a group of andesitic central cones erupted.

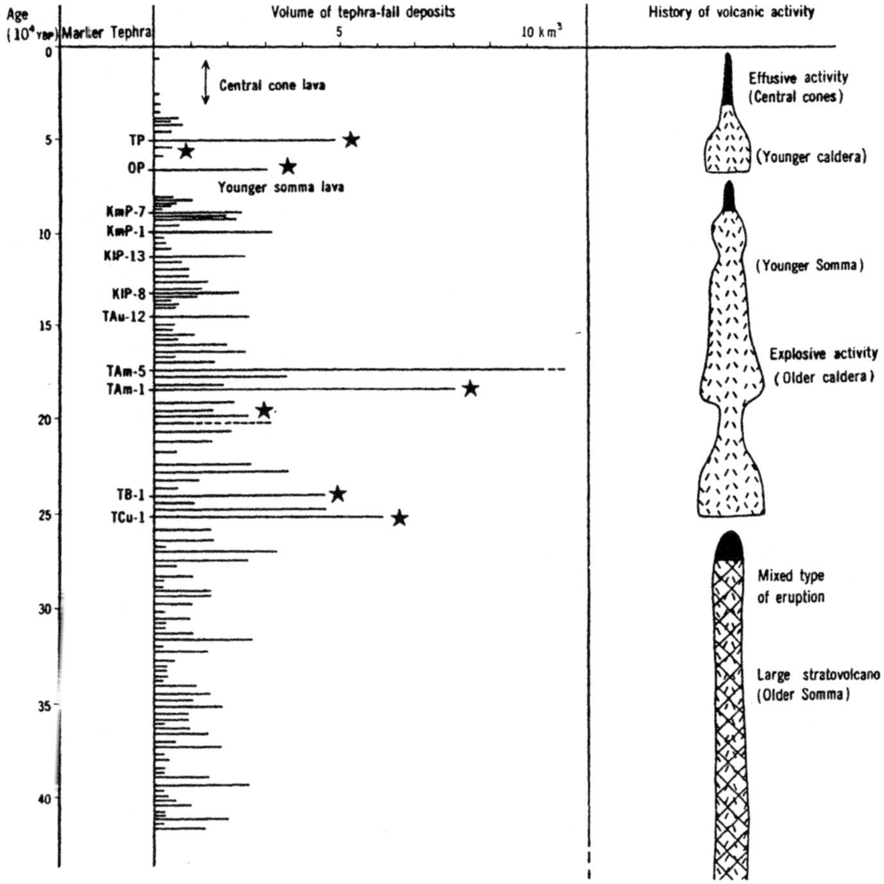

Fig. 2. Eruptive history of the Hakone Volcano in relation to tephras. An asterisk shows eruptions also producing tephra-flow deposit. The bulk volume of tephra in these eruptions amounts to several times that of eruptions producing only tephra-fall deposits.

Numerous Pleistocene tephra layers of andesite-dacite pumice are found mantling wide areas of southern Kanto, chiefly below the Fuji tephra group. From their distribution and lithofacies most of them were assigned to the eruption of Hakone. Among these tephras, thirteen layers have been dated by the fission-track method (26,34). Figure 2 shows the eruptive history of the volcano in terms of changes in tephra volume. The age scale of 300,000 to 450,000 years has been inferred from the stratigraphic sequence in which the Sakura ash, an important widespread marker of about 400,000-450,000 years BP, is interspersed with the lowest part of the Hakone tephra group.

The growth history of the volcano is shown diagrammatically on the right-hand column of Figure 2. The stratigraphic relation between cone-forming material and tephra during the past 250,000 years (after the latest activity of the Older Somma), can be directly confirmed in the field. The activity of the Older Somma stage is inferred from the sequence of the tephra itself.

The eruptive phase of Older Somma, in which frequent but comparatively less explosive activity occurred, started about 400,000-450,000 years ago and continued until immediately before the voluminous pumice eruption, TCu-1, about 250,000 years ago. After that, there was a period of voluminous tephra eruptions consisting of two groups, older and younger. The older period of activity may be further subdivided into two sub-stages, both of which fall into the destructive period of the older stratovolcano. The activity ejecting the younger tephra layers from OP to TP (Figure 2) caused the latest caldera collapse.

Immediately before the eruption of each large tephra, such as TCu-1, TAm-1 and OP, there was a quiescent period that ranged perhaps from several thousand years to ten thousand years. At the initial phase of each quiescent period, enormous amounts of lava were effused forming part of the Older Somma, as well as the Younger Somma and the Central Cones.

Formerly it was argued that caldera collapse occurred in the last stage of activity of a stratovolcano. However, as seen from the Hakone tephra sequence (Figure 2), it can be argued that after a period of quiescence, vigorous explosive activity ejected an enormous volume of felsic tephra and formed a caldera. This introduced a new stage of activity into the history of volcanism at Hakone, suggesting repeated extrusions of new magma.

4.3 Aira caldera and Sakurajima Volcano

Kagoshima Bay, located in southern Kyushu, was formed as a large volcanic rift valley. In this rift valley, large explosions have frequently occurred since the beginning of the Quaternary, with every explosion the ground subsided to form depressions. Aramaki and Ui (12) discriminated between and described at least thirty units of voluminous tephra deposits by electron microprobe analysis.

The latest eruptions are represented by the Ito pyroclastic flow deposits of about 21,000-22,000 years ago. The source is assigned to Aira Caldera, located at the head of the Kagoshima Bay (Figure 1). The eruption started with ejection of a great amount of airfall pumice (more than 20 km^3 in volume), followed by the Tsumaya pyroclastic-flow deposit (less than 6 km^3) and finally the large-scale Ito pyroclastic-flow deposit (ca. 150 km^3)(46). The widespread marker-tephra, Aira-Tn ash is now confirmed to be an airfall equivalent of the Ito pyroclastic-flows (24).

This big eruption should have been responsible for forming the Aira Caldera as it is today with a diameter of 20 km, although a smaller caldera is known to have existed in the western part of the present caldera at the beginning of late Pleistocene some 100,000 years ago. The resumption of activity inside the caldera occurred about 14,000 years ago, resulting in the formation of the present volcano, Sakurajima, and other volcanic cones. Sakurajima Volcano is an active stratovolcano which has ejected more than thirteen tephra units and lavas. The latest large scale activity was in 1914 A.D. and activity still continues at present but on a smaller scale.

Figure 3 shows the stratigraphy and chronology of marker-tephras from Kyushu to Kanto. This enables us to interpret the eruptive history of principal volcanoes. Among the stratovolcanoes of the Mt. Fugi type, such active volcanoes as Sakurajima, Kirishima, and the central cones of Aso and Asama, have grown only in the last several tens of thousands of years of less. The larger the size of the cone the longer the period of activity and the greater the number of tephra sheets. The eruptions were not generally highly explosive but occurred at frequent intervals, so that the tephras of any one cycle of eruption have a comparatively small volume. For composite volcanoes with a caldera (Mt. Hakone type), the larger the size, the longer the period of activity, and in many cases these date back to the middle Pleistocene, as in the cases of Daisen, Akagi, Yatsugatake and Ontake. However, they have not been very active in the Holocene. The widespread marker-tephras in Japan were formed by great explosions usually resulting in caldera collapse. Moreover, the large caldera volcanoes of the Aira type have a history of a number of repeated large-scale eruptions after the middle Pleistocene, such as in the cases of Kikai, Ata, Aira, Kakuto and Aso.

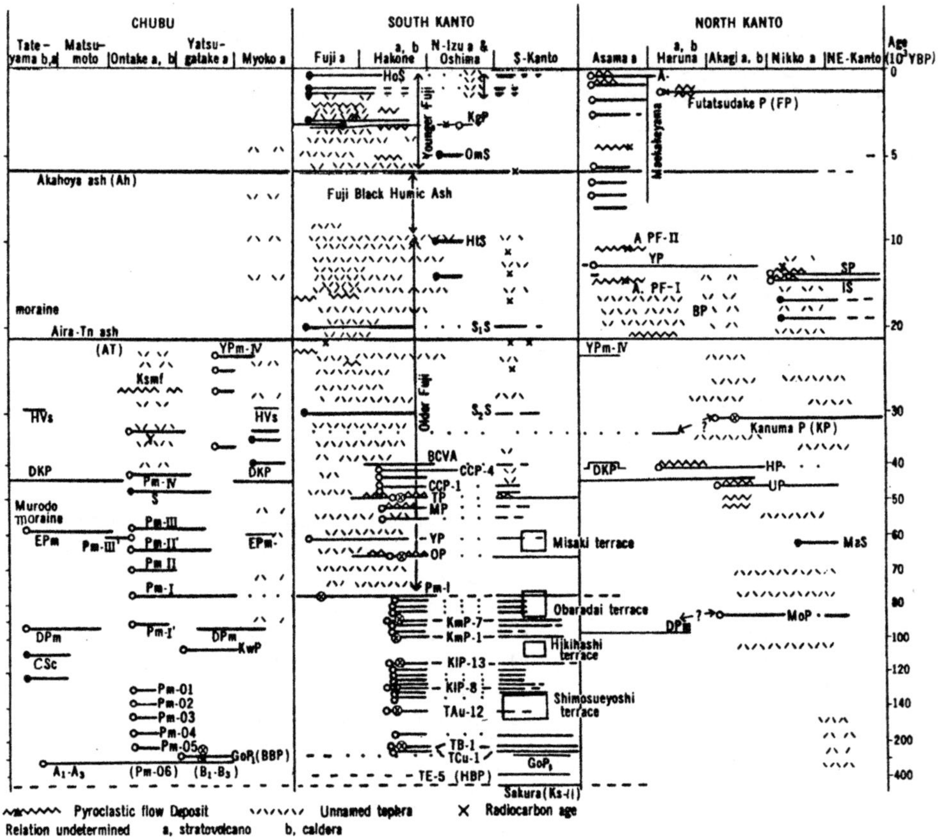

Figure 3. Time-space diagram of late-middle Quaternary marker tephras from Kyushu to Honshu.

5. ARCHAEOLOGICAL APPLICATIONS OF TEPHROCHRONOLOGY

Many eruptions of tephra in the Holocene have long been dated with the aid of stratigraphic relations to human remains, whose chronology was established by typological studies on artifacts. Since the beginning of the 1970's, recognition of the widespread marker-tephras in archaeological sites, especially the Aira-Tn and Akahoya ashes, has compelled us to revise some former views on archaeological chronology, both in the neolithic and paleolithic ages. In the Kanto district, a number of archaeological assemblages were recovered form the volcanic ash soil below the Aira-Tn ash (21,000-22,000 YBP), with an approximate age of 30,000-22,000 YBP. Diversity in technological traditions and rapid cultural change, however, appear to begin immediately after the eruption of the Aira-Tn ash, about 20,000 YBP. The Neolithic Jomon ceramic culture began about 10,000 YBP, during

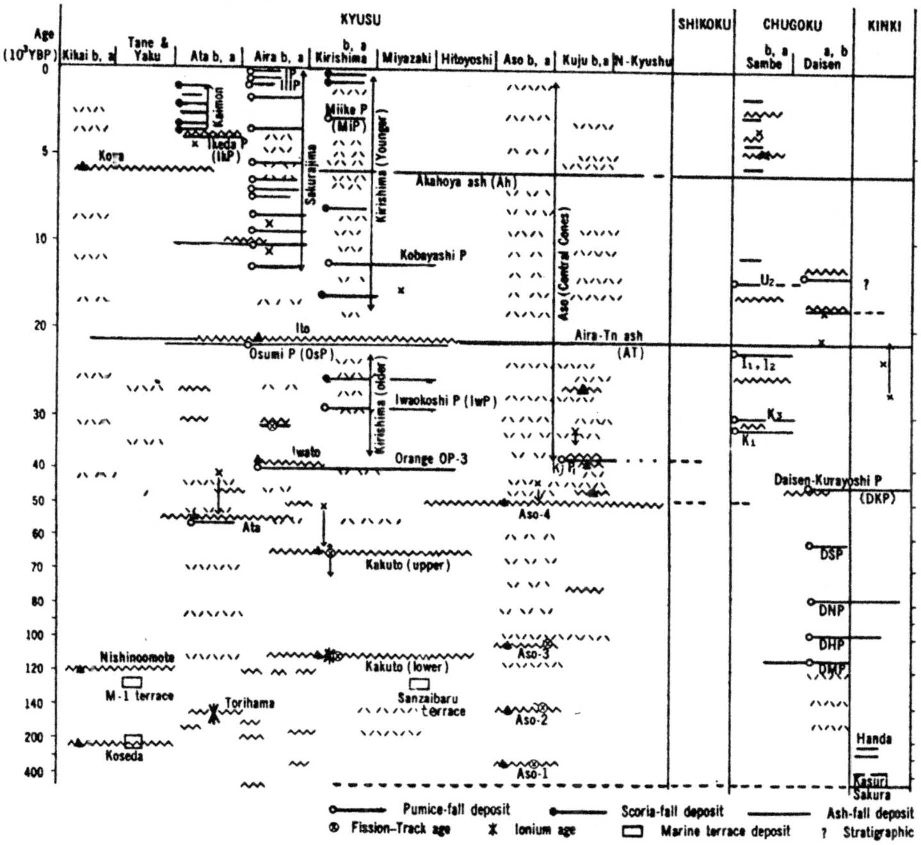

which the Akahoya ash (6,300 YBP), provides the definite datum plane in the archaeological sequence.

As shown in Figure 3, many other Holocene tephras are found in archaeological sites, especially in areas adjacent to active volcanoes and are useful time-markers in correlating artifacts from place to place. This is especially so in the case of tephra eruptions with approximate dates of 1,000-2,000 YBP which have not yet been definitely found mentioned in historical documents, such as the Towada-a eruption and the Haruna-Futatsu-dake eruption. The relationship between the tephra sheet and human remains is very significant in determining its age and also for checking the date obtained by other methods.

One of the most important problems in Japanese archaeology is the age and process of the initiation of human occupation. While the Japanese islands were repeatedly connected to the continent during most of the glacial ages, and human occupation of Japan in the last 30,000 years is well documented, the evidence for earlier occupation of the islands, suggested by some 100 assemblages, is still the subject of controversy. Of these, a few assemblages were recovered by controlled excavation from the tephra formations whose chronology was well established. At the same time, for most of these assemblages, lithic specimens themselves are yet to be fully analysed and described.

6. SEA LEVEL CHANGES RECOGNISED BY TEPHRA STUDIES

6.1 Holocene sea level

Researchers reporting sea level changes in the Holocene of Japan have been obtaining heights of paleosea-levels primarily from beach angles of marine terrace and fossiliferous intertidal deposits, and obtaining radiocarbon ages from organic materials contained in the deposits (e.g. 47,48). Studies on sea level changes using dated tephras of Holocene age as time-markers had not been implemented despite the advantages in tracing the datum plane over extensive regions.

With the recent discovery of the Akahoya ash, which erupted during the hypsithermal age and covered an extensive area from southwestern to central Japan, studies have started on sea level changes in the Holocene using this tephra (10). The Akahoya ash is found in alluvium at various localities in southwestern and central Japan. Investigations are being conducted in these localities to see if the ash accumulated during the culmination of the transgression or immediately before it. In such areas where tectonic uplift is considerable (more than 3 mm/year), the Akahoya ash occurred after the emergence of the coast, whilst on the stable or subsiding coasts, the ash was usually found within marine sediments with an elevation of +2m to 40m. In Japan, Holocene sea level data are obtained more often from uplifting coasts than from stable and subsiding coasts. It is believed that the use of the Akahoya ash may provide more data on stable and subsiding coasts and, accordingly, elucidate the regional differences in Holocene crustal movements.

A number of Holocene tephras other than the Akahoya ash are distributed on the east coast of southern Kyushu, on the northeastern part of Tohoku and in the Ishikari lowland in Hokkaido. Further sutdies on sea level changes using tephras may be fruitful.

6.2 Sea levels in the middle-late Pleistocene

Generally, areas in which it is favorable to compile a history of sea level changes are those where uplifting coral reefs are developing, such as the Barbados (49) or New Guinea (50), because in such areas the present elevation of paleosea-levels and their radiometric ages can be determined. In Japan, a detailed study (51) was made of Kikai-shima Island, Ryukyu Island Group, a coral island where uplift is active.

In middle to high-latitude regions where coral reefs do not develop, the height of paleosea-levels can be measured but ages cannot be determined. In Japan, however, there are some localities where one can use tephras as time-markers in the Pleistocene sequences. In such localities marker tephras cover marine terraces or are interspersed within marine deposits, as around Kagoshima Bay in southern Kanto and on the coast in the northeastern part of the Tohoku district. Studies are in progress at those localities and the author will now discuss results of a compilation of sea level changes in southern Kanto by tephrochronology, mainly from about 450,000 years ago to the present (34).

In the Kanto Plain, the largest in Japan, Quaternary marine and fluviatile deposits and terraces are well developed. Furthermore, a large number of tephras from a group of volcanoes situated in the western and northern part of the plain, cover those terraces or are interbedded with the strata. Marine terraces and deposits are distributed in southern Kanto, including Tokyo and Yokohama (Figure 4). Also, tephra formations mainly from Fuji and Hakone volcanoes are considered to be important in tracing the geomorphological development and history of sea level changes.

In southern Kanto, the rate of tectonic movement differs with locality. Therefore, a danger of error exists if correlation of marine terraces is made by relying solely upon their elevation or the degree of dissection. Hence, correlation by means of tephra is regarded as indispensable. For example, because of tephra layers it has been recognised that the Shimosueyoshi terrace, which is widespread throughout this locality, formed in the last interglacial age at the summit level of the Ooiso Hills (100-200 m), where strong uplift was undergone. The same terrace is also found at a lower elevation (5-10 metres above sea level) in a locality which was scarcely uplifted, and is almost at the level of Holocene deltas.

Also in this district, a distinction can be made between two zones, with Tokyo Bay as the border: one is on the west side of the Bay, i.e. Yokohama, the Miura Peninsula and Ooiso Hills, where the surface is tending to uplift and where several marine terraces and older shorelines are cut: the other is on the east side of the Bay which is the northern part of the Boso Peninsula. Here the surface was a seabed during most of the middle Pleistocene and was later uplifted, resulting in a well developed sequence of marine sediments. In the former zone, a number of traces of paleosea-levels can be distinctly observed, whilst in the latter zone temperature and depths of water for various ages can be determined from faunal and floral fossils and other data in the strata. The correlation of a number of

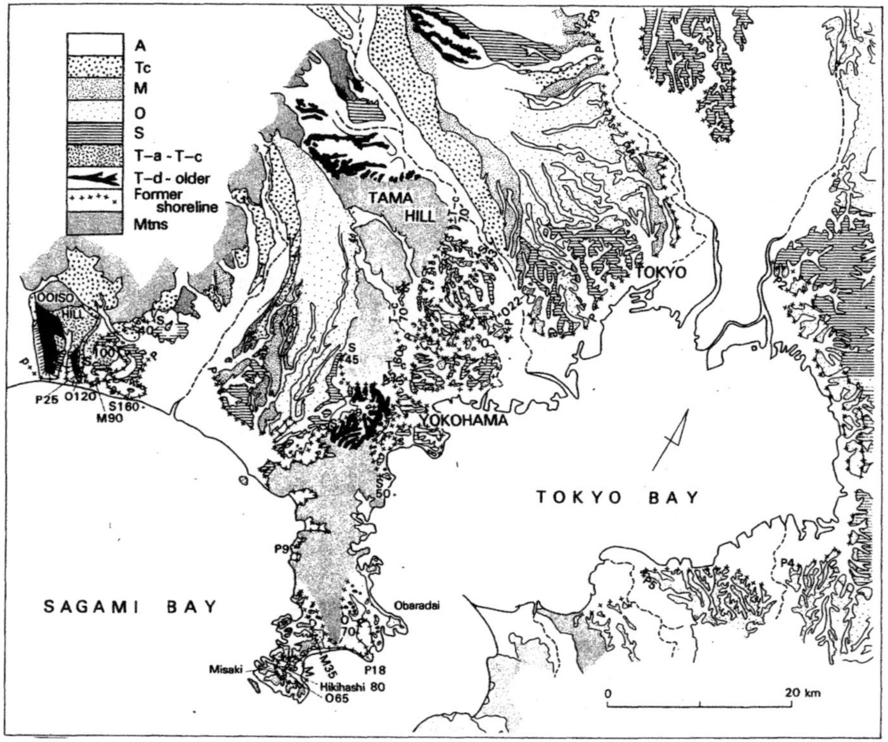

Fig. 4. Terraces of southern Kanto. A, Alluvial plain; Tc, Tachikawa terrace; M, Misaki terrace; O, Obaradai terrace; S, Shimosueyoshi terrace; T-a∿T-c, Tama-a to Tama-c terraces; T-d∿older, Tama-d and other older terraces. The heights of former shorelines are shown in metres above sea level. P, Postglacial shoreline.

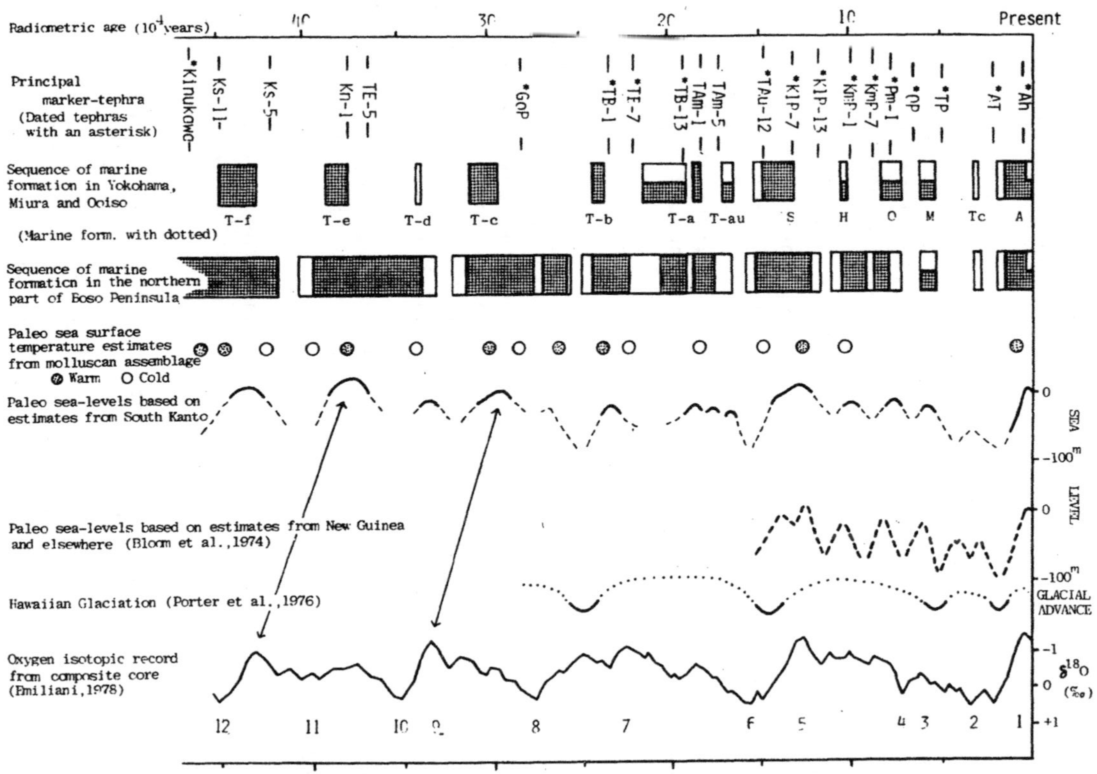

Fig. 5. The marine sequence in southern Kanto showing interstratified tephra used for analysis and induced sea level changes (see text for discussion).

marker-tephra formations distributed in southern Kanto is possible. Detailed petrographic features were described for about 120 tephras representing a period of 400,000 years. The accurate geomorphic development, sea level changes and history of crustal movement in this region can be clarified (Figure 5), (52,22,53).

The major results of such research can be summarised as follows:

1) Despite the different rate of uplift with different localities, the times of emergence and submergence of each marine sediment layer were the same for every locality in the area west of Tokyo Bay. Namely, each marine layer constitutes its own terrace surface and the surfaces are covered with air-fall tephra which becomes a time-marker. Also, within the strata of transgressive periods, there exist marker-tephras which can be traced from place to place, endorsing the fact that the transgression had been progressing at almost the same time. These facts suggest that sea level changes were repeated periodically at a rate greater than that of tectonic uplift, while considerable uplift progressed at a constant rate peculiar to each locality. Eleven culmination periods of high sea level can be distinguished during the past 450,000 years as shown in Figure 5.

2) In the area west of Tokyo Bay, each period for which evidence of high sea level is found also corresponds to a period of higher water temperature. Namely, some marker-tephras found within, or immediately above, the marine terrace deposits in the uprising area are those falling within the horizon yielding molluscan assemblages, which indicate higher water temperature in the subsiding area (Figure 5). Also it has been discovered that tephras accumulating on the terraces during the period of regression fell within the fossil bed indicating lower water temperature in the latter area (53).

(3) In order to mutually evaluate the magnitude of repeated high sea levels during the past 450,000 years, it is necessary to obtain as much information as possible and to synthesise it. First, for the section in the northern part of Yokohama where marine terraces are typically developed Machida (34, pp. 219-220), determined the elevations of older sea levels at each of the higher sea level periods by applying the average uplift rate of 0.3 mm/year from the difference between the present elevation and the estimated elevation of the original paleosea level at Shimosueyoshi terrace (130,000 years ago). This was assumed to be +7m, and that of the Holocene transgressive peak (about 6,000 years ago) was taken as +2m. The results show that they were highest in the Shimosueyoshi peak (34, see S on Figure 5).

There is a problem, however, if an assumption of constant uplift rate is applied to the middle Pleistocene by extrapolation. Accordingly, the second criteria for evaluation, information on paleotemperatures estimated from the faunal and/or floral assemblages on the eastern coast of Tokyo Bay, should be taken into consideration. Third, it is necessary to mutually compare the distribution of marine terraces and marine deposits for each period. The sea level curve in Figure 5 is a result of integration of those data and information. Four major higher sea levels can be recognised for the last 400,000 years, each indicating a full interglacial period, i.e. Tama-e (T-e, ca. 370,000 years ago); Tama-c (T-c, ca. 300,000 years ago); Shimosueyoshi (S, ca. 130,000 years ago) and Holocene (from ca. 7,000 years ago to the present). Consequently, it is also considered that Tama-b (T-b, ca. 240,000-230,000 years ago), Tama-a (T-a, ca. 190,000-180,000 years ago) and Obaradai (O, ca. 80,000 years ago) were comparatively higher sea level periods.

4) Determination of the ages of lower sea levels, indicating full glacial periods, and evaluation of their magnitude is more difficult than those of higher sea levels. However, on the eastern coast of Tokyo Bay (northern part of the Boso Peninsula), problems regarding low sea level periods can be approached by investigating terrestrial sediments and buried valley bottoms. In addition, comparison of fossil assemblages between periods and investigation of the strata yielding large continental vertebrates (in particular, Palaeoloxodon naumanni), indicate that a land bridge existed between the continent and Japan (53). It is considered that among several periods of maximum lowering of sea levels, those of about 260,000-240,000 years ago (immediately before higher sea level period, Tama-b) and of about 150,000 years ago (immediately before the Shimosueyoshi high sea level period), are the lower sea level periods (glacial ages) on the scale. They are equal to, or greater than, the latest glacial age with its peak about 20,000 years ago (Figure 5).

5) Widespread marker-tephras also play an important role in compiling the chronology of sea level changes before about 450,000 years ago. It has recently become known that the vitric tephra called Kasamori 11 (Ks-11), which is sandwiched in the Kasamori marine formation below the lower part of the Tama-e formation below the lower part of the Tama-e formation in the northern Boso Peninsula and other localities in southern Kanto, can be identified as a marker-tephra of the Osaka Group called the Sakura ash. This ash can be recognised in the vicinity of Osaka (14), resulting in the establishment of an important datum plane in the middle Pleistocene (23). Because the nature of this ash resembles the abovementioned Aira-Tn ash, it was thought that it was of the co-ignimbrite airfall type, which could have originated from one of the large calderas in Kyushu. The

estimated age of this tephra, about 450,000 years, is based on its stratigraphic relationship to the underlying dated tephra, Kinukawa ash (29), and with the overlying Tama-e formation. Stratigraphically, it is included in the deposits of the Byobugaura transgression (Tama-f) in southern Kanto, and within the deposits immediately below the transgressive peak Ma 7 of the Osaka Group sequence in Kinki. During the period from the Brunhes-Matuyama boundary to the horizon of this tephra at least three interglacial-glacial cycles were recorded in both the Osaka Group and the Kazusa Group in southern Kanto.

6) The sea level curve obtained from tephrochronology of marine terraces in wouthern Kanto can be compared with other curves derived from various parts of the world. On Figure 5, the sea level curve in New Guinea and other places (50), and the variation curve of oxygen isotopic ratios obtained from foraminifera in deep sea cores from various places (54), and the dated glacial advances in Hawaii (55) are also shown. All these curves resemble one another for the last 150,000 years. For the period older than 150,000 years, the resemblance is also remarkable between the sea level curve of southern Kanto and the variation curve of oxygen isotopic ratios. Principal high sea level periods and the numbered oxygen isotopic stages could be correlated as follows: T-e to the peak at stage 11; T-c to the peak at stage 9; T-b and T-a to the peaks among stage 7; S to stage 5; and P to stage 1.

Recently there has been increasing discussion on the cause of climatic changes in the Quaternary, relating to seasonal and latitudinal distribution of solar radiation, due to variation in relation to the earth's orbit (e.g. 56,54). One of the problems in testing this theory is the uncertainty in geological chronology. For the past 150,000 years, however, considerably more accurate ages have been determined and it can be said that a chronology has almost been established. For older ages, the chronology is still insufficient. Even for the oxygen isotopic ratio curve by Emiliani (54), there were only three fixed points, i.e. the present, 125,000 years and 700,000 years. The curve was drawn by interpolation between these fixed points. Under such circumstances, the tephrochronology of southern Kanto, where several dates were determined between 140,000 and 450,000 years, has great significance. However, radiometric dating of marker-tephras between 300,000 and 400,000 years, which is missing, must be supplied and important marker-tephras should be dated by various methods for crosschecking.

7) Although the assumption that uplift was proceeding at an equal rate in each locality of southern Kanto is meaningful as a first approach, further detailed examination is required

for more quantitative discussion, (e.g. for the calculation of original elevation of past sea levels). Machida (57) compared the degree of tilt of various groups of marine terrace surfaces, for which ages were determined by tephrochronology, and noted that they had been mainly tilted at a constant rate. However, in the coastal areas of the southern part of the Boso Peninsula, the Miura Peninsula and the Ooiso Hills (which were not dealt with in the above discussion), the tendency of recent accelerating uplift is clearly observed. On the Ooiso coast, for instance, the mean uplift rate in the recent 6,000 years was about 3mm/year, while that in the past 60,000 years was about 1.2 mm/year, and that in the last 130,000 years was about 1.2-1.5 mm/year.

Such rapid uplifting and tilting in the Holocene is not noted in the middle part of the Kanto Plain, so that this movement seems to be confined to the abovementioned coastal areas; the coasts are bounded by the seabed, which becomes suddenly deeper close to the plate boundary called the Sagami Trough. In order to explain this, it is necessary to conduct tectonic studies and to consider hydro-isostatic influences accompanying the rise of sea level in the Holocene.

7. MOUNTAIN GLACIATION AND RELATED PHENOMENA STUDIED BY TEPHRA

It goes without saying that sea level changes in the Quaternary were associated with the advance and retreat of glaciers. During the glacial ages in Japan small mountain glaciers formed. So far hardly any detailed studies on geomorphologic change during glaciation and the influence of glaciers on rivers have been made. As suggested by Kobayashi, et al. (58), it is possible, and successful, to compile a history of glacial fluctuation in terms of marker-tephras in some localities in Japan (Figure 6). In the next section, recent results of tephrochronology of the glaciers at Mt. Tateyama in the northern Japan Alps are described.

7.1 Glacial chronology of Mt. Tateyama

Adjoining glaciated peaks (e.g. Mt. Tateyama, 3,015 m) consisting of granite and gneiss is Tateyama Volcano, which was active during the last 150,000-200,000 years. After forming a stratovolcano (first stage of activity) Tateyama ejected a great deal of tephra and formed a pyroclastic plateau and a caldera (second stage). The volcano further extruded tephra and lavas (third stage) (59). Fukai (60) investigated the glacial chronology of this region and the relationship with volcanic activity. He showed that the glacial till called the Murodo gravel layer, deposited at the maximum glacial extent, occurs above the pyroclastic flow deposits of the second stage as well as the airfall tephra of the third stage, and is covered by one of the lava flows of the third stage.

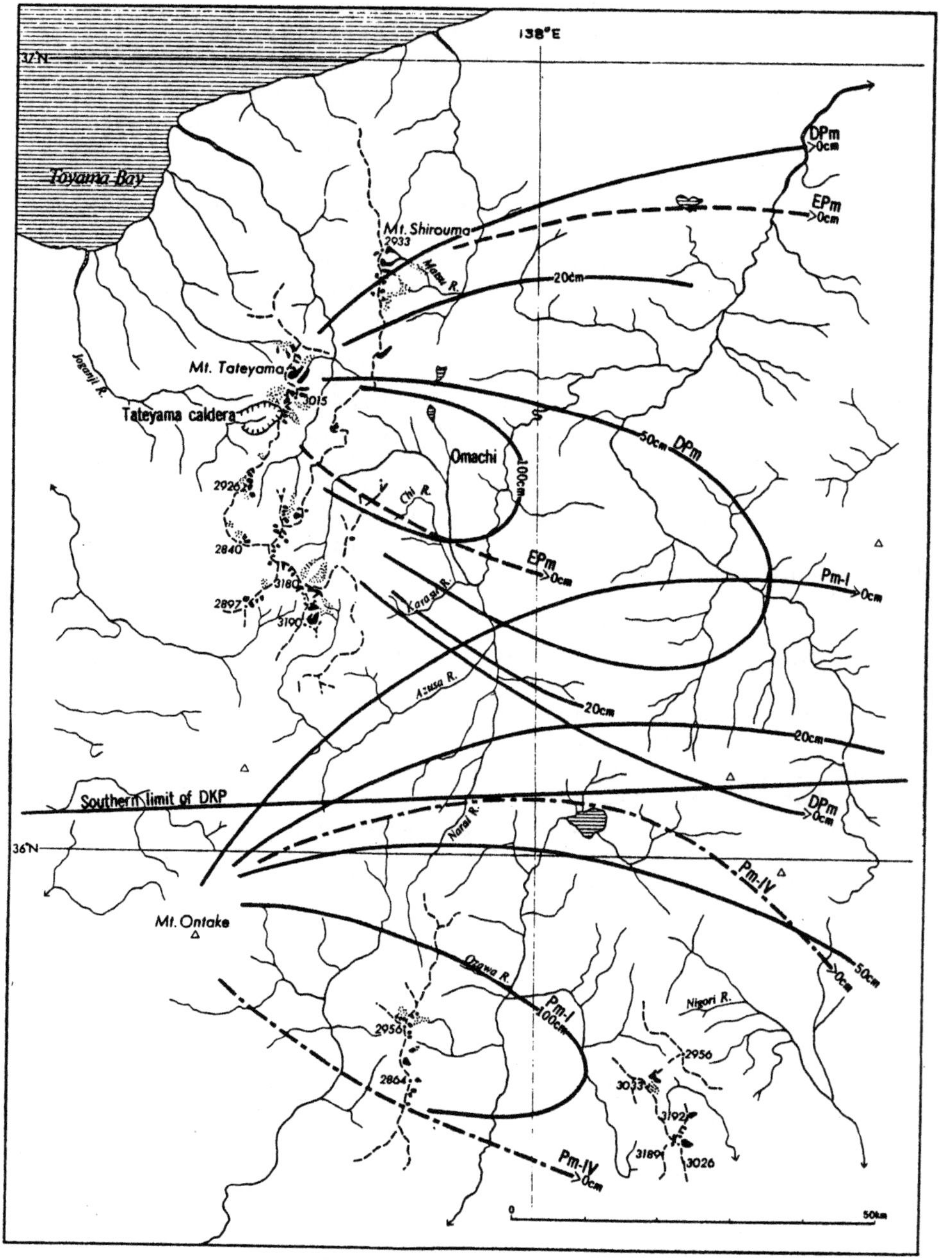

Fig. 6. Glaciated areas in the Japan Alps and distribution of principal marker-tephras. Glaciated areas in the late Wurmian age are indicated by solid decoration and those in the middle or early Wurmian age by dots.

Machida and Arai (32) studied six marker-tephras of Pleistocene age around the northern Japan Alps, all of which are important for the establishment of a glacial chronology at Mt. Tateyama. The stratigraphic relationship between glacial deposits and tephras is shown on Figure 7. The pyroclastic flow deposits of the second stage are correlated on the basis of petrography to Omachi DPm (61), which mantles the wide area to the east of the

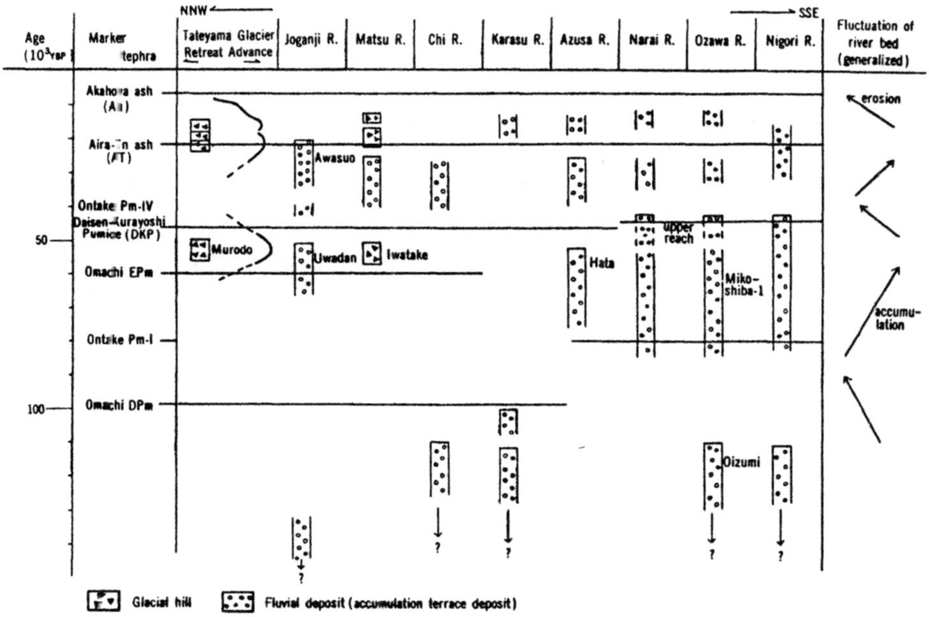

Fig. 7. Stratigraphic relationship between moraine deposits, river terrace deposits and tephras around the Japan Alps. Rivers (R) marked with an asterisk indicate that the summits of source mountains were glaciated. For location of rivers, see Figure 6.

northern Alps occurring immediately below the marker-tephra, Ontake Pm-I (80,000 years old). The tephra-fall of the third stage is assigned to Omachi EPm (61), with an approximate age, estimated from the stratigraphic position between radiometrically dated tephras, around 60,000 years. Furthermore, the lava flow of the third stage covering the Murodo glacial till is stratigraphically below Daisen Kurayoshi pumice (DKP), one of the widespread tephras from Daisen Volcano. Although DKP has not yet been accurately dated by radiometric methods, its stratigraphic relation with other dated tephras in northern Kanto confirms its age within a limited range, approximately 45,000-

47,000 years (Figure 3). Accordingly, the age of Murodo glacial advance is estimated to have been occurred between 47,000-60,000 years ago, presumably around 50,000-55,000 years. The terminal moraines of the maximum advance, though partly eroded, were situated at an elevation not higher than 2,250 m.

The younger group of moraines and cirques situated on slopes higher than 2,400 m represent the Hida glacial stage proposed by Fukai (60). It can only be said that the age of the glaciation is about 20,000 years or slightly later, as an excellent marker tephra such as Aira-Tn ash (21,000-22,000 years) cannot be found above the younger moraines.

7.2 Accumulation terraces around the Japan Alps

It has so far been considered that river terraces in Japan were formed mainly by tectonic influences. Recently, however, as the result of tephrochronological studies and observations of river terrace deposits, it was discovered that there were cases where rivers experienced cyclic cut and fill in the upper and middle reaches, with no influence of sea level change. The periods of cut and fill of various rivers coincide fairly well with each other, suggesting a climatic influence on the regime of rivers (Figure 7).

In the upper course of the Jōganji River, the source of which is the glaciated Mt. Tateyama, outwash deposits did not fully develop due to the existence of a gorge. On the other hand, along the river lower than an elevation of 600 m, several terraces are found. They are classified as the Uwadan and Awasuno terraces, both of which are fill-top terraces. The Uwadan terraces were formed in the valley which cut into pyroclastic flow deposits of the second stage of the Tateyama Volcano and are covered with DKP airfall tephra. Accordingly, the time of terracing is assumed to have occurred about 50,000-60,000 years ago and the culmination of its accumulation is approximately at the period of glacial advance represented by the Murodo glacial deposit. The Awasuno terrace, a remarkable deposit that attains a thickness of 140 m, was formed when downcutting of the river started about 20,000 years ago. This is interpreted from the fact that the marker tephra, Aira-Tn ash, is interbedded within the uppermost part of the terrace deposits, and can also be correlated with the peak of the abovementioned Hida glacial advance.

Taking advantage of the modes of occurrence of the tephra layers as key beds for correlation, valley filling around the Japan Alps might perhaps have taken place simultaneously with the glaciations (Figure 7), as in the case of the Joganji River. The accumulation period of the rivers prior to the stage of

Uwadan terrace formation is slightly older than the marker-tephras, Omachi DPm or Ontake Pm-I, which can be recognised in various localities in central Japan (Figure 7). If a rough estimate is made, those older fill-top terraces could fall either within the low sea level period that is considered to have occurred between 125,000 years and 105,000 years (62), or the period about 150,000 years (probably the last-but-one, large glacial period).

The primary causes of active valley filling in those periods differ from the case of higher latitude regions where glaciers developed on a large scale. Glaciers in Japan were small, therefore the influence of outwash is considered to have also been rather small. On the other hand, in the drainage basin shown in Figure 7, one half to two thirds of the area became periglacial, so that the increase in supply of detritus under periglacial conditions is one of the causes of valley filling. In addition, decrease in precipitation and/or discharge in the glacial age is a factor which cannot be ignored. Analysis of those causes is a subject for future studies.

7.3 Problems concerning glacial chronology of the late Pleistocene

In the last full glaciation around 20,000 years ago, glaciers advanced in various parts of the world and the sea was lowered to its maximum level. However, with respect to the ages and magnitudes of the preceding glaciations and sea level changes, it is hardly possible with the data available to arrive at any definite conclusions.

In the region of the eastern margin of the Laurentide ice sheet, two glacial substages, Cherrytree and Guildwood, are recognised to have existed during the Wisconsin glacial age which are older than 20,000 years. The former, dated at about 35,000 years, was a relatively small-scale glacial period, while the latter is considered to have been on a large-scale. As for the ages of the latter, Dreimanis and Karrow (63) estimated about 50,000 years, but Dreimanis and Raukas (64) considered that it was older (about 70,000 years). On the South Island of New Zealand, the last glacial advance period (Early Otira Glacial), precedent to the last glacial substage (Late Otira Glacial, about 20,000 years ago), is estimated to have occurred about 50,000-75,000 years ago. During this time range, a high sea level stage (one of the interstadials) around 60,000 years was recorded in Barbados, New Guinea, Kikai and southern Kanto. Accordingly, whether the abovementioned glacial advance period is before or after 60,000 years constitutes an important chronological problem. The record of deep-sea cores has not provided significant information on this problem. In the generalised curve of marine oxygen isotope variations (Figure 5), stage four should fall between 70,000 and 75,000 years. This data has often been used in discussing the chronological problems of the late Pleistocene, providing the basis for the discussion of the age of the beginning of the Wisconsin glaciation (65). However, this has not been estimated by accurate radiometric dating but by assuming a constant deposition rate on deep-sea floor.

A glacial advance which could be correlated with the period represented by the Murodo glacial deposit in the Japan Alps is the Mauna Kea Volcano in Hawaii (55). Here, the age of the last but one advance, the Early Makanaka Glacial, was determined to be about 55,000 years based on K-Ar dating on a lava from a vent beneath the ice cap. Both of these two cases were similar in that the scale of glacial advance that occurred about 50,000-55,000 years ago was by no means smaller than the last event which happened about 20,000 years ago.

According to information from tephra studies in south Kanto, sea level around 50,000-55,000 years ago should have been remarkably low. One of the marker-tephras from Hakone Volcano, the Tokyo pumice, dated at about 50,000 years by fission-track method, directly covers peaty layers with cold pollen assemblages in southwestern Kanto. River terraces along the Sagami and Tama Rivers in southern Kanto, directly covered by this tephra, are considered to have been formed at a low sea level period because of their steep profiles. In addition, in the southern part of the Miura Peninsula, where uplift is rapid, airfall terrace Tokyo pumice covers the scarp of the Misaki marine terrace to an elevation at least 5 m above sea level. This has an age of about 60,000 years (the present elevation of the paleo-shoreline is 35 m. Since the average uplift rate in this locality during the past 60,000 years is estimated to be about 1mm/year, sea level at the time of eruption of the Tokyo pumice must have been at a lower position than -55 m. Thus, evidence is obtained that there was a remarkably low sea level period at the time of the post-Misaki terrace. However, during the interim period between the Obaradai and Misaki high sea levels (80,000-60,000 years), no evidence has been obtained to show substantial lowering of sea level.

On the other hand, the sea level curve obtained from New Guinea (50) is partly different from the information from southern Kanto or from deep-sea cores, which presents a clearly indented shape (Figure 5). Almost all of the interim periods between times of high sea level had remarkably low sea levels.

Problems concerning the age and scale of glacial substages and low sea level periods in the chronology of the late Pleistocene still remain. There is still a possibility of finding glacial substages older than 50,000-55,000 years and later than the long interglacial stage of 130,000-125,000 years. Because the scales are different in various districts, more data will be needed from each area. Nevertheless, it can be said that the data concerning the glacial substage of 50,000-55,000 years obtained from Hawaii and the Japan Alps, suggest a possibility of modifying the peak of stage four of marine oxygen isotope variations to having occurred 50,000-55,000 years ago, rather than 70,000-75,000 years ago.

CONCLUSIONS

This paper has indicated that studies utilising tephras are producing a number of fruitful results in a variety of fields including the development of geomorphic features and a refinement of Quaternary chronology. The eruptive history of several volcanoes in Japan, glacio-eustatic sea level changes, the advance and retreat of mountain glaciers, and the history of erosion of rivers can all be interpreted in detail using tephras. It can also be noted that the latter three are mutually linked in deep causal relationships.

The causal relationship between volcanic activity and climatic changes has been discussed by a number of researchers, but no conclusion has been arrived at. Certainly, a large-scale eruption is effective in producing aerosols in the atmosphere for a relatively long period of time and thus weakening solar radiation. On the other hand, it is possible that climatogenetic sea level changes may produce isostatic influence over the earth's crust and upper mantle, and hence affect the rise and fall of underground magmatic activities. The question as to whether there is any causal relationship between volcanic activity and climatic or sea level changes cannot be concluded unless data are obtained on a global basis.

REFERENCES

1. Uragami, K., Yamada, S., and Naganuma, Y.: 1933, Bull. Volcanol. Soc. Japan, 1, pp. 44-60.
2. Suzuki, T.: 1887, Tohoku Region Quat. Res. Group: 1969, Quaternary system of Japan, pp. 37-83.
3. Nakamura, K.: 1960, 1961, Sci. Paper Coll. General Education, Univ. Tokyo, 10, pp. 125-145, 11, pp. 281-319.
4. Harada, M.: 1943, Rept. Inst. Soil and Manure, Tokyo Imp. Univ. 3, pp. 1-140.
5. Sugihara, S.: 1954, Sundai Shigaku 4, pp. 1-5.
6. Committee on nomenclature of the pyroclastic deposit in Hokkaido: 1974, Distribution of the late Quaternary pyroclastic deposit in Hokkaido, Japan.

7. Katsui, Y.: 1959, Bull. Volcanol. Soc. Japan, ser. 2, 4, pp. 33-48.
8. Katsui, Y.: 1963, J. Fac. Sci., Hokkaido Univ., ser. 4, 11, pp. 631-650.
9. Kanno, I.: 1961, Bull. Kyushu Agr. Expt. Station 7, pp. 1-185.
10. Machida, H. and Arai, F.: 1978, Quat. Res. Japan 17, pp. 143-163.
11. Aramaki, S. and Ui, T.: 1966a, J. Geol. Soc. Japan 72, pp. 337-349.
12. Aramaki, S. and Ui, T.: 1976, Bull. Earthq. Res. Inst. 51, pp. 151-182.
13. Ono, K.: 1965, J. Geol. Soc. Japan 71, pp. 541-553.
14. Yokoyama, T.: 1969, Mem. Fac. Sci. Kyoto Univ., ser. Geol. Miner. 36, pp. 19-85.
15. Yoshikawa, S.: 1976, J. Geol. Soc. Japan 82, pp. 497-515.
16. Itihara, M., Shimoda, C., Itihara, Y.: 1978, J. Geosci. Osaka City Univ. 21, pp. 27-36.
17. Nakamura, K., Aramaki, S., and Murai, I.: 1963, Quat. Res. Japan 3, pp. 13-30.
18. Kobayashi, K.: 1969, Etude sur le Quat. dans le Monde, pp. 981-984.
19. Kobayashi, K. and Momose, K.: 1969, Etude sur le Quat. dans le Monde, pp. 909-962.
20. Arai, F.: 1972, Quat. Res. Japan 11, pp. 254-270.
21. Yokoyama, T.: 1972, Quat. Res. Japan 11, pp. 247-253.
22. Arai, F., Machida, H. and Sugihara, S.: 1977, Quat. Res. Japan 16, pp. 19-40.
23. Machida, H., Arai, F. and Sugihara, S.: 1980, Quat. Res. Japan 19 (in press).
24. Machida, H. and Arai, F.: 1976, Kagaku 46, pp. 339-347.
25. Nishimura, S. and Sasajima, S.: 1970, Chikyu Kagaku 24, pp. 222-224.
26. Machida, H. and Suzuki, M.: 1971, Kagaku 41, pp. 263-270.
27. Nishimura, S. and Miyachi, M.: 1973, J. Japan Assoc. Min. Petr. Econ. Geol. 68, pp. 225-229.
28. Nishimura, S. and Yokoyama, T.: 1973, Proc. Japan Acad. 49, pp. 615-618.
29. Nishimura, S. and Yokoyama, T.: 1975, Paleolimnology of Lake Biwa and the Japanese Pleistocene 3, pp. 138-142.
30. Kigoshi, K.: 1967, Sci. 156, pp. 932-934.
31. Fukuoka, T. and Kigoshi, K.: 1970, Bull. Volcanol. Soc. Japan 15, pp. 111-119.
32. Machida, H. and Arai, F.: 1979, J. Geogr. Japan 88, pp. 313-330.
33. Itihara, M., Yoshikawa, S., Inoue, K., Hayashi, T., Tateishi, M. and Nakajima, K.: 1975, J. Geosci. Osaka City Univ. 19, pp. 1-29.
34. Machida, H.: 1975, Quaternary Studies, pp. 215-222.

35. Horn, D.R., Delach, M.N., Horn, B.M.: 1969, Geol. Soc. Amer. Bull. 80, pp. 1715-1724.
36. Furuta, T.: 1976, Marine Geol. 20, pp. 229-237.
37. Arai, F. and Machida, H.: 1980, Bull. Volcanol. Soc. Japan ser. 2 (in print).
38. Aramaki, S.: 1963, J. Fac. Sci. Tokyo Univ. sec. 2, 4, pp. 229-443.
39. Nakamura, K.: 1964, Bull. Earthq. Res. Inst. 42, pp. 649-788.
40. Machida, H.: 1964, J. Geogr. Japan 73, pp. 293-308, 337-350.
41. Katsui, Y.: 1955, J. Geol. Soc. Japan 61, pp. 481-495.
42. Fukuyama, H.: 1978, J. Geol. Soc. Japan 84, pp. 309-316.
43. Koaze, H., Sugihara, S., Shimizu, F., Utsunomiya, Y., Iwata, S. and Okazawa, S.: 1974, Sundai Shigaku 35, pp. 1-86.
44. Machida, H.: 1967, Geogr. Rept. Tokyo Metrop. Univ. 2, pp. 11-19.
45. Kuno, H.: 1950, J. Fac. Sci. Univ. Tokyo, sec. 2, 7, pp. 257-279.
46. Aramaki, S. and Ui, T.: 1966b, Bull. Volcanol. 29, pp. 29-47.
47. Fujii, S. and Fuji, N.: 1967, J. Geosci. Osaka City Univ. 10, pp. 43-51.
48. Pirazzoli, P.A.: 1978, Quat. Res. 10, pp. 1-29.
49. Mesollella, K.J., Matthews, R.K., Broecker, W.S. and Thurber, D.L.: 1969, J. Geol. 77, pp. 250-274.
50. Bloom, A.L., Broecker, W.S., Chappell, J.M.A., Matthews, R.K. and Mesolella, K.J.: 1974, Quat. Res. 4, pp. 185-205.
51. Konishi, K., Omura, A. and Nakamichi, O.: 1974, Proc. 2nd Intern. Coral Reef Sympo., pp. 595-613.
52. Machida, H., Arai, F., Murata, A. and Hakamata, K.: 1974, J. Geogr. Japan 83, pp. 302-338.
53. Sugihara, S., Arai, F. and Machida, H.: 1978, J. Geol. Soc. Japan 84, pp. 583-600.
54. Emiliani, C.: 1978, Earth Planet. Sci. Lett. 37, pp. 349-352.
55. Porter, S., Stuiver, M. and Yang, I.C.: 1977, Sci. 195, pp. 61-63.
56. Hays, J.D., Imbrie, J. and Shackleton, N.J.: 1976, Sci. 194, pp. 1121-1132.
57. Machida, H.: 1973, J. Geogr. Japan 82, pp. 53-76.
58. Kobayashi, K. and Shimizu, H.: 1966, J. Fac. Sci. Shinshu Univ. 1, pp. 97-113.
59. Yamasaki, M., Nakanishi, N. and Miyata, K.: 1966, Sci. Rept. Kanazawa Univ. 11, pp. 73-92.
60. Fukai, S.: 1974, Bull. Educ. Toyama Univ. 22, pp. 119-133.
61. Nakaya, S.: 1972, Quat. Res. Japan 11, pp. 305-317.
62. Steinen, R.P., Harrison, R.S. and Matthews, R.K.: 1973, Geol. Soc. Amer. Bull. 84, pp. 63-70.
63. Dreimanis, A. and Karrow, P.F.: 1972, Intern. Geol. Congr. 24th Sess., sec. 12, pp. 5-15.
64. Dreimanis, A. and Raukas, A.: 1975, Quaternary Studies, pp. 109-120.
65. Suggate, R.P.: 1974, Quat. Res. 4, pp. 246-252.

QUATERNARY TEPHRA OF NORTHERN CENTRAL AMERICA

W. I. Rose, Jr., G. A. Hahn[1], J. W. Drexler, M. L. Malinconico[2], P. S. Peterson[3], and R. L. Wunderman
Michigan Technological University, Houghton, Michigan 49931, U.S.A.

ABSTRACT. Silicic Plinian tephra units representing more than 30 Quaternary eruptions blanket Guatemala and El Salvador. They were erupted mainly from 5 principal sources, all of them calderas. Several of the eruptions were accompanied by ash flows. These eruptions also have the most extensive tephra deposits. The total volume of material erupted is equivalent to 300-500 km^3 of dense rock. A major uncertainty is the volume of tephra scattered very far from the source. The volume of silicic magma erupted in the Quaternary is similar to the volumes of mafic lava produced at the volcanic front. The basaltic and andesitic cones of the volcanic front parallel the offshore Middle America trench and the active underthrust zone. The five caldera sources form a trend parallel to the volcanic front, on the side opposite the trench, where the older continental crust abuts the volcanic zone. The ages of silicic volcanism precede and overlap with the age of mafic volcanic front, which is largely younger than 50,000 years. All of the calderas have multiple eruptions which span at least many tens of thousands of years. Between the calderas the interfingering of ashes has allowed a network of relative ages to be established. We used a variety of techniques to characterize these units. They can be readily distinguished from units from many other provinces, but considerable effort is required to distinguish among the local units. Standard field and petrographic observations (stratigraphic data, thicknesses, grain size, lithic content, mineralogy) establish the critical frame-

---

[1]Present address: Noranda Exploration, Salmon, ID 83467 U.S.A.
[2]Present address: Dartmouth College, Hanover, NH 03755 U.S.A.
[3]Present address: Climax Molybdenum, Climax, CO 80401 U.S.A.

S. Self and R. S. J. Sparks (eds.), Tephra Studies, 193–211.

work which disallows most erroneous correlations. Geochemical analysis, particularly trace elements, provide a rapid means of ruling out many more possible correlations. Quantitative mineralogical analysis by electron microprobe of hornblende and Fe-Ti oxides was a very effective last resort for correlation.

1. INTRODUCTION

Five calderas of Quaternary age have been recognized in northern Central America (Table 1). They form a WNW-trending line which parallels both the volcanic front and the Middle-America Trench (Figure 1). Each of these centers has produced a series of silicic pyroclastic eruptions, forming a complex sequence of airfall and ash-flow deposits. This report describes these pyroclastic units and the methods used to distinguish and correlate them.

Table 1. Quaternary Calderas of Northern Central America.

	NAME	AREA, Km^2	ASSOCIATED PYROCLASTIC UNITS
A	Atitlân	500	W, Los Chocoyos (H), I?
B	Amatitlân	220	R?, L, T, E, J
C	Ayarza	10	Mixta, Pinos Altos (B), Tapalapa
D	Coatepeque	50	XVIII - I
E	Ilopango	100	Tierra Blanca, Tierra Blanca Joven

2. PREVIOUS AND CURRENT WORK

Koch and McLean's (1) work established the physical and petrographic basis for tephra studies in the Guatemalan Highlands. They established the stratigraphy in the Guatemala City area. Hahn et al. (2) demonstrated that the youngest ash flow deposits (Los Chocoyos Ash) in the many basins of the Guatemalan Highlands were correlative, using bulk chemical analyses. He also documented the lateral chemical changes of this ashflow and established the correlation of the Los Chocoyos Ash with the Worzel D layer in the Equatorial Pacific. Rose et al. (3) described the Los Chocoyos ash-flow member in the Quezaltenango Valley and showed that its physical characteristics were consistent with a source from the Atitlân caldera.

Drexler et al. (4) extended the correlation of the Los Cho-

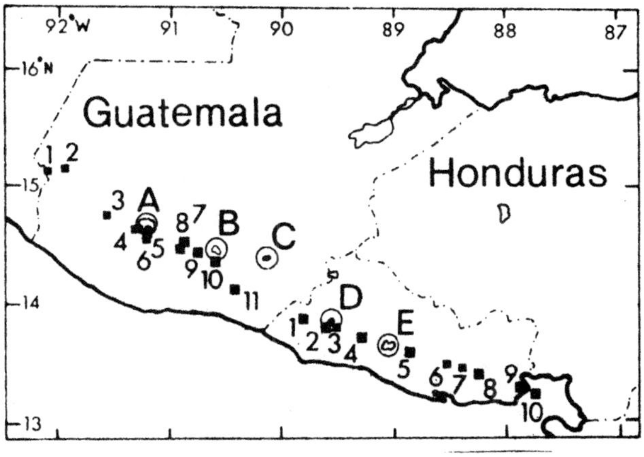

Figure 1. Location map for the volcanic front volcanoes of Northern Central America (numbered squares) and the five Quaternary calderas discussed in this report. A = Atitlán, B = Amatitlan, C = Ayarza, D = Coatepeque, E = Ilopango.

coyos Ash to the Y8 layer in the Gulf of Mexico. They also compiled chemical data on many other Guatemalan tephra units and deep sea ashes. The Y8-Los Chocoyos correlation established the 85,000 ± 5,000 yr. B.P. age, and allowed a minimum volume of 200 km^3 (dense rock equivalent) to be estimated.

Peterson (5) described the tephra units of Ayarza and established a correlation of younger Ayarzan tephra (Pinos Altos) with some sites of the B-tephra in the Guatemala City region, and with ash in Pacific deep sea core. The Pinos Altos ash was dated by C^{14} study of a log in the overlying consanguinous Tapalapa ashflow at 23,100 yr. B.P.

Meyer (6) described the stratigraphic units of Coatepeque caldera and made an estimate of the total volume of Coatepeque ashes.

Steen-McIntyre and Sheets (7) and Hart and Steen-McIntyre (8) have established the physical and petrographic characteristics of the Tierra Blanca joven tephra and correlated it throughout western El Salvador where it is closely associated to protoclastic Mayan archaeological sites.

3. BRIEF DESCRIPTION OF THE UNITS

The rocks are calc-alkalic, ranging from dacite to rhyolite. A significant fraction of many of the units is more mafic, usually

Table 2. Representative analyses of pyroclastic rocks erupted from five calderas in northern Central America.

	ATITLÁN →					
Wt. %	1	2	3	4	5	6
SiO_2	74.9	53.4	73.4	74.4	73.7	70.7
Al_2O_3	14.1	20.5	13.0	11.9	12.8	13.8
Fe_2O_3*	1.9	9.34	0.82	0.83	1.66	1.65
MgO	0.37	3.80	0.21	0.20	0.40	0.41
CaO	2.15	8.73	0.88	0.80	1.9	1.62
Na_2O	2.83	3.25	3.06	3.50	3.9	3.60
K_2O	2.25	1.11	3.94	4.30	2.7	3.48
TiO_2	0.17	0.85	0.13	0.12	0.21	0.26
P_2O_5	0.02	0.18	0.03	nd	nd	0.05
TOTAL	98.69	101.16	95.47	96.05	97.27	95.57
ppm						
Sr	335	594	146	109	269	303
Rb	66	25	128	124	76	103
Zr	118	177	67	59	147	161
Y	3	14	13.3	nd	nd	10
Ba	810	273	910	1135	970	879
Sc	2	23	2	2.8	3.9	3
La	19	10	19.5	20	22	21

Legend:

1. W-tephra, Km 131, CA1 rhyolite component (SOL-2).
2. W-tephra, Km 131, CA1 basalt component (SOL-1).
3. H-tephra, mean of 18 samples.
4. Los Chocoyos ash flow, high K blocks, mean of 15 samples (3).
5. Los Chocoyos ash flow, low K blocks, mean of 16 samples (3).
6. Late Atitlán ash (758e), from n. shore of Lake Atitlán.

basaltic. The bulk chemistry of selected samples is given in Table 2. These rocks are the most silicic subset of modern Central American volcanic rocks (Figure 2). The degree of crystallization and mineralogy is summarized in Table 3. Nearly all

Table 2. Representative analyses of pyroclastic rocks erupted from five calderas in northern Central America (cont.)

	AMATITLÁN →				AYARZA →	
Wt. %	7	8	9	10	11	12
SiO_2	75.1	67.5	70.0	72.1	72.9	51.5
Al_2O_3	14.1	16.9	16.3	14.4	13.7	18.1
Fe_2O_3*	1.1	3.3	3.0	2.39	1.28	10.4
MgO	0.21	0.85	0.88	0.5	0.20	3.8
CaO	0.92	3.52	3.15	1.90	0.63	8.0
Na_2O	3.42	4.18	3.91	3.9	3.8	3.7
K_2O	3.96	3.17	2.40	3.5	4.23	1.12
TiO_2	0.15	0.42	0.33	0.30	0.13	1.28
P_2O_5	0.01	0.15	0.17	0.06	0.04	0.25
TOTAL	98.97	99.99	100.14	99.05	96.94	98.15
ppm						
Sr	138	435	428	nd	64	610
Rb	130	82	62	105	128	17
Zr	78	257	205	nd	106	194
Y	16	12	7	nd	nd	nd
Ba	820	854	815	992	620	290
Sc	3.1	6	3.7	3.9	0.7	26
La	21.6	20.8	16	22.2	20	11

Legend:

7. T-tephra from near Palencia (23-T).
8. E-tephra from Rio Molino (9-E).
9. C-tephra from Rio Molino (9-C).
10. J_1 tephra from south of Amatitlán (J_1-IV).
11. Mixta tephra, rhyolite component mean of 3 analyses (5).
12. Mixta tephra, basalt component (5).

units are quartz and plagioclase-bearing and contain iron-titanium oxides. Most rocks are 70-99% glass by weight. The ferromagnesian and accessory mineralogy is quite variable and these data were useful in correlation (see also (1)).

Table 2. Representative analyses of pyroclastic rocks erupted from five calderas in northern Central America (cont.)

	⟶		COAT. ⟶	ILOPANGO	
Wt. %	13	14	15	16	17
SiO_2	71.7	73.3	69.9	66.0	69.2
Al_2O_3	15.8	14.3	14.7	15.2	15.4
Fe_2O_3*	2.12	1.53	2.7	3.5	3.2
MgO	0.3	0.3	0.09	1.06	0.84
CaO	1.31	1.12	1.42	4.81	4.35
Na_2O	3.7	3.2	3.0	4.1	4.3
K_2O	3.5	3.8	4.2	2.0	2.1
TiO_2	0.23	0.18	0.11	0.36	0.33
P_2O_5	0.04	0.03	0.04	0.16	0.13
TOTAL	98.70	97.76	96.16	97.19	100.03
ppm					
Sr	126	nd	67		318
Rb	119	135	180		55
Zr	157	nd	216		169
Y	16	nd			6
Ba	793	734			848
Sc	2.6	2.5			6
La	24.4	25.3			15

Legend:

13. Pinos Altos, tephra, mean of 5 analyses of tbj ashflow block, Colonia Layco (TB-4) (5).
14. Tapalapa ashflow, mean of 3 analyses (5).
15. Coatepeque, tephra, 5m thick, reverse graded unit at Km 48, CA1.
16. Bulk phreatoplinian tbj tephra, Colonia Layco (TB-6).
17. tbj ashflow block, Colonia Layco (TB-4).

All of the units are characterized by vesicular pumice of tan to white color. Most units are typical Plinian airfall deposits. Several have associated non-welded ashflows. Volume estimates of many of these units and their sources are compiled

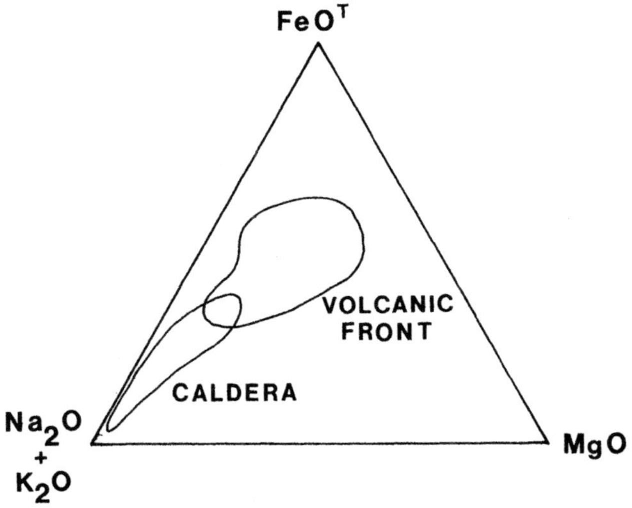

Figure 2. AFM plot showing generalized fields of compositions for caldera and volcanic front rocks.

in Table 4. Together the rocks represent at least 300 km^3 of dense rhyolite. The volcanic front represents 1,000 km^3 of basalt and andesite lavas. Together the recent volcanic rocks of northern Central America are a bimodal association.

The relationship of the Guatemalan units, as currently understood, is shown in Figure 3. To date we have not been able to extend this framework to El Salvador. The volcanic front activity seems to largely post-date the Los Chocoyos Ash in the Guatemalan Highlands.

4. VOLUME ESTIMATION

Volumes of tephra deposits were estimated using the method of Rose et al. (9) adapted to a reiterative computer program called ASHVOL. Inputs are isopach thickness and area, as is usual with such calculations, but the integration performed allows inference of volumes which occur as unmeasured near-source on distal deposits. The interactive program, written in BASIC by R.L. Wunderman is available on request from the senior author.

The calculations (Table 4) establish clearly the volume of dominance of the Los Chocoyos unit. They also form a framework for a comparison of the other tephra units.

Table 3. Mineralogy of Tephra and Ash Flows, northern Central America.

Light Minerals		Dark Minerals	
Plag (An_{15-45})	0.5-20%	Hbd	0-8%
Qtz	0- 8%	Biot	0-3
Accessories sometimes:		Opx	0-2
Cummingtonite, Olivine		Mag	tr-3
Cpx, Allanite, Zircon		Ilm	tr

Thickness of an airfall unit in its near-source exposure is not necessarily a good volume criteria, as shown in Figure 3, which contrasts several of the units. Some ash deposits, such as W or E, are anomalously thick near source with respect to their total volume, but thin rapidly with distance.

5. THE LOS CHOCOYOS ASH

Because it is formed throughout the Guatemalan Highlands (Figure 3) and because of its volume dominance (Table 4) this unit is of greatest significance. It represents the largest recent eruption of the Atitlán caldera (10).

The Los Chocoyos Ash consists of two distinct members representing the Plinian airfall and subsequent ashflow eruption. The airfall deposit is called the H-tephra member. It is very glass-rich and homogeneous in composition (Table 2, no. 3), consisting of biotite-bearing rhyolite. Its thickness reaches a maximum of about 4 m near Lake Atitlán, but is found almost everywhere in the highlands of Guatemala. The unit consists of a single ungraded bed to the east of Atitlán but becomes 3 fall units to the west. Near Lake Atitlán pyroclastic surges overlie the H-tephra member; everywhere else the ashflow member overlies the H-tephra directly and with no trace of a weathering break or any erosion.

The ash-flow member is largely confined to topographic lows including many of the basins of the Guatemalan Highlands (see map in (2)). Though its thickness exceeds 200 m in several of the basins nowhere does it show even a trace of welding. Flow units are absent or obscure in most exposures, but near Guatemala City the ashflow consists of three well defined flow units. Because its patchy distribution requires the overrunning of many topo-

Table 4. Volumes and Sources of Major Silicic Pyroclastic Units of Northern Central America.

Unit(s)	Volume, Km^3 DRE	Source
1902 Dacite	2.5	Santa María
Siete Orejas	>2.	Siete Orejas
Los Chocoyos	>200.	Atitlán
W-tephra, ashflow	4.	Atitlán
R-tephra, ashflow	>10?	?
X-tephra	1.6	Acatenango Area
Y-tephra	1.5	Acatenango Area
S-tephra	0.8	Acatenango Area
T-tephra, ashflow	>16.	Amatitlán
L-tephra, ashflow	>17.	Amatitlán
C-tephra	0.8	Sabana Grande(?)
E-tephra	2.5	El Durazno
Pinos Altos tephra/ Tapalapa ashflow	>2.0	Ayarza
Mixta, tephra	0.1	Ayarza
I-XVIII, tephra, ashflows	35(?)	Coatepeque
tbj; tephra	>3.5	Ilopango
TOTAL ESTIMATED	>300.	

graphic barriers, because its lack of welding implies low temperatures of deposition and because of its very great dispersion (Figure 4) we have inferred that the deposits resulted from collapse of an exceptionally high column.

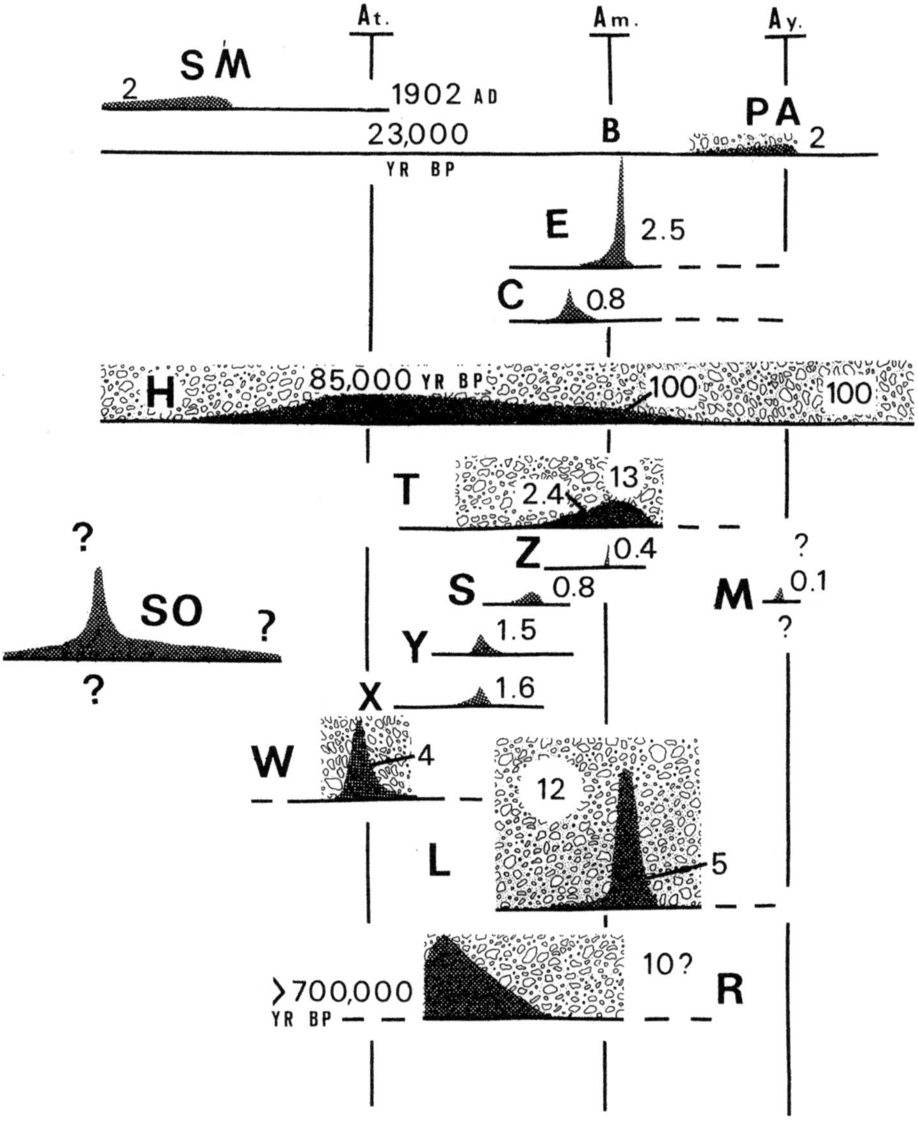

Figure 3. Composite cross-sectional diagram of Quaternary tephra units from NW to SE across Guatemala. Letter designations correspond to usage in text. At, Am and Ay denote positions of Atitlán, Amatitlán and Ayarza calderas. Numbers indicate ages and dense rock equivalent volumes for units. Vertical scale is highly exaggerated, but indicates relative thicknesses of units. Ash flow units are shown schematically with their associated airfall, though their distribution is strongly influenced by topography.

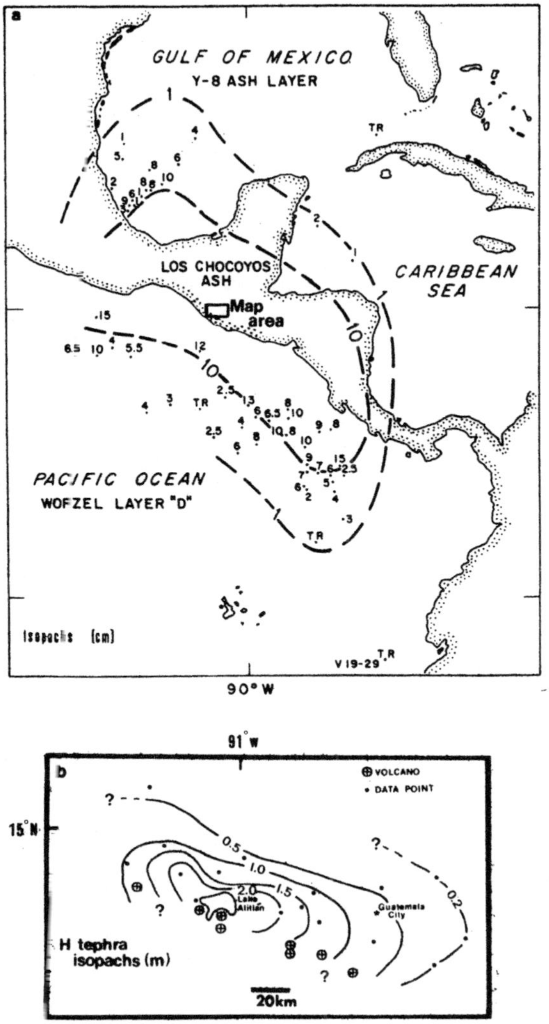

Figure 4. Isopach maps of the Los Chocoyos Ash (4). Area shown in b is indicated by "map area" in a.

The shift in activity from airfall to ashflow was accompanied by a change in composition of the magma. Throughout the ashflow there are two populations of pumice rocks: a biotite-bearing rhyolite (high K) similar in composition to the H-tephra and a much more heterogeneous less potassic hornblende-bearing rhyolite (low K). A few of the hornblende-bearing blocks appear to reflect physical mixing. As the ashflow member was deposited, the proportions of less potassic blocks increased; overall the mixture is about 60% high K: 40% low K (3).

Correlation of the Los Chocoyos Ash with tephra units in dated deep sea cores (2, 4) from the Equatorial Pacific, the Gulf of Mexico and the Caribbean Sea have enabled the dating of the Los Chocoyos eruption at 85,000 yr. B.P. We do not know whether the deep sea ashes are associated with the H-tephra member, the ash-flow member or both. Low K rhyolite shards are found by microprobe study of the deep sea ashes, but they make up a very small fraction of the total population.

6. OTHER UNITS

In describing other silicic tephra and ashflow units, we will follow a geographic traverse from northwest to southeast, beginning in the area of Quezaltenango, Guatemala and continuing to Lake Ilopango, in El Salvador.

Ashes of the Quezaltenango Region. The 1902 tephra from Santa Maria Volcano represents the only historic Plinian eruption in northern Central America. Details of the eruption were published by Sapper (11, 12) and many other authors and summarized by Rose (13). The Plinian deposits were examined by Williams (14). The eruption produced about 2.5 km^3 of dense dacite magma, distributed from a high (>25km) eruptive column over a broad region. Although its composition is slightly more mafic, the tephra is physically similar to many of the other deposits. Because of the prevailing winds at the time of eruption its distribution is important only to the west of Santa María.

The climatic eruptions of Siete Orejas Volcano, just west of Santa María, produced a voluminous Plinian tephra deposit which is more than 10 m thick on the south side of Quezaltenango and at San Juan Ostuncalco (15). This unit has been tentatively correlated with the lower part of tephra stratigraphy found at Nahualá and westward toward Quezaltenango. The pumice is rhyolitic in composition and represents an eruption of signficant volume (>2 km^3 DRE). Exposures near San Juan Ostuncalco show that the Siete Orejas eruption predates the Los Chocoyos event (3).

Mapping by Johns (16) in the Cerro Quemado area, describes sequences of mixed tephra units that post-date the Los Chocoyos unit. These are very minor volumetrically and consist of continuous sequences of alternating basalt and rhyolite tephra. Their source vents are unknown.

Ashes of the Atitlán Caldera. Besides the Los Chocoyos Ash, two other tephra units have been recognized as part of the Atitlan eruptive sequence. The W-tephra unit is a Plinian deposit which is more than 9 m thick near the Caldera but thins rapidly both toward the east and west. In poor exposures west of Chimaltenango it is found interfingering with the X and L tephra (see Figure 3).

Several tephra units post-dating the Los Chocoyos eruption with sources from Atitlán have been described. Hahn (17) identified the I-tephra in the area to the south of Lake Atitlán. Units of silicic Plinian tephra overlie the flows of San Pedro Volcano on its northern flanks and Rose et al. (18) suggested that these units may correlate with the I sequence. A late sequence of ashes consisting of a thin Plinian deposit sandwiched between phreatoplinian layers is found along the north shore of Lake Atitlán at the caldera rim and inside it, along the shore (19). Chemical study of these units is incomplete, but all are in the rhyodacite range, and it is likely that correlations among them will be found. No correlations with tephra units to the east have been verified.

Ashes with Unknown Sources Between Atitlán and Amatitlán. Several Plinian tephra units with volumes ranging from 0.5 - 2.0 km^3 form part of the Sumpango Group succession (20). In stratigraphic order (oldest to youngest) these are X, Y, S, Z_1, and Z_3. All predate the Los Chocoyos. These tephra are well exposed in roadcuts around Chimaltenango. They seem to have sources in the area of Acatenango.

Amatitlán Caldera Succession. In the Guatemala City area is the most complex tephra sequence. Several of the units have been shown to have local sources (1, 21). These are from oldest to youngest: R, L, Z_2, Z_4, Z_5, T, C, E, J. R predates the Bruhnes/Matuyama reversal, is poorly exposed and may not be related to the Amatitlán caldera. L and T probably represent the caldera-forming events: they both predate the Los Chocoyos eruption. E and J are later Plinian tephra units coming from craters within the Amatitlán caldera. C tephra comes from a crater to the northeast of Agua Volcano, outside of the Caldera. All of the units are silicic ranging from dacite to rhyolite in composition. This sequence is the subject of current work (21).

Ayarza Caldera Sequence. Ayarza is a small (7 x 4 km) double caldera (22, 5). Two ash sequences, assigned to the formation of the two parts of the caldera have been recognized. The older ash is the Mixta Plinian deposit, a spectacular mixed magma of very small volume (<0.2 km^3) and extent, but of considerable petrologic interest. The younger units, associated with formation of the western Ayarza caldera, are the cogenetic sequence of Pinos Altos tephra, Tapalapa ashflow and Sabana Redonda tephra. These units are part of the same eruptive sequence. The B-tephra of the Guatemala City succession (1) is correlated with this sequence which is dated by C^{14} at 23,100 ± 500 yr. B.P. The Pinos Altos is also correlated with ash in core RC-12-32 in the Pacific.

Coatepeque Caldera Sequence. The sequential pyroclastic units of the Coatepeque caldera were described (6) as eighteen dacite to rhyodacite tephra units, interrupted by 3 soil units. From his description, and from direct field observation some of the units are ashflows. The entire section is under reinterpretation (23). We conclude that there are at least 3 eruptive episodes. Meyer (6) estimated the total volume of all units to be in excess of 35 km^3 DRE. The age of the sequence is placed between 10,000 and 45,000 yr. BP based on a study of the interbedded fossil soils. The Coatepeque ashes have not been traced to the west toward Ayarza, so that we do not yet know the stratigraphic relationships of Coatepeque and Ayarza ashes. The youngest ashes of Ilopango, the Tierra Blanca joven tephra and ashflow are clearly younger than all of the Coatepeque sequence.

Ilopango Caldera Sequence. The pyroclastic rocks associated with Ilopango caldera represent at least 3 distinct periods of activity. The entire sequence is presently under study (24). The latter eruption, resulting in the Tierra Blanca joven (tbj) tephra and associated ashflows has been described in detail (8). It has been dated by C^{14} at about 300 A.D. The pyroclastic rocks of Ilopango are dacite to rhyodacite in composition. The volumes of older pyroclastic units of the Tierra Blanca are not known, but the tbj totals at least 3.5 km^3 DRE. Because the tbj is phreatoplinian, the caldera lake was probably present at the time of its eruption.

In summary the five known calderas and several vents outside the caldera have produced in excess of 30 silicic pyroclastic units which total more than 300 km^3 of dense dacitic to rhyolitic magma. The ages of the units are not all established, but most, including some units from each caldera, are younger than 100,000 yr. B.P.

7. METHODS USED FOR IDENTIFYING AND CORRELATING.

The most important methods used in recognizing and correlating were stratigraphic and physical ones. Since most of these units are thin blankets, many exposures allow examination of sequential units. Physical criteria such as thickness, grain size, shape, grading were important field criteria. Mineralogy, particularly the ferromagnesian minerals, was important as well.

We obtained whole rock major and minor element geochemistry and glass and mineral analyses by microprobe on many samples. All of this data was useful for correlation. We will describe each method in turn.

Whole Rock Geochemistry. We made whole rock analyses of tephra and ashflow units by X-ray fluorescence and neutron activation. Table 5 shows the list of elements and the methods used.

Table 5. List of Trace Elements Analysed For Use in our Correlation Studies.

XRF:	Fe, Mg, Ca, Ti, P, Mn, Sr, Zn, Cu, Ni, Co, Cr, S, V, Pb, Zr, Rb, Sn, La, Y
NAA:	Ce, La, Nd, Sm, Eu, Tb, Yb, Lu, Hf, Ta, Th, U, Sc, Cr, Co, Ni, Zn, Cd, As, Sb, Se, Br

We used neutron activation on samples of which we had only limited quantities, particularly the deep sea samples. Whole rock geochemistry works best in glassy units, such as the Los Chocoyos, or in units with abundant large pumices where crystal/glass atmospheric fractionation is inhibited. Table 6 shows a comparison of glass separate and whole rock analyses for several trace elements. Unit E is much more crystal-rich than Unit T, and shows significant discrepancies in trace elements. Units with more than about 10% crystals can become difficult to recognize chemically if crystal/glass atmospheric fractionation is occurring. The problem can be avoided in a variety of ways: 1) use glass

Table 6. Results of Trace Element Analyses of Glasses and Whole Rocks From Two Tephra Units

		E-tephra		T-tephra	
		WR	GL	WR	GL
ppm	Sr	400	123	180	108
	Pb	22	14	17	11
	Zr	276	185	160	175
	Y	10	16	15	15
	Rb	73	125	120	120
	Sc	8	<2	2.5	<2
	Ba	840	854	900	932
	La	25	25	24	28

separates only, 2) use pumice blocks only, 3) use elements which are not captured by phenocryst phases.

For correlation purposes we used the statistical method of variation analysis (25, 26). The calculations reduce a comparison of two possible correlative units from as many as 30 or more elemental concentrations to 2 or 3 numbers. We calculated %CV, $\bar{X}$ and the coefficient of similarity for each comparison. By repeated trials of samples known to be correlative we established reliable cutoffs for these parameters. These values and the definitions of the parameters are given in Table 7. If more than 10 elements are known, these comparisons become very helpful in correlation. A computer program, called Coefvar, is written in Univac BASIC and is available on request from the senior author.

Mineral Analyses by Microprobe. These were very helpful correlation criteria. All of the units examined contained both titanomagnetite and ilmenite. These minerals are easily separated and are amenable to precise energy dispersive microprobe analysis. With coexisting Fe-Ti oxide phases, the Buddington and Lindsley (27, 28) T-fO_2 estimates can be made. Figure 5 shows the silicic pyroclastic units of northern Central America plotted by Fe-Ti oxide temperatures. There is range of temperatures, from less than 700°C to more than 1000°C. Lower temperature magmas usually are biotite-bearing while higher temperature ones contain hornblende. There is not a simple relationship between these temperatures and the bulk composition of the rocks (28).

Hornblende compositions were helpful in correlation. Changes in amphibole composition appear to be related to temperature (29).

Table 7. Statistical Calculations Used in Trace Element Fingerprinting (25, 26).

PARAMETER	DEFINITION	EMPIRICAL CUT-OFFS
$\bar{X}$	mean of $\frac{Cx}{Co}$	1.00 ± 0.15
%CV	$\sigma*100/\bar{X}$	<25
COEF. SIMILARITY	mean of $\frac{Cx}{Co}$ or $\frac{Co}{Cx}$ whichever is <1.	>.90

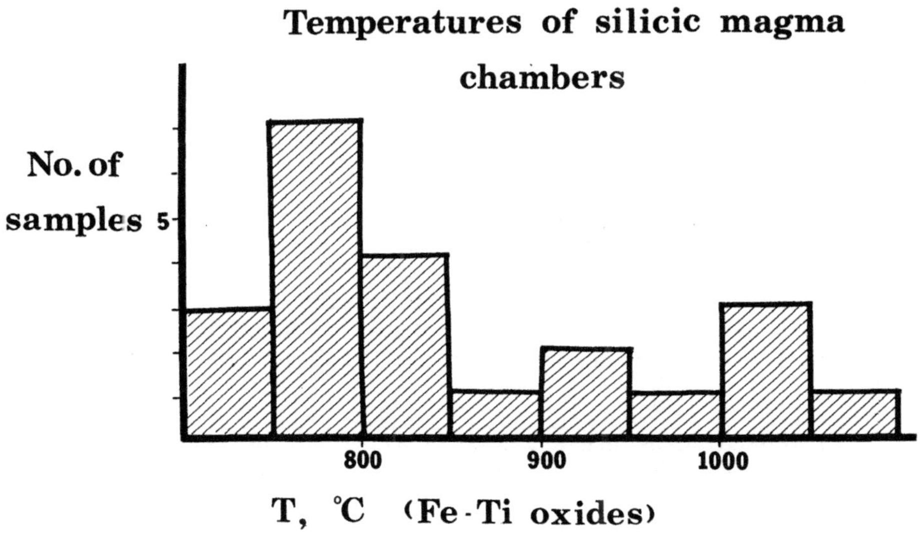

Figure 5. Fe-Ti oxide temperatures of silicic tephra units of northern Central America (29).

The composition of biotite and glass were less useful for correlation. Glass compositions, as determined by microprobe, are insufficiently distinct when dealing with so many similar units. Biotites are subject to vermiculitization and the compositions we determined in the tephra and ash flows did not reflect the primary composition (5).

In practice we used all of the correlation data in the following way. Field and petrographic observations were of first rank. These are, in order, stratigraphy, grain size, thickness, lithic content and mineralogy. Such data eliminated most erroneous correlations. The bulk chemical data served to eliminate many more. Quantity of sample available and its importance helped dictate whether we used X-ray fluorescence or neutron activation or both. Fe-Ti oxide and amphibole compositions, determined by microprobe, were a very effective last resort in this work.

The methodology we used stands in contrast to deep sea ash correlation studies, e.g. which emphasize microprobe determinations of glass compositions. We believe that our own approach is much more successful in the near source regions where one is confronted with a large number of deposits of very similar composition. In the deep sea cores, often smaller amounts and numbers of units are found. Often samples from different provinces are found, which are more easily distinguished on the basis of major element chemistry.

We found morphological study of ash and refractive index determinations of glass to be of less utility in our correlation. This is also because so many units are indistinguishable by these criteria (30).

8. CONCLUSIONS

1) Quaternary silicic volcanism has occurred chiefly from 5 calderas in northern Central America.

2) The volumes of silicic rocks erupted in the last 10^5 yrs. (>300 km^3) are comparable to the volumes of basalts and andesites (1000 km^3) erupted along the volcanic front during the same period.

3) The calderas are aligned parallel to the volcanic front, slightly to the north, and spaced more widely than the volcanic front.

4) Activity at the calderas began before the volcanic front activity but continued and has overlapped with it.

5) All of the calderas have multiple eruptions, and have produced deposits which are roughly proportional to their size.

6) Correlation of on-land silicic pyroclastic sequences based on major and minor element chemical analyses of bulk samples and minerals, as well as traditional field and petrographic methods can be successful in establishing a complex, near-source stratigraphy of very similar units.

REFERENCES

1. Koch, Allan and McLean, Hugh: 1975, GSA Bull. 86, pp. 529-541.
2. Hahn, G.A., Rose, W.I., Jr., and Meyers, T.: 1979, GSA Spec. Paper 180, pp. 101-112.
3. Rose, W.I., Jr., Grant, N.K. and Easter, J.: 1979, GSA Spec. Paper 180, pp. 87-100.
4. Drexler, J.W., Rose, W.I., Jr., Sparks, R.S.J. and Ledbetter, M.T.: 1980, Quat. Res. 13, pp. 373-391.
5. Peterson, P.S.: 1980, unpubl. M.S. thesis, Michigan Tech. University.
6. Meyer, J.: 1964, N. Jb. Geol. Palaont. 119, pp. 215-246.
7. Steen-McIntyre, V. and Sheets, P.D.: 1978, Geol. Soc. Amer. Abst. with Prog. 10, p. 497.
8. Hart, W.J. and Steen-McIntyre, V.: 1980, in: Volcanoes and Human Activity, ed. by P. Sheets, in press.

9. Rose, W.I., Jr., Bonis, S., Stoiber, R.E., Keller, M. and Bickford, R.: 1973, Bull. Volc. 37, pp. 338-354.
10. Newhall, C.G., Self, S. and Paull, C.K.: 1980, Trans. A.G.U. 61, p. 69.
11. Sapper, K.: 1903, Centralblatt f. Miner. Palaont., pp. 33-44, pp. 71-72.
12. Sapper, K.: 1904, N. Jb. Miner. Geol., Palaont., pp. 39-103.
13. Rose, W.I., Jr., Bull. Volc. 34, pp. 29-45.
14. Williams, S.N.: 1980, unpubl. M.S. thesis, Dartmouth College.
15. Gierzycki, G.A.: 1976, M.S. thesis, Michigan Tech. Univ.
16. Johns, G.W.: 1975, M.S. thesis, Michigan Tech. Univ.
17. Hahn, Gregory A.: 1976, M.S. thesis, Michigan Tech. Univ.
18. Rose, W.I., Jr., Drexler, J.W., Penfield, G. and Larson, P.B.: 1980, Bull. Volcanol. 43, no. 1, pp. 131-153.
19. Newhall, C.G.: 1980, pers. comm.
20. McLean, Hugh: 1970, Ph.D. thesis, Univ. of Washington.
21. Wunderman, R.L.: 1980, Abstract volume, NATO Adv. Study Inst. Iceland, June 1980.
22. Peterson, P.S. and Rose, W.I., Jr.: 1980, Trans. A.G.U. 61, p. 66.
23. Williams, S.N.: 1979, unpub. MS thesis, Dartmouth College.
24. Hart, W.J.: 1980, NATO Adv. Study Inst., Abst. Vol., Iceland, June 1980.
25. Borchardt, G.A., Harward, M.E. and Schmitt, R.A.: 1971, Quat. Research 1, pp. 247-260.
26. Sarna-Wojcicki, A.M., Bowman, H.W. and Russel, P.C.: 1979, USGS Prof. Paper 1147.
27. Buddington, A.F. and Lindsley, D.H.: 1964, Jour. Petrol. 5, pp. 310-357.
28. Lindsley, D.H. and Rumble, D., III: 1977, Amer. Geophys. Union Trans. 58, p. 519.
29. Love, M.A., Rose, W.I., Jr. and Wunderman, R.L.: 1980, Trans. A.G.U. 61, p. 66.
30. Drexler, J.W.: 1978, unpubl. M.S. thesis, Michigan Tech. U.

A TEXTURAL STUDY OF BASALTIC TEPHRAS FROM LOWER TERTIARY DIATOMITES IN NORTHERN DENMARK

A.K. Pedersen and K.A. Jørgensen

Geologisk Museum, Øster Voldgade 5-7, DK-1350 København K., Denmark

ABSTRACT. More than 200 tephra layers deposited in early Tertiary diatomite and clay in northern Denmark record explosive volcanism in the Skagerrak and in the North Sea. The early activity is basaltic, rhyolitic, and mafic and salic alkaline while later activity is exclusively tholeiitic Fe-Ti-basaltic. Several of the younger basaltic tephras exceed 4 km^3 by volume. The morphology and grain size of the tephras indicate both subaerial and phreatic activity. The Fe-Ti-basaltic tephras are distinctly surtseyan in type, which may indicate volcanism in a shallow sea environment.

1. INTRODUCTION

Tephras deposited in a 60 m thick unmetamorphosed diatomite ('mo-clay') in northern Denmark (Fig. 1) record a long and complex volcanic episode in the North Sea - Skagerrak area (1-5). The diatomite was deposited in a marine shelf environment under varying oxic and anoxic conditions (6-8). It is uppermost Paleocene to lower Eocene in age (9-11) and it probably represents a time span of a few million years. Parts of the tephra sequence are well preserved, in particular when embedded within concretions of manganoan calcite, while other parts have suffered notable alteration by reaction with saline and fresh water, but even altered tephra layers have retained their grain morphology. Out of the more than 200 tephra layers 179 have been numbered (1) to include a lower negative series (layers -39 to -1) and an upper positive series (layers 1 to 140). The tephra layers are from less than 1 mm to 18 cm thick and all pyroclastic particles are less than 1 mm in size. Some characteristic numbered tephra

S. Self and R. S. J. Sparks (eds.), Tephra Studies, 213–218.

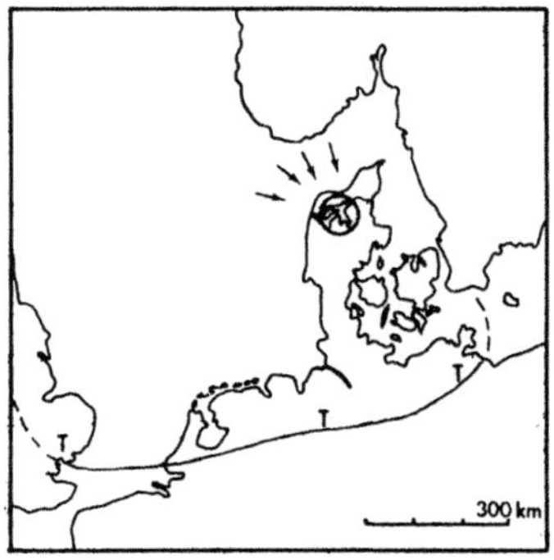

Fig. 1. Enclosed by a circle are shown the diatomite localities in northern Denmark. Arrows indicate the inflow direction of tephras from sources in the North Sea and Skagerrak. Line marked T indicate the southern limit of extension of the early Tertiary tephras (11, 12).

layers have been recognized in clay deposits of similar age in other parts of Denmark (1, 2) and some extend into northern Germany (12). Studies of thickness variations (1, 2, 13) show that the basaltic ashes were deposited by air fall from the west and north from within a distance of a few hundred kilometres from the diatomite localities in Denmark. Very voluminous tholeiitic basaltic tephras are not a common phenomenon, and as the volcanoes are not exposed we shall try to characterize the volcanism by the tephra morphology, mainly in layers which previously have been described in terms of chemistry and mineralogy (5, 14).

2. VOLCANIC EVOLUTION IN THE NORTH SEA - SKAGERRAK AREA

Recent microprobe work on glasses and crystals from selected tephra layers have enabled the identification of at least 4 distinct stages of volcanism in the tephra sequence.

Stage 1 (layers -39 to -21). Here tholeiitic basaltic tephras dominate, several rhyolitic ones are present including the thickest layer of that part of the sequence (-33). Several intermediate tephras are also found, but they are poorly characterized at present. Composite layers, formed from an early salic and a later mafic ash shower, also occur.

Stage 2 (layers -20 to -14). Comparatively wide spacing between tephra layers in the sediment points to a low volcanic activity at this stage. Basaltic, mafic alkaline and salic alkaline tephras are found. One layer (-19b) contains lithic clasts of a large variety of alkaline rocks and the presence of perovskite, phlogopite and geikelite-rich ilmenite (MgO = 10.1 wt %) indicates that the eruption was nephelinitic and carried high pressure megacrysts. The salic alkaline layers are characterized by light coloured glass, sanidine, alkali amphiboles and Ti-aegirine (14). Stage 2 represents a typical continental rift zone volcanism.

Stage 3 (layers -13 to +19). Here tholeiitic basaltic layers dominate but several rhyolitic tephras are also present. A voluminous rhyolitic tephra (+19) marks the final termination of salic volcanism in the North Sea - Skagerrak area and is the thickest tephra layer (18 cm) in the diatomite sequence.

Stage 4 (layers +2+ to +140). This stage is represented by a monotonous sequence of tholeiitic Fe-Ti-basalts which seems to represent a tephra equivalent to Tertiary Fe-Ti-basalt lava plateau such as the oldest exposed lavas on the Faroes (15). Many of the tephras exceed several km^3 by volume (as solid basalt) even with very conservative estimates.

3. TEPHRA SIZE

Grain size analyses of five Danish tephras, four Fe-Ti-basaltic and one rhyolitic, from (16) are shown on the Md_ϕ-σ_ϕ diagram (Fig. 2) which has been used (17) to discriminate tephras from strombolian and surtseyan eruptions. All the tephras are comparatively well sorted and very fine-grained, most of all the rhyolite. The tephras have been affected by prolonged aeolian differentiation, followed by sorting through the settling in saline water, and their grain size distribution gives no direct evidence of the explosive mechanisms. Surtseyan tephras are generally more widely dispersed and more fine-grained than strombolian or hawaiian types (17, 18).

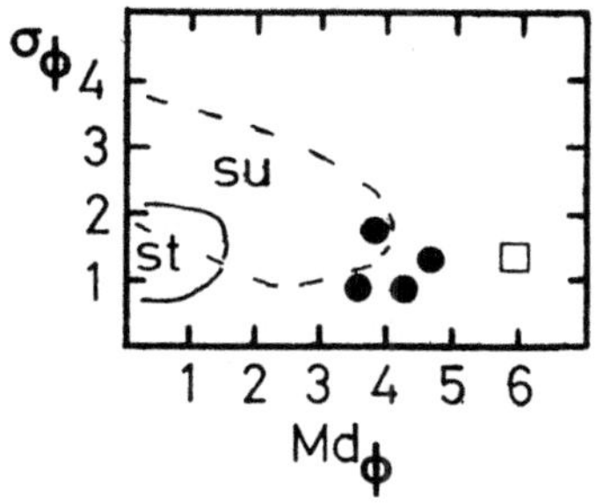

Fig. 2. Md_ϕ-σ_ϕ diagram (17) used to discriminate strombolian (st) and surtseyan (su) eruptions. Dots: Fe-Ti-basalt tephras (layers +1, +22, +114, +118) from stage 3 and 4. Open square: Rhyolitic tephra (layer +19) from stage 3. Grain size analyses from (16).

4. TEPHRA MORPHOLOGY

Basalts from stage 1 and 4. Five Fe-Ti-basalt tephras from stage 4 (layers +30, +31, +51, +79 and +102) and one from stage 1 (layer -35) were selected for study. The typical tephra contains both dark brown sideromelane and tachylite (Fig. 3) but only

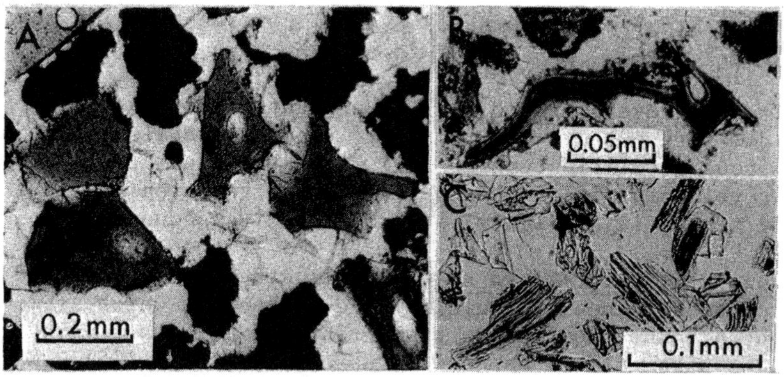

Fig. 3. A: Fe-Ti-basaltic layer +31. Sideromelane and tachylite shards embedded in calcite matrix. The glass morphology is typical of surtseyan-type eruptions. B: Fe-Ti-basaltic layer +30. A typical achnelith in basaltic tephra. C: Rhyolite layer -32. Pipevesicular shards of rhyolitic glass. The glass morphology is typical of plinian-type eruptions.

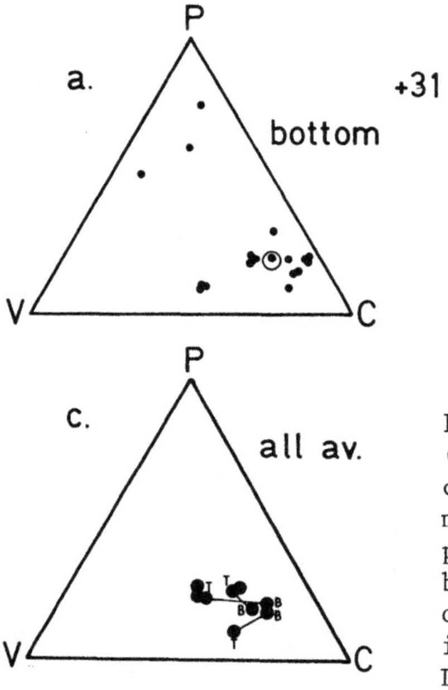

Fig. 4. P (planarity) - V (convexity) - C (concavity) diagrams (19) which show grain morphology of 6 basaltic tephras. a: bottom of layer +31, b: top of same layer, c: mean of 20 grains in each of the investigated sample areas. Lines connect T (top) and B (bottom) in 3 layers.

few lithic fragments. The grains are dominated by planar and concave outlines as shown in the P-V-C-diagrams (Fig. 4) which have been used to discriminate hyalotuffs and hyaloclastites (19).

The grain morphologies in the Danish basalt tephras are characteristic for surtseyan type eruption (17, 20). In several tephras grains with convex forms (Fig. 3B) become increasingly common upwards in the layer (Fig. 4a, b). Such "achneliths" (17) are formed in abundance in strombolian and hawaiian type eruptions (17, 20) and could indicate a minor ash-contribution from pure subaerial basaltic craters.

Rhyolitic and salic alkaline tephras from stage 1 and 2. Fig. 3C shows the grain morphology typical of the investigated salic tephras from stage 1 and 2 (layers -33, -32, -18, -17). The grains are strongly vesiculated thin walled pumice glass shards. Such ashes are characteristic of plinian type eruptions (20) and indicate violent eruptions from craters standing well above sea level.

Rhyolitic tephra from stage 3. The rhyolite (layer +19) which was produced in the latest salic eruption is characterized by pipevesicular shards and in addition it contains equant glass shards. This could indicate that the eruption was at least partly of a phreatoplinian type (20) and formed when a salic gas rich magma erupted in shallow water.

Mafic alkaline (nephelinitic) tephra from stage 2. The mafic alkaline tephra (layer -19b) is dominated by a large variety of lithic fragments of rock and crystals while glassy grains are subordinate. Such a tephra could originate from an ultraexplosive pipe-forming eruption.

5. CONCLUSIONS

The tephras in the lower Tertiary diatomites of Denmark are air fall tuffs deposited in saline water a few hundred kilometres from the eruption sites. The early rhyolitic and the salic alkaline tephras originate from plinian-type eruptions from volcanoes above sea level. The single mafic alkaline tephra has been formed in an ultraexplosive eruption. The latest salic tephra as well as all investigated basaltic tephras are produced by phreatomagmatic activity. The abundant surtseyan eruptions point to a widespread upwelling of Fe-Ti-basaltic magma into a shallow sea and explain why so many very voluminous basaltic eruptions have left so little geological evidence in their source areas in the North Sea and in the Skagerrak.

REFERENCES

1. Bøggild, O.B.: 1918, Dan. geol. Unders., raekke 2, 33, 159 p.

2. Andersen, S.A.: 1937, Dan. geol. Unders., raekke 2, 59, 53 p.
3. Andersen, S.A.: 1937, Geol. Fören. Stockh. Förh., 59, pp. 317-346.
4. Gry, H.: 1940, Meddr dan. geol. Fören., 9, pp. 586-627.
5. Pedersen, A.K., Engell, J. and Rønsbo, J.G.: 1975, Lithos, 8, pp. 255-268.
6. Gry, H.: 1935, Dan. geol. Unders., raekke 2, 61, 171 p.
7. Bonde, N.: 1979, Meded. Werkgr. Tert. Kwart. Geol., 16(1), pp. 3-16.
8. Pedersen, G.K.: 1980, Sedimentology, (in press).
9. Perch-Nielsen, K.: 1976, Bull. geol. Soc. Denmark, 25, pp. 27-40.
10. Hansen, J.M.: 1979, Bull. geol. Soc. Denmark, 27, pp. 89-91.
11. Knox, R.W. O'B. and Harland, R.: 1979, J. Geol. Soc. Lond., 136, pp. 463-470.
12. Andersen, S.A.: 1938, Z. Geschiebeforsch. Flachldgeol., 14, pp. 179-207.
13. Norin, R.: 1940, Geol. Fören. Stockh. Förh., 62, pp. 31-44.
14. Rønsbo, J.G., Pedersen, A.K. and Engell, J.: 1977, Lithos, 10, pp. 193-204.
15. Noe-Nygaard, A. and Rasmussen, J.: 1968, Lithos, 1, pp. 286-304.
16. Pedersen, G.K. and Surlyk, F.: 1977, Sedimentology, 24, pp. 581-590.
17. Walker, G.P.L. and Croasdale, R.: 1972, Bull. Volcanol., 35, pp. 303-317.
18. Walker, G.P.L.: 1973, Geol. Rundsch., 62, pp. 431-446.
19. Honnorez, J. and Kirst, P.: 1975, Bull. Volcanol., 39, pp. 1-25.
20. Heiken, G.: 1974, Smithsonian Contr. Earth Sci., 12, 101 p.

RELATIONS BETWEEN TECTONICS AND VOLCANISM IN THE ROMAN PROVINCE, ITALY

Johan C. Varekamp

Department of Geology, Arizona State University,
Tempe, Arizona 85281, U.S.A.

ABSTRACT. Tephrochronology is useful for correlation of tectonic disturbances in volcanic areas because of the usually rapid sedimentation of pyroclastic sequences. The Vulsini volcano in central Italy provides an example. Three regional tectonic phases can be distinguished during the past million years: 1. formation of a large graben, 2. sinking of segments of the volcano complex, 3. faulting, accompanied by phreatomagmatic volcanism. Each event was followed by small volume eruptions of basic, unevolved potassic magma. Subsequent eruptions produced more differentiated rocks. The tectonic activity created room for magma at shallow level, where crystallization and minor magma mixing took place.

1. INTRODUCTION

Comparison of the age of tectonic disturbances in a chain of volcanoes can furnish evidence of synchronous tectonic development in different areas, which helps in the distinction between local and regional tectonics. What field evidence of tectonic activity can we find in young volcanic areas? Several criteria, judiciously applied, may aid in identifying this activity.

1. Offsets in volcanic rock sequences, overlain by unconformable material.
2. Slumps, mudflows and coarse clastic fans.
3. Steep valleys, indicated by large thickness variations in the infilling ash flows.
4. Linearly arranged chains of scoria cones.
5. Phreatomagmatic deposits.
6. Linear zones of alteration.

S. Self and R. S. J. Sparks (eds.), Tephra Studies, 219–225.

Most of these criteria are best applied to volcanoes with low aspect ratios, which lack the high erosion rates commonly associated with an increase in relief during sedimentation. Tectonic uplift causes a steep relief, resulting in steep valleys and clastic fans.

In central Italy, magma reservoirs are situated mainly in the thick calcareous Mesozoic series, which contain several good aquifers. When these aquifers are offset, water can reach the magma reservoirs, resulting in phreatomagmatic eruptions (see (1) for a detailed discussion on the difference in topographic morphology resulting from deep and shallow phreatomagmatic eruptions). Correlation of tectonic phases is possible using offsets in marker ash beds, mantling of newly formed relief and other cross cutting stratigraphic relationships. Using these criteria, as well as absolute age determinations and tephra correlations, three tectonic phases have been identified during the past million years.

2. TECTONIC PHASES IN THE VULSINI AREA

The Vulsini area is located about 100 km north of Rome. The regional tectonic setting is shown in Figure 1. The area is situated in the southern continuation of the Radicofani-Tiber graben, with the mid-graben horst structure of Monte Cetona. This large structure formed at the beginning of the Pleistocene, and was accompanied by the initiation of potassic volcanism in the area.

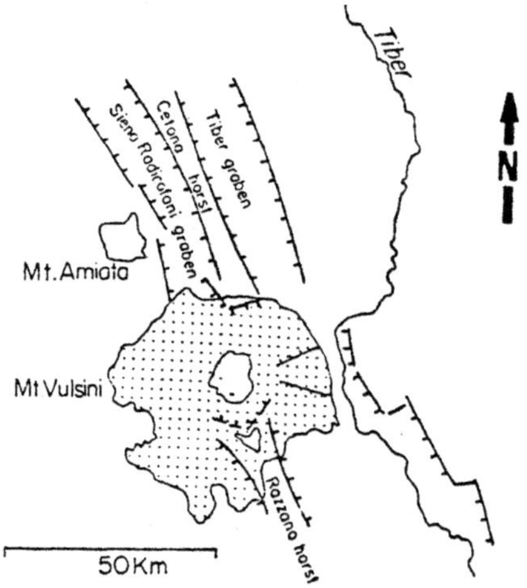

Fig. 1. Regional tectonic setting of the Vulsini area. Stippled area is Quaternary volcanics; Lake Bolsena in centre.

Detailed stratigraphic, structural and petrologic information on the Vulsini area is given by (2) and (3,4).

During early development the whole basin sank as a part of the large graben structure. Rapidly subsiding troughs became filled with tuffs and epiclastic material as well as lacustrine deposits, including diatomites (tectonic phase 1, 1.0-0.7 M.Y. ago). At the end of this period a new wave of tectonic disturbance affected the area and a graben formed NE of Orvieto (Figure 2), which was formerly a higher area. E-W faults were active east of Montefiascone and the central part of the area began to subside. The newly formed depression was partially filled with ash flows. After these eruptions, tectonic activity increased and the central part of the area continued its collapse, forming the Bolsena depression. E-W faults were active near Benano (tectonic phase 2, 0.4-0.35 M.Y.). This tectonic activity was followed by basic potassic volcanism in the east part of the area.

The Latera volcano then started its cycle of activity in the west. After the eruption of a series of large ash flows the Latera caldera formed, accompanied by, or followed directly by, tectonic activity in the whole area. This activity is

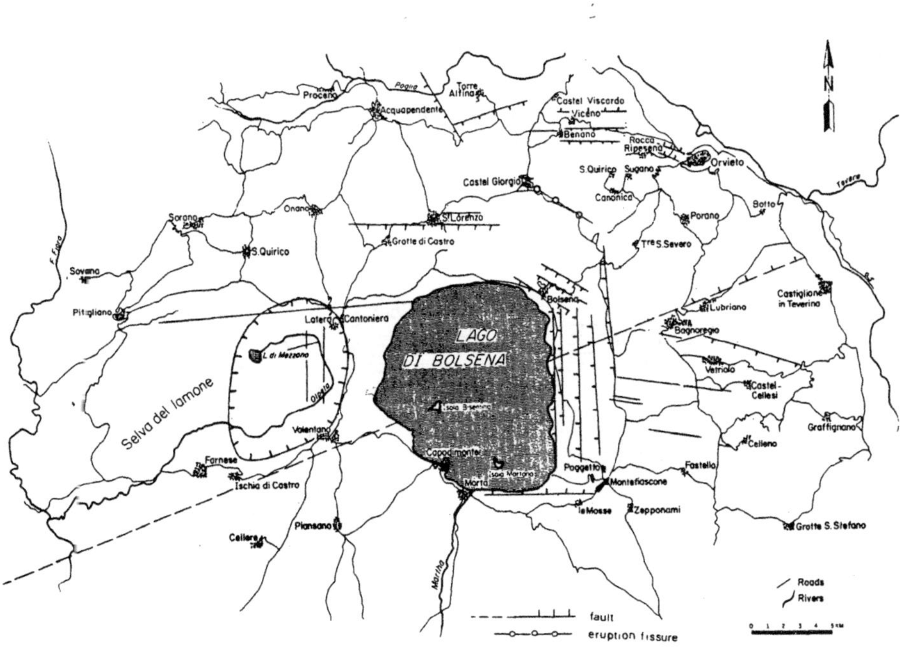

Fig. 2. Tectonic map of the Vulsini area (3).

indicated by large amounts of phreatomagmatic deposits in the Montefiascone area, which mantle the relief of the Bolsena depression. Secondary centres line up along NE-SW faults, and the islands in the Lake Bolsena formed by Surtseyan-type eruptions (tectonic phase 3, 0.1-0.5 M.Y.).

A tectonic map is given in Figure 2. During the first tectonic phase, faults of the Apennine system, running NNW-SSE were of importance, but during phase 2 N-S and E-W faults also became active. During phase 3 the Tyrrhenian fault system (NE-SW) was active. The subsidence of the central part of the area during phase 2 is considered to be related to regional tectonic movements, which also caused offsets in other parts of the Vulsini area. The term volcano-tectonic depression is thus more appropriate than the term caldera for the Bolsena depression. The volume of the basin is difficult to reconcile with a hypothesis of collapse due to withdrawal of magma, since the volume of the ash flows erupted before its collapse is far less than the volume of the depression (3).

3. RELATIONS BETWEEN VOLCANISM AND TECTONICS

The periods of strong subsidence (phase 1 and 2) were both immediately followed by periods of basic potassic volcanism. Later more silicious melts were erupted as ash flows (phonolites and trachytes).

Phase 3 is distinctly different. Following this phase a variety of magmas were erupted including trachybasalts, trachytes and leucitites. The tectonic character of this phase was also different, as transverse fault directions became more important and vertical offsets were less pronounced.

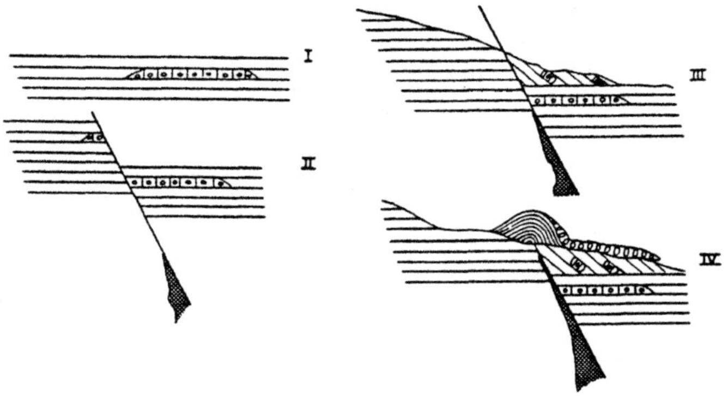

Fig. 3. Development of a scoria cone on a faulted and subsequently eroded tephra section.

Of particular interest is the time lag between the fault movements of phase 2 and the subsequent magmatism, as illustrated by two examples from the Vulsini area. Some of the Scoria cones on the east side of the Bolsena Lake are built over a fan of debris, related to the erosion of a N-S trending fault scarp (Figure 3).

Some of the lavas related to the scoria cones cover lacustrine deposits from the Bolsena Lake. The Bolsena "low" was, therefore, already filled with water when these eruptions occurred, there was a hiatus between the tectonic disturbance and the following volcanic activity. Another example is found along the road from Marta to Viterbo, near the junction to Montefiascone (Figure 4). Here an ash flow deposit and pumice series were cut by a fault, and the deposits dissected into a sharp relief by erosion. The gullies later became filled with fluvially reworked volcanics and these deposits are covered by local, leucitic lava flows and scoria. A period of erosion and fluvial deposition followed the faulting, in turn followed by magmatic activity.

The tectonic character of the phase 2 movements is normal faulting along steep fault planes. Such normal faulting, with large blocks sinking like keystones, inhibits direct rise of magma. The period of repose between tectonics and volcanism is difficult to estimate but is probably in the order of hundreds of years or less. The highly mafic and primitive compositions of the rocks erupted after phase 2 suggest that the faulting is of fundamental nature and reached deep magma reservoirs, perhaps in the lower crust or upper mantle.

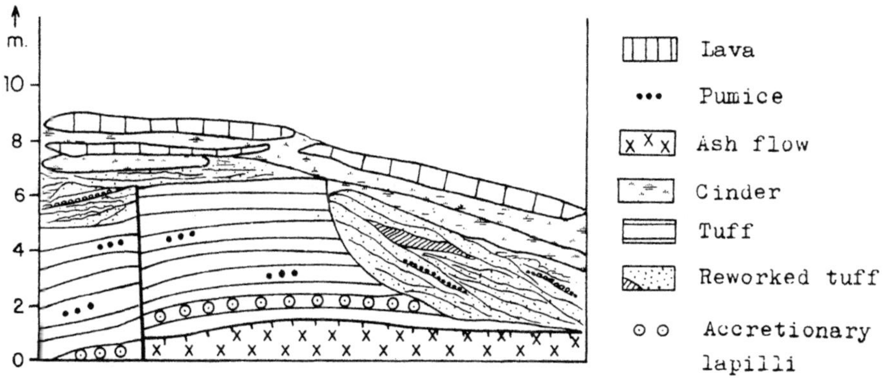

Fig. 4. Road section east of Marta.

If we compare the dated tectonic phases in the Vulsini area with tectonic disturbances in the other areas of the Roman province, similarities are obvious (5). Phase 1, the initial graben formation, can be recognized in all these areas. Phase 2 is less obvious in the Vico area, but strong subsidences occurred in the Sabatini area about 0.3 M.Y. ago. In the Alban Hills a large caldera-like structure formed about 0.4 M.Y. ago. Phase 3 is present in all the areas and thick phreatomagmatic deposits are ubiquitous, particularly in the Alban Hills and Sabatini.

4. DISCUSSION

How can we explain the observed features, given the constraints on the sequence of events and timing placed by tephrostratigraphy. Periods of strong subsidence, akin to rift tectonics (vertical subsidence in the Radicofani-Tiber graben is up to 1000 meters) may be induced by a linear hot spot zone below Italy. Rifting creates room for rising magma, but at shallow levels the normal faulting closes the pathways for the magma. It takes some time before the magma pressure opens these faults, allowing eruptions to occur.

The time lag between subsidence and related volcanic activity is thus explained. During this repose the primitive magmas of deep origin temporarily reside in shallow reservoirs, where mixing with remnant batches of more differentiated potassic magma from an earlier cycle may occur. This is indicated by non-equilibrium phenocryst assemblages found in several of these lavas (3).

A comparable model of rifting, followed by injection of a more primitive magma into an older magma reservoir that contains differentiated melts, has been discussed by (6). A period of inflation follows the rifting episode until eruptions can occur.

ACKNOWLEDGEMENTS

I thank M.F. Sheridan and J. Fink for constructive reviews of this manuscript.

REFERENCES

1. Funiciello, R., Locardi, E., Lombardi, G. and Parotto, M: 1978, Proc. Conf. Therm. Waters Geothermal Energy and Volc. of Medit. area, pp. 227-240.
2. Sparks, R.S.J.: 1975, Geol. Rundschau 64, pp. 497-523.
3. Varekamp, J.C.: 1979, Geol. Ultraectina 22, 384 p. (PhD. thesis).
4. Varekamp, J.C.: 1980, Bull. Volc. 43, 3, in press.

5. Locardi, E., Funiciello, R., Parotto, M. and Lombardi, G.: 1976, Geol. Romana 15, pp. 279-300.
6. Sigurdsson, H. and Sparks, R.S.J.: 1981, J. Petrol. 22, pp. 41-84.

DEEP-SEA TEPHRA STUDIES

QUATERNARY TEPHROCHRONOLOGY IN THE MEDITERRANEAN REGION

Jörg Keller

Mineralogisches Institut der Universität
7800 Freiburg, West Germany

ABSTRACT. Quaternary volcanism in the Mediterranean region was highly explosive, especially in the Roman-Campanian province, the Aeolian Islands and the Hellenic island arc. Pantelleria and Etna were less explosive. Tephra layers originating from all of these volcanoes have been encountered in Quaternary sediments. A summary of tephra layer occurrences in the Mediterranean Quaternary record is given in this paper. Correlation results and petrographic characterization techniques are described.

Examples for widely distributed tephra layers come mainly from sediment coring in the deep sea, but include examples on land several hundred kilometers from the inferred source volcanoes. Identified sources are mainly from the Campanian province, including Ischia, the Phlegrean volcanic area and Somma-Vesuvius. Others are from Pantelleria, Lipari, Etna and Santorini.

1. INTRODUCTION

Volcanic ash layers in Mediterranean deep-sea sediments were first described in piston cores taken by the R.V. Albatross during the Swedish Deep-Sea Expedition 1947-48 (1-3). Later coring, e.g. by the vessels Vema, Robert Conrad, Chain, Meteor and Trident and deep-sea drilling by Glomar Challenger showed that tephra layers are widespread and frequent in the central and eastern Mediterranean. They represent valuable stratigraphic markers for intercore correlations (4-6).

As many as 25 different tephra layers have been described and intercorrelated within the Upper Quaternary sequence of the

S. Self and R. S. J. Sparks (eds.), Tephra Studies, 227–244.

Eastern Mediterranean Sea (5). Petrographical tephra characterization based on chemical composition of the volcanic glass, the refractive index, the volcanic mineral assemblage and specific mineral compositions leads in most cases to an unambiguous identification of the magma type, yielding constraints on the volcanic source province. In several cases highly distinctive physical and chemical parameters allow the correlation with a specific source volcano or with one specific eruption of a particular source volcano (5-11). This is best achieved if, in addition to complete identification of petrological parameters, the exact stratigraphic position can be determined.

The framework of the stratigraphy of Quaternary sediments in Mediterranean piston cores was refined in recent years using basin-wide stagnation layers ("sapropels"), their radiometric ages and an oxygen-isotope stratigraphy, which compares the Mediterranean isotope record with the open oceans (4,12). The resulting time-scale was used by Keller et al. (5), Thunell et al. (10) and McCoy (6) in similar ways to interpolate stratigraphic ages of tephra layers between well dated isotope-stage boundaries.

In recent years two problems of Mediterranean tephrochronology have received particular attention. 1. The identification and the distribution pattern of the so-called Minoan Santorini tephra (9,13) and the possible influence of tephra-fall on the Minoan world in Crete and the Aegean. 2. The distinction of very similar tephra layers originating from recurrent highly explosive trachytic eruptions in the Campanian province (including Ischia and the Phlegrean Fields). Volcanic tephra layers of this composition are generally referred as "Campanian trachytes". A Y-zone tephra of this composition is the most widely distributed ash in piston cores from the Eastern Mediterranean. This is the Y-5 tephra, after Keller et al. (5).

2. THE IONIAN SEA

As many as 15 airborne tephra layers have been identified in the Upper Quaternary sequence of cores from the Ionian Sea (5). The stratigraphic interval containing volcanic layers covers the last 200,000 years.

Core RC 9-191 (Fig. 1) contains the most complete tephra sequence with 10 layers in one continuous profile. Of these, seven layers originate from the Roman-Campanian region. Calc-alkaline composition and east-west dispersal patterns suggest a Hellenic arc origin for two layers. These are termed X-1 and V-3 according to our stratigraphic terminology (5). A silicic peralkaline ash occurring in the Y-zone (Y-6) originated from

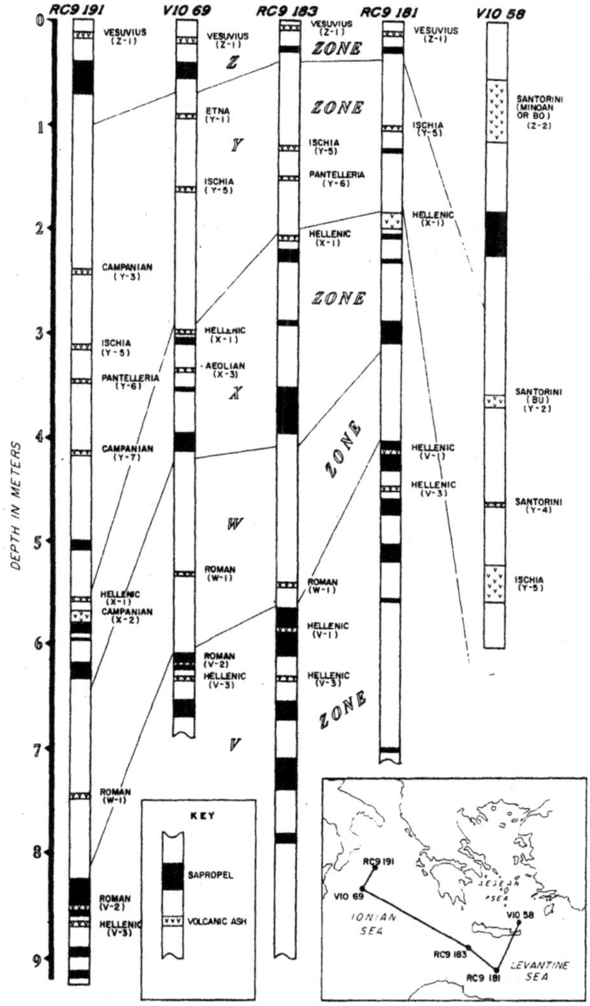

Fig. 1. Lithostratigraphic correlation and tephrochronology of representative cores from the Ionian, Levantine and Aegean Seas; after (5).

explosive activity on the Island of Pantelleria about 45,000 years B.P. This layer dates the "Green Tuff" eruption of Pantelleria (J. Wright, pers. comm. 1980).

The combined stratigraphic results from cores Robert Conrad 9-191 and Meteor M 22-60 bring the total number of Roman-Campanian tephras in the Ionian Sea to 9 different layers. They occur in Z-, Y-, X-, W- and V-zones. The Z-zone Vesuvius leucite tephritic

Table 1. Mineralogical composition of Eastern Mediterranean deep-sea tephra-layers.

Layer	Core	Olivine	Augite	Aegirine - Augite	Acmite	Hypersthene	Hornblende	Biotite	Apatite	Sphene	Zircon	Melanite	Melilite	Na - Amphibole		Plagioclase	Sanidine	Sodalite	Leucite	Refractive index n	Petrographic type
X-1	RC-9-181,190		xxx			xxx			x							xxx				1.546 ± 0.004	calc-alkaline, andesitic
V-1	RC-9-181,410		xxx			xxx	x		x							xxx				1.509 ± 0.002	calc-alkaline, rhyodacitic
V-3	RC-9-183,632		xxx			xxx			x							xxx				1.517 ± 0.003	calc-alkaline, dacitic
Z-1	RC-9-191, 15	x	xxx				xx	x	x	x		x				xxx	xxx		xxx	1.511 ± 0.003	leucite-tephritic
Y-3	RC-9-191, 245	x		xxx			x	xxx	x							xx	xxx			1.520 ± 0.002	trachytic
Y-5	RC-9-191, 310			xxx			x	xxx	x	x						xx	xxx	x		1.520 ± 0.004	trachytic
Y-6	RC-9-191, 345			xxx					xx					xx		xx	xxx			1.525 ± 0.002	peralkaline
Y-7	RC-9-191,415			xxx	xx			xx	x	x	x					xx	xxx			1.519 ± 0.002	trachytic
X-2	RC-9-191,575			xxx			x	xx	x	x						xx	xxx	x		1.521 ± 0.002	trachytic
W-1	RC-9-191,750		xxx					xx	x							xx	xxx		x	1.531 ± 0.004	tephritic
V-2	RC-9-191,850		xxx					xx	x				x			xx	xxx		x	1.526 ± 0.002	melilite-tephritic
X-3	V10-69, 335	x	xxx			xxx	xxx									xxx				1.55 ± 0.01	calc-alkaline, andesitic
Y-8	22-M-60,481		xxx			xxx	x		x		x					xxx				1.55 ± 0.01	calc-alkaline, andesitic
X-4	22-M-60,708	x	xx				xx		x							xxx				1.56 ± 0.01	alkalibasaltic
X-5	22-M-60, 755			xxx				xx	x	x						xxx	xxx			1.522 ± 0.003	trachytic
X-6	22-M-60, 775			xxx			x	xx	x	x						xx	xxx			1.520 ± 0.003	trachytic

ash and the Roman W-1 and V-2 tephrites are highly distinctive in their chemical composition and in their mineral content (Table 1). Major difficulties still exist in distinguishing trachytic tephras. Six layers in the Ionian Sea (Y-3, Y-5, Y-7, X-2, X-5, X-6) are of very similar composition which is referred as "Campanian alkali-trachyte type".

Petrographic characterization and correlation of the different Campanian alkali-trachytes was attempted using diagnostic phases among the phenocrysts, and chemical composition. Chemical methods use two different approaches to derive representative results. X-ray fluorescence analysis can be done on a mechanically separated glass shard fraction, whereas microprobe analysis gives single glass shard compositions. Both methods present inherent problems. For Campanian alkali-trachytes amongst the most diagnostic chemical parameters are total alkali content and K_2O/Na_2O ratio (Fig. 2). Loss of alkalies during microprobe analysis heavily affects these parameters. For microprobe work on volcanic glass shards the major problem is the uncontrolled volatilization of sodium. Low probe currents, large beam diameter and short counting time can contribute to overcome this problem. Figure 3 gives an example for the effects of sodium loss. It also shows that sodium contents can be considered reliable if a large scanning area on a SEM or a large beam diameter on an

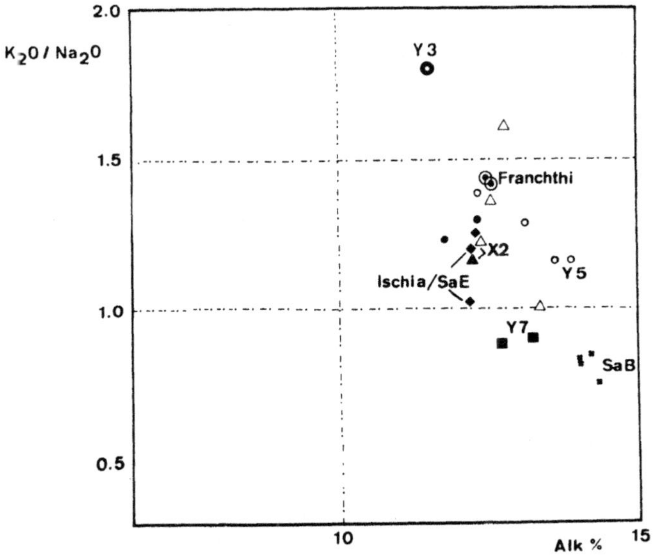

Fig. 2. K_2O/Na_2O vs. alkalies diagram of various "Campanian trachyte type"-tephras. Tephra designations as in Figs. 1 and 5. Open symbols for microprobe data, full symbols for XRF analyses.

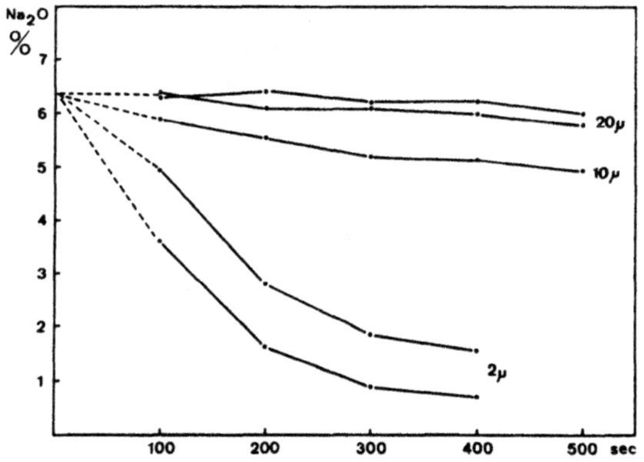

Fig. 3. Apparent sodium contents in microprobe analysis depending on beam diameter (scanning area) and counting time for alkali trachytic volcanic glass shards. Analytical conditions: scanning electron microscope with Li(Si) energy dispersive analytical equipment (15kV/1nA).

electron microprobe are used. Table 3 compares XRF analytical data on a bulk sample with energy dispersive microprobe results on single glass shards. It is shown that both methods give comparable results.

Figure 4 uses a graphic representation proposed by Thunell et al. (10), which avoids the consideration of sodium. The FeO-K_2O ratio distinguishes between an Ischia-group and a Campanian Ignimbrite group. In the latter field, however, available data for many more "Campanian alkali-trachytes" also fall.

The use of microprobe analyses of phenocrysts in these trachytic tephras may provide additional discriminants for this problematic group, which constitutes the dominant material in Mediterranean tephrochronology. Such studies are currently under way (Morche and Keller, unpubl.).

3. THE TYRRHENIAN SEA AND ADRIATIC SEA

Central Italy is the source for the majority of widespread tephra layers in Quaternary sediments of the Mediterranean area. Accordingly, the Tyrrhenian and Adriatic basins, which are close to the source volcanoes, must bear important tephrochronological information.

However, in this review of the state-of-the-art in Mediterranean tephrochronology it must be pointed out that detailed

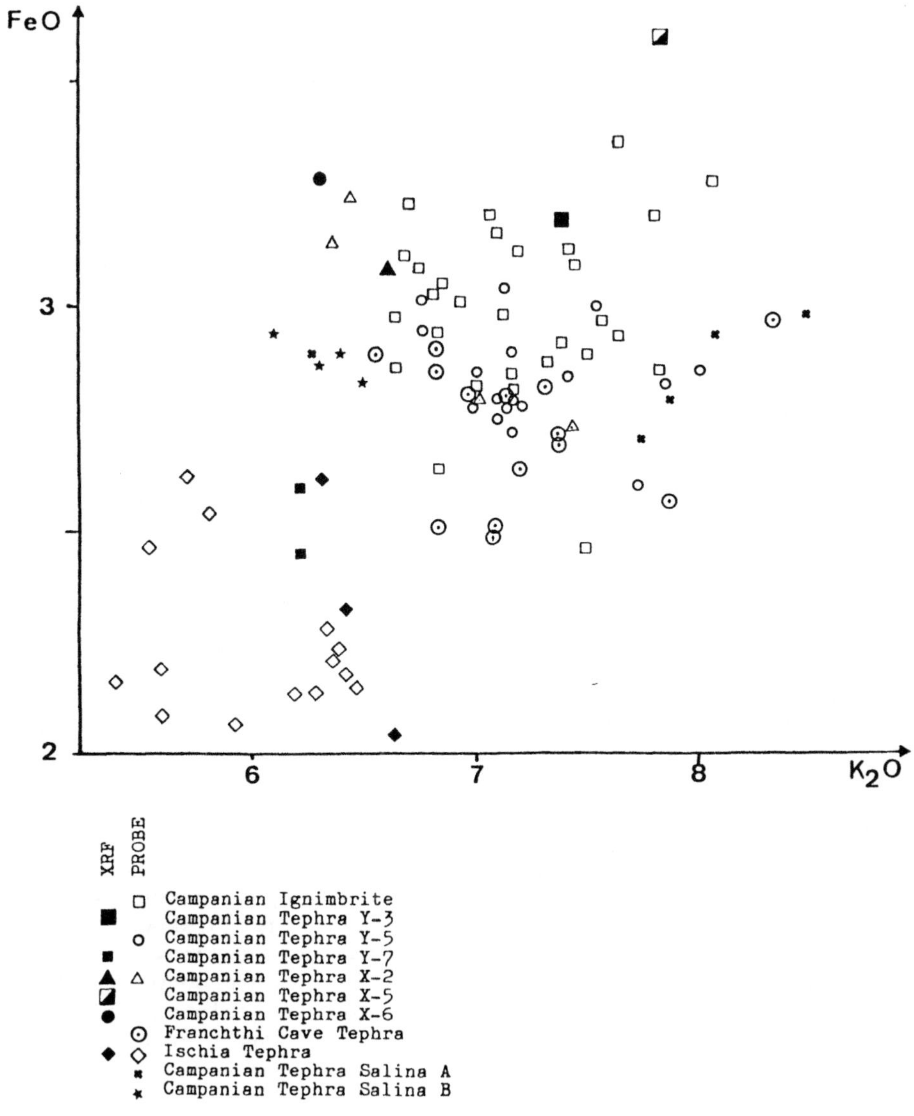

Fig. 4. Chemical comparison of glass composition in Campanian alkali-trachyte tephras using the FeO (total iron) versus K_2O plot proposed by Thunell et al. (10). This diagram discriminates between a "lower Fe/lower K" group which includes samples from Ischia, the Ischia-tephra of the Aeolian Islands and tephra Y-7, and a "higher Fe/higher K" group, which includes the Grey Campanian Ignimbrite and various similar tephra layers.

results are still awaited from these areas. High sedimentation rates in the Tyrrhenian and the Adriatic Sea are the reason that piston cores seldom reach a comparable stratigraphic depth as in the Ionian cores. Cores containing ash layers have been reported from the Adriatic Sea by van Straaten (14), and for the Tyrrhenian Sea by Norin (2), Ryan (4) and Ryan et al. (15). Very limited petrographic data exist. Cita et al. (12) have attempted on a purely lithostratigraphic basis, to correlate piston cores from the Central Tyrrhenian Sea and the DSDP Site 132-profile with the Ionian sequence. Their Fig. 12 shows up to seven tephra layers in a stratigraphic sequence spanning the last 100,000 years.

A detailed petrographic characterization of all these tephras remains to be established. However, available indications (2, 16, unpublished results) point again to a Campanian origin for almost all of the Tyrrhenian and Adriatic tephras. A few individual layers in the Southern Tyrrhenian Sea are different and may correlate with the Aeolian Islands and with the great submarine volcanoes such as Marsili Seamount (7).

Tephrochronological studies on land were undertaken on the Aeolian Islands, because these islands have a position halfway between the Campanian source volcanoes and the important tephrochronological sequences in the Ionian deep-sea cores. Five different layers of "Campanian trachytes" so far have been detected in the Upper Pleistocene of these islands (Fig. 5). Mineralogical and chemical compositions are currently being determined and correlations with Ionian cores such as RC 9-191 are being established (W. Morche, unpublished).

A prominent layer on the Aeolian Islands is the alkalitrachytic Ischia-Tephra (22) with estimated age brackets of approximately 25,000 and 40,000 years B.P. Fig. 5 shows the Ischia-Tephra (Sa E) within the idealized stratigraphy of the island of Salina. The stratigraphic relationship between this widespread Ischia tephra and the Y-5 layer, correlated with the Grey Campanian Ignimbrite (17,25), remains a major problem for tephrochronological studies in the Mediterranean.

4. AEGEAN AND LEVANTINE SEA TEPHROCHRONOLOGY

Eastern Mediterranean tephrochronology is based mainly on cores V 10-58 (13), RC 9-181 (4,5) and TR 172-19 and 22 (10,17). These cores contain the most complete tephra sequences. Fig. 6 summarizes the stratigraphy of tephra layers in the Eastern Mediterranean and Aegean deep-sea sediments. At present the following layers have been recognized and characterized:

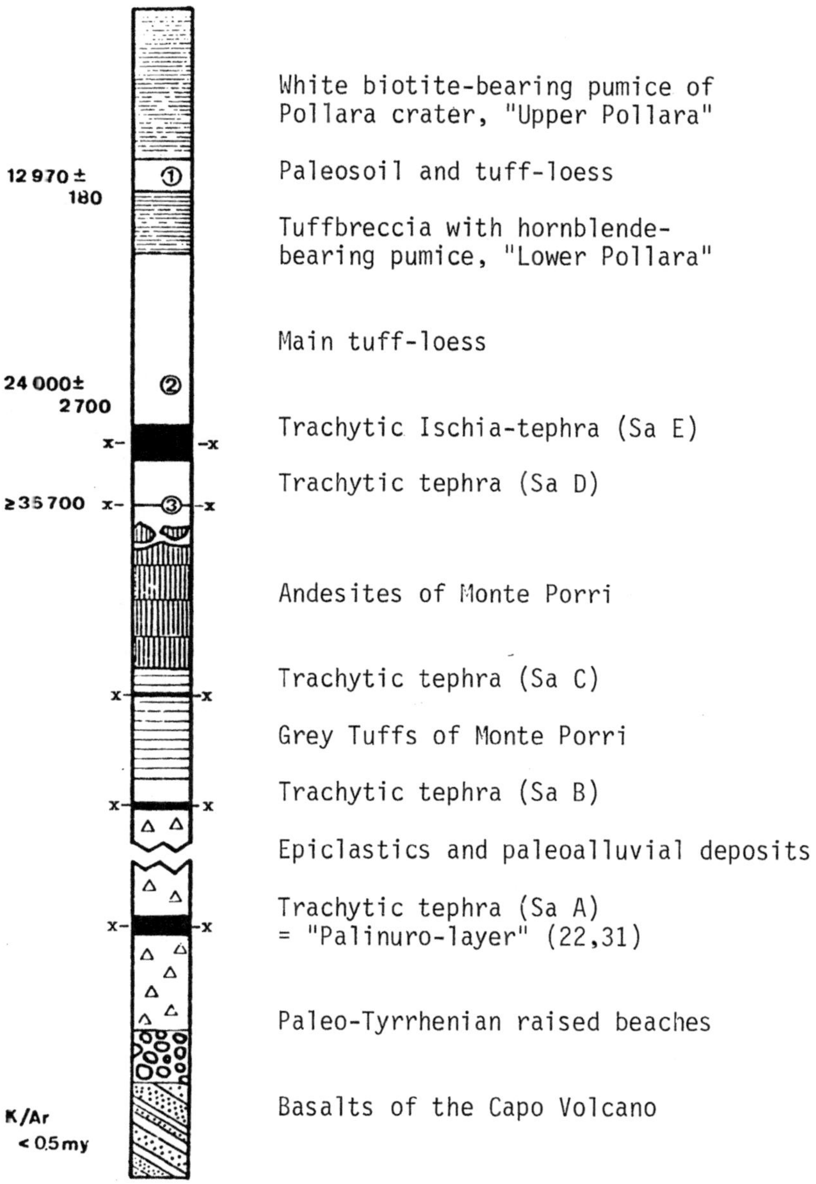

Fig. 5. Alkali-trachytic tephra layers, "Campanian trachytes" within the local volcanic stratigraphy of Salina, Aeolian Islands (Southern Tyrrhenian Sea); after (22) with additional unpublished data of Morche and Keller. Numbers are C-14 ages B.P.

The post-glacial Vesuvius ash (Z-1) is widespread in Ionian cores and still present south of Crete in core RC 9-181. This ash is characterized by its tephritic composition and its distinctive mineralogy with leucite and melanite (Table 7).

The Minoan Santorini (or Thera) tephra (Z-2) has received major attention in Aegean tephrochronology because of its possible influence on the Minoan world about 3,500 years ago. Recent studies of Watkins et al. (11), Keller (18) and McCoy (8) reviewed the distribution pattern and agree in a strictly easterly dispersal axis extending at least 600 km from the source volcano. Occurrences on land have been identified on the islands of Kos (18) and Rhodes (19), both approximately 200 km from Santorini. Correlation of the Minoan ash is now based upon the most complete spectrum of petrographical criteria, including refractive index, chemical bulk composition, trace elements, glass shard composition and phenocryst content. An example of tephra correlation and distinction with the aid of mineral compositions is given in Fig. 8. These data help to discriminate between such chemically similar ashes as the Upper and Lower Pumice Series on Santorini (20,21) (Fig. 7 and Tab. 2).

Two Santorini layers of minor thickness and dispersion occur within the Y-zone. These are named Y-2 and Y-4 in the terminology of Keller et al. (5). Y-2 can be correlated with the 18,000 years old Akrotiri ignimbrite on Santorini (10,17,20). Y-4 has a limited core V 10-58. Its extrapolated age on the oxygen-isotope scale is clase to 30,000 years B.P. In addition, Thunell et al. (10) and Federman and Carey (17) report a Y-zone pumice tephra from Yali in the Eastern Mediterranean deep-sea sediments. The age of this "YALI C" tephra is 31,000 years (17). Its silicic composition differs from the more andesitic characteristics of the Santorini Y-4. The Yali tephra and Y-4 were not found in the same core. Thus, the relative positions in the schematic profile of Figure 6 is a tentative suggestion.

The alkali-trachytic tephra Y-5 is of Campanian origin (22,23). Keller (23) suggested a source volcano on the Island of Ischia, but Thunell et al. (10) showed a closer match of chemical composition with the Grey Campanian Ignimbrite (25). The Franchthi Cave deposit from the Peloponnesus (24) is very close in composition (Table 3) and is correlated with Y-5 tephra (26). The age of Y-5 on the oxygen isotope time-scale is 38,000 years (5,10).

The distribution and composition of a brown tephra, called Hellenic Andesite X-1, is discussed in Keller et al. (5). A tephritic tephra, W-1, which is correlated with the Roman potassic province (5), is only found in cores from the Western Levantine and the Ionian Sea.

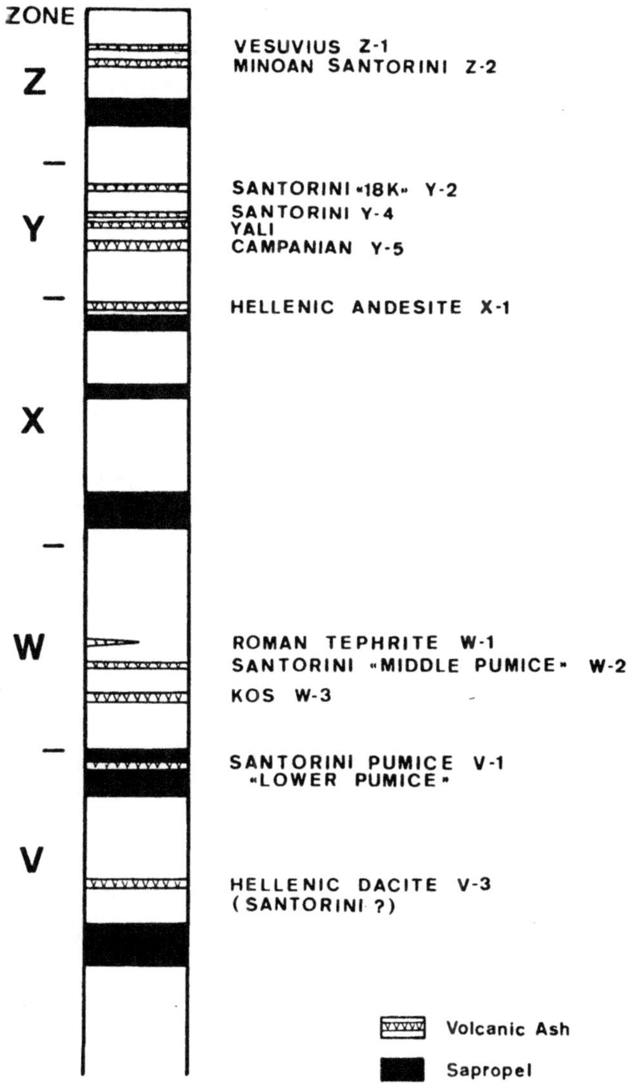

Fig. 6. Schmatic summary of the tephra-layer sequence in deep-sea sediments from the Eastern Mediterranean and Aegean Sea; data mainly from (5,6,10,13,17).

The W-2 layer was found by Federman and Carey (17), in only one core of the Trident cruises (TR 172-19-398 cm). The position of this tephra corresponds to approximately 100,000 years on the oxygen-isotope stratigraphy. According to its compositional characteristics this layer is correlated with the Middle Pumice Series of Santorini (17).

Table 2. Microprobe analyses of glass shards from Santorini Thera tephras and the V-Zone tephra 1, from the Levantine Sea. Analysis with TPD-microprobe and ORTEC energy-dispersive detector at R.S.E.S., Australian National University, Canberra. Column 1 by courtesy of N.G. Ware, Canberra.

	Minoan Thera Pumice	Lower Thera Pumice (BU)		RC-9-181 410 cm
	1	2	3	4
SiO_2	70.76	67.8	67.26	67.95
TiO_2	0.31	0.25	0.26	0.21
Al_2O_3	13.70	14.22	14.11	14.17
FeO	2.06	2.02	2.36	2.33
MgO	0.33	0.28	0.43	0.35
CaO	1.43	1.22	1.41	1.22
Na_2O	4.64	4.12	4.32	4.08
K_2O	3.34	3.51	3.16	3.16
Cl	0.36	0.34	0.36	0.30
Total	96.93	93.76	93.65	93.77
H_2O (by difference)	3.0	6.0	6.0	6.0

1 Minoan Pumice, Fira quarries, Thera (22).

2 Lower Thera pumice, main chaotical pumice unit near C. Alai. Average of 3 analyses.

3 Lower Thera pumice, basal air-fall layer, C. Alai. Average of 4 analyses.

4 Tephra layer in core RC-9-181, 410 cm, V-1 tephra, intercalated in W-V zone sapropel (5). Average of 5 analyses.

The Kos Plateau Pumice results from one of the most powerful eruptions in the Aegean during the Pleistocene (22). A deep-sea tephra in the W zone (W-3) is correlated with this rhyolitic eruption (6,17). Also Thunell et al. (10) refer to a Kos Plateau Pumice tephra in their Eastern Mediterranean cores. Federman and Carey (17) give an extrapolated age of 120,000 years B.P. for the Kos eruption.

The deepest piston cores in the eastern Mediterranean penetrated into the V-zone, about 200,000 years B.P., according to the time-scale adopted (5). Two tephra-layers, V-1 and V-3, occur in the deepest interval (Fig. 1). Both layers have silicic calc-alkaline compositions (rhyodacitic and dacitic). These characteristics show a Hellenic arc volcanic source. McCoy (6) tentatively correlated V-1 with the Lower Thera Pumice ("Unterer Bimssteinhorizont"). This discussion is possible since Seward et al. (27) provided a new fission track date for the Lower Pumice on Thera. Their age of about 100,000 years is much older than former estimates. However, there is still a stratigraphic discrepancy, since the estimated age of V-1 is about 160,000 years on the time scale of the deep-sea stratigraphy. The chemical composition of V-1 glass shards is compared with the Lower Thera Pumice in Figure 7 and Table 2. On the basis of these data a correlation between the V-1 layer and the Lower Thera Pumice seems possible. The composition of V-3 points again to Santorini as the source. However, it is not yet established whether V-3 correlated with one of the smaller pumice layers, which occur below the Lower Pumice (BU) in the Santorini caldera-wall profiles, or with an unexposed larger unit. Based upon the oxygen-isotope data, an age of about 180,000 years B.P. is suggested for V-3.

5. TEPHRA LAYERS FROM DSDP-DRILLING DURING LEG 42A IN THE MEDITERRANEAN

Only preliminary data are published by Nesteroff et al. (16), for up to 40 layers of volcanic material found in the complete Quaternary sequence cored at site 132 in the Central Tyrrhenian Sea during Leg 13. Tephra layers are frequent in the upper part of the Quaternary sequence and frequency diminishes down hole. During Leg 42 at site 373 in the Central Tyrrhenian Sea, one core was taken in Quaternary sediments from 96.5 to 106.0 m below sea bottom (28). This core (core 373-1) represents a stratigraphic interval in the middle part of the Pleistocene; unfortunately a more precise stratigraphic interval was not obtained. The sediments encountered in Core 1 contained 4 tephra layers with the following sample position (each section is 150 cm long):

373: Core 1, section	1	108 - 110 cm	= Tephra 373-1
	1	118 - 120 cm	= Tephra 373-2
	3	88 - 90 cm	= Tephra 373-3
	3	140 - 142 cm	= Tephra 373-4

Chemical and mineralogical data (Table 4) define two very similar rhyodacitic ashes in Section 1 and two potassic trachytes (with $K_2O/Na_2O>2.0$) in section 3. If a hypothetical age of

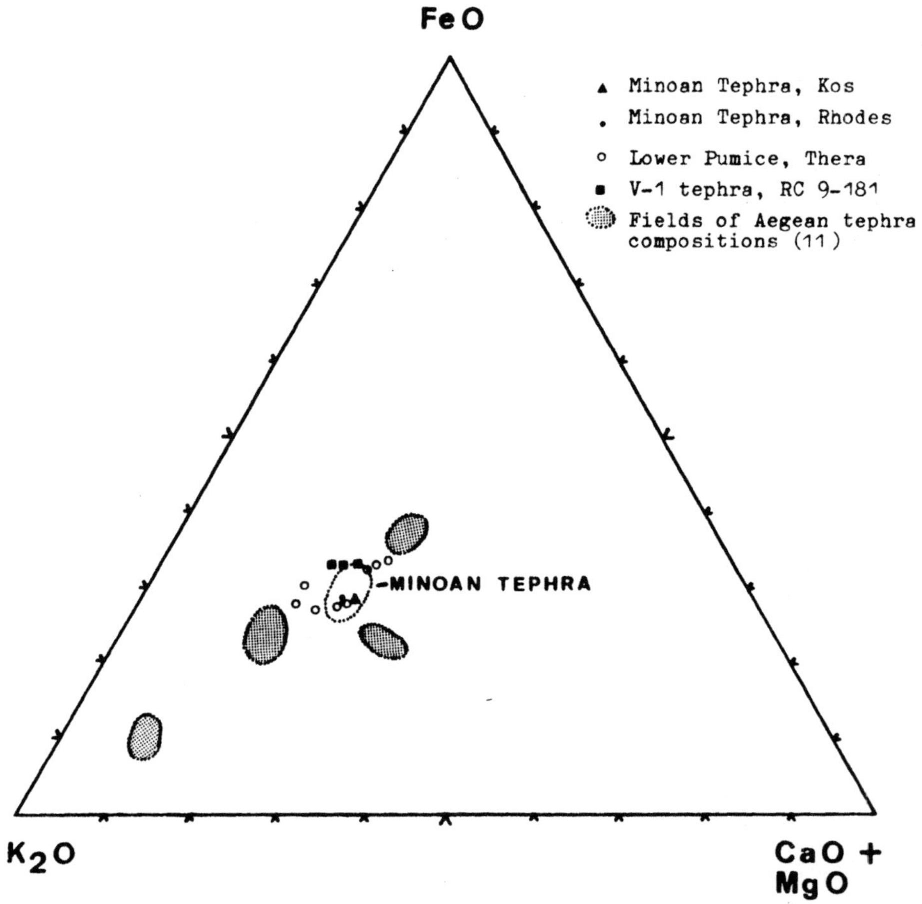

Fig. 7. K_2O-FeO-(CaO+MgO) plot of analyses of glass shards from Aegean tephra layers. Various fields of Aegean tephra compositions are taken from (11). Tephra layers from Kos and Rhodes are identical with the Minoan Thera pumice. In this diagram, the Lower Thera pumice and tephra V-1 from core RC-9-181 are similar and plot close to the Minoan tephra field.

approximately 500,000 years (28) is taken for this cored "Middle Pleistocene" sequence, which is about 100 meters below the sea bottom, the silicic calk-alkaline ashes could belong to the latest activity in the Tuscany province. A climactic period occurred contemporaneously in Roman province potassic volcanism (29), to which the potassic ashes in section 3 can be related.

Site 376 was drilled on the Florence Rise to the west of Cyprus (30). Ash-layers were recovered in Core 1, which penetrated

Table 3. Comparison between analytical results of two different methods for the Franchthi Cave tephra.

	1	2	
SiO_2	57.55	57.6	57.08 - 58.54
TiO_2	.4	.28	0.22 - 0.32
Al_2O_3	17.7	17.9	17.81 - 18.04
Fe_2O_3	3.33	3.1	2.87 - 3.15
MnO	.18	.12	0.13 - 0.20
MgO	.48	.37	0.33 - 0.40
CaO	2.23	1.73	1.66 - 1.83
Na_2O	5.15	5.1	4.19 - 5.73
K_2O	7.36	7.35	6.99 - 7.92
Cl	--	.77	0.63 - 0.84
	94.38	94.22	
H_2O (by difference)	5.0	5.0	

1) XRF-analysis on a separated glass shard concentrate.
2) Energy dispersive microprobe analysis on single glass shards (average and range of 4 determinations. Analyses by N. Ware, Canberra)

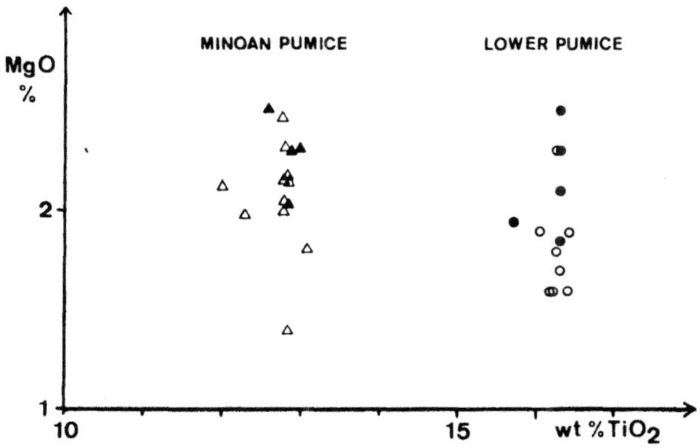

Fig. 8. Magnetite chemistry discriminating between Minoan Thera pumice and Lower Thera pumice series. Open symbols are for magnetites from basal air-fall layers, full symbols for main chaotical pumice units.

9 m below sea bottom. The core is thought to be Upper Pleistocene in age. Samples of volcanic ashes were received from the following sections:

376:	Core 1, section 5	89 - 91 cm:	Tephra 376-1
	Core 1, cc, about 9 meters sub-bottom:		Tephra 376-2
	Core 3, section 1	129-131 cm:	Tephra 376-3

Results of chemical and petrographic analysis are reported in Table 4. Tephra 376-3 which occurs about 18 m sub-bottom

Table 4. Composition of tephra-layers found in DSDP Leg 42A at Sites 373 (Tyrrhenian Sea) and 376 (West of Cyprus). Averages of 5-12 analyses, analytical equipment as in Table 2.

Microprobe analyses normalized to 100%, volatile-free, total iron as FeO

	373.1	373.2	373.3	373.4	376.1	376.2
SiO_2	71.2	70.8	59.9	62.6	71.6	77.5
TiO_2	.51	.54	.17	.26	.42	.03
Al_2O_3	14.6	14.53	20.7	19.24	15.06	13.27
FeO	3.63	3.82	2.54	2.57	3.04	.45
MgO	.84	.88	.3	.56	.67	.39
CaO	2.5	2.62	3.6	2.18	1.89	.71
Na_2O	3.0	3.14	3.95	4.55	4.2	3.2
K_2O	3.6	2.6	8.7	8.0	3.0	4.6
Inferred H_2O:	5%	5%	5%	5%	3.5%	8%
Refractive index:	1.518	1.521	1.520	1.517	1.512	1.492
Phenocrysts:	rare Fsp, Cpx	Fsp, Cpx, Opx, Bi	Fsp, Cpx, Tit, Amph, Bi	Fsp, Cpx, Tit, Amph, Bi	Bi, Amph, Cpx, Opx Ap, Zr	Bi, Amph, rare Cpx Opx Ap, Zr

Phenocryst abbreviations: Fsp = feldspars, Cpx = augite, Opx = hypersthene, Bi = biotite, Amph = amphibole, Ap = apatite, Tit = sphene, Zr = zircon.

is phenocryst-free and too fine-grained for glass analysis. The refractive index of 1.492 ± 0.003 shows a rhyolitic composition. All analyzed layers have calc-alkaline, silicic compositions. This is in accordance with a possible origin in the Hellenic arc. None of the layers has a Santorini mineralogy. Tephra 376-2 is close in composition to the Plateau Tuff of Kos (22). However, definitive correlation with this tephra would require a precise stratigraphic attribution of both source eruption and tephra-layer. Correlation of a W-zone tephra (W-3) with the Kos plateau-tuff was also suggested in piston-core studies (6, 10,17).

REFERENCES

1. Mellis, O.: 1954, Deep-Sea Research 2, pp. 89-92.
2. Norin, E.: 1958, Reports Swedish Deep-Sea Exped. 1947-48, 8, pp. 4-136.
3. Olausson, E.: 1961, Reports Swedish Deep-Sea Exped. 1947-48, pp. 337-391.
4. Ryan, W.B.F.: 1972, in: The Mediterranean Sea (D.J. Stanley, ed.), pp. 149-169.
5. Keller, J., Ryan, W.B.F., Ninkovich, D. and Altherr, R.: 1978, Geol. Soc. America Bull. 89, pp. 591-604.
6. McCoy, F.W.: 1980-81, Marine Geology, in press.
7. Keller, J. and Leiber, J.: 1974, METEOR-Forschungsergebnisse C, 19, pp. 62-76.
8. McCoy, F.W.: 1980, in: Thera and the Aegean World (C. Doumas, ed.) 2, pp. 57-78.
9. Ninkovich, D. and Heezen, B.C.: 1965, Colston-Paper 17, pp. 413-453.
10. Thunell, R.C., Federman, A.N., Sparks, R.S.J. and Williams, D.F.: 1979, Quaternary Research 12, pp. 241-253.
11. Watkins, N.D., Sparks, R.S.J., Sigurdsson, H., Huang, T.C., Federman, A.N., Carey, S. and Ninkovich, D.: 1978, Nature 271, pp. 122-126.
12. Cita, M.B., Chierci, M.A., Ciampo, G., Moncharmont Zei, M., D'Onofrio, S., Ryan, W.B.F. and Scorziello, R.: 1972, Initial Reports of the DSDP 13, pp. 1263-1339.
13. Keller, J. and Ninkovich, D.: 1972, Zeitschrift Deutsche Geol. Ges. 123, pp. 579-587.
14. Straaten, L.M.J.U.: 1967, Revue Géogr. Phys. et Géol. Dynam. 9, pp. 219-240.
15. Ryan, W.B.F., Workum, F. and Hersey, J.B.: 1965, Geol. Soc. America Bull 76, pp. 1261-1282.
16. Nesteroff, W.D., Wezel, F.C. and Pautot, G.: 1972, Initial Reports of the DSDP 13, pp. 1034-1035.
17. Federman, A.N. and Carey, S.N.: 1980, Quaternary Research 13, pp. 160-171.
18. Keller, J.: 1980, in: Thera and the Aegean World (C. Doumas, ed.), 2, pp. 49-56.
19. Doumas, C. and Papazoglou, L.: 1980, Nature 287, pp. 322-324.
20. Günther, D. and Pichler, H.: 1973, Neues Jahrb. Geol. Paläont. Monatshefte 1973, pp. 394-415.
21. Friedrich, W.L. and Pichler, H.: 1976, Nature 262, pp. 373-374.

22. Keller, J.: 1971, in: Acta Intern. Scientific Congr. on the Volcano of Thera (S. Marinatos, ed.), pp. 152-167.
23. Ninkovich, D. and Heezen, B.C.: 1967, Nature 213, pp. 582-584.
24. Farrand, W.R.: 1977, Geol. Soc. America Abstracts with Programs 9, p. 971.
25. Barberi, F., Innocenti, F., Lirer, L., Munno, R., Pescatore, T. and Santacroce, R.: 1978, Bull. Volcanol. 41, pp. 10-31.
26. Vitaliano, C.J., Farrand, W.R., Jacobsen, T.W. and Taylor, S.R.: 1980, this volume.
27. Seward, D., Wagner, G.A. and Pichler, H.: 1980, in: Thera and the Aegean World (C. Doumas, ed.) 2, pp. 101-108.
28. Hsü, K.J. et al.: 1978, Initial Reports of the DSDP 42, pp. 151-174.
29. Evernden, J.F. and Curtis, G.H.: 1965, Curr. Anthropology 6, pp. 343-364.
30. Hsü, K.J. et al.: 1978, Initial Reports of the DSDP 42, pp. 219-304.
31. Lirer, L., Pescatore, T. and Scandone, P.: 1967, Atti Accademia Gioenia Sci. Nat., Catania 18, pp. 85-115.

AREAL DISTRIBUTION, REDEPOSITION AND MIXING OF TEPHRA WITHIN DEEP-SEA SEDIMENTS OF THE EASTERN MEDITERRANEAN SEA

Floyd W. McCoy

Lamont-Doherty Geological Observatory, Palisades, New York 10964

ABSTRACT

Regional distributions of tephra deposits within Upper Quaternary pelagic sediments of the eastern Mediterranean and southern Aegean Seas have been determined from cores of deep-sea sediments. Reconstructions of the extent and thickness of individual ash layers required an evaluation of physical and biological redepositional processes, as well as possible mixing effects during sampling. Slumping or other mass-wasting processes on the sea-floor have significantly altered tephra distributions, as illustrated by regional variations in isopachs for the Z-2 Minoan and Y-5 Campanian Tuff ash layers; local variability can be equally significant as indicated by thickness variations of the Y-5 Campanian Tuff ash layer in a small area of the Mediterranean Ridge. Additional dispersal or mixing due to oceanic currents and biological activity, the latter primarily by benthic burrowing organisms, have also been important physical processes in both pre- and post-burial modification of ash layers.

INTRODUCTION

Within the Upper Quaternary stratigraphic sequence of deep-sea sediments in the eastern Mediterranean and southern Aegean Seas, twenty-five layers of volcanic ash are known. They are a remarkable documentation of volcanic activity during the past 200,000 years in this area, a record that extends into man's cultural history through the possible devastation of the Minoan culture by a ca 1400 B.C. eruption of Thera (Santorini) and the burial of Pompeii by a 79 A.D. eruption of Vesuvius. Each tephra deposit is a characteristic stratigraphic horizon; their

S. Self and R. S. J. Sparks (eds.), Tephra Studies, 245–254.

tephrochronology forms a valuable stratigraphic framework - particularly in conjunction with the sapropel chronology - for correlating and interpreting the complexities of sedimentation within the various physiographic basins that form the eastern Mediterranean and southern Aegean sea-floor.

A generalized sapropel and tephrochronology is summarized in Figure 1 as interpreted from numerous piston cores from the Ionian and western Levantine Seas (1) and from the eastern Levantine basin (2) in association with corresponding paleotemperature variations and oxygen-isotope stages identified from $\delta^{18}O$ measurements. Sapropel nomenclature follows that of (2); age designations in Figure 1 are inferred from ^{18}O stages (1), radiometric (fission track) dates (2), and marine planktonic foraminifera zones (1).

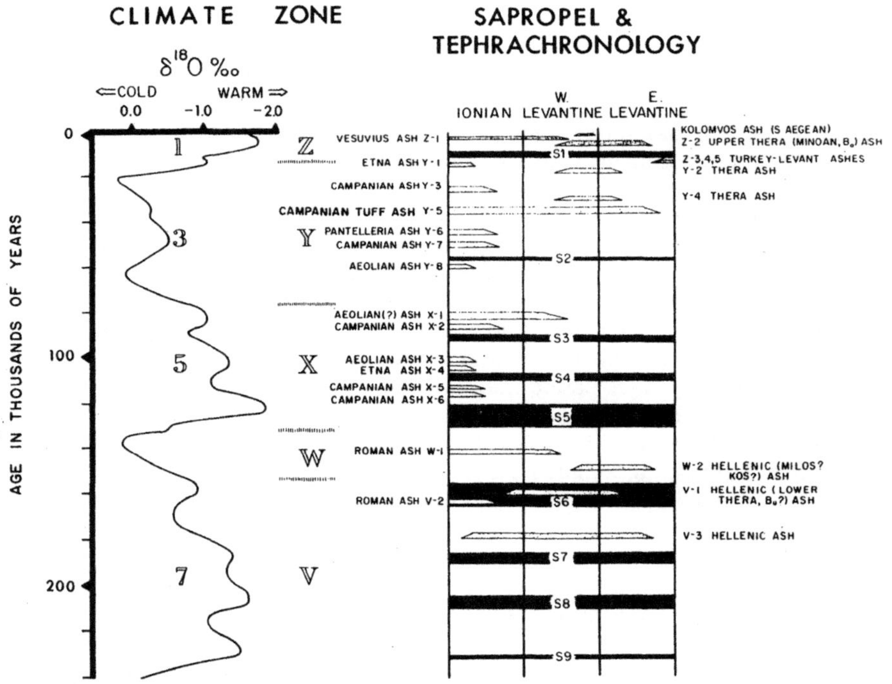

Figure 1. Sapropel chronology and tephrochronology for the Ionian, western Levantine and eastern Levantine basins in the eastern Mediterranean Sea (modified after 1, 2).

AREAL DISTRIBUTION OF TEPHRA

The areal distribution of ash layers found in cores is outlined in Figures 2 and 3 with accepted alpha-numeric designations assigned from their chronostratigraphic position within planktonic foraminifera zones (1, 3). The two deposits with the largest areal extent are the Z-1 Vesuvius ash (the 79 A.D. eruption that buried Pompeii), and the Y-5 Campanian Tuff ash (4) (formerly identified as the "Ischia" ash).

Many of these ash deposits have been inadequately sampled in piston cores and thus remain poorly studied: the Z-3 and Z-4 tephra south of Turkey, the Y-8 and X-3 ash layers probably derived from eruptions in the Aeolian islands area, the Y-6 deposit possibly correlative with the Green Tuff on Pantelleria (5), the X-4 tephra apparently derived from Etna, and older tephra that have been recovered in only a few long cores such as the W-2 or V-2 ashes. Few are known in outcrop, with the notable exception of the Z-1 Vesuvius, Z-2 Minoan, Y-4 Thera and Y-5 Campanian Tuff ash layers (4,6,7,8,9). Only two layers, the Z-2 Minoan and Y-5 Campanian Tuff ashes, have been studied in detail from deep-sea cores (1,2,3,4,10,11,12) including the now classic account of the Minoan (Santorini) tephra (13). Subsequent studies of the Minoan tephra have been stimulated by both geological and archeological interests, most recently by a second international conference on Thera in 1979 (14).

Available information for estimating the tephra layer distributions given in Figures 2 and 3 clearly indicate the wide areal extent of these layers and their potential as stratigraphic marker beds. Suggested boundaries for each ash layer are approximate and obviously subject to modification as each deposit receives additional study. Studies in progress on the Y-1 Etna tephra sampled in cores from the Ionian Sea, for instance, suggest that this deposit may extend further to the east onto the Mediterranean Ridge than shown in Figure 2 (15).

REDEPOSITION AND MIXING OF TEPHRA

The portrayal of tephra deposition in Figures 2 and 3 has been established from only those cores that presumably recovered stratigraphic intervals of appropriate geologic age. Definition of depositional limits is, and must be, based upon the absence of ash layers in a number of cores rather than in only one or two cores because of possible pene- and post-depositional layer modification due to slumping, biological activity, current dispersal, mechanical mixing by the coring process, and other redispersal processes. These effects can be significant. Considering the high topographic relief in the Hellenic trench area (16) and in the irregular "cobblestone" topography along much of the Mediterranean Ridge where slopes may be 40° or more, in addition to benthic mixing processes (17) and potential coring

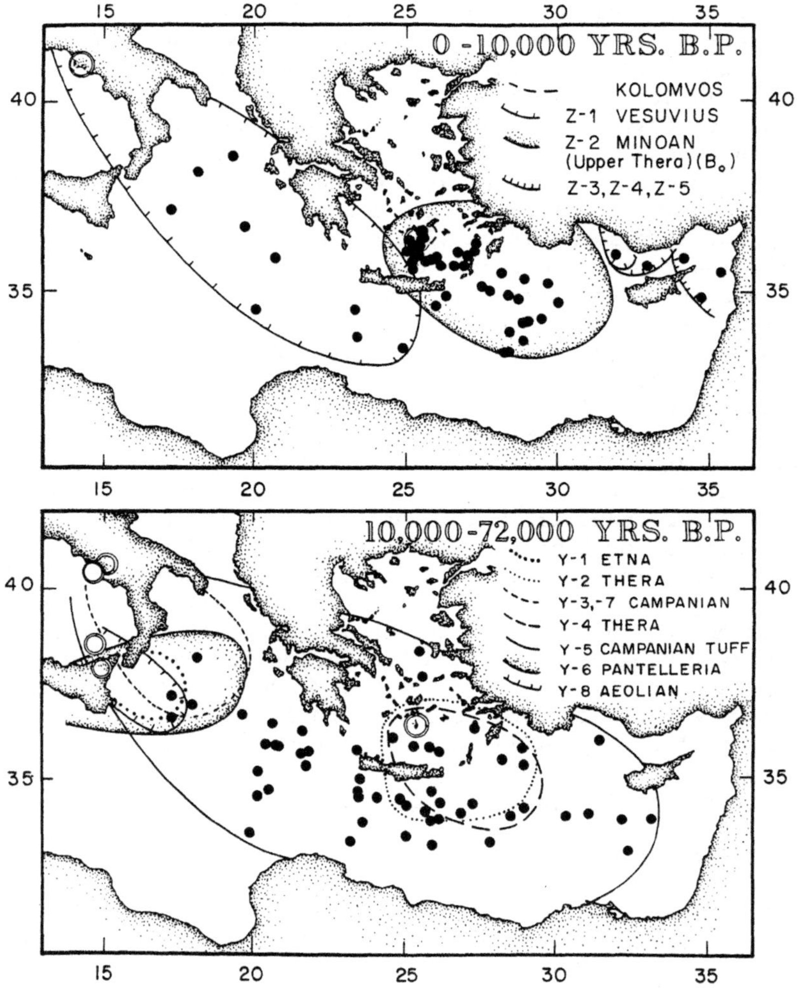

Figure 2. (A) Areal distribution of tephra layers deposited between 0 and 10,000 years B.P. ("Z" planktonic foraminifera zone). Concentric circles mark probable or known volcanic sources of these layers.

(B) Areal distribution of tephra layers deposited between 10,000 and 72,000 years B.P. ("Y" planktonic foraminifera zone). Concentric circles mark probable or known volcanic sources of these layers.

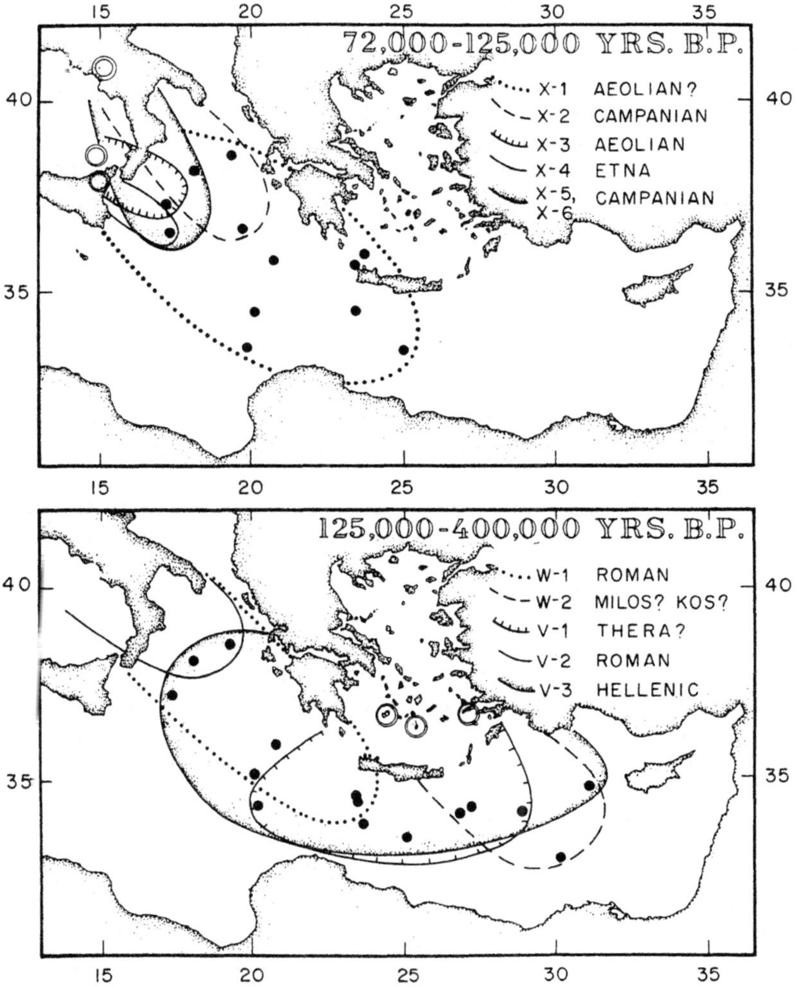

Figure 3. (A) Areal distribution of tephra layers deposited between 72,000 and 125,000 years B.P. ("X" planktonic foraminifera zone). Concentric circles mark probable or known volcanic sources.

(B) Areal distribution of tephra layers deposited between 125,000 and 400,000 years B.P. ("W", "V", and "U" planktonic foraminifera zones). Concentric circles mark probable or known volcanic sources.

problems (18,19), it is indeed remarkable that identifiable tephra deposits remain intact at all.

Isopachs of the Z-2 (Minoan) and Y-5 (Campanian Tuff) ashes based upon measurements of ash layer thicknesses recovered in cores illustrates the magnitude of redeposition and disturbance of these layers. As has been noted elsewhere (3), variations in present-day thicknesses of the Minoan tephra downwind along the presumed primary dispersal axis can vary between a 1 cm thickness only 40 km from the volcanic source on Santorini, to nearly 200 cm thick 160 km from Santorini within the Cretan Trough due to redeposition of ash into this topographic depression (presuming equitable coring disturbances, if any, to cores). Similar comparisons for the Y-5 (Campanian Tuff) ash between layer thicknesses actually measured in cores and the presumed original thickness, shown in Figure 4, illustrates the regional variability resulting from redispersal processes on the sea-floor (again presuming equitable coring disturbances, if any).

Local variability in thickness can be significant as indicated by studies of cores taken only a mile or so apart in two small areas along the Mediterranean Ridge known as Cobblestone areas 3 and 4, where surveying and sampling programs were conducted using precise navigational techniques (areas are outlined by small black boxes in Figure 4). In Cobblestone area 3 (Figure 5), redeposition of tephra into small basins has formed thicker layers due to removal of material by slumping from surrounding high areas; this is particularly evident in plotting layer thickness variations of the Y-5 (Campanian Tuff) ash (20). Original thickness of this ash in the Cobblestone 3 area is presumed to have been about 7 cm (see Figure 4) but within this 90 km^2 area, the thickness varies twofold between nearby cores due to redeposition of ash into depressions from surrounding steep slopes and redispersal or mixing of displaced volcanic particles into nearby pelagic sediments.

Studies are in progress on post-burial mixing of tephra resulting from benthic biological activity. The best indication of the mixing intensity that might be expected is the work done on the Toba ash layer in Indian Ocean cores (Figure 6) where nearly one-third of the original ash layer has been mixed upwards by animal activity since deposition of the ash approximately 75,000 years ago (17,21). Mechanical mixing effects could also result from the coring process (18,19). Any interpretation of the presumed original (undisturbed) thickness of ash layers, such as that depicted for the Y-5 (Campanian Tuff) ash in Figure 4, must also consider the 2:1 relationship between undisturbed layer thicknesses on land in outcrop and those in deep-sea sediments (presuming a pyroclastic fall deposit) (12).

These are important considerations in any study of tephra, but are especially appropriate to studies of pyroclastic fall deposits forming distinctive layers on the sea-floor because of

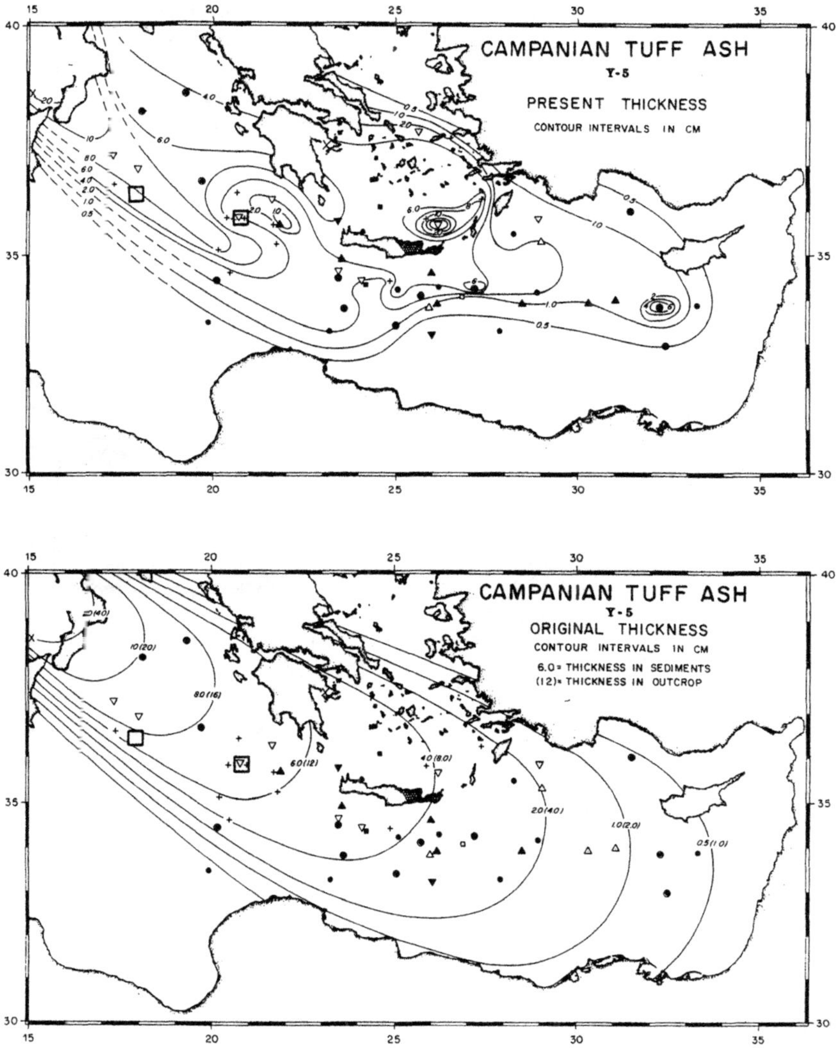

Figure 4. Thickness and areal distribution of the Y-5 ash layer: comparison between present-day layer thickness measured in moist cores and the inferred original thickness using these core data. Isopachs are in cm; numbers in parenthesis are equivalent layer thickness in outcrop using technique in (12). No correction has been made for compaction of layers during coring. Symbols identify cores; symbol identification in (3). Cross-hatching on Crete delineates the area where glass shards apparently derived from this eruption have been identified in soils. Small boxes in the Ionian Sea outline Cobblestone area 3 (right) and Cobblestone area 4 (left).

their remoteness from direct observation and because of possible complications associated with their sedimentation and sampling processes. In the eastern Mediterranean, an area that has become a classic locale for tephra studies, these factors can be significant and must be considered.

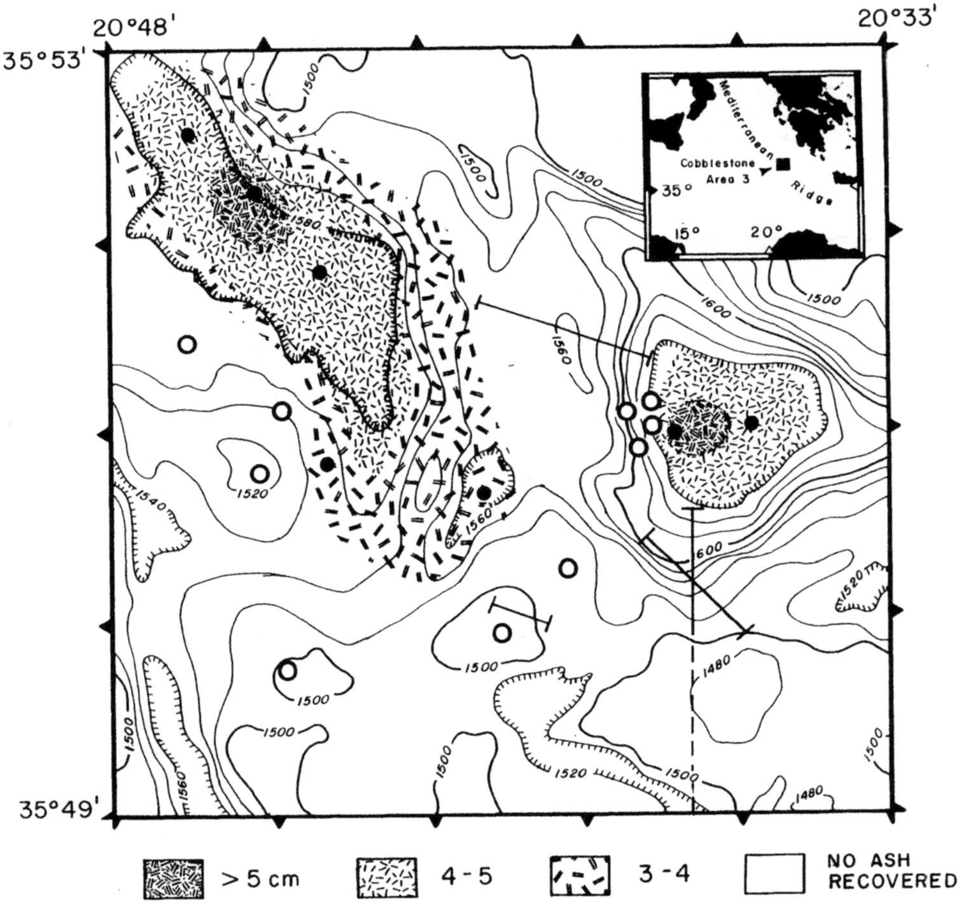

Figure 5. Variability in thickness of the Y-5 ash layer in Cobblestone area 3 (20). Open circles are cores without this tephra; solid circles represent cores with the tephra.

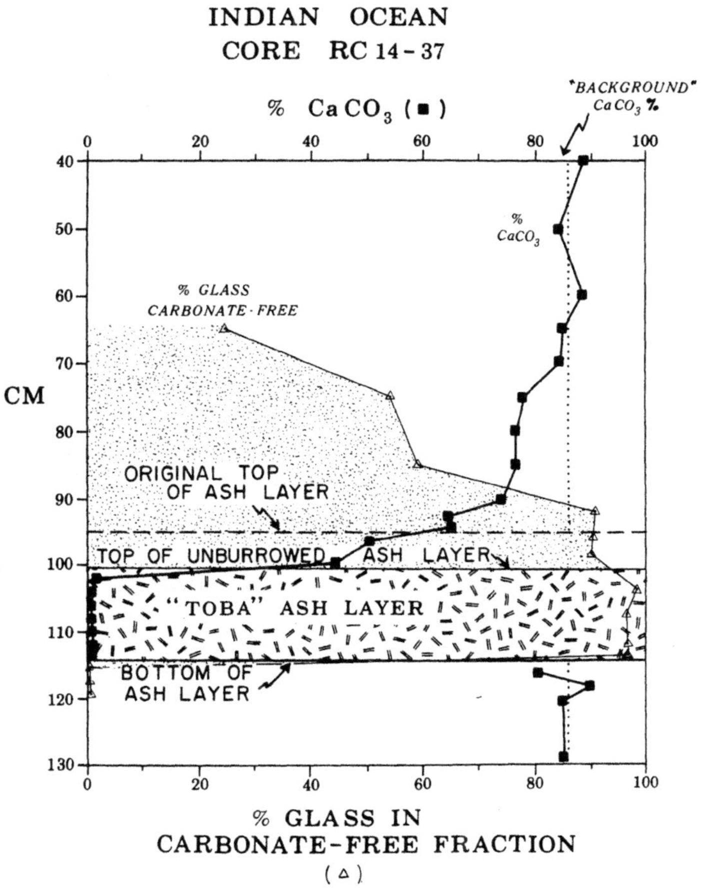

Figure 6. Mixing of the Toba ash into pelagic sediments by burrowing. The presumed original thickness of the layer (20 cm) has been reduced considerably by this biologic activity with mixing obvious up to 30 cm above the present-day upper contact of the tephra (17).

ACKNOWLEDGEMENTS

Comments from D. Ninkovich (now at Cities Service Co.), L. Burckle and S. Coughlin (Lamont-Doherty Geological Observatory), G. Heiken (Los Alamos Scientific Laboratory), S. Self (Arizona State University) and S. Sparks (University of Cambridge) are appreciated. LDGO curating facilities are supported by the ONR (contract N00014-75-C-0210-Scope E) and NSF (contract OCE-78-25448). This is LDGO Contribution No. 3172.

REFERENCES

1. Keller, J., Ryan, W.B.F., Ninkovich, D., and Altherr, R.: 1978, Geol. Soc. Amer. Bull., 89, pp. 591-604.
2. McCoy, F.W.: 1974, Ph.D. thesis, Cambridge, Mass., Harvard Univ., 132 p.
3. McCoy, F.W.: 1980, Thera and the Aegean World, 2, pp. 57-78.
4. Thunell, R., Federman, A., Sparks, S., and Williams, D.: 1979, Quat. Res., 12, pp. 241-253.
5. Wolff, J.A., and Wright, J.V.: Bull. Volcanol., in press.
6. Bond, A., and Sparks, R.S.J.: 1975, J. Geol. Soc. Lond., 131, pp. 1-15.
7. Farrand, W.R.: 1977, Geol. Soc. Amer., Abst. with Prog., 9, p. 971.
8. Keller, J.: 1980, Thera and the Aegean World, 2, pp. 49-56.
9. Keller, J.: 1971, Acta, 1st Int. Sci. Cong. Volcano Thera, pp. 152-169.
10. Keller, J., and Ninkovich, D.: 1972, Z. Deutsch. Geol. Ges., 123, pp. 579-587.
11. Richardson, D., and Ninkovich, D.: 1976, Geol. Soc. Amer. Bull., 87, pp. 110-116.
12. Watkins, N.D., Sparks, R.S.J., Sigurdsson, H., Huang, T.C., Federman, A., Carey, S., and Ninkovich, D.: 1978, Nature, 271, pp. 122-126.
13. Ninkovich, D., and Heezen, B.C.: 1965, Proc. 17th Symp. Colston Res. Soc., pp. 413-453.
14. Thera and the Aegean World: 1978, 1, pp. 1-822; 1980, 2, pp. 1-427.
15. Kidd, R., personal communication.
16. Olausson, E.: 1971, Opera Bot., 30, pp. 29-39.
17. Ninkovich, D., and Ruddiman, W.: 1977, Proc., X Inqua Cong., p. 326.
18. Olausson, E.: 1960, Rept. Swedish Deep-Sea Exped. 1947-48, 8, pp. 287-334.
19. McCoy, F.W.: 1980, Mar. Geol., 38, pp. 263-282.
20. McCoy, F.W., and Coughlin, S.: in prep.
21. Ninkovich, D., Shackleton, N.J., Abdel-Monem, A.A., Obradovich, J.D., and Izett, G.: 1978, Nature, 276, pp. 574-577.

MARINE TEPHROCHRONOLOGY AND QUATERNARY EXPLOSIVE VOLCANISM IN THE LESSER ANTILLES ARC

Haraldur Sigurdsson and Steven N. Carey

Graduate School of Oceanography
University of Rhode Island
Kingston, Rhode Island 02881 USA

ABSTRACT. Explosive volcanism in the Lesser Antilles arc has been tackled by combined tephrochronologic studies of land deposits and 100 deep-sea piston cores from the adjacent Atlantic and Grenada Basins. Volcanic production in the arc during the late Quaternary has been estimated from deep-sea core data for a 10^5 year period as 527 km^3 (285 km^3 D.R.E.), with the majority of this material (445 km^3) deposited in the marine environment. Marine tephra deposits are principally of two types: ash-fall layers and a variety of subaqueous pyroclastic gravity flow deposits. Ash-fall layers and associated air-borne dispersed ash represent about 1/3 of total production. They have been deposited almost exclusively in the Atlantic east of the arc and their dispersal is entirely controlled by westerly high-altitude winds. A variety of pyroclastic sediment gravity flow deposits account for approximately 2/3 of the volcanic production. Most important are deposits which form as a result of the entry of ignimbrites directly into the sea during major eruptions. They have travelled up to 250 km from source along the back-arc basin and form massive, poorly graded deposits up to 5 m thick. Pyroclastic gravity flow deposits occur predominantly in the Grenada Basin, west of the arc and their distribution is aided by the steep (9°) slopes of the western arc flank. Microprobe analyses of glass shards in tephra deposits from the deep-sea and land show that rhyodacites are the dominant products of volcanism in this arc.

1. INTRODUCTION

Explosive volcanism is a characteristic feature of island arc evolution. One might therefore expect that the pyroclastic

S. Self and R. S. J. Sparks (eds.), Tephra Studies, 255–280.

rocks would feature most prominently in geologic studies of island arcs. This is not the case, however, particularly in petrologic studies, which focus generally on lavas and other holocrystalline rocks. In some cases the poor preservation of pyroclastic deposits in the land environment has hindered their study. These low-density and friable deposits are easily eroded and those which survive erosion are subject to rapid weathering because of their porous and glassy character.

The products of explosive volcanism are, however, excellently preserved in the marine environment adjacent to island arcs, as tephra layers in deep-sea sediments. Sampling of these deposits is a simple process by standard piston coring techniques, which can yield cores up to 15 m length. This sampling is expensive, however, with the cost of core running at approximately $300 per meter. The study of tephra in the marine environment has many advantages over the study of land-based deposits, but the two field methods should ideally be combined. The marine environment provides biostratigraphic control and thus dating of volcanic events. The good chemical preservation of glassy tephra in sediments enables chemical correlations of ash layers between cores and to land sources by use of the correlation techniques described in this paper, thus making wide spread deep-sea tephra layers time-stratigraphic markers of unique value to other geologic studies.

This paper is a summary of the application of marine tephrochronology to the study of Quaternary volcanism in the Lesser Antilles island arc. It is based on data from 100 cores collected by R/V GILLISS and R/V ENDEAVOR in 1976 and 1978 and extensive field studies on the volcanic islands. Much of the work discussed here has been published elsewhere. Carey and Sigurdsson [1,2] have documented the deep-sea deposits from two explosive eruptions in the arc and Sparks et al. [3,4] present field and theoretical studies on the entrance of pyroclastic flows into the sea. The type and distribution of volcanogenic sediments around the arc is discussed by Sigurdsson et al. [5].

The goals set for this study included these outstanding questions: What is the proportion of pyroclastics in arc volcanism? How is the pyroclastic material transported into the adjacent basins and what is its volumetric importance in sedimentation in the forearc and backarc basins? What distance can pyroclastic debris flows travel under water and what is the lithology of their deposits? Is volcanism episodic in the arc? What is the lifespan of individual centers? Does activity migrate along the arc or is it random?

Answers to most of these questions have been obtained as an outcome of the deep-sea coring program. This detailed information on the history of volcanism in the Lesser Antilles arc only relates to the Quaternary, however, but deeper sampling programs are under way to extend this record back to the period of Tertiary evolution of this arc.

1.2 Geologic Setting

Volcanism in the Lesser Antilles occurs along a 800 km long arc, associated with slow subduction (2 cm/yr) of the Atlantic plate. Activity in the Quaternary has taken place in centers on eleven islands and one submarine volcano. The arc consists of three geochemically and structurally distinct linear segments, which may, in turn, reflect segmentation of the subducting slab (Fig. 1). The northern segment, from Saba to Montserrat, is

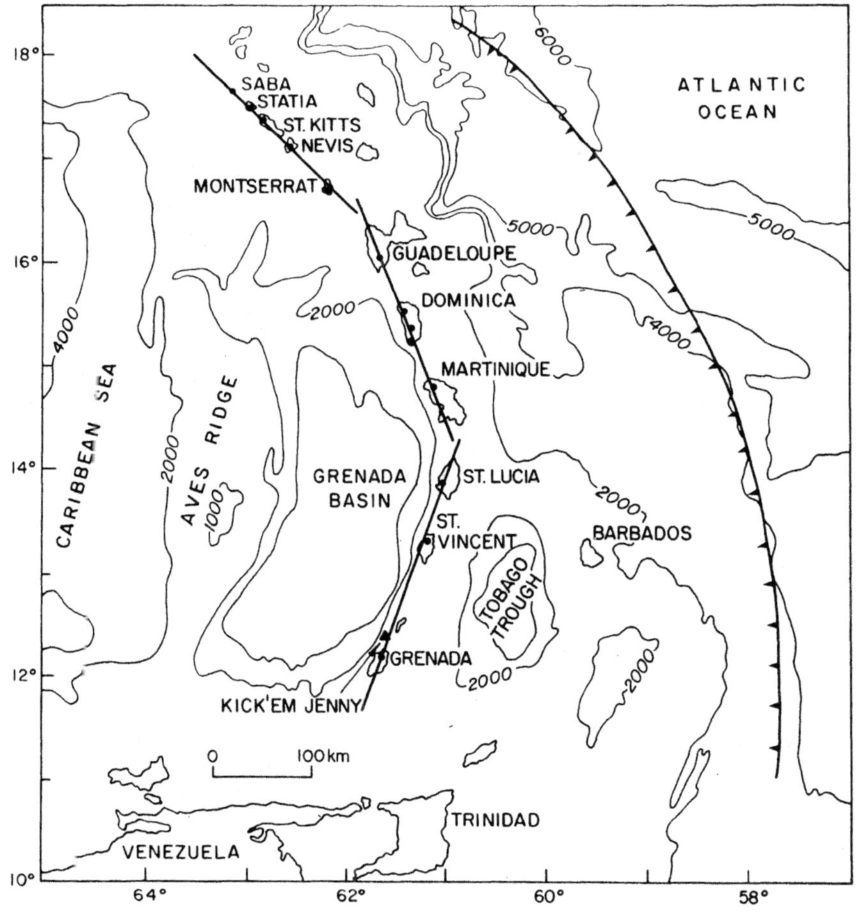

Fig. 1 Segmentation of the Lesser Antilles volcanic arc, showing the location of young volcanic centers.

characterized by a magma suite of the island-arc tholeiitic type, showing relative iron-enrichment and low potash [6]. This segment is also notable for shallow focal depths of earthquakes (<70km) and only 17° angle of dip of the seismic zone [7]. Its upper surface is at a depth of 100 km below the volcanoes.

The central segment of the arc, from Guadeloupe to Martinique, produces magmas of the calc-alkalic suite [6] and has been the most productive segment in the Quaternary. The seismic zone under this segment increases in dip from 20° to 50° and extends to 190 km depth. Its top lies at about 120 km below the volcanic arc [7]. The southern segment, from Grenada to St. Vincent, erupts an alkalic suite of magmas associated with highly undersaturated lavas enriched in incompatible elements and transition elements [6]. The seismic zone dips at 30° under the arc and its upper surface is at depth of 100 km below the volcanic axis.

The style of volcanism in the arc is dominantly explosive. Large magnitude explosive eruptions in the silicic centers of Dominica, Martinique, St. Lucia and Guadeloupe have produced extensive plinian pumice fall deposits and ignimbrites. Growth of domes is another prominant style of volcanism in these silicic centers, often associated with explosions and formation of block-and-ash flows [8].

Lava flows occur in more basic centers, such as on St. Vincent, but are still subordinate in volume to pyroclastic deposits. The basic centers, such as the Soufriere of St. Vincent, produce principally two types of deposits: pyroclastic flow and ash-fall deposits. This activity ranges from vulcanian to phreatomagmatic eruptions and the type of activity in the basic centers appears to be sensitive to the state of the vent, and in particular to the presence of crater lakes.

Volcanic mud-flow and flash-flood deposits, which occur on all the islands, testify to the rapid remobilization of loose pyroclastic deposits during and after explosive eruptions. They grade into coastal fans of coarse breccias and conglomerates, which reach thicknesses of 100 and 200 m on the west coast of Dominica and the east coast of Martinique. The accumulation of these deposits is probably episodic and related to individual volcanic events, rather than representing a steady-state sedimentation process. Primary and secondary pyroclastic flow deposits accumulated rapidly at the west coast of St. Vincent during the 1902 Soufriere eruption and built out an extensive fan [9]. The slumping of this fan into the Grenada Basin during the climax of the eruption represents an important type of volcanic sedimentation process [1].

There are three processes which transport volcanogenic sediments into the sea. Firstly, pyroclastic flows which enter the sea directly may either continue into deep water as subaqueous debris flows, or in two stages, involving accumulation at the coast or shelf break, followed by slumping of oversteepened deposits and subsequent turbidite transport to deep water. The second process of volcanogenic sedimentation involves direct fallout of tephra from high eruption plumes. The third process is the erosion of loose pyroclastics on land immediately following an eruption.

2. COMPOSITION AND CORRELATION OF TEPHRA LAYERS

Glass compositions of the major Quaternary tephra deposits on each volcanic island are predominantly rhyodacitic or rhyolitic with the exception of andesitic and dacitic types from the southern islands of St. Vincent and Grenada (Fig. 2B). In general the glasses become more alkalic from north to south along the axis of arc when compared at equal silica values. This trend is in accord with the north to south compositional variation demonstrated by Brown et al. [6] for samples consisting predominantly of lavas.

When compared with whole rock analyses of lavas and pyroclastics (Fig. 2A) the glasses from the major pyroclastic units are distinctly more siliceous (Fig. 2B). The higher silica values reflect both the difference in analytical technique (whole rock vs. glass) and the generally more evolved nature of the pyroclastics relative to the lavas. There is however, an excellent agreement between the glass compositions of the major land based pyroclastic deposits and the glass compositions of tephra layers in deep sea sediments surrounding the arc. This corresponce in glass compositions forms the basis for the correlation of marine tephra layers and the establishment of the late Quaternary tephrochronology for the Lesser Antilles region.

Despite similarities in silica content, glasses from each volcanic island exhibit distinct ranges in oxides such as K_2O, FeO^*, CaO and MgO enabling discrimination of individual centers on a triangular plot such as Figure (3). Using this plot and biostratigraphic dating, deep-sea tephra layers can be correlated to the most probable source island and in some instances to individual land-based units [1,2].

This correlation technique is only feasible, however, by use of the electron microprobe, where chemical analyses of individual glass shards can be obtained. Except for the analysis of sodium, the method is routine. We have found that the great majority of glassy tephra suffer extreme sodium loss during microprobe analysis and conventional analyses yield results which are often 50% too low. We have therefore found it necessary to analyse sodium separately by a microprobe technique which determines the

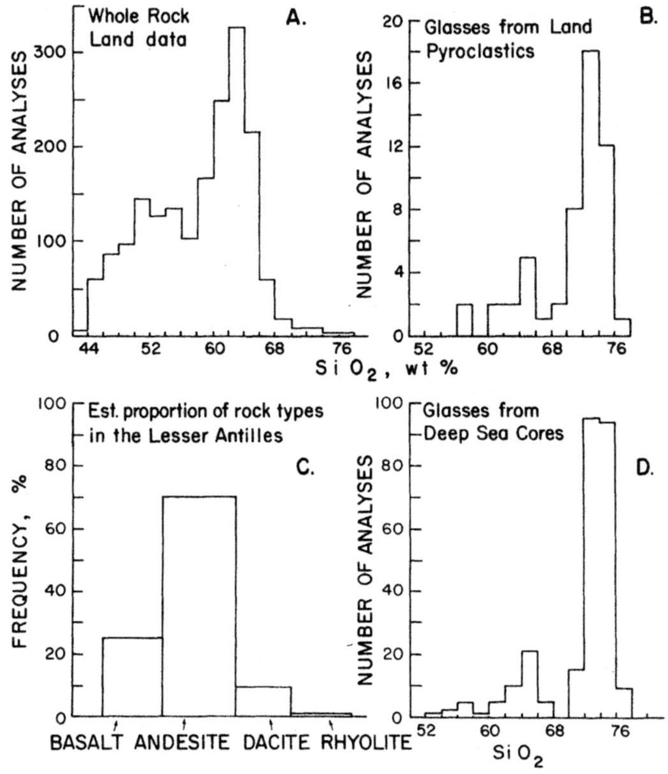

Fig. 2 The relative proportions of volcanic rock types in the Lesser Antilles arc, as a function of silica content. A: silica frequency histogram of 1518 whole rock analyses from 15 islands of the arc [6]. B: Microprobe analyses of glass shards in fifty-three Quaternary pyroclastic deposits on the volcanic islands. C: Baker's [20] estimate of the proportions of rock types in the arc. D: Microprobe analyses of glass shards in tephra layers from deep-sea cores of the Grenada Basin and the Atlantic.

initial sodium concentration of glass from the rate of sodium loss during electron bombardment [10].

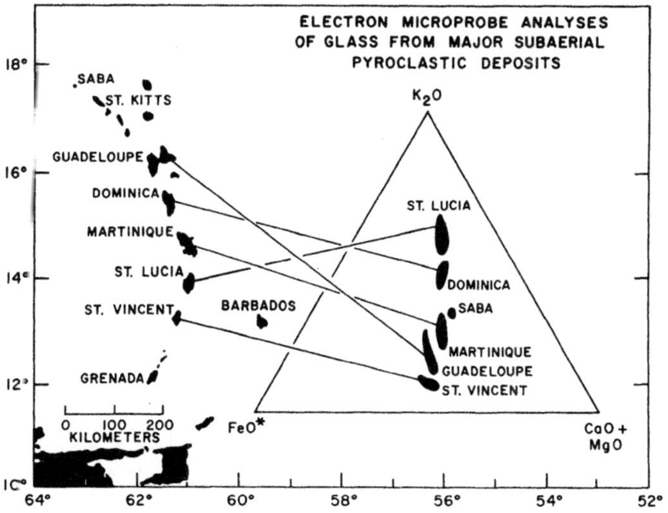

Fig. 3 The chemical "finger-printing" of glasses in tephra deposits of the major volcanic centers of the arc.

3. VOLUME AND DISTRIBUTION OF PYROCLASTIC DEPOSITS

Tephra from the arc has been deposited in the back-arc Grenada Basin to the west and in the Atlantic up to 800 km to the east. The evidence from deep-sea cores shows that the distribution of tephra is highly asymmetrical around the arc, with the bulk of pyroclastics occurring in the Grenada Basin, where volcanogenic sediments form 34% of all sediments cored, in contrast to only 4% in the Atlantic. These volume relationships are a direct consequence of the high influx rate of pyroclastic flows into the Grenada Basin, where pyroclastic debris flow deposits and ash-turbidites make up 98% of all volcanogenic sediments (Fig. 4). This has led to an average sedimentation rate of 15 cm/10^3 years in the Grenada Basin. In contrast, ash-fall deposits form 99% of the volcanogenic sediments on the Atlantic floor, where pyroclastic debris flow deposits are rare. Dispersed ash typically makes up 3 to 7% of all sediments, but is more important in the Atlantic (Fig. 4).

A quantitative assessment of late-Quaternary explosive activity and volcanogenic sedimentation in and around the arc has been carried out on the basis of deep-sea coring and land-based studies [5]. Currently available sampling of the sedimentary basins is representative of the last 10^5 years and the volume

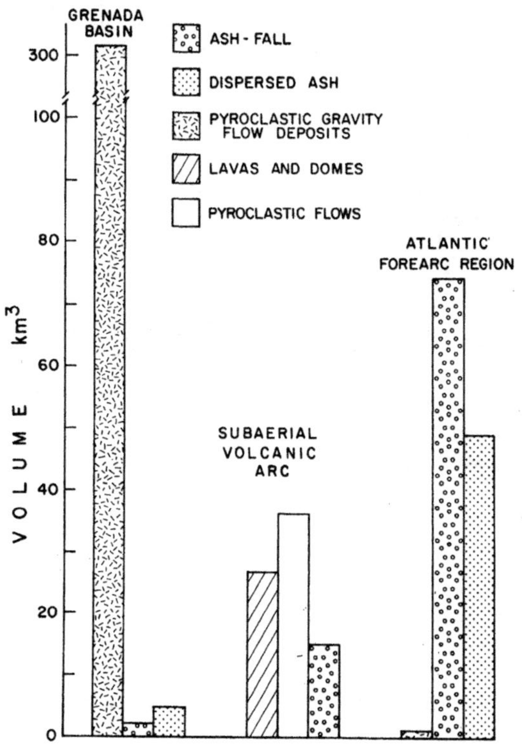

Fig. 4 The volumetric distribution of volcanic deposits in the back-arc Grenada Basin, in the subaerial arc and in the Atlantic fore arc region during the past 10^5 years.

estimates discussed here apply to production of the arc during this period (Table 1). The volume estimates (Fig. 5) indicate that total volcanic production in the arc is 527 km^3 (285 km^3 dense-rock equivalent) in the last 10^5 years. The vast majority of this production is now in the marine environment, or 445 km^3.

Field studies show that the bulk of volcanism in this period has taken place in the central islands of the arc, with over half the preserved subaerial volcanic products on Dominica (53 km^3). Other productive islands include Martinique, St. Lucia, Guadeloupe and St. Vincent (Table 1). The major deep-sea ash layers can be chemically correlated with deposits on these islands.

The easterly dispersal of air-fall tephra from the arc has affected a wide region of the Atlantic (Fig. 5), but represents a surprisingly small fraction of total production (24%). In the

Table 1

Volume and distribution of volcanics produced in the Lesser Antilles arc during the last 100,000 years

	apparent vol.*	dense-rock equivalent**
Subaerial volcanics		
Dominica	53 km^3	39.8 km^3
Martinique	10	7.5
St. Lucia	9	6.8
Guadeloupe	8	6.0
St. Vincent	5	3.8
Total on land	85 km^3	63.9 km^3
Megascopic volcanogenic layers		
Atlantic	75	37.5
Grenada Basin	313	156.5
Dispersed Ash		
Atlantic	49	24.5
Grenada Basin	5	2.5
Total marine	442 km^3	221 km^3
Total volcanic production	527 km^3	285 km^3

*Volume estimates for deep-sea volcanic deposits are based on coring evidence (see fig. 6). Estimates of volumes of subaerial volcanics are based on unpublished field data.

**Dense-rock equivalent was calculated by multiplying volcanogenic sediment volumes by a factor of 0.5 and subaerial volcanics by a factor of 0.75. Densities of volcanogenic sediments measured in the deep-sea cores range from 1.1 for ash-fall deposits to 1.3 for pyroclastic flow deposits. Given a whole-rock density of 2.4 to 2.6, a conversion factor of 0.5 gives a close approximation of dense-rock equivalent.

Atlantic, the air-fall tephra consists of two components: discrete ash layers and dispersed ash. The dispersed ash volume is surprisingly large or nearly equivalent to the volume of discrete ash layers [11].

4. ASH-FALL DEPOSITS

Most ash-fall layers are thin (0.5 to 10 cm) pale colored and consist predominantly of fine-grained (<250 μm) glass shards, with minor proportions of lithics and crystals. Most of the grey or white dacite or rhyolite layers are virtually lithic-free, whereas dark-grey basaltic andesite tephra, such as from the phreatomagmatic 1979 Soufriere eruption on St. Vincent contain 20 to 65% of lithics (Fig. 6).

The crystal content of a given layer varies in a systematic manner with distance from the arc, reflecting downwind sorting in the ash-plume (Fig. 7). The crystal/glass ratio in the ash-fall

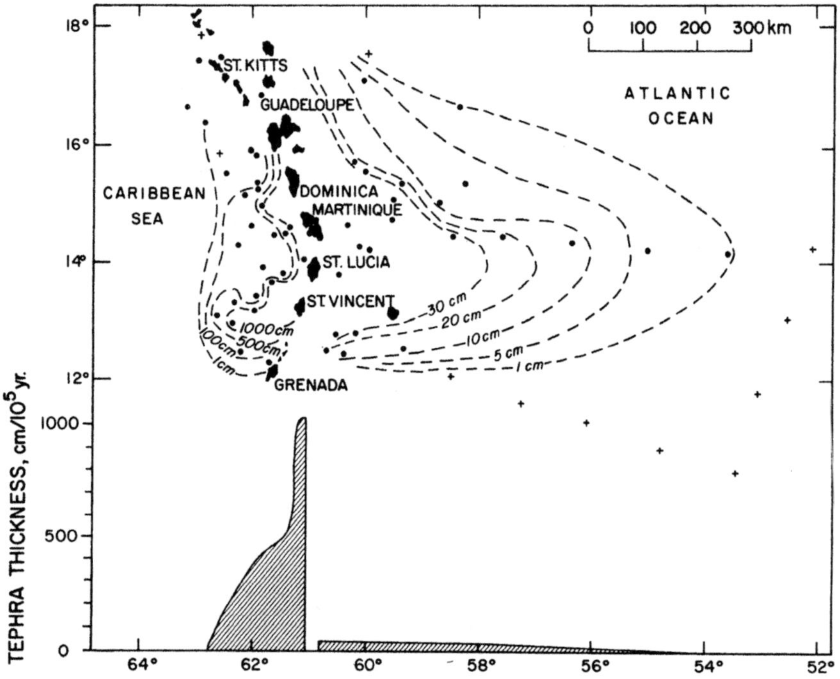

Fig. 5 The cumulative thickness of tephra and other volcanogenic sediments (subaqueous pyroclastic turbidites, volcanic sands) deposited in the last 10^5 years around the Lesser Antilles arc. Circles indicate location of piston cores containing volcanogenic layers; crosses are cores devoid of megascopic tephra layers. Lower part of figure is an E-W cross section at 14°N through the cumulative tephra-deposits.

layer from the 1902 Soufriere eruption decreases, for example, from 1.9 at a distance of 110 km to 0.6 at 200 km distance [1]. Aeolian fractionation of glass and crystals also occurred in the Roseau ash-fall layer from Dominica [2]. Modal analyses of pumices indicate an original crystal/glass ratio of 0.25 to 0.33 in the erupting magma. In contrast, the ratio rises abruptly along the dispersal axis, reaching a maxim of 3 approximately 100 km east of Dominica, before tapering off with distance as the glass fraction becomes a dominant component. The Roseau tephra layer shows a depletion in glass relative to the source up to a distance of 510 km down wind. The high enrichment of crystals near the source requires that a major portion of the glass fraction was removed further down wind.

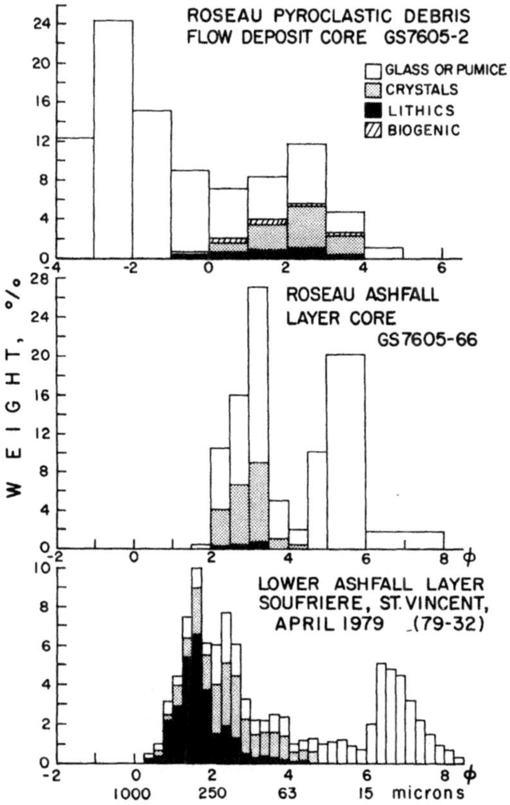

Fig. 6 Examples of grain size and modal histograms for ash fall and pyroclastic flow deposits from the Lesser Antilles arc.

The crystal/glass ratio of the ash-fall layers is a useful parameter in quantitative estimates of aeolian fractionation of an eruption plume. Mass balance calculations for the Roseau tephra, based on the observed crystal/glass fractionation, have shown, for example, that a glass fraction with a minimum volume of 12.2 km^3 was transported beyond the observed fall-out envelope [2], thus only 52% (13.3 km^3) of the erupted tephra from the ash-fall phase of this eruption is preserved as a discrete ash layer. This missing glass fraction must have been deposited outside the observed fall out region as dispersed ash.

Most of the ash-fall layers which are visible in Quaternary deep-sea sediments near the arc are from large-magnitude explosive eruptions from the silicic volcanic centers of Dominica, Martinique and St. Lucia. These eruptions have produced total volumes in excess of 1 km^3 and deposited ash layers with a minimum thickness

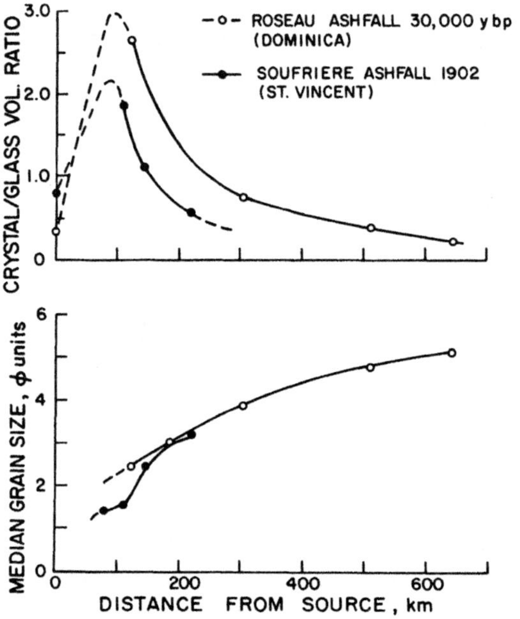

Fig. 7 Variations in crystal/glass ratio as a result of aeolian fractionation of airborne tephra from the Quaternary Roseau eruption [2] and tephra from the 1902 eruption of the Soufriere of St. Vincent [1].

of 0.5 to 1 cm. Ash-fall layers usually have a sharp base but a gradational upper contact with the overlying sediment due to bioturbation. Bioturbation has obliterated many ash-fall layers with original thickness less than 1 cm. Such volcanic events can, however, be located in the core by studies of the variation in the concentrations of dispersed ash in the adjacent sediment (Fig. 8).

Small magnitude eruptions, such as most of the historical explosive eruptions in the arc (Mt. Pelée, 1902; Soufriere, 1979) do not form a megascopic ash-fall layer and merely contribute to the dispersed ash component of the sediment. The thin ash-fall layers of recent small magnitude eruptions can, however, be recovered by coring techniques which do not disturb the surface sediment. A series of gravity cores to the east of St. Vincent thus located a young surface layer of dark-brown tephra with a maximum thickness of 0.5 cm. This layer has been chemically correlated with the 1902 Soufriere eruption [1]. The same study found that tephra compositionally identical to the 1902 eruption of Mt. Pelée is, on the other hand, dispersed in the surface sediment east of Martinique.

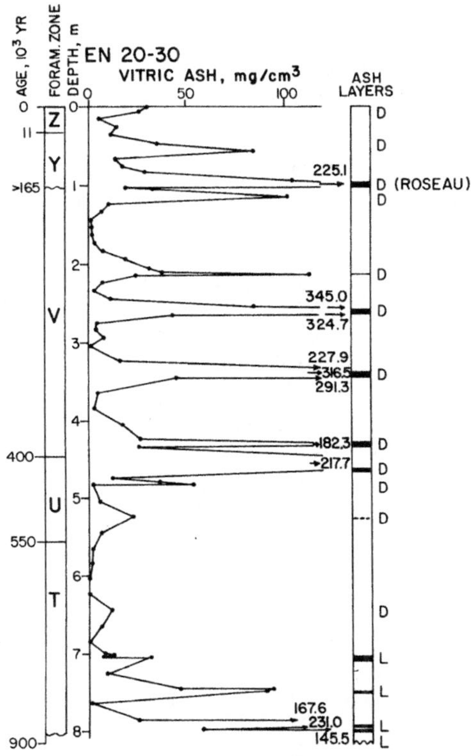

Fig. 8 The concentration of vitric ash (>11μm) and core log for core EN20-30, approximately 50 km east of Martinique. The diagram illustrates the variable content of volcanic ash dispersed throughout the sediment, in mg/cm^3, on basis of a 10 cm sampling interval. Major peaks in ash concentration occur adjacent to megascopic volcanic ash layers (see right-hand column) whereas minor peaks (50-100 mg/cm^3) mark ash deposition from minor explosive eruptions and ash layers which have been dispersed by bioturbation. The subaerial sources of megascopic ash layers and dispersed ash zones have been determined by microprobe analyses of glass shards and correlation with island tephra deposits. D and L identify tephra from Dominica and St. Lucia, respectively.

Both historical and deep-sea evidence shows that ash-plumes from explosive eruptions in the arc are carried to the east (Fig. 9). The ash plumes of the 1902 eruptions of Mt. Pelée and the Soufriere were thus carried over 1200 km east into the Atlantic at minimum velocity of 13 to 18 m/sec. Similarly, virtually all ash-fall layers in Quaternary cores occur east of the arc, whereas only three layers of ash-fall origin have been found in the Caribbean.

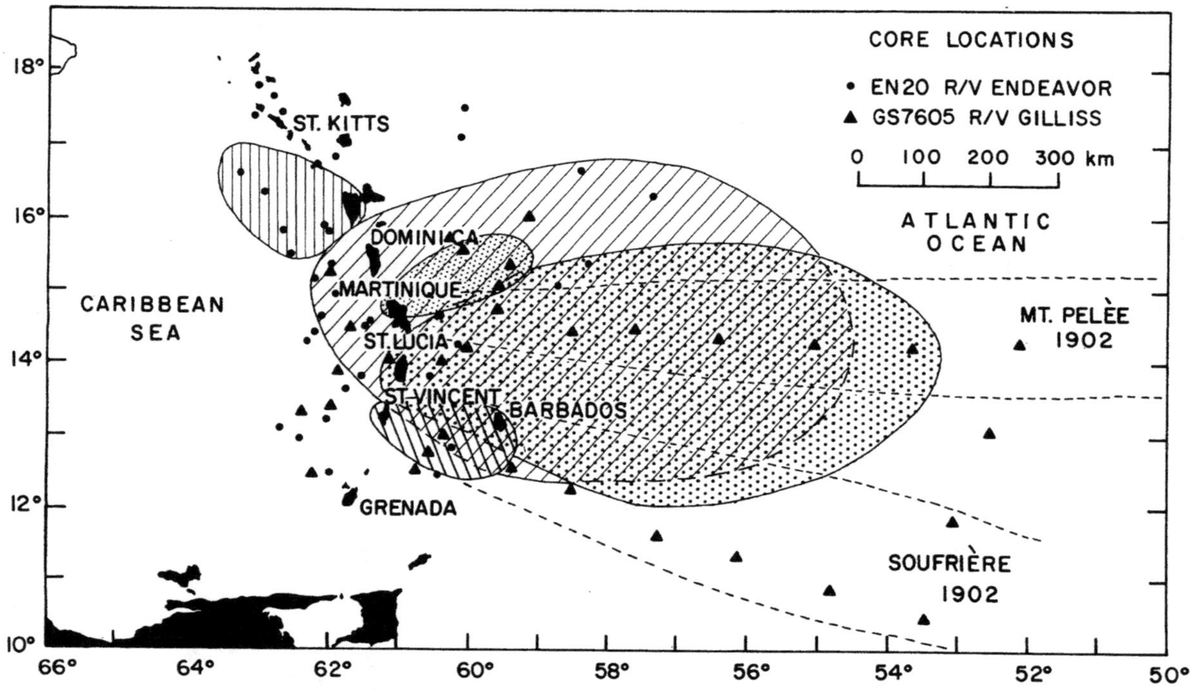

Fig. 9 The Quaternary fall-out patterns of air-borne tephra from the Lesser Antilles arc, as deduced from piston coring evidence. Also shown are the ash plumes of the 1902 eruptions of the Soufriere of St. Vincent and Mount Pelee in Martinique on basis of reports from passing ships [1].

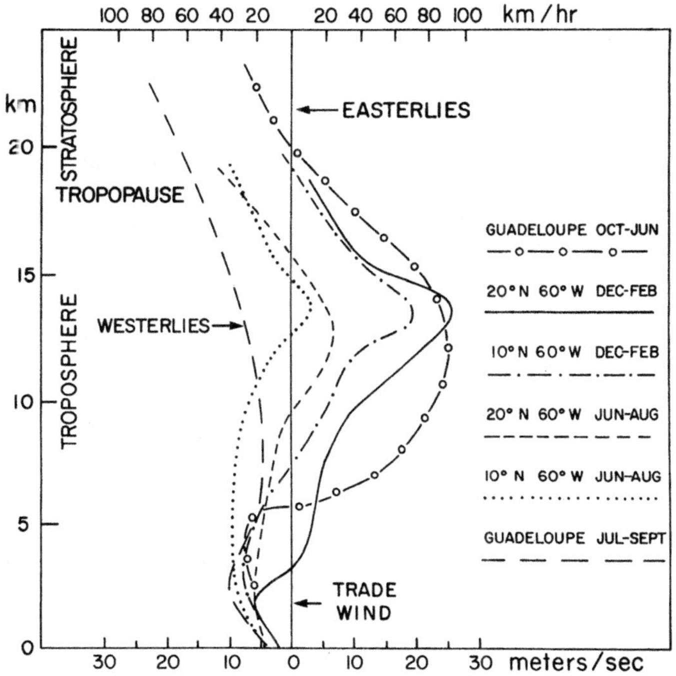

Fig. 10 Seasonal and height variation in wind circulation above the Lesser Antilles arc, based on data from Newell et al. [21] and H. Westercamp (personal communication).

This distribution can be accounted for by the general wind distribution above the arc. The easterly trades, which are the dominant surface winds in the region, are replaced by westerlies above 5 to 8 km height, averaging 14 to 25 m/sec. (Fig. 10). The principal factor which determines the height of an eruption plume above the arc is the height of the tropopause where a marked thermal inversion occurs, separating the cooling trend of the tropospheric atmosphere from the warming trend in the stratospheric atmosphere. The tropopause fluctuates between 16 and 18 km and during the explosions of the Soufriere on 14 April 1979 the tropopause was at 18.1 km at temperature of -80.5° (A. F. Krueger personal communication).

The buoyant rise of eruption columns as thermals will occur along the negative tropospheric thermal gradiant. At the tropopause the column loses its positive buoyancy, spreads laterally and is carried as a plume with the prevailing wind. How much of the plume penetrates the tropopause and enters the stratosphere is debatable, but in either case its subsequent transport is dominated by the prevailing westerlies which straddle the tropopause in the Caribbean region.

The ash-fall layers provide indirect evidence that Lesser Antilles eruption clouds do not penetrate much above the tropopause. The wind structure of the tropical lower stratosphere undergoes a quasi-biennial or 26-month oscillation, with westerlies alternating with easterlies [12]. If eruption plumes rose to elevation of 20 km or above in the tropical stratosphere, then the effect of the quasi-biennial wind oscillation should lead to an even distribution of Quaternary ash-fall layers to east and west of the arc. On the contrary, the observed distribution shows transport to the east for virtually all air-borne tephra. We conclude that either the Quaternary tephra did not reach the stratosphere above the tropopause or that observed quasi-biennial oscillation of to-day did not exist during the Quaternary.

5. PYROCLASTIC DEBRIS FLOWS, TURBIDITES AND RELATED DEPOSITS

Large quantities of coarse pyroclastic material are transported to the deep sea (>1500 m) surrounding the Lesser Antilles volcanic arc by a variety of sediment gravity flow mechanisms. In contrast to the dispersal and deposition of airfall tephra the distribution of deposits formed by these flows is strongly influenced by the bathymetry of the arc flanks adjacent to the flow source. Requirements for the generation of pyroclastic sediment gravity flows include the accumulation of pyroclastic material in the marine environment and a condition of instability leading to sediment mobilization. These requirements may be met by a number of different volcanological and sedimentological processes operating in an active island arc such as 1) direct entry of subaerial pyroclastic flows into the sea with continued subaqueous movement downslope, 2) slumping of rapidly accumulating pyroclastics in shallow water derived from ash-fall, pyroclastic flow or remobilized subaerial deposits and 3) slumping of reworked pyroclastic material (epiclastic) which has been accumulated by current transport. The first two processes are considered to occur contemporaneous with major explosive eruptions while the third may occur independently of eruptions. It is likely, however, that significant hybridization of the processes, especially multistage histories, can occur both in time and space.

Pyroclastic deposits which form in the deep sea from different types of sediment gravity flows reflect both the nature of their source material and the mechanism by which it was transported and deposited. Our interpretation and classification of these deposits in the deep-sea cores is based upon the sediment gravity flow models presented by Middleton and Hampton [13]. In the next sections we present specific examples of pyroclastic sediment gravity

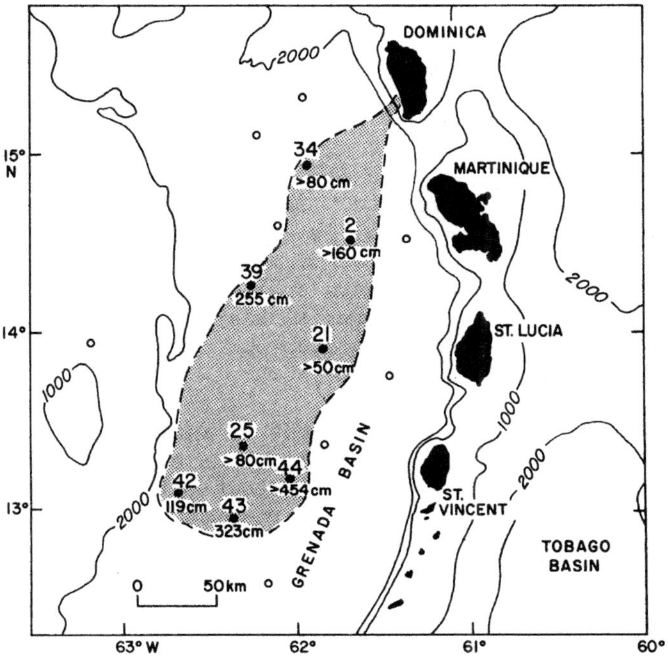

Fig. 11 Distribution of the Roseau subaqueous pyroclastic flow deposit in the Grenada Basin. Solid dots are location of cores which sampled the deposit, with layer thickness in cm [2].

flow deposits which have been formed by the various processes described above.

5.1 Subaqueous pyroclastic debris flow deposits

Pyroclastic flow deposits form a high proportion of the volcaniclastic units exposed on many of the Lesser Antilles islands. These deposits include basaltic andesite scoria deposits of St. Vincent type, andesitic and dacitic block-and-ash flows of Mt. Pelée type and andesitic and dacitic ignimbrites. In many instances these deposits outcrop with significant thicknesses along the coasts of islands indicating that some flows have directly entered the sea. This has been confirmed by observations of pyroclastic flows entering the sea during the historic eruptions of Mt. Pelée on Martinique and the Soufriere of St. Vincent [9]. The entry of pyroclastic flows into the sea is thus an important mechanism for transporting large quantities of pyroclastics to the marine environment and a likely mechanism for initiating pyroclastic sediment gravity flows.

A good example of a deposit formed by the entry of a large volume rhyodacitic pyroclastic flow into the sea is provided by the Roseau subaqueous pyroclastic debris flow deposit [2]. This deposit forms a major stratigraphic unit over an area of $1.4x10^4$ km^2 in the Grenada Basin west of the volcanic arc (Fig. 11) and has a volume of approximately 30 km^3. A high proportion of the deposit (>80%) is volcanogenic material consisting of rhyolitic glass shards, pumice, crystals of plagioclase, hypersthene, augite, hornblende, and titanomagnetite. Non-volcanic components include varying amounts of pelagic clay (both as distinct clasts and admixed with tephra), tests of foraminifera, radiolaria and petropods, and charcoal wood fragments up to 2 cm in length. The unit has been correlated to the subaerial Roseau ignimbrite (30,000 y. B.P.) on Dominica by electron microprobe analyses of glass shards and mineral phases.

The deposit is typically present in the cores as a single poorly sorted massive unit with grain size ranging from clay size to pumice clasts up to 6.5 cm in diameter (Fig. 6). Thicknesses vary from 0.5 m to at least 5 m and in some instances piston coring could not penetrate the entire unit because of its thickness and coarse grain nature. Both normal and reverse grading is present but grading is most prominent in cores which are furthest from source.

The Roseau pyroclastic debris flow deposit resulted from the entry of hot subaerial pyroclastic flows into the sea along the west coast of Dominica. This interpretation is supported by the presence of thick (>30 m) outcrops of the subaerial Roseau pyroclastic flows along the west coast near the capital city of Roseau [14] (Fig. 12). As no significant shelf area is present along the west coast the flows were able to continue down the steep submarine flank (ave. 9°) to the floor of the Grenada Basin. Transport occurred mainly as a high concentration debris flow with the supporting interstitial matrix being formed by a mixture of seawater, fine ash, and eroded marine sediment. The high concentration of coarse grains and large pumices in these deposits has greatly contributed to the buoyancy and mobility of the debris flow [15]. The presence of grading and eroded clay clasts in some of the units however, suggests that the flow mechanism may have been transitional to a more turbulent lower concentration flow as dilution with seawater increased. Deposition of the deposit resulted from decreased flow velocities on the gently sloping floor of the Grenada Basin. The flow was apparently able to travel over 250 km from its source on slopes less than 1 degree. This remarkable mobility of subaqueous debris flows was confirmed by Embley [16] who reported a subaqueous debris flow west of the Canary Islands which had travelled 700 km from its source over a slope as low as 0.1°. Other pyroclastic gravity flow deposits of the same type as the Roseau deposit have been found in the Grenada Basin and chemically correlated with St. Lucia and Martinique.

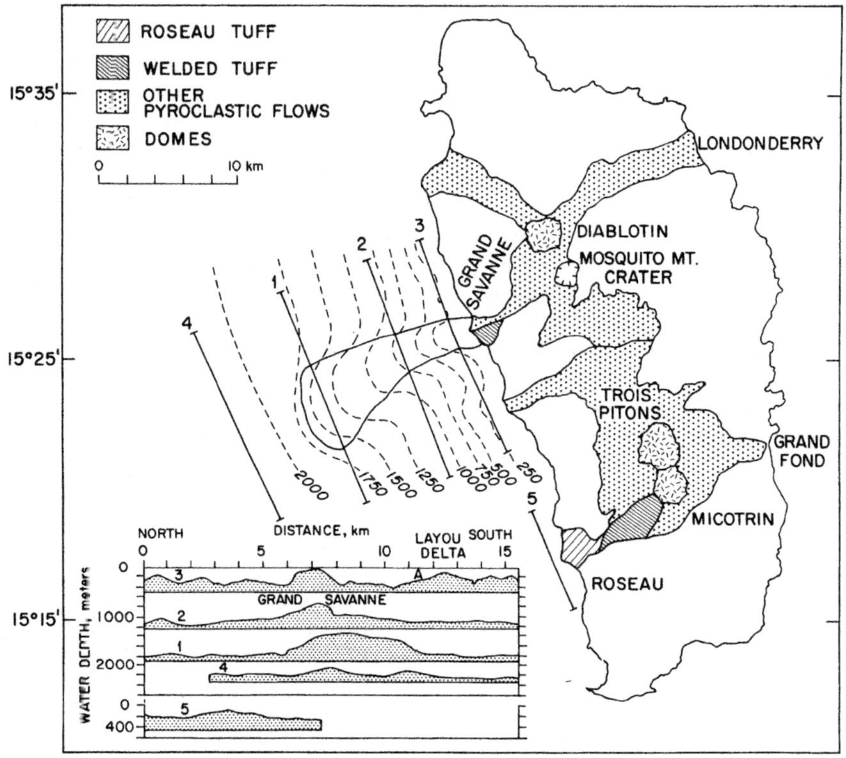

Fig. 12 Pyroclastic deposits on the island of Dominica and adjacent sea floor. Off-shore profiles 1 to 4 (insert) show the prominent submarine grand Savanne ridge, believed to represent accumulation of subaqueous pyroclastic flows with relatively low mobility [3]. Profile 5 off Roseau shows that such a ridge is not associated with the pumice-rich and presumably more mobile Roseau subaqueous pyroclastic flow [2].

In contrast to the large volume, extensive, subaqueous pyroclastic debris flow deposits of the Roseau type, smaller, more localized deposits are also present along the western flank of Dominica. North of the Roseau pyroclastic flow deposits along the west coast of Dominica is the broad fan-like apron of the Grand Savanne (Fig. 12). This feature consists entirely of pyroclastic flow deposits with minor ash-fall layers. The source of these units is likely to be the dacitic dome complexes of Diablotin and Mosquito Mountain 6 km to the east.

The ages of these deposits are not well constrained, but a sample of carbon from the uppermost fine pisolitic ash-fall deposit yielded an age of >22,000 yrs. In general, the freshness of the deposits and the youthful appearance of the apron's morphology indicates that a majority of the complex is late Quaternary in age. The thickness of these units at the coast (some as much as 50 meters) demonstrates that these flows also entered the sea like the Roseau pyroclastic flows to the south. There is however a striking difference in the types of submarine deposits which formed off the Grand Savanne [3].

Extensive 3.5 kHz seismic profiling parallel to the west coast of Dominica revealed the existence of a major east-west trending ridge extending from sea-level at Grand Savanne to a distance of 13 km offshore in water depths of 800 m (Fig. 12). Near shore the ridge is 2 km wide and 250 m thick but widens to 4 km and a thickness of 400 m at its extremity. This ridge represents the submarine extension of the subaerial Grand Savanne sequence and was formed by the entry of pyroclastic flows into the sea.

In the south, seaward of the subaerial Roseau pyroclastic flows no such ridge was observed. Shallow water profiles did show some constructional features off the mouths of the Layou and Roseau rivers but these are interpreted as only small fluvial deltas. It is apparent that the pyroclastic flows from the Grand Savanne had limited mobility underwater compared to the flows which entered the sea in the south.

The morphology of the subaqueous Grand Savanne ridge is consistent with the build-up of 8-20 individual debris flow deposits. We feel that the limited mobility of these flows may be related to the type and volume of pyroclastic flows produced near Grand Savanne compared to the south. There are at least two types of pyroclastic flow deposits found on the Grand Savanne apron: block-and-ash deposits and ignimbrite. The block-and-ash flows consist of blocks, some up to 10 meters, of dense, poorly vesicular hornblende and plagioclase bearing dacite set in an ash matrix. These deposits form the base of the exposed section and are up to 40 meters thick [3]. Overlying the block-and-ash flows are welded and non-welded crystal-rich dacitic ignimbrite flow units with a distinct absence of pumice clasts. In contrast, the Roseau pyroclastic flows to the south contain abundant low density rhyodacitic pumice set in an ash matrix [14] and are of larger volume than units of the Grand Savanne.

One of the factors affecting the mobility of debris flows is yield strength of the interstitial supporting matrix [13]. Upon entrance into the sea, the dense pumice-poor pyroclastic flows of the Grand Savanne may have had a limited ability to mix with seawater thus retaining a high total yield strength and limited subaqueous mobility. On the other hand, the lower density pumice-rich pyroclastic flows such as the Roseau flow in the south were able to mix more extensively with seawater and sediment to form a supporting matrix with a lower yield strength. This enabled the flow to move greater distances over similar slope angles. Alternatively, the limited mobility of the Grand Savanne deposits could be attributed to subaqueous welding. Sparks, et al., [4] presented theoretical arguments suggesting that welding of pyroclastic flow deposits may be enhanced in the submarine environment because of the marked viscosity changes in glass exposed to vaporized seawater.

Entrance of pyroclastic flows into the sea can thus produce a variety of submarine pyroclastic deposits and is an efficient mechanism for transporting significant volumes of coarse volcanogenic sediment to the deep sea surrounding active island arcs.

5.2 Pyroclastic turbidites

Explosive eruptions often result in the rapid progradation of island coastlines by the deposition of pyroclastic flows, ash-fall material, lahars, and the fluvial transport of freshly eroded pyroclastics [9,17]. The sequences are commonly unstable and are quickly modified by wave and current action leading to the accumulation of pyroclastic material in shallow water. If bathymetric conditions are suitable and a triggering mechanism supplied, this material may slump and move into deeper water via sediment gravity flows.

During the May 7, 1902 eruption of the Soufriere volcano of St. Vincent, areas of the west coast, particularly at the mouth of the Wallibou river valley, were expanded by the deposition of pyroclastic flows. Eyewitness accounts report that sections of this newly formed area up to 2 km long and 100 meters wide subsequently slumped into the sea at the climax of the eruption. Sediment cores recovered from the Grenada Basin west of St. Vincent were found to contain a thin pyroclastic turbidite (ave. 5 cm) forming the surface sediment and consisting of volcanic ash identical in composition to the material produced during the May 7, 1902 eruption. This deposit exhibits Bouma divisions A and E [18] with grain sizes of < 5 to 250 microns. We interpreted this layer as a pyroclastic turbidite generated by the slumping of freshly deposited pyroclastics along the west coast of St. Vincent during the eruption [1].

Similar thin (1-20 cm), localized turbidites and grain flows, such as the one produced by the 1902 activity of the Soufriere, are common in the sediment cores from the Grenada Basin especially near the break in slope along its eastern margin. A good example is core GS7605-24 taken 60 km offshore west of St. Vincent. This core contains 21 such layers with an average thickness of 8 cm and constituting 39% of the 5 meter core. Sedimentary features of

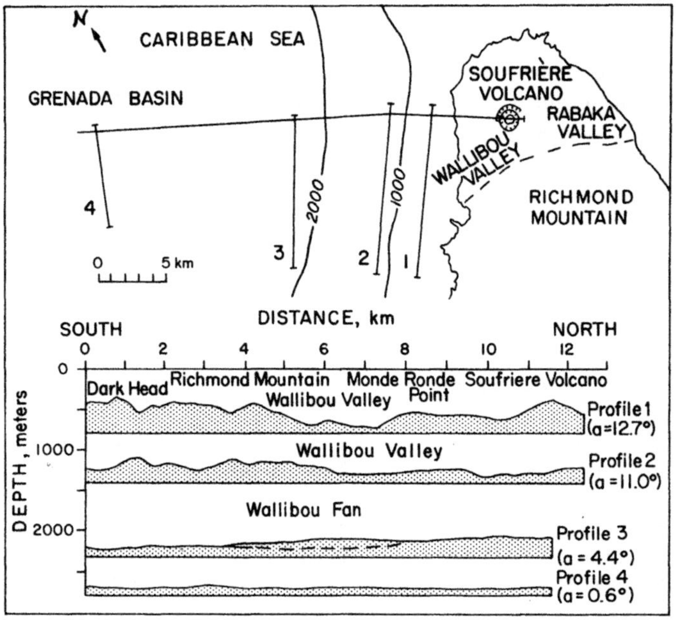

Fig. 13 Bathymetric profiles off the west coast of St. Vincent, showing the submarine extension of the Wallibou Valley and its mergence with the Wallibou pyroclastic sediment fan in region of lower slope (profile 3).

these deposits indicate deposition by turbidity currents, fluidized sediment flows and grain flows.

The accumulation of these deposits at the base of the western flank of the arc produces localized submarine fan deposits whose provenance can be established by the correlation techniques outlined earlier. Using seismic profiles and core lithologies it is thus possible to define fan deposits for different islands in the arc. St. Vincent for example has a thin fan sequence which begins at 2200 meters along the west flank of the Grenada Basin and extends out 100 km. Sediment influx to this fan appears to be directed through a submarine channel on the west flank of St. Vincent (Fig. 13, profiles 1 and 2) which is a continuation of the subaerial Wallibou River valley.

Other islands in the arc such as Martinique and St. Lucia also have localized fans developed at the base of their western flanks. Occasionally these fans are intercalated as shown by the occurrence of layers from different sources in a single core.

The origin of the pyroclastic turbidites and grain flows which contribute to the growth of these fans is undoubtedly complex and likely to be multistage in nature. We believe that the majority of the deposits are a direct consequence of the episodic major explosive eruptions of the arc such as some of the examples outlined above.

Independent of the volcanic processes however, is the continuous erosion of the subaerial arc complexes producing an additional source of sediment. This sediment is transported to the marine environment mainly by fluvial processes and then redistributed by ocean currents. In the Lesser Antilles, surface currents are dominated by an east to west flow through the narrow island passages. Current velocities exceeding 150 cm/sec have been reported in the Grenada and St. Vincent passages [19]. Any sediment supplied by fluvial transport is thus redistributed to the Caribbean side of the arc where it may subsequently accumulate and slump into deeper water.

6. DISCUSSION AND CONCLUSIONS

Marine tephrochronology has greatly augmented our knowledge of the Quaternary volcanic history of the Lesser Antilles arc and provided a basis for quantitative assessment of the rate of volcanism in this arc. Deep-sea cores and land studies show that two eruption styles have formed the bulk of the volcanogenic deposits. Firstly, large-magnitude (5 to 50 km^3) explosive eruptions from dacite or dacitic andesite centers are periodic and occur at intervals of ten to twenty thousand years. During the Quaternary we know of 8 such events in Dominica, 10 events in St. Lucia, 2 from Martinique and 1 from Guadeloupe which are recorded in deep-sea cores (Fig. 14). These large-scale explosive events of dacitic magma form high eruption columns which lead to ash-fall up to 800 km east of the arc. Pyroclastic flows generated during such eruptions have entered the sea and produced several types of submarine volcanogenic sediments up to 5 m thick. Low-density pyroclastic flows, with high concentration of pumice and high flow rate have been buoyant and mobile in the marine environment. The youngest of these, the Roseau ash from Dominica, has travelled as a subaqueous pyroclastic debris flow 250 km distance along the Grenada Basin. Higher-density flows, such as block-and-ash flows and crystal-rich ignimbrites are, on the other hand, less mobile and have been deposited directly near shore and on the steep submarine flank of the arc.

The second eruptive style involves basaltic andesite and basaltic centers, characterized by frequent small (0.1 to 2 km^3) eruptions involving both extrusion of lava, ash-fall and pyroclastic flows. This style of activity is now typified by the Soufriere

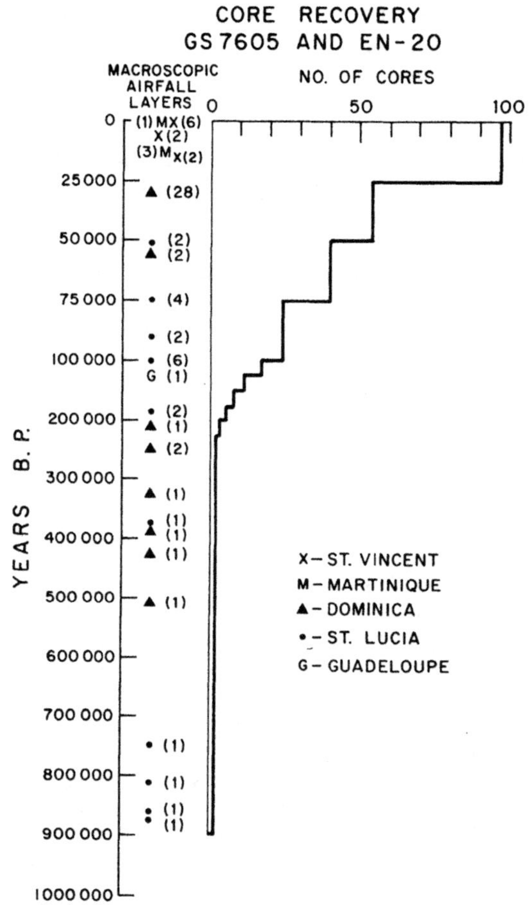

Fig. 14 Core recovery and stratigraphic column of Quaternary Lesser Antilles tephra in deep-sea cores, based on biostratigraphy. Numbers in brackets indicate the number of cores which sample a given tephra layer.

of St. Vincent and Kick'em-Jenny submarine volcano, where small-volume eruptions occur on a two-to 100-year time scale. Primary deposits from this type of activity are largely restricted to the arc flanks and are not important in the deep-sea. They include thin ash-fall layers and dispersed ash, and thin (2-20 cm) ash turbidites which have e.g. developed a fan west of St. Vincent at the break in slope between the arc and the Grenada Basin.

The deep-sea coring and land evidence show that total volcanic production in the arc has been 527 km^3 in the last 10^5 years

(285 km^3 D.R.E.), mostly from five islands in center of the arc. The vast majority of this material or approximately 84% has been deposited in the deep marine environment. The distribution of the marine tephra layers (ash-fall and pyroclastic debris flow deposits) is, however, highly asymmetrical around the arc, both in volume and in type. Subaqueous pyroclastic flows and related gravity flow deposits are the dominant type of tephra and account for 2/3 of all volcanic production. The pyroclastic sediment gravity flows are almost exclusively deposited in the back-arc Grenada Basin and form a very minor component of Quaternary sedimentation in the forearc region, such as in the Tobago Trough or on the Atlantic abyssal plain to the east. Ash-fall deposits are, on the other hand, less than 1/3 of total volcanic production, but are the dominant tephra type in the Atlantic, where ash-fall layers and related dispersed ash form 99% of all volcanogenic sediment.

The two factors which most strongly influence the distribution of tephra from the arc during major explosive eruptions are the prevailing atmospheric circulation and the bathymetry or slope of the arc flanks. The easterly dispersal of air-borne tephra reflects the important role of the westerly winds, which prevail above 8 km in the region, up to the tropopause at 18 to 20 km height. The distribution of subaqueous pyroclastic debris flows to the west of the arc correlates well with steep slopes (average 9°) of the submarine western flank of the arc, leading from the subaerial volcanoes to the flat basin floor at 2700 m depth. Slopes on the eastern flank are, in contrast, gentle and average 1 1/2°. Pyroclastic flows were discharged into water on both sides of the arc during large explosive eruptions. Flows associated with high eruption rates would rapidly pass through the critical region at the coastline and flow down the steep western flanks of the arc into the Grenada Basin. Flows advancing into the Atlantic on east side of the arc would, on the other hand, enter relatively shallow water and a region of low slope. The combined effects of low slope and hydromagmatic explosions resulting from the interaction of hot flow and sea water in shallow water may have resulted in disintegration of many of the flows on east side of the arc. The products resulting from disintegration of pyroclastic flows in shallow water east of the arc are volcanogenic sands, which are transported by the strong westerly flowing currents through the island passages and into the Caribbean.

Systematic microprobe analyses of glass shards in all deep-sea tephra layers cored by us gives a new quantitative view of the relative proportions of magma types erupted from this arc. It is found that the great majority of the major pyroclastic deposits on land and at sea have rhyodacite to rhyolite glass compositions, with exception of tephra from southern centers such as on St. Vincent and Grenada, which range from andesitic to dacitic types. Based on volume, the rhyodacitic and rhyolitic glasses are by far the dominant types. These compositional relations are in good accord with the findings of Brown et al. [6] based on 1518 whole-rock analyses from the arc, but differ markedly from Baker's [20] estimates, where andesite and basalt were regarded as the dominant rock types.

REFERENCES

1. Carey, S.N. and Sigurdsson, H.: 1978, Geology 6, pp. 271-274.
2. Carey, S.N. and Sigurdsson, H.: 1980, J. Volcanol. Geotherm. Res. 7, pp. 67-86.
3. Sparks, R.S.J., Sigurdsson, H. and Carey, S.N.: 1980a, J. Volcanol. Geotherm. Res. 7, pp. 87-96.
4. Sparks, R.S.J., Sigurdsson, H. and Carey, S.N.: 1980b, J. Volcanol. Geotherm. Res. 7, pp. 97-106.
5. Sigurdsson, H., Sparks, R.S.J., Carey, S. and Huang, T.C.: 1980, Volcanogenic sedimentation in the Lesser Antilles arc (submitted to J. Geology).
6. Brown, G.M.B., Holland, J.G., Sigurdsson, H., Tomblin, J.F. and Arculus, R.J.: 1977, Geochim. Cosmochim. Acta. 41, pp. 785-801.
7. Tomblin, J.F.: 1975, The Lesser Antilles and Aves Ridge, in The Ocean Basins and Margins, vol. 3: The Gulf of Mexico and the Caribbean, ed. A.E.M. Nairn and F.G. Stehli, pp. 467-500.
8. Roobol, M.J. and Smith, A.L.: 1976, Geology 4, pp. 521-524.
9. Anderson, T. and Flett, J.S.: 1903, Pt. 1, Royal Soc. Phil. Trans. Ser. A-200, pp. 353-553.
10. Nielsen, C. and Sigurdsson, H.: 1980, Quantitative methods for electron microprobe analysis fo sodium in natural and synthetic glasses (submitted to Amer. Mineral.).
11. Huang, T.C., Sigurdsson, H. and Carey, S.: 1980, Dispersed tephra in Lesser Antilles Arc (in prep.).
12. Wallace, J.M.: 1973, Rev. Geophys. Space Phys. 11, pp. 191-222.
13. Middleton, G.V. and Hampton, M.A.: 1973, Sediment gravity flows: mechanics of flow and deposition, in Turbidites and deep water sedimentation. SEPM short course.
14. Sigurdsson, H.: 1972, Bull. Volcanol. 36, pp. 148-163.
15. Hampton, M.A.: 1979, J. Sed. Petrol. 49, pp. 753-758.
16. Embley, R.W.: 1976, Geology, vol. 4, pp. 371-374.
17. Kuenzi, W.D., Horst, O. and McGehee, R: 1979, Geol. Soc. Amer. Bull., vol. 90, pp. 827-838.
18. Bouma, A.: 1962, Sedimentology of some Flysch Deposits, Amsterdam, Elsevier Publ. Co., 168 p.
19. Stalcup, M.C. and Metcalf, W.: 1972, Jour. Geophys. Res., vol. 7, pp. 1032-1049.
20. Baker, P.E.: 1968, Bull. Volcanol. 32, pp. 189-206.
21. Newell, R.E., Kidson, J.W., Vincent, G. and Boer, G.J.: 1972, The general circulation of the tropical atmosphere, Cambridge, Mass. Inst. Tech. Press, vol. 1, 258 p.

TEPHROCHRONOLOGY AT DSDP SITE 502 IN THE WESTERN CARIBBEAN

Michael T. Ledbetter

Department of Geology
University of Georgia
Athens, Georgia 30602

ABSTRACT

A composite core consisting of four Hydraulic Piston Cores at DSDP Site 502 in the western Caribbean were analyzed for dispersed tephra. Peaks in dispersed tephra abundance were combined with megascopic ash layers to produce a tephrochronology which is based on the magnetostratigraphy and biostratigraphy for the section recovered. The frequency of tephra falls at Site 502 increased in the late Miocene and Quaternary; this may be related to subduction rates of the Cocos Plate under Central America, or to major changes in wind direction over Central America.

1. INTRODUCTION

Central America is characterized by bimodal volcanism, consisting of basaltic andesite volcanics and rhyolitic ignimbrite and airfall deposits. The widespread airfall ashes on land have marine tephra equivalents in the three ocean basins surrounding Central America (Hahn et al. (1); Drexler et al. (2)). The discovery of widespread marine tephra in the equatorial Pacific was accomplished using seismic reflectors, identified as white ash by Worzel (3). Over a dozen discrete ashes within the Worzel ash were geochemically fingerprinted and source areas for the largest volume tephra were suggested (Bowles et al.) (4). Of these, the Worzel Layer D is one of the largest (Rose et al.)(5), and may be correlated to the Los Chocoyos Ash with a source in the Lake Atitlán caldera, Guatemala (Hahn, et al.) (1).

S. Self and R. S. J. Sparks (eds.), Tephra Studies, 281–288.

Recent studies of deep-sea tephra have enhanced terrestrial pyroclastic studies by providing information on the areal distribution of airfall ashes (e.g., Watkins et al. (6); Ninkovich et al., (7)), the eruption mechanics (e.g., Wilson et al. (8); Huang et al. (9); Ledbetter and Sparks (10)), and the frequency of explosive eruptions (Kennett and Thunell (11)). Additionally, the age of major volcanic eruptions can be determined readily by tracing an ash layer into well-dated marine cores (e.g., Ninkovich and Shackleton (12); Drexler et al. (2)).

The recovery of a nearly complete stratigraphic section DSDP Site 502, close to many of the major volcanic centers in Central America and northern South America (Fig. 1), provides an opportunity to establish a tephrochronology at an important convergent plate boundary. The frequency of explosive eruptions may be determined by the presence of dispersed and megascopic tephra within a closely sampled section. Fluctuations in the eruption frequency may be compared to spreading rate changes on the East Pacific Rise in order to test the hypothesis that eruption frequency reflects changes in the subduction rate of the Cocos Plate under the Caribbean Plate.

2. METHODS

A composite section was compiled from four holes at Site 502 in order to sample a complete stratigraphic section at each site. A 20 cm sampling interval was employed in the Quaternary section; a 50 cm interval was used in the older section. The sample spacing in years (0.1-15 KYRS) is approximately the same throughout the composite section.

Approximately 1 gram samples were sieved through a 62 μm screen and treated with HCL. A qualitative compositional analysis of the dispersed tephra abundance was performed and the relative amount of tephra was classified as either absent, trace, common, abundant or a layer. The compositional data was plotted with depth in the composite section (Fig. 2).

Each peak in dispersed tephra abundance and each megascopic ash was numbered and resampled with a larger (approximately 9g) sample. After sieving with a 38 μm screen, the percentage of tephra in the coarse fraction was plotted with age in the section (Fig. 3).

3. DISCUSSION

Thirteen megascopic and at least 63 dispersed tephra were identified in the composite section at Site 502 (Fig. 2). The

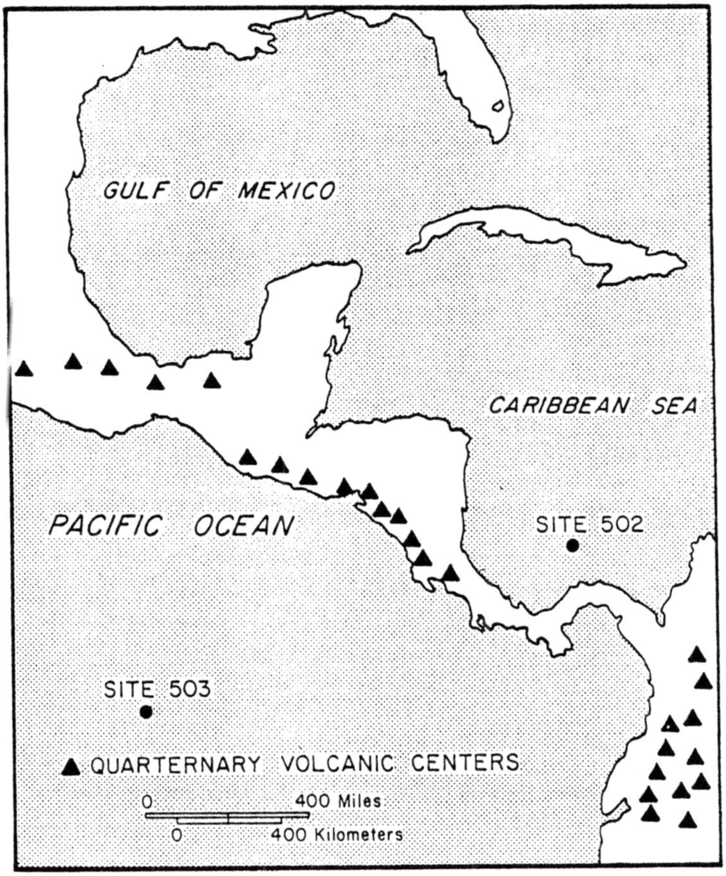

Figure 1. Locations of DSDP Sites 502 and 503 with explosive volcanic centers of Central America and northern South America shown for reference (from Ninkovich and Shackleton, 1975). Site 503 was too far from source to receive megascopic ash layers.

tephra abundance (mg greater than 38 μm g^{-1} total sample) for each ash is shown with an assigned age (Fig. 3) determined from magnetostratigraphy and biostratigraphy (Prell and Gardner et al. (13)). The relative magnitude of peaks in tephra abundance is a function of distance to source, wind direction, eruptive column height, and total volume of eruptive products. Relative tephra abundance at a site, therefore, cannot be used as a measure of eruption magnitude alone.

The eruption frequency record at Site 502 clearly shows that the late Miocene and the Pleistocene are characterized by increased volcanic eruptions (Fig. 3). These two periods of increased volcanism correspond to the global pattern of increased volcanic activity proposed by Kennett and Thunell (11) and Kennett et al. (14), which has been disputed by Ninkovich and Donn (15). The latter attribute the Quaternary increase to the approach of core-sites into the ash-fall zone adjacent to the convergent plate margins. Site 502 is on the Caribbean Plate however, which has not moved with respect to the Central American volcanic centers (Fig. 1). Therefore, the increased tephra frequency during the Quaternary at Site 502 may represent a true increase in eruption frequency, not simply an approach of the site toward the ash-fall zone surrounding the source region.

The increased eruption frequency of Central American volcanoes in the late Miocene and Quaternary may represent an increase in subduction rates during those periods (Kennett and Thunell (11)). This hypothesis is consistent with the increased spreading rates on the East Pacific Rise (EPR) during the Quaternary (Rea and Scheidegger, (16)). If the subduction rate is equivalent to the spreading rate on a rigid plate, then the Quaternary increase in volcanism may be due to an increased subduction rate under Central America.

The episodic nature of the volcanism which produced the ash layer at Site 502 is similar to that observed throughout the circum-Pacific region (Nobel et al. (17); McBirney et al. (18); Hein, et al. (19)). The late Miocene and Quaternary increases in volcanism may represent increased spreading rates and a concomitant increase in subduction rates in the Pacific basin. The increased subduction rates may also play a role in

---

Figure 2. Tephra abundance as a composite of four Hydraulic Piston Cores at Site 502 is plotted versus corrected depth. The concentration of tephra in the >38 μm non-carbonate fraction is qualitatively divided into four categories for dispersed tephra in addition to megascopic layers. Each peak in abundance is numbered for reference to Figure 3.

SITE 502 TEPHRA ABUNDANCE

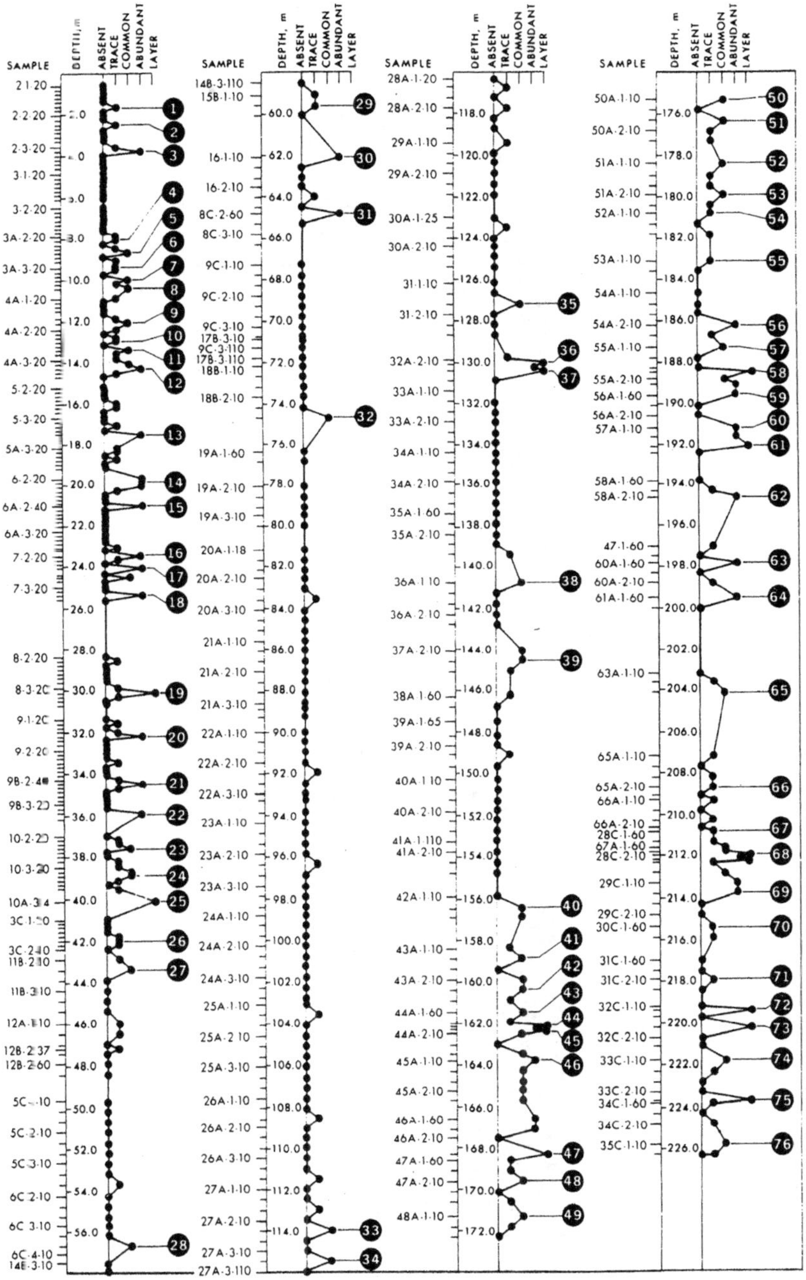

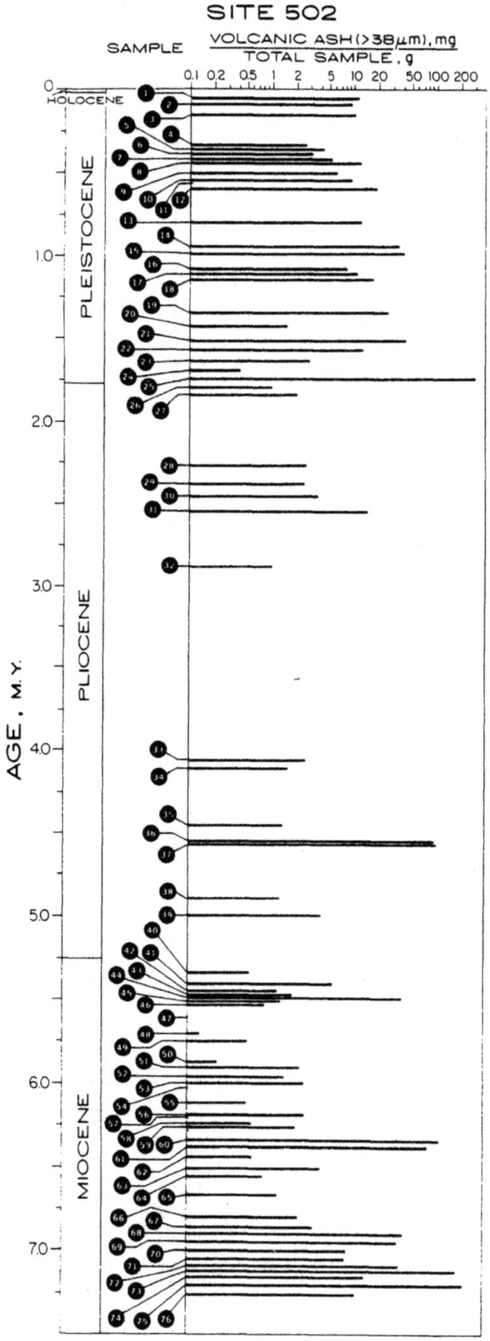
SITE 502
SAMPLE
VOLCANIC ASH (>38μm), mg
TOTAL SAMPLE, g
0.1 0.2 0.5 1 2 5 10 20 50 100 200
HOLOCENE
PLEISTOCENE
PLIOCENE
MIOCENE
AGE, M.Y.
0
1.0
2.0
3.0
4.0
5.0
6.0
7.0

the evolution of calc-alkaline magmas at convergent plate boundaries (Scheidegger and Kulm, (20)). The episodic nature of spreading and subduction in the Paleogene (Kennett and Thunell (11)) cannot be determined at Site 502, but could be tested with additional sites which penetrate older sediment on the western Caribbean Plate.

An alternative explanation for the Quaternary and late Miocene peaks in tephra abundance at Site 502 may be a shift in prevalent wind directions. If winds changed direction during the Pliocene so that volcanic ash was carried to the west instead of east into the Caribbean, then the low ash frequency in the Pliocene could be due to that change rather than to a low eruption frequency. Since the large ash falls during the late Pleistocene are found on the Pacific side of Central America (Bowles et al. (4); Drexler et al. (2)), the predominant wind direction must have been to the west. Therefore, Site 502 has recorded ash falls in spite of a predominantly upwind location and no change of wind direction in the Pliocene could account for the marked decrease. It is proposed therefore, that the tephra frequency has recorded the eruption frequency of Central American volcanoes.

REFERENCES

1. Hahn, G.A., Rose, W.I., Jr. and Meyers, R.: 1979, in: Ash Flow Tuffs, W. Elston and C. Chapin (Eds.), Geol. Soc. Am. Special Paper 180, pp. 100-112.
2. Drexler, J.W., Rose, W.I., Jr., Sparks, R.S.J. and Ledbetter, M.T.: 1980, Quat. Res. 13, pp. 327-345.
3. Worzel, J.L.: 1959, Proceedings of the National Academy of Science 45, pp. 349-355.
4. Bowles, F.A., Jack, R.N. and Carmichael, I.S.D.: 1973, Geol. Soc. Am. Bull. 84, pp. 2371-2388.
5. Rose, W.I., Jr., Grant, N.K. and Easter, J.: 1979, in: Ash Flow Tuffs, W. Elston and C. Chapin (Eds.), Geol. Soc. Am. Special Paper 180, pp. 87-100.
6. Watkins, N.D., Sparks, R.S.J., Sigurdsson, H., Huang, T.C., Federman, A., Carey, S., Ninkovich, D.: 1978, Nature 271, pp. 122-126.

Figure 3. Quantative measurements of tephra concentration for larger samples at each horizon denoted in Figure 2 are plotted with age at Site 502. Increased frequency of ash fall from explosive eruptions in the Central American region (Fig. 1) occurred in the late Miocene and Pleistocene.

7. Ninkovich, D., Sparks, R.S.J. and Ledbetter, M.T.: 1978, Bull. Volcanologique 41, pp. 1-13.
8. Wilson, L., Sparks, R.S.J., Watkins, N.D. and Huang, T.C.: 1978, J. Geophys. Res. 83, pp. 1829-1838.
9. Huang, T.C., Watkins, N.D. and Wilson, L.: 1979, Geol. Soc. Am. Bull. Part II 90, pp. 235-288.
10. Ledbetter, M.T. and Sparks, R.S.J.: 1979, Geology 7, pp. 240-244.
11. Kennett, J.P. and Thunell, R.C.: 1975, Science 187, pp. 479-503.
12. Ninkovich, D. and Shackleton, N.J.: 1975, Earth Planet. Sci. Letters 27, pp. 20-34.
13. Prell, W.L. and Gardner, J.V. et al.: in press, Init. Repts. Deep Sea Drilling Project 68.
14. Kennett, J.P., McBirney, A.R. and Thunell, R.C.: 1977, J. Volcanol. and Geotherm. Res. 2, pp. 145-163.
15. Ninkovich, D. and Donn, W.L.: 1976, Science 194, pp. 899-906.
16. Rea, D.K. and Sheidegger, K.F.: 1979, J. Volcanol. and Geotherm. Res. 5, pp. 135-148.
17. Noble, D.C., McKee, E.H., Farrar, E. and Peterson, U.: 1974, Earth Planet. Sci. Letters 21, pp. 213-220.
18. McBirney, A.R., Sutter, J.F., Naslund, H.R., Sutton, K.G. and White, C.N.: 1974, Geology 2, pp. 585-589.
19. Hein, J.R., Scholl, D.W. and Miller, J.: 1978, Science 199, pp. 137-141.
20. Scheidegger, K.F. and Kulm, L.D.: 1975, Geol. Soc. Am. Bull. 86, pp. 1407-1412.

DEEP-SEA RECORD OF CENOZOIC EXPLOSIVE VOLCANISM IN THE NORTH ATLANTIC

Haraldur Sigurdsson and Benny Loebner

Graduate School of Oceanography
University of Rhode Island
Kingston, Rhode Island 02881 USA

ABSTRACT. Cenozoic explosive volcanism associated with rifting and continued opening of the Norwegian and Greenland Seas has produced two compositional series of silicic tephra. These are preserved in deep-sea sediments. A high-potash and high-alumina series of glasses range from quartz-trachytes to alkali rhyolites and associated comendites. A low-potash series, ranging from icelandites through dacites and rhyolites, is characterised by low alumina and enriched in Fe and Ca. Both series have been erupted throughout the Cenozoic. The deep-sea cores indicate five apparent episodes of Cenozoic explosive volcanism. The first is a late Paleocene episode recorded in North Sea exploration wells and stems from volcanic activity in the British Isles. A middle Eocene episode is attributed to volcanism on the subaerial Iceland-Faeroes ridge. The middle Oligocene episode coincides with the Kialineq plutonic event in East Greenland. The subsequent lull in rhyolitic explosive volcanism during late Oligocene coincides with minimum activity of the hotspot, low spreading rates and absence of known subaerial volcanic sources. An early to middle Miocene episode is attributed to rejuvenation of the Iceland hotspot and emergence of subaerial Iceland, whereas the Plio-Pleistocene increase in explosive silicic volcanism coincides with rift-jumping between the Icelandic volcanic zones. Ash-fall dispersal in the region is dominated by westerly winds, except during glacial stages, when ice-rafting leads to tephra deposition south of Iceland.

1. INTRODUCTION

Tephra in deep-sea sediments is generally attributed to explosive volcanism at converging plate margins, whether of the

S. Self and R. S. J. Sparks (eds.), Tephra Studies, 289–316.

island arc or Andean type. The formation of extensive tephra deposits is not, however, restricted to this type of tectonic environment. Sediments in the North Atlantic and adjacent Norwegian and Greenland Seas contain abundant tephra deposits formed by explosive volcanism at rifted plate margins from 50 Ma to the present time. Recent drilling on Glomar Challenger Leg 38 and piston coring in the region has now provided an extensive sampling of this explosive volcanism, thereby creating the opportunity for a reconstruction of the history of Cenozoic explosive volcanism in the North Atlantic.

There are clear benefits from the study of these deep-sea tephra. Firstly, temporal variations in volcanism can be established in the framework of biostratigraphic dating, which is particularly important in light of the often poorly preserved, poorly dated and incomplete record of Cenozoic explosive volcanism on land. Secondly, chemical correlation of deep-sea ash layers with land-based deposits provides important correlations between the biochronologic and radiometric time scales. Thirdly, the glassy nature of tephra in the deep-sea sediments provides samples of Cenozoic magmas which are often unavailable by other means and equivalent to exposed plutons in the deeply eroded source regions. The possibility of sampling the glassy equivalents of the magmas of Skaergaard, Skye and other classical Tertiary igneous centers in the deep-sea sediments is an exciting prospect.

In this paper we compile a record of tephra layers in deep-sea sediments of the North Atlantic during the Cenozoic era, document the compositional types of glasses, analyse possible episodicity, dispersal patterns and, finally, speculate on their sources. The complex history of repeated generation of rhyolitic magmas and explosive volcanism in this region is discussed in the framework of continental rifting, subaerial sources on the Iceland-Faeroes Ridge and repeated rift-jumping and plate-tectonic adjustments.

1.1 Geologic Setting

The occurrence of rhyolitic ash layers in North Atlantic deep-sea sediments throughout the Cenozoic requires explosive volcanism from subaerial sources, either at the rifted continental margins, from an elevated hotspot on the mid-ocean ridge (e.g. Iceland) or from oceanic island volcanoes (e.g. Jan Mayen). A second requirement is the availability of rhyolitic magmas in the subaerial source regions, whether generated by crustal fusion or fractional crystallisation processes. Present knowledge of the evolution of the Norwegian and Greenland Seas indicates that such sources have been available sporadically during the Cenozoic.

The first plate tectonic events in the region were associated with spreading at the Ran Ridge and opening of the Labrador Sea. Activity at the Ran Ridge ceased at about 40 Ma [1]. Opening of the Norwegian Sea can be traced back to anomaly 24 time [2]. This oldest identifiable magnetic anomaly in the region has an age of 56 Ma on the time scale of LaBrecque et al. [3], in good agreement with radiometric dates of magmatic activity at the rifted continental margins. The British Tertiary igneous province evolved during the period 50 to 60 Ma [4,5] and the Tertiary igneous rocks of East Greenland have radiometric ages which cluster between 50 and 56 Ma [6,7,8].

Opening continued uninterrupted on the Reykjanes Ridge and Mohns Ridge throughout the Tertiary and up to present time, with the exception of adjustment in spreading direction from northwesterly to westerly direction at anomaly 13 time, at 36 Ma [2].

The principal complexities in the evolution of this ocean basin occur in the region bounded by the Jan Mayen fracture zone to the north and the Iceland-Faeroes Ridge in the south (Fig. 1). Initial evolution of this region is associated with the Aegir Ridge, where spreading from anomaly 24 time to anomaly 7 (25.5 Ma) led to the formation of the Norway Basin. According to Talwani and Eldholm [2] the Aegir Ridge became extinct by anomaly 7 time and the spreading axis jumped west to the Greenland continental margin. This estimate of the time of extinction of the Aegir Ridge is uncertain, however, and may be too low, judging from the presence of middle Oliogocene sediments above basement at site 337 near this ridge axis.

The westward rift jump resulted in the separation of the Jan Mayen Ridge as a microcontinent from East Greenland in the Scoresby Sund region. This event is potentially very important for the history of explosive rhyolitic volcanism in the region as it provides subaerial volcanic sources at a rifted continental margin, analogous to the British Eocene and Paleocene igneous province and the East African rift today. The rifting of the continental Jan Mayen ridge from Greenland occurred during spreading on the Kolbeinsey Ridge. Vogt <u>et al</u>. [9] have shown that this process began at least 24 Ma and their tentative identification of anomaly 7 east of the Kolbeinsey Ridge indicates that initial rifting of the Jan Mayen ridge may be pre-25.5 Ma.

This Oligocene rifting of the East Greenland margin was probably accompanied by formation of igneous centers. The early to middle Oligocene plutonic and volcanic rocks in the Mesters Vig and Kialineq regions of East Greenland are, however, the only known occurrences of Oligocene magmatism in the region and have been dated at 28 to 39 Ma [10,11]. Future dating of other plutonic centers north of Scoresby Sund may further establish the presence of an Oligocene East Greenland magmatic episode.

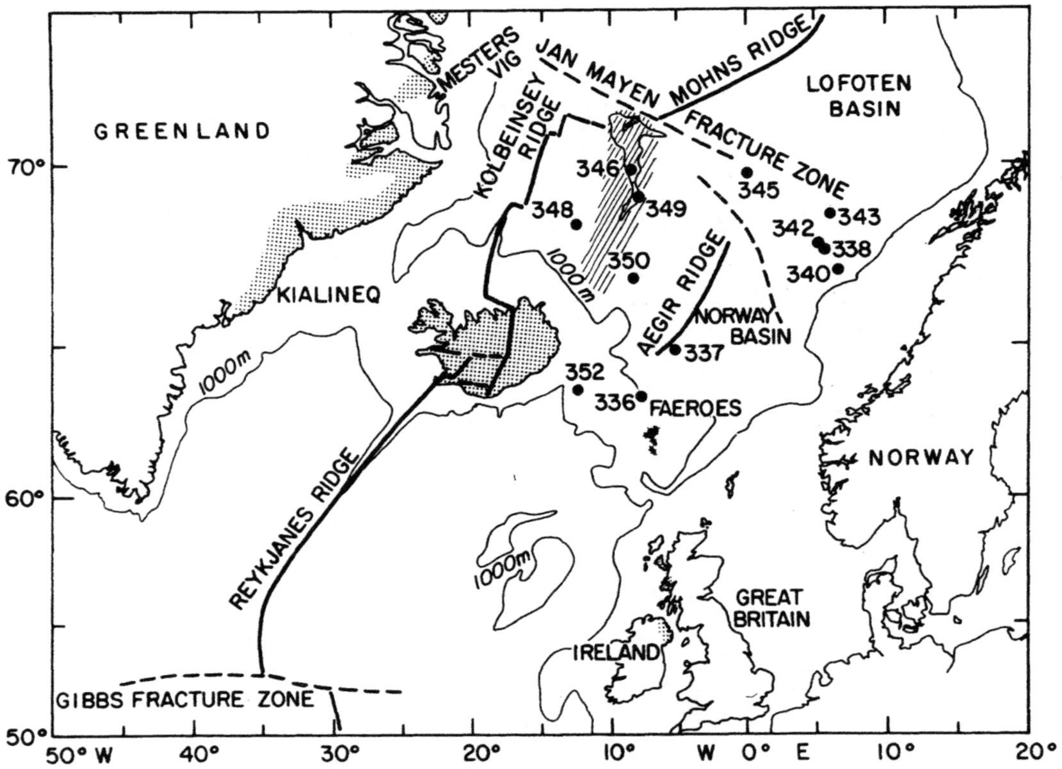

Fig. 1 Structure and bathymetry of the Norwegian and Greenland Seas, showing location of Leg 38 drilling sites. Ruled area represents Jan Mayen continental fragment. Stippled areas represent known Tertiary igneous centers.

The evolution of the Iceland-Faeroes Ridge and the emergence of Iceland above sea level are important elements in the history of explosive volcanism of the region. Voppel _et al._ [12] have put forward the most recent model for the evolution of the Iceland-Faeroes Ridge. They propose a spreading center on the ridge up to anomaly 22 time (53 Ma), connected by right-offset fracture zones to the Reykjanes Ridge to the south and Aegir Ridge to the north. By anomaly 21 time (50 Ma) the Iceland-Faeroes Ridge spreading center was abandoned, and gave way to a northward propagation of the Reykjanes Ridge onto the western part of the Iceland-Faeroes Ridge. The details of subsequent rift jumping on the transverse ridge are not known and inferences about subsequent evolution assume symmetrical spreading about the Reykjanes Ridge. The Iceland-Faeroes Ridge remained above sea level until mid-Oligocene (30 Ma); [2,13].

Two prominent "waists" in the morphology of the aseismic ridge, as defined e.g. by the 1000 m contour, separate the Iceland platform from the aseismic ridge on the Greenland margin and from the Iceland-Faeroes Ridge to the north-east. The physiographic bulge of the Iceland Platform has been interpreted as reflecting an abrupt increase in basalt discharge of the Iceland hotspot [14]. The magnetic anomaly interpretation of Voppel _et al._ [12] indicates formation of the "waists" and minimum activity of the hotspot at about anomaly 18 time (42 Ma). Vogt _et al._ [9], on the other hand, place anomaly 13 on the Greenland side of the Denmark Strait, implying the period 25 to 35 Ma for the lull in hotspot activity. The timing of subsequent increase in hotspot activity and emergence of the Iceland platform are still subject to speculation and our only constraints are the radiometric dates of 16 Ma for the oldest exposed subaerial lavas on Iceland [15]. Vogt _et al._ [9] speculate that the increase occurred ca. 25 Ma. If this estimate is correct, then the mid-ocean ridge system in the region was probably submerged from the time of submergence of the Iceland-Faeroes Ridge in middle Oligocene (30 Ma) until the emergence of the Iceland platform (ca. 20 Ma). This model predicts a lull in subaerial explosive volcanism in late Oligocene. Subsequent Miocene explosive volcanism in the region can be attributed to the Iceland hotspot, where scores of rhyolite-producing central volcanoes have been identified.

2. QUATERNARY TEPHRA

2.1 Pleistocene

The record of Pleistocene explosive volcanism in the region is preserved in piston cores as discrete tephra layers or as ash zones, i.e. sediment units with pronounced increase in the concentration of dispersed volcanic ash. Three ash zones have been

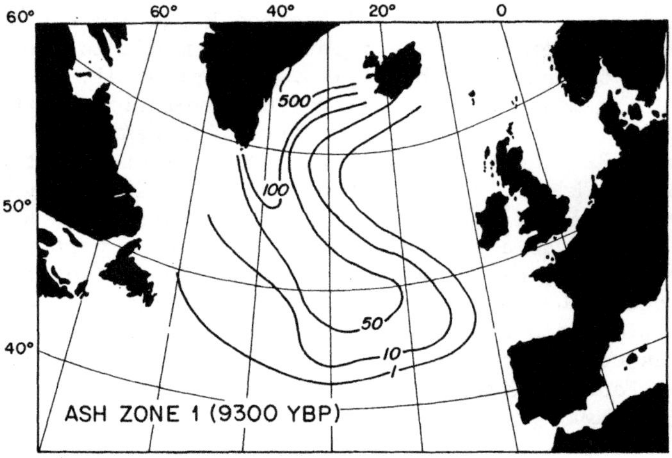

Fig. 2 Dispersal pattern of ice-rafted volcanic ash zone 1. The abundance of sand-sized glass shards (> 63 μm) for a 1-cm^2 column through the ash zone is indicated by 10^3 contours [60].

identified in the North Atlantic Pleistocene sediments [16,17]. The youngest has been dated as 9,300 y.b.p. and is present in 30 cores between Iceland and 45° (Fig. 2). This 20 to 30 cm thick zone occurs at depths of 20 to 60 cm below the sea-floor. It is marked by an increase in glass shards up to a maximum of 10,000 shards per gram of sediment, but nowhere does the volcanic ash component exceed 1% of total sediment. The glass shards are rhyolitic in composition and up to 1000 μm in length. Ruddiman and Glover [16] estimate a total ash volume of 0.94 km^3 for ash zone 1, or equivalent to an ash layer of 0.9 mm thickness with areal extent of 4.8 x 10^6 km^2. Land-based Icelandic tephrochronology has not yet been extended back far enough to allow correlation with ash zone 1. In the Faeroes, however, Waagstein and Johansen [18] identify a 9,400 y.b.p. ash layer in soils, which could well be the land equivalent of ash zone 1.

Ash zone 2 dates from 64,700 ± 3,500 y.b.p. [19] and is also prominent in sediments south of Iceland, as far south as 45°N. This ash zone is present at 364 to 382 cm depth in a core 275 km west of Iceland [20]. The total volume of ash zone 2 is 7.59 km^3; equivalent to a layer of 1.4 mm thickness, and 5.5 x 10^6 km^2 in extent [16].

Ash zone 3 was deposited approximately 340,000 y.b.p. in the North Atlantic south of Iceland. It is comparable to zone 2, with a volume of 6.27 km^3 and equivalent to a 1.1 mm ash layer with surface area of 5.5 x 10^6 km^2.

Ruddiman and Glover [16] suggest Icelandic volcanoes as a source for these ash zones. They point out, however, that the coarse grain size, great distance from Iceland (up to 1800 km) and southerly transport all speak against atmospheric dispersal. Instead, they propose that the ash distribution was controlled by ice-rafting and by the counter-clockwise circulation pattern of pack ice in the North Atlantic during the Pleistocene.

Although the ash zones have not yet been correlated to sources, it seems very likely that they are the products of Pleistocene explosive eruptions in Iceland volcanoes or possibly Beerenberg volcano on Jan Mayen. The North Atlantic had extensive ice cover north of 45°N during most of the Pleistocene. Deposition of the ash on to ice delayed ash sedimentation on the sea-floor near Iceland and lead to southerly ice-rafting of much of the ash, to a zone of melting and deposition of ice-rafted debris at 45 to 50°N. The dominance of this pattern of ash dispersal during most of the Pleistocene has consequently blurred the record of explosive volcanism in North Atlantic sediments.

The distribution of volcanic ash in Recent and Pleistocene sediments of the Norwegian and Greenland Seas has been documented by Kellogg [21,22,23] and Kellogg et al. [20]. The ash distribution in surface sediments, as determined by analysis of 169 trigger core tops, shows two major lobes emanating from Iceland. One lobe trends NNE from Iceland and past Jan Mayen Island; the other trends ESE from Iceland (Fig. 3). Fourteen piston cores from the Norwegian and Greenland Seas contain a total of 36 volcanic ash layers and zones ranging in thickness from 2 to 18 cm [21]. These cores generally fall within the volcanic ash region in surface sediments. Most of these ash layers probably represent major eruptions, as they exhibit thicknesses of 5 to 15 cm at a distance of 900 to 400 km from Iceland.

Approximate age estimates of these ash layers, obtained from Kellogg's calculated sedimentation rates and radiometric dates, are given in Figure 4. This incomplete Quaternary record shows two distinct peaks in the frequency of ash layers. One peak occurs in the Holocene and the other during the Eemian interglacial stage (115,000-124,000 y.b.p.). Other ash layers appear randomly distributed throughout the sampled record, which extends to 360,000 y.b.p.

The increase in explosive volcanism during the Eemian and Holocene is probably more apparent than real and could be attributed to better preservation of ash due to lack of ice cover in the Norwegian and Greenland Seas during these interglacial stages. The scarcity of ash layers and the more frequent occurrence of dispersed ash zones in the record from 80,000 y.b.p. up to the Holocene is, on the other hand, probably a reflection of extensive

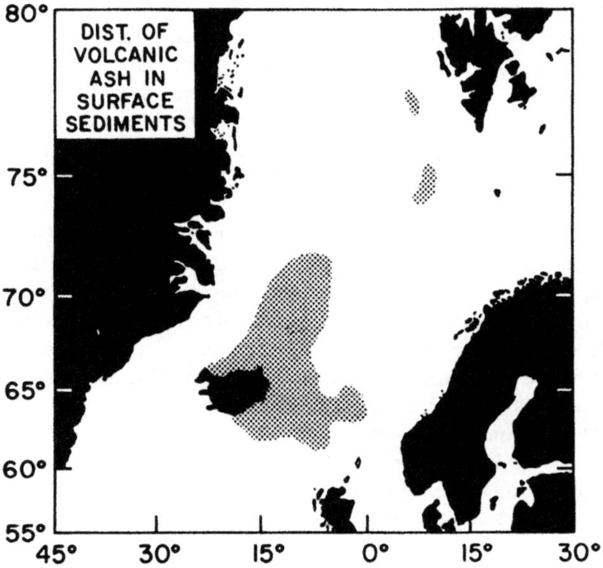

Fig. 3 The distribution of volcanic ash in surface sediments in the North Atlantic region, determined by analysis of 169 trigger core tops [21].

ice cover, resulting in ice rafting and wider dispersal of the ash, both in space and time. Following the same line of reasoning, it is possible that the concentration of ash layers observed between 225,000 to 230,000 y.b.p. (Fig. 4) is associated with ice-free conditions during the interglacial stage which initiated climatic cycle C, 225,000 y.b.p. [24].

2.2 Holocene

The postglacial record of explosive volcanism in the region comes primarily from Icelandic tephrochronology and from the occurrence of drifted pumice from raised beaches in Europe and North America. Production of widespread tephra in this period has been dominated by the activity of the central volcanoes Hekla, Öraefajökull, Snaefellsjökull, Torfajökull and Askja. Four major rhyolitic to dacitic eruptions in Hekla have produced the tephra layers H_1, H_3, H_4 and H_5, dated at 846, 2,900, 4,500 and 7,000 y.b.p respectively [25,26,27]. The well-studied Hekla tephra include the largest-volume explosive events (H_3 with 12 km^3 tephra, 2.2 km^3 dense-rock equivalent) and thus provide convenient guidelines in interpreting the deep-sea ash layers. The Hekla tephra have been dispersed in a northeasterly direction over Iceland and the Norwegian Sea. The thickness of each tephra layer in the fallout region decreases with distance from Hekla in a systematic manner

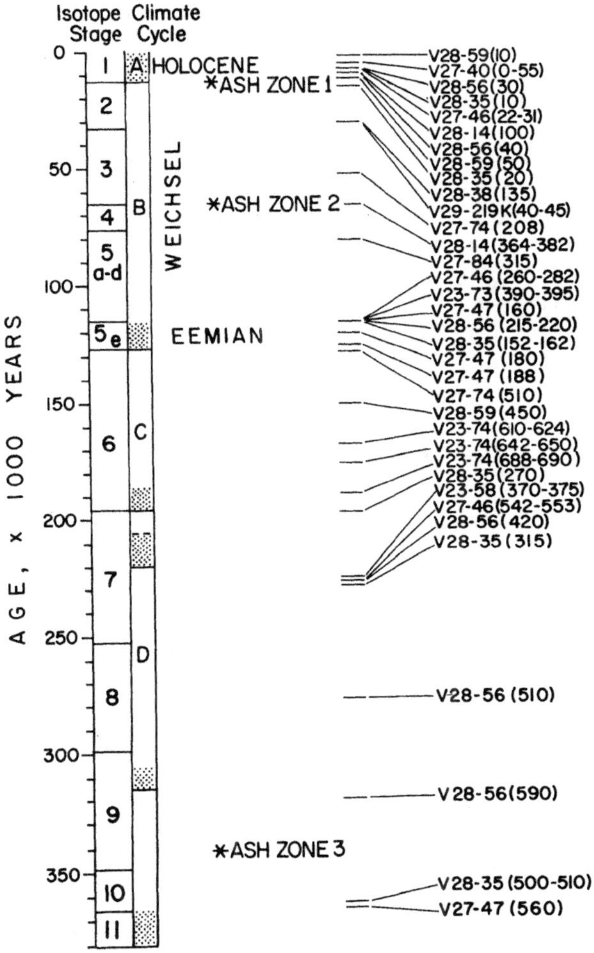

Fig. 4 Distribution of volcanic ash layers in North Atlantic piston-cores, showing ice-rafted ash zones [16] and discrete volcanic ash layers [21]. Isotopic stages are from Shackleton and Opdyke [59] and climatic cycles after Ruddiman and McIntyre [24].

(Fig. 5). In none of these layers, however, does the 5 cm isopach extend beyond the coast of Iceland and we can surmise that deep-sea tephra layers from the post-glacial Hekla eruptions and other eruptions of similar magnitude will be no more than 1 to 2 cm thick. The 10 to 18 cm thick deep-sea layers reported by Kellogg [21] in Norwegian Sea cores must consequently represent eruptions which are several orders of magnitude larger than the documented post-glacial events.

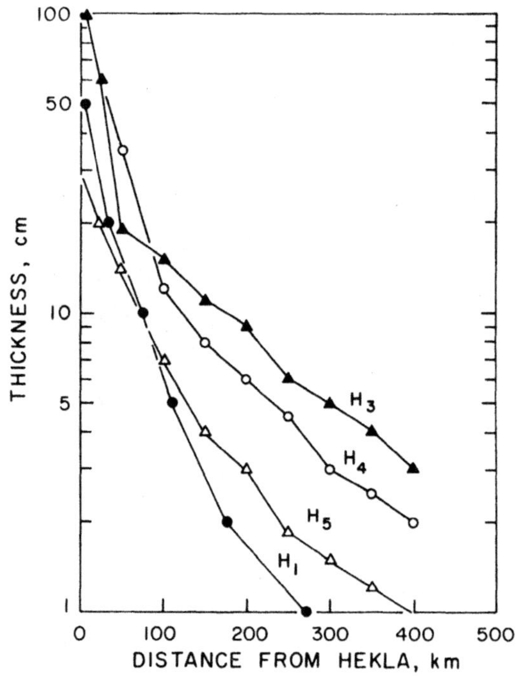

Fig. 5 Thickness of Holocene Hekla tephra layers as a function of distance from source [26].

Other major post-glacial explosive eruptions in Iceland are probably all inferior to H_3 in volume. They include the eruption of Öraefajökull in 1362, which produced 6 km^3 of tephra dispersed east over an area of 280,000 km^2 within the 0.1 cm isopach [28]. Two eruptions of Snaefellsjökull in western Iceland produced silicic tephra layers dated 1,750 and 3,960 y.b.p. [29]. Finally, the 1875 eruption of Askja produced 0.8 km^3 of tephra (0.21 km^3 dense-rock equivalent) which was transported east as far as Scandinavia [30,31].

Pumice transported by surface currents occurs widely on post-glacial raised beaches in the North Atlantic region. The occurrence of drifted pumice has been documented in Britain, Scandinavia, Svalbard, Greenland and the Canadian Arctic, notably around Baffin Bay. The use of these widespread pumice horizons in correlating raised beaches has spurred efforts to date these levels [32,33, 34]. Many of the pumice lumps are 3 to 5 cm in diameter and predominantly of dacitic to rhyolitic composition. Both Blake and Binns indicate Iceland as the most likely source, on compositional and geologic grounds.

Radiometric dates of associated organic matter show that pumice drifting has occurred in the period 1,000 to 7,000 y.b.p. Many of the European occurrences are in the range 6,000 to 7,000 y.b.p. (Tapes I) and could represent pumice derived from the Hekla eruption of H_5 tephra 7,000 y.b.p. Similarly, several pumice horizons in the Canadian Arctic and elsewhere, dated 4,000 to 5,000 y.b.p. [32] could be derived from the 4,500 y.b.p. H_4 event [26]. Further detailed work on correlation of these drifted pumice levels with Icelandic events is required.

2.3 Controls of Quaternary dispersal

Tephra from explosive eruptions which have been large enough to deposit an identifiable ash-layer in deep-sea cores must have been injected above the tropopause (8 to 10 km) and transported dominantly in the stratosphere. The empirical evidence from historical eruptions [30] and the distribution of volcanic ash in surface sediments (Fig. 3) shows that stratospheric transport is dominantly to the north-east and east of Iceland. This is also consistent with the general stratospheric circulation in the region. West winds dominate in the stratosphere during winter, spring and fall at the latitude of Iceland, ranging from 15 to 30 m/sec at 10 km to over 100 m/sec at 60 km [35,36]. In mid-summer, however, weaker west winds (0-10 m/sec) prevail up to 20 to 30 km, with east winds of 5 to 60 m/sec at higher elevations [37]. A small percentage of Icelandic tephra would therefore be dispersed to the west.

Ocean currents during interglacials, e.g. the Eemian and Holocene, do not significantly affect the dispersal of tephra, since most of the glass shards in deep-sea cores are sufficiently large (200 to 500 μm) to have settling velocities far in excess of current velocities. The principal effect of ocean currents on tephra dispersal in the deep-sea relates to the drift of sea-ice. Ruddiman and Glover [16] have shown that the transport of volcanic ash carried by glacial sea-ice was controlled by a counterclockwise current in the North Atlantic south of Iceland (Fig. 2). The absence of the warm Norwegian current from the Norwegian and Greenland Seas during glacial stages probably permitted year-round ice cover of the entire Norwegian Sea [22]. The transport of ice-rafted volcanic ash from this region during Quaternary glacial stages thus probably largely occurred southward through the Denmark Strait.

3. TEPHRA IN DSDP CORES

Seventeen sites drilled during Leg 38 of the Deep-Sea Drilling Project are the principal source of information about Tertiary explosive volcanism in the North Atlantic region. We have sampled and analysed 81 ash layers from these cores.

The Glomar Challenger penetrated 5278 m of sediment during Leg 38. A total of 1808 m of core was recovered, representing core recovery of only 34%. These sections contain sediments ranging in age from Eocene to Pleistocene and provide a fairly representative record spanning the period of opening and evolution of the Norwegian and Greenland ocean basins [38]. Discrete ash layers are common in the succession and, in addition, dispersed volcanic ash is often an important constituent of the general lithology. The Initial Report for Leg 38 documents 259 ash layers and ash pockets in these cores. Our study shows, however, that only 81 of these are well-preserved ash layers, consisting principally of volcanic glass shards. The remainder of the reported layers were found to consist of zeolites, potassium feldspar and clay minerals, presumably representing the breakdown products of volcanic glass, or were sand layers, composed of quartz, feldspar, lithic fragments of volcanic rocks and other terrigenous components. These latter layers are believed to represent distal turbidites and ice-rafted debris, particularly in the Plio-Pleistocene sediments.

Approximately 80% of the layers are rhyolitic, consisting of 90 to 95% colorless to pale-grey bubble-wall glass shards (300 to 600 μm) with a trace of feldspar or quartz. Many of the rhyolitic layers also contain pale-brown to brown glass shards of dacitic to basaltic composition and are the products of mixed magma eruptions. The remainder of the tephra layers are basaltic in composition, with vesicular, dark-brown and glassy fragments (200 to 300 μm) and generally finer-grained than the rhyolitic tephra. The layers range in thickness from 1 to 9 cm with the exception of two layers of 14 and 18 cm thickness.

The largest sample of ash layers comes from early to mid-Tertiary sediments at sites 338 to 343 on the Voring Plateau (26 layers). Site 348 on the Iceland Plateau yielded 21 layers of late-Tertiary and Quaternary age, and fourteen layers were recovered from sites 346 (5), 349 (6) and 350 (3) on the adjacent Jan Mayen Ridge, ranging from Miocene to Recent. Site 337 on the Aegir Ridge in the Norway Basin yielded ten layers and six ash layers came from sites 336 and 352 on the Iceland-Faeroes Ridge. Site 345 yielded four ash layers from the Lofoten Basin.

3.1 Glass chemistry

The major-element composition of glass shards in Leg 38 ash layers has been determined with the microprobe. The rhyolitic glasses are highly susceptible to thermally-induced sodium loss during analysis, generally resulting in a 50% decrease in count rate during the first ten seconds of analysis. Sodium was therefore determined separately by the decay-curve technique [39].

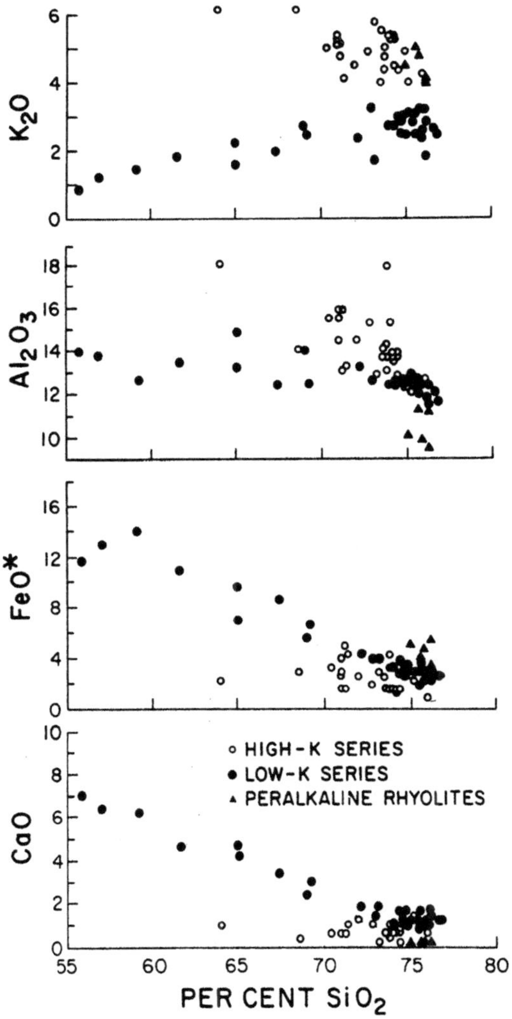

Fig. 6 Oxide variation diagram of glasses in North Atlantic deep-sea tephra.

The analysed volcanic glasses span the range from basalts to rhyolites, but intermediate types are rare, while rhyolites are the dominant type. There are two major rock series, shown on variation diagrams in Figure 6. One series spans the range from basalts to rhyolites and is characterized by low potassium, low alumina and high total iron and calcium content (low-K series). The other series is only represented by trachytes and rhyolites, with high potassium and alumina but low iron and calcium (high-K series). The glasses have been divided into compositional classes,

Table I: Major element classification of glassy tephra (water-free basis)
Oxide values in wt. %

	Type	SiO_2	K_2O	Al_2O_3	FeO^*	CaO
Basalt	A-1	<55	<0.30			
Basalt	A-2	<55	>0.30			
Icelandite	B-1	55-66	<2.5	<15	>7	4-7
Trachyte	B-2	55-66	>4	>16	<4	<2
Dacite	C-1	66-70	<3	<14	>5	>2
Quartz trachyte	C-2	66-70	>4	>14	<4	<2
Rhyolite	D-1	>70	<3.8	11.5-14	2-6.5	1-3
Alkali rhyolite	D-2	>70	>3.8	12-17	1-5	<1.7
Comendite	D-3	70		<11.2	>3	<0.25

as shown in Table I. We have used potassium as the principal discriminant, in view of the dichotomy between the high-potassium and low-potassium series throughout the compositional range. The two series can also be discriminated on basis of FeO*, CaO and Al_2O_3 up to 74% SiO_2, but in the high-silica glasses the concentrations of these three oxides in the two series tend to merge. MgO, Na_2O and TiO_2 trends are not significantly different in the two series.

Peralkaline rhyolites, with <11% Al_2O_3, are the third type of acid glasses (D-3 type, Table I). These comendites have 4 to 5% FeO*, very low CaO and are generally high in K_2O. They are probably related to the high-K series, judging from the common association of quartz trachytes, alkali rhyolites and comendites in volcanic series on some oceanic islands (Easter Island) and continental rifts (Oslo province). Microprobe analyses of representative samples are given in Table II.

Glasses of intermediate composition are associated with the low-potassium series. They occur either as mixed tephra, containing minor andesite and dacite low-K glass shards in a predominantly low-K rhyolite ash layer, or, more rarely, occur as homogenous andesite layers. The low-K andesites are typically iron-rich (10 to 14% FeO*) and are compositionally equivalent to icelandites.

Rhyolitic volcanic rocks equivalent to the high-K and low-K rhyolitic deep-sea tephra series, as well as the peralkaline rhyolitic tephra, are found both on the North Atlantic continental margins and on Iceland. The high- and low- alkali rhyolite types recognised on Iceland [40] correspond to the high-K and low-K series, respectively. Most Icelandic rhyolites are of the low-K type, generally high in iron and contain plagioclase as the principal phenocryst phase. They are the dominant rhyolite type of the Icelandic Tertiary central volcanoes and are field-associated with tholeiitic basalts. In contrast, several Quaternary central volcanoes on Iceland have erupted high-K rhyolites with phenocrysts

TABLE II: Microprobe Analyses of Glassy Tephra from Leg 38

	Low-potash series						High-potash series				
	1	2	3	4	5	6	7	8	9	10	11
SiO_2	58.02	63.81	65.28	67.87	70.55	72.74	63.23	66.73	68.83	72.57	73.61
Al_2O_3	12.42	14.49	12.00	12.16	12.27	11.94	17.87	13.68	13.98	13.33	9.46
FeO^*	13.59	7.00	8.45	6.59	3.71	2.74	2.37	2.97	2.82	1.70	4.45
MgO	1.14	1.41	0.04	0.01	0.10	0.04	0.39	0.02	0.02	0.10	-
CaO	6.06	4.19	3.34	2.97	1.45	1.16	0.95	0.45	0.49	0.77	0.13
Na_2O	3.53	4.07	5.20	5.45	5.08	5.02	6.90	7.04	4.42	4.49	4.67
K_2O	1.45	2.15	1.92	2.45	3.12	2.81	6.08	6.00	5.15	4.33	4.58
TiO_2	1.37	0.84	0.58	0.48	0.27	0.13	0.76	0.16	0.17	0.17	0.20
MnO	0.41	0.18	-	0.21	0.15	-	0.22	0.12	0.15	-	-
	97.99	98.14	96.81	98.19	96.70	96.58	98.77	97.17	96.03	97.46	97.10
Norms											
Q	10.30	17.04	17.16	18.28	24.84	29.68	-	13.17	20.80	27.57	35.61
Or	8.74	12.95	11.72	14.74	19.07	17.19	36.38	36.49	31.69	26.25	27.87
Ab	30.48	35.09	45.45	46.97	44.45	43.98	53.03	38.04	38.95	38.98	23.85
An	14.04	15.20	3.85	1.51	1.51	1.81	-	-	-	3.52	-
Ne							3.12	-	-	-	-
Ns							0.08	5.41	-	-	3.90
Di	14.58	5.14	11.81	12.03	5.25	3.69	4.11	2.05	-	0.35	0.59
Ol							1.66	-	-	-	-
Hy	19.20	12.96	8.87	5.54	4.35	3.13	-	4.53	5.44	2.98	7.76
Il	2.65	1.63	1.14	0.93	0.53	0.25	1.46	0.31	0.34	0.33	0.39

1. 337-10-1(71-77)
2. 348-9-5(128-131)
3. 346-8-5(43-45)
4. 346-8-5(35-40)
5. 348-7-5(111-113)
6. 337-1-2(100-101)
7. 346-4-5(113-115)
8. 346-4-5(122-125)
9. 343-5-6(122-126)
10. 342-3-3(34-36)
11. 338-20-3(102-104)

of anorthoclase or sanidine and quartz [40]. They are field-associated with alkalic or transitional basalts. Glasses of the high-K series span the range from quartz-trachytes (64 to 66% SiO_2) to alkali rhyolites. Their equivalents can also be found in many of the Tertiary alkaline intrusions of East Greenland, where quartz syenites to alkali granites abound. The composition of the mid-Miocene trachyte tephra from site 346 is, for example, strikingly similar to the average composition of the Kangerdlugssuaq intrusion [41].

Major-element compositions of Scottish Tertiary rhyolitic glasses are, on the whole, analogous to the high-K rhyolitic tephra, although considerable variability exists [42,43]. They are, however, characterized by the presence of two feldspars, a plagioclase and a sanidine [44].

The petrogenesis of the two deep-sea tephra rhyolite series must await further study of the coexisting minerals and trace-element data. The present evidence indicates that the compositional differences between the two series can be attributed to the type of phenocryst phases which coexisted in these liquids during their

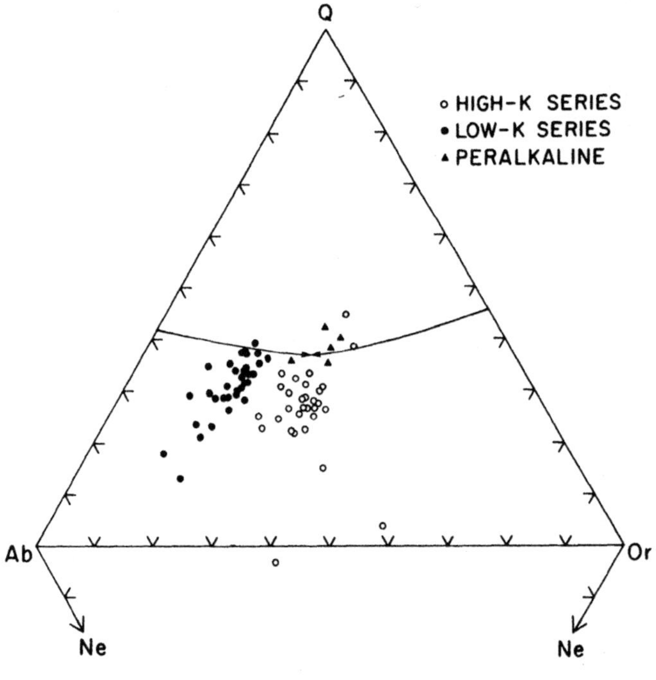

Fig. 7 Normative projection of North Atlantic tephra in the "granite" system. The quartz-feldspar boundary curve is for P_{H_2O} = 1000 kg/cm^2.

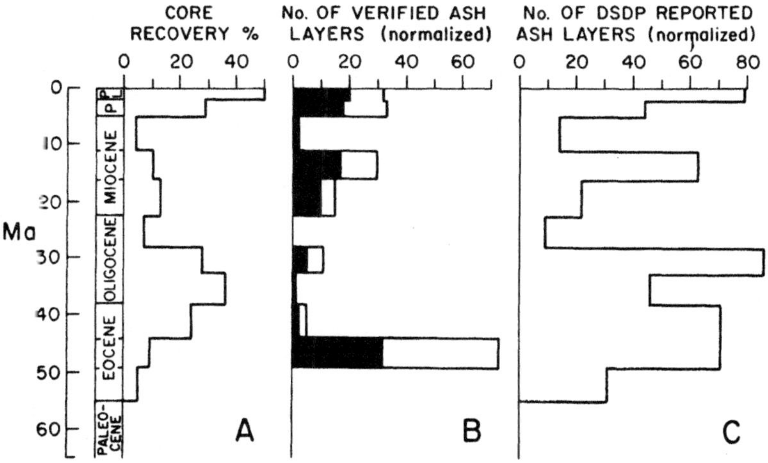

Fig. 8 Frequency of DSDP Leg 38 tephra layers and core recovery in the Cenozoic stratigraphic stages. A: Average core recovery, by stratigraphic stage, for all sites [38]. B: Frequency of tephra layers for all sites, normalized for core recovery. Low-K series are shaded; high-K series unshaded. Only those ash layers which have been verified and analysed by us are included. C: Number of ash layers and ash pods reported in core logs of Leg 38 DSDP Initial Reports, normalized for core recovery.

evolution, with alkali feldspar and quartz, with or without plagioclase, in the high-K rhyolites, and plagioclase alone or with pyroxene in the low-K series (Fig. 7).

Figure 8 shows that the two compositional series have co-existed in the region during the Tertiary and Quaternary periods and both series are present in the four apparent episodes of explosive volcanism in middle Eocene, middle Oligocene, middle Miocene and the Plio-Pleistocene. The rare peralkaline rhyolites (D-3 type) are, on the other hand, entirely restricted to the first two episodes. Tephra of the low-K series are, on the whole, twice as abundant as the high-K series.

3.2 Temporal variations

Marine tephrochronology is potentially the best tool in the search for temporal episodes of explosive volcanism, regional or world-wide [45]. The validity of the use of the observed frequency of volcanic ash layers in deep-sea cores as an index of episodicity is, however, subject to the sampling interval, i.e. the core recovery. Volcanic episodicity should be reflected as a cyclic variable in frequency of ash layers down core and our ability to resolve a cycle is dependent on the sampling interval, the period (duration of a cycle) and sedimentation rate. Ledbetter and Ellwood [46] have shown that a minimum of four samples per cycle are required for satisfactory resolution of the variable being analysed. During Leg 38 a total of 5278 m of Cenozoic sediments were penetrated and total core recovery was 1800 m, or 34%. Given an average sample size of 9.5 m (standard DSDP core length), the total sample number is 190 and average sample interval 28 m. The minimum cycle length which this sample interval is capable of resolving is then 112 m, if we adopt 4 samples per cycle as a minimum sample density. Is this average sample interval sufficient to resolve cycles at the level of stratigraphic stages in the Cenozoic (e.g. early, middle and late Oligocene)? The thickness of stratigraphic stages in Leg 38 cores is quite variable, ranging from a few meters to 185 m, and the available sampling is thus being stretched to its very limit in trying to resolve volcanic episodes at the level of stratigraphic stages.

Mindful of these limitations, we have attempted to assess the variation in relative frequency of ash layers with time throughout the Cenozoic in the Norwegian and Greenland Seas. Using biostratigraphic ages assigned in the DSDP Initial Reports for Leg 38 [38] we have compiled two histograms of ash-layer frequency (Fig. 8). In one approach we have counted the ash layers and pods occurring in each stage of the Cenozoic epochs, as reported in the Initial Reports, and normalized this number to observed core recovery. In the other approach only those layers were counted which have been verified by us, by sampling, to consist of rhyolitic tephra, thus excluding ice-rafted debris, turbidite layers, basaltic layers and layers consisting of authigenic minerals of uncertain origin. The number of rhyolitic layers has likewise been normalized for core recovery in Fig. 8.

Both methods indicate four apparent maxima in the frequency of explosive volcanism in the Cenozoic: in middle to late Eocene, middle Oligocene, early to middle Miocene and Plio-Pleistocene. This episodicity may be more apparent than real, however, as the data is subject to two uncertainties. Firstly, the biostratigraphic scale based on planktonic foraminifers is not applicable at high latitudes and ages assigned to some Leg 38 sediments are somewhat uncertain [47]. Secondly, the magnitude of the normalizations, carried out to correct for the poor core recovery, is so large that spurious frequency peaks may have been generated. The average core recovery for all stratigraphic stages of the sites containing tephra layers is only 18%.

There is a problem in equating the tectonic and spreading history of the region with the record of the volcanic ash layers in the sediments. In most cases, the age of the oldest datable sediment at a given site, determined from the fauna, is significantly greater than the radiometric age of underlying basement (see e.g. (2), Table 3, sites 343 and 345). At site 343 the early Eocene sediments appear to be 20 to 25 m.y. older than basement and at site 345 the early Oligocene to late Eocene sediments exceed the radiometric basement age by 7 to 16 m.y.

Furthermore, basement ages inferred from interpretation of magnetic anomalies (2) are in several cases significantly higher than the assigned faunal age of the oldest datable sediment at a given site. At sites 338 and 343, for example, the basement ages inferred from magnetic anomalies are about 10 and 5 m.y. older than the oldest overlying sediment (see (2), Table 3). Such discrepancies seem too large to be due to a hiatus but may rather be attributed to the observation that the Heirtzler et al. [48] time scale generally gives ages that are too great in early Tertiary time by 5 to 10 m.y. [49]. Subsequent interpretations and correlations of plate tectonic history and explosive volcanism presented in this paper are consequently subject to the uncertainties in the correlations of the paleontologic, geomagnetic and radiometric time scales. In this paper we have adopted an age of 56 Ma for anomaly 24 after LaBrecque et al. [3] instead of the 60 Ma age of Heirtzler et al. [48].

3.3 Eocene

Ash layers of Eocene age are present in sediments at three sites. The Eocene ashes at site 336 and early Eocene layers at site 340 are completely altered to zeolites, potassium feldspar and clay aggregrates whereas middle to late Eocene ash layers at sites 340 and 343 are composed of over 90% clear, colorless glass shards which appear unaffected by alteration. Preserved Eocene ash layers are thus restricted entirely to the Vøring Plateau.

The Eocene ashes are entirely of rhyolitic composition and layer 340-4-3 (105-113) is of peralkaline composition (type D-3). The high-K and low-K series are equally represented among the Eocene rhyolitic ashes. The high-K series include a 14 cm thick rhyolite layer and several other layers at the 150 m level in cores from site 343.

3.4 Oligocene

Sites on the Iceland-Faeroes Ridge (336), Norway Basin (337) and Vøring Plateau (338) contain ash layers of Oligocene age. Site 337 is near the Aegir Ridge spreading axis which became extinct in late Oligocene. Ashes at this site are, with one exception, of basaltic composition and represent the greatest concentration of basaltic tephra observed in Leg 38 cores. These seven layers, 1 to 4 cm thick, are composed of vesicular basaltic glass and are most likely the products of local explosive basaltic volcanism on the active ridge axis in early middle Oligocene, prior to the westward jump of the ridge axis from the Norway Basin [2]. One major Oligocene ash layer (6 cm) at site 337 consists predominantly of colorless low-potash rhyolitic glass shards (type D-1) together with minor dark-brown shards of icelandite glass. The composition of this layer matches very closely with that of an Oligocene rhyolitic mixed 7 cm layer at nearby site 336 and the two layers may represent the same volcanic event. Further north, the middle Oligocene ash layers also include two peralkaline rhyolite layers at site 338, which are comparable to other comendite ashes recovered from the Vøring Plateau, such as the Eocene layers described above. Tephra of the high-K series are dominant in the Oligocene.

3.5 Miocene

Ash layers of Miocene age are found at five sites. They are particularly abundant on the Jan Mayen Ridge (sites 346 and 350), on the Iceland Plateau (site 348) and in the Lofoten Basin to the east (345). Only one Miocene ash layer has been deposited as far east as the Vøring Plateau (342).

The Miocene is dominated by low-alumina, low potash and iron-rich rhyolitic ash layers (type D-1) of very uniform composition, which frequently contain minor amounts of brown low-alumina dacitic glass shards (C-1) and basaltic glass. These mixed ash layers are thickest (18 cm) at site 350 some 350 km ENE of Iceland and range between 5 and 1 cm at sites on the Jan Mayen Ridge and Lofoten Basin north-east of Iceland. Their composition is remarkably similar to rhyolitic Miocene obsidians from eastern Iceland (see e.g. [50], Table 1, anal. 2A).

The Miocene low-potash rhyolitic ashes fall within the north-easterly dispersal pattern of Holocene Icelandic tephra [21]; (Fig. 3 this volume). This fact, together with the close chemical match with Icelandic Tertiary obsidians, points to an Icelandic source for these ash layers.

The Miocene sediments also contain three acid high-potash ash layers. Two of these are 2 to 3 cm high-alumina dacitic layers of middle Miocene age from the north end of the Jan Mayen Ridge (site 346). The third is a rhyolitic 1 cm layer of early Miocene age on the Vøring Plateau (site 342). Basaltic tephra (1 to 3 cm) are present in early to middle Miocene sediments on the Iceland Plateau (site 348).

3.6 Pliocene

We have identified a total of 11 ash layers of Pliocene age, ranging in thickness from 1 to 6 cm. They are best preserved at site 348 on the Iceland Plateau (8 layers) and also occur at site 337 in the Norway Basin and at site 350. Most common (7 examples) are low-potash rhyolites of D-1 type, which are present throughout the Pliocene at site 348 and are the only type recovered from site 337. Two of these rhyolite layers are mixed, with low-alumina, low-potash, brown colored icelandite glass shards.

The upper Pliocene sediments at site 348 contain four pods of high-potash, low-iron rhyolitic ashes of uniform composition (D-2 type). These ash pods are distributed over a 5 m section and may represent a single major ash layer which has been disturbed during coring. Six basaltic ash layers (1 to 3 cm) are distributed through the early Pliocene sediments at site 348.

3.7 Pleistocene

Only nine ash layers from the Pleistocene epoch are preserved, due to the poor core recovery of Pleistocene sediments from Leg 38. All but two of these layers occur on the Jan Mayen Ridge and Iceland Plateau at sites 349 and 348, north-east of Iceland, with a solitary layer at site 337 in the Norway Basin, and at site 352. Six of these are low-alumina, low-potash rhyolites (D-1 type), ranging from 1 to 9 cm in thickness and displaying a very minor range in chemical composition. Basaltic glass shards are mixed in two of these layers. Two layers (4 and 5 cm thick) at site 349 and one layer at site 352 are high-alumina rhyolites of D-2 type.

4. DISCUSSION AND CONCLUSIONS

The deep-sea record shows that Cenozoic explosive volcanism in the North Atlantic region has produced two compositional tephra series: a low-potash series ranging from basaltic to rhyolitic types, characterized by low alumina, high iron and calcium: and a high-potash series ranging from trachytes to alkali rhyolites characterized by high alumina, low iron and calcium. Comendite tephra are associated with the latter series. Both series have been erupted in the Cenozoic, but tephra of the low-K series are always the dominant type.

The major oxide compositions of tephra cannot be used to identify unequivocally the source regions and evidence from land shows that magmas comparable to both series have been erupted or intruded both in the rifted continental margin environment (Greenland, Britain) and in the hot-spot environment (Iceland). We can safely assume, however, that none of the Tertiary tephra of the high-K series are derived from Iceland, since alkalic rhyolites and other acid rocks comparable to the high-K series are only known from the Icelandic Quaternary [40]. Further work in progress on the mineralogy and trace element abundances of the deep-sea tephra are likely to constrain the source regions.

The distribution of volcanic ash layers in deep-sea sediments cored during Leg 38 indicates four episodes of explosive volcanism in the region during the Tertiary: in middle Eocene, middle Oligocene, early to middle Miocene and Plio-Pleistocene (Fig. 8). There is a general decline in ash layer frequency from early to middle Tertiary and an apparent increase in the late Tertiary and Quaternary. There is no evidence of late Oligocene explosive volcanism in the North Atlantic. We caution, however, that these observations on temporal variations are subject to considerable error due to the poor core recovery from Leg 38.

The principal input parameters for our model of the history of Cenozoic explosive volcanism in the North Atlantic region are summarized in Fig. 9. They include spreading rates, active rift axes, land-based evidence of subaerial volcanism on the rifted margins and on the hot-spot generated transverse ridge, and the ash-layer frequency as an index of explosive activity. There are two pre-requisites for the formation of deep-sea acid ash-layers. The first is the presence of subaerial volcanic vents, either at the rifted margins or on elevated parts of the active mid-ocean ridge. The second requirement is the generation of acid magmas.

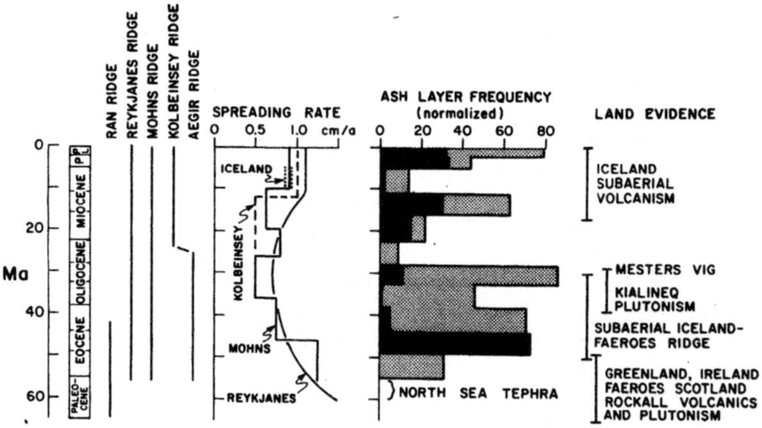

Fig. 9 Cenozoic episodes of explosive volcanism, plate tectonic evolution and subaerial volcanic sources in the North Atlantic. Ash layer frequency is shown for both analysed and verified tephra (solid) as well as DSDP-reported layers (shaded). North Sea tephra after Jacque and Thouvenin [52]. Spreading rates for Kolbeinsey Ridge from Vogt et al. [9], for Mohns Ridge from Talwani and Eldholm [2], for Reykjanes Ridge from Vogt and Avery [61].

With exception of the Mesters Vig and Kialineq igneous activity (28 to 39 Ma; [10,11]), known subaerial sources at the rifted continental margins were active only in the period 50 to 65 Ma [4,5,6,51]. The bulk of volcanism in East Greenland and northwest British Isles thus pre-dates the middle Eocene episode of explosive volcanism.

This explosive volcanism which resulted from the igneous activity in the early Tertiary of the British Isles is not recorded in DSDP cores from Leg 38. There are two reasons for this. Firstly, most of the cores do not penetrate into the Paleocene and secondly, dispersal patterns of ash-plumes from Scotland and Northern Ireland volcanoes were probably easterly and thus deposited south of the Leg 38 cores. These tephra have been found in North Sea oil exploration wells, where the British Tertiary explosive volcanism marks an upper Paleocene (Sparnacian) episode [52].

Within the opening basin the Iceland-Faeroes ridge remained above sea level up to 30 Ma [13]. There is no evidence as yet of acid igneous activity on this ridge, but the presence of large-amplitude circular magnetic anomalies on the ridge have been interpreted as the eroded cores of Tertiary central volcanoes [53]. The most likely source of the middle Eocene episode thus appears to be the Iceland-Faeroes ridge.

The general decline in explosive volcanism during middle Tertiary was interrupted in middle Oligocene time by an episode characterized by high-potash and peralkaline acid tephra. We propose that this episode stems from renewed activity at the East Greenland margin, such as the contemporaneous activity in the Kialineq region, where alkali granites and quartz-syenites were intruded [11]. Some contribution to this episode may also have come from the Iceland-Faeroes ridge.

The total absence of tephra layers in late Oligocene represents a lull in explosive volcanism which coincides with the absence of subaerial sources, decreased hot-spot activity and spreading rates, and jump of the spreading axis from the Aegir Ridge to Kolbeinsey Ridge. Spreading rates on the Mohns Ridge [2] and Reykjanes Ridge show a decrease from 1.25 cm/a in the Eocene to 0.5 cm/a in the Oligocene. This general decrease was accompanied by subsidence of the Iceland-Faeroes ridge by late Oligocene [13], marking the disappearance of the only known subaerial volcanic source.

The increase in hot-spot activity, which began at 25 Ma [9], led to the build-up of the Iceland platform and the early and middle Miocene episode in explosive volcanism. Tephra from this episode are predominantly of the low-potash type and closely comparable to the rhyolitic products of the early Icelandic central volcanoes. Extensive coastal erosion has levelled off older parts of the Icelandic lava pile in the east and west and the present base of the succession exposed above sea level is 16 Ma [15]. The submarine remnants of older central volcanoes, such as those revealed by magnetic surveys on the Iceland shelf [54], are the likely sources of the early Miocene deep-sea tephra, whereas the middle Miocene tephra are contemporaneous with the numerous rhyolite-producing central volcanoes which occur in the exposed Icelandic Tertiary lava succession.

Approximately a third of the early and middle Miocene tephra are of the high-K series and have no known compositional equivalents in the Icelandic Tertiary. The magmatic activity associated with the rifting of the continental Jan Mayen Ridge from the East Greenland margin is a probable source of these tephra. The westward jump of the spreading axis from the Aegir Ridge to the Greenland margin occurred at 24 Ma [9] and led to the formation of the Kolbeinsey Ridge. Volcanism in the rift during the early stages of separation of the continental Jan Mayen Ridge may have been subaerial and produced high-K acid magmas of the type which are associated with the earlier Eocene magmatism of the East Greenland margin.

The marked decline in deep-sea tephra layers in late Miocene is a feature which is not mirrored by the tectonic or volcanic history of the region. On the contrary, the late Miocene is a period of increased spreading rates on the Kolbeinsey Ridge [9] and Mohns Ridge [2] and marks an abrupt increase in accumulation rate of the north Iceland basalt lava pile, from 1 km/m.y. to 4 km/m.y. [55]. The apparent late Miocene decline in explosive volcanism may, however, only be a function of the very poor core recovery (4%) from this stratigraphic stage.

The Plio-Pleistocene acid tephra peak coincides with the jump of the Iceland spreading axis from a central volcanic zone in western Iceland to its present position in the eastern volcanic zone 4 m.y. ago [56]. It has been proposed that this rift-jumping onto older crust has resulted in increased acid volcanism due to fusion and remobilisation of pre-existing plagiogranites and leucocratic intrusions in the Icelandic crust [57].

The record of explosive volcanism in the Quaternary is very well preserved in piston cores, but only selective aspects of this material have been studied so far. The Vema cores contain several thick tephra layers (10 to 22 cm) which indicate eruptions orders of magnitude larger than any of the Holocene rhyolitic events in Iceland. Microprobe studies in progress on the glassy tephra indicate that the Quaternary layers are predominantly low-K rhyolites, including the three ash zones of Ruddiman and Glover [16]. Icelandic central volcanoes seem the only plausible sources, as the Beerenberg volcano on Jan Mayen has exclusively erupted basalts and trachybasalts [58].

The Quaternary tephra give a revealing picture of dispersal patterns and are promising indicators of conditions of open water alternating with ice cover in the Norwegian and Greenland Seas and the adjacent North Atlantic. There is an impressive concentration of discrete ash layers during the interglacial Holocene and Eemian stages, whereas the glacial stages are characterized by lower frequency of discrete ash layers but common dispersed ash zones. These relationships indicate deposition of tephra from easterly-borne ash plumes onto open water during interglacial stages, resulting in settling and good preservation of discrete tephra layers and development of ash zones during the glacial stages is likely on the other hand to reflect the deposition of ash onto ice cover in the Norwegian Greenland Seas. Subsequent ice rafting of tephra through the Denmark Strait led to deposition of glacial tephra as dispersed ash in the North Atlantic south of Iceland.

Acknowledgements

We thank Andrew Davis for assistance with microprobe analyses. This work was supported by the National Science Foundation.

ADDENDUM

After the preparation of this report, W.L. Donn and D. Ninkovich kindly provided us with a preprint of their paper on "Cenozoic explosive volcanism in the North Atlantic Ocean", (Journal of Geophysical Research, in press). Their analysis of DSDP cores indicates major volcanic episodes in Middle Eocene and Pliocene, with minor episodes in Pleistocene, Miocene and Oligocene. Their interpretation regarding volcanic sources, however, differs from that presented in this paper. Donn and Ninkovich attribute all the Cenozoic ash layers to subaerial volcanism in Iceland.

REFERENCES

1. Srivastava, S.P.: 1978, Geophys. J. R. Astr. Soc. 52, pp. 313-357.
2. Talwani, M. and Eldholm, O: 1977, Geol. Soc. Amer. Bull. 88, pp. 969-999.
3. LaBrecque, J.L., Kent, D.V. and Cande, S.C.: 1977, Geology 5, pp. 330-335.
4. Macintyre, R.M., McMenamin, T. and Preston, J.: 1975, Scott. J. Geol. 11, pp. 227-249.

5. Mussett, A.E., Brown, C.G., Eckford, M. and Charlton, S.R.: 1973, Geophys. J. R. Astr. Soc. 30, pp. 405-414.
6. Beckinsale, R.D., Brooks, C.K. and Rex, D.C.: 1970, Bull. Geol. Soc. Denm. 20, pp. 27-37.
7. Brooks, C.K. and Gleadow, A.J.W.: 1977, Geology 5, pp. 539-540.
8. Larsen, H.C.: 1978, Nature 274, pp. 220-223.
9. Vogt, P.R., Johnson, G.L. and Kristjansson, L.: 1980, J. Geophys. 47, pp. 67-80.
10. Gleadow, A.J.W. and Brooks, C.K.: 1979, Contr. Min. Petrol. 71, pp. 45-60.
11. Brown, P.E., Breemann, O., Noble, R.H. and MacIntyre, R.M.: 1977, Contrib. Mineral. Petrol. 64, pp. 109-122.
12. Voppel, D., Srivastava, S.P. and Fleischer, U.: 1979, Deutsch. Hydrogr. Zeitscher. 32, pp. 154-172.
13. Nilsen, T.F.D.: 1978, Contr. Min. Petrol. 67, pp. 63-78.
14. Vogt, P.R.: 1974, The Iceland phenomenon: imprints of a hot spot on the ocean crust and implications for flow below the plates, in: Geodynamics of Iceland and the North Atlantic Area, L. Kristjansson, ed., NATO Adv. Study Inst. Ser., Reidel, pp. 49-62.
15. Moorbath, S., Sigurdsson, H. and Goodwin, R.: 1968, Earth Planet. Sci. Lett. 4, pp. 197-205.
16. Ruddiman, W.F. and Glover, L.K.: 1972, Bull. Geol. Soc. Amer. 83, pp. 2817-2836.
17. Ruddiman, W.F. and McIntyre, A.: 1973, Quat. Res. 3, pp. 117-130.
18. Waagstein, R. and Johansen, J.: 1968, Dan. Geol. Foren., Medd. 18, pp. 257-264.
19. Ruddiman, W.F.: 1977, Science 196, pp. 1208-1211.
20. Kellogg, T.B., Duplessy, J.C. and Shackleton, N.J.: 1978, Boreas 7, pp. 61-73.
21. Kellogg, T.B.: 1973, Late Pleistocene climatic record in Norwegian and Greenland Sea deep-sea cores, Ph.D. Thesis, Department of Geology, Columbia University, 545 p.
22. Kellogg, T.B.: 1976, Geol. Soc. Amer. Memoir 145, pp. 77-110.
23. Kellogg, T.B.: 1977, Mar. Micropaleontol. 2.
24. Ruddiman, W.F. and McIntyre, A.: 1977, J. Geophys. Res. 82, pp. 3877-3887.
25. Thorarinsson, S.: 1967, Soc. Sci. Islandica.
26. Larsen, G. and Thorarinsson, S.: 1978, Jokull 27, pp. 28-46.
27. Thorarinsson, S., Einarsson, T. and Kjartansson, G.: 1959, Geogr. Am. 41, pp. 135-169.
28. Thorarinsson, S.: 1958, Acta Nat. Isl. 2, pp. 1-100.
29. Steinthorsson, S.: 1968, Natturufr. 37, pp. 236-238.
30. Thorarinsson, S.: 1944, Tefrokronologiska studier pa Island, E. Munksgaard, Kobenhavn, 217 p.
31. Sparks, R.S.J., Wilson, L. and Sigurdsson, H.: 1980, Phil. Trans. Roy. Soc. Lond. (in press).
32. Blake, W., Jr.: 1970, Canad. J. Earth Sci. 7, pp. 634-664.

33. Binns, R.E.: 1972, Scott. J. Geol. 8, pp. 105-114.
34. Binns, R.E.: 1972, Bull. Geol. Soc. Amer. 83, pp. 2303-2324.
35. Craig, R.A.: 1965, The upper atmosphere; meterology and physics, New York Academic Press, 509 p.
36. Palmen, E.H.: 1969, Atmospheric circulation systems: their structure and physical interpretation, New York, Academic Press, 603 p.
37. Reiter, E.R.: 1971, Atmospheric transport processes, Part 2: chemical tracers, U.S. Atom. Energy Commission, 382 p.
38. Talwani, M., Udintsev, G., et al.: 1976, Initial Reports of the Deep Sea Drilling Project, 38, Washington, 1256 p.
39. Nielsen, C. and Sigurdsson, H.: 1980, Quantitative methods for electron microprobe analysis of sodium in natural and synthetic glasses, (submitted to Amer. Mineral.).
40. Sigurdsson, H.: 1971, Feldspar relations in Icelandic alkalic rhyolites, Mineral. Mag. 38, pp. 503-510.
41. Kempe, D.R.C. and Deer, W.A.: 1976, Lithos 9, pp. 111-123.
42. Carmichael, I.S.E. and McDonald, A.: 1961, Geochim. Cosmochim. Acta. 25, pp. 189-222.
43. Carmichael, I.S.E.: 1962, Geol. Mag. 99, pp. 251-264.
44. Carmichael, I.S.E.: 1963, Q. J. Geol. Soc. Lond. 119, pp. 95-131.
45. Kennett, J.P. and Thunell, R.: 1975, Science 187, pp. 497-503.
46. Ledbetter, M.T. and Ellwood, B.B.: 1976, Geology 4, pp. 303-304.
47. Hailwood, E.A., Bock, W., Costa, L., Dupeuble, P.A., Muller, C. and Schmitker, D.: 1979, Chronology and biostratigraphy of Northeast Atlantic sediments, DSDP Leg 48, Initial Reports of the Deep Sea Drilling Project, 48, pp. 1119-1142.
48. Heirtzler, J.R., Dickson, G.O., Herron, E.M., Pitmann, W.C. and Lepichon, X.: 1968, J. Geophys. Res. 73, pp. 2119-2136.
49. Berggren, W.A., McKenna, M.C., and Hardenbol, J.: 1978, J. Geol. 86, pp. 67-81.
50. Walker, G.P.L.: 1962, Geol. Soc. London, Q. J. 118, pp. 275-293.
51. Myers, J.S.: 1980, Earth Planet. Sci. Lett. 46, pp. 407-418.
52. Jacque, M. and Thouvenin, J.: 1975, Lower Tertiary tuffs and volcanic activity in the North Sea, in: Petroleum and the continental shelf of North-West Europe, Vol. 1, Geology, Ed. A.W. Woodland, John Wiley and Sons, pp. 455-465.
53. Bott, M.P.H. and Gunnarsson, K.: 1980, J. Geophys. 47, pp. 221-227.
54. Kristjansson, L.: 1976, Mar. Geophys. Res. 2, pp. 285-289.
55. Saemundsson, K., Kristjansson, L., McDougall, I. and Watkins, N.D.: 1980, J. Geophys. Res. (in press).
56. Saemundsson, K.: 1974, Bull. Geol. Soc. Amer. 85, pp. 495-504.
57. Sigurdsson, H.: 1977, Nature 269, pp. 25-28.

58. Hawkins, T.R.W. and Roberts, B.: 1970, Norsk Polarinst., pp. 19-41.
59. Shackleton, N.J. and Opdyke, N.D.: 1973, Quat. Res. 3, pp. 39-55.
60. Ruddiman, W.F. and Glover, L.K.: 1975, Quat. Res. 5, pp. 361-389.
61. Vogt, P.R. and Avery, O.E.: 1974, J. Geophys. Res. 79, pp. 363-389.

STUDIES ON INDIVIDUAL VOLCANOES OR TEPHRA LAYERS

NEW ZEALAND CASE HISTORIES OF PYROCLASTIC STUDIES

George P.L. Walker

Department of Geology, University of Auckland,
Private Bag, Auckland, New Zealand.

ABSTRACT. Studies on the rhyolitic pyroclastics of the Taupo Volcanic Zone, hitherto mainly stratigraphic, are now being directed at determining the nature of the eruptions, and the results of some recent research are reviewed. Aspects covered include the location of the source vent of fall deposits (which in very powerful eruptions may not be where the deposit is thickest); the use of crystal contents to determine the volumes of plinian deposits; the distinction between magnitude, dispersive power, intensity, and violence of explosive eruptions; features shown by water-scavenged ashes and evidence for water being erupted with the ash and causing fluvial erosion; and the special features of low-aspect ratio ignimbrites. The review ends by correlating the form of a volcano with the intensity of its eruptions, volcanoes having high-intensity eruptions being flat or (like Taupo) inverse in form.

1. INTRODUCTION

Eight great explosive eruptions took place during the past 42,000 years from the two main active rhyolitic volcanoes (Taupo and Okataina) of the Taupo Volcanic Zone in New Zealand, as well as many lesser ones, and studies of the pyroclastic products of these great eruptions are yielding new insights into the nature of explosive volcanism besides extending the scale and range of known volcanic phenomena. These eruptions included plinian, phreatoplinian and ignimbrite-forming types, and some also generated extensive air-fall ashes of co-ignimbrite type.

The extensive spreads of young rhyolitic pumice and ash in

S. Self and R. S. J. Sparks (eds.), Tephra Studies, 317–330.

the centre of the North Island have long aroused interest. The pioneer studies were by Grange, Baumgart, and Healy (1-3), and intensive studies have been made in the past two decades by Pullar, Birrell, Vucetich, Howorth, Nairn and others (4-11). The stratigraphic study of the pyroclastic deposits has been done meticulously and in great detail. Few volcanological interpretations have however been attempted until Self and Sparks described two phreatoplinian deposits (12) and Nairn described the Rotomahana deposits of 1886 (13).

The author's studies during the past two years were directed at finding out what could be learned from the deposits about the nature of the eruptions. New isopach maps have been drawn, and also maps of the average maximum diameters of the three largest pumice clasts (MP) and the three largest lithic clasts (ML) measured at each site. Maximum clast size measurements are very useful for correlation purposes, and the MP and ML maps are useful when seeking the vent position and studying eruption dynamics. In addition extensive suites of samples were collected for granulometric and component analyses. The following account summarises the results of some of this research.

2. VENT POSITION DETERMINATION

The rhyolitic volcanoes lack any prominent cone or obvious central vent, and one very basic problem is how to determine vent positions. Two plinian pumice deposits from the Taupo volcano, called Waimihia and Taupo (respectively 3400 and 1800 y old) illustrate the nature of the problem. Both deposits show a low rate of thinning outwards from where they are thickest. The Waimihia pumice is between 6 and 7 m thick along 30 km of its dispersal axis, extending from the shore of Lake Taupo eastwards, and previous workers considered that the vent was situated about 20 km east of the lake shore (3). The Taupo pumice has a maximum known thickness of 1.8 m within an area centred about 8 km east of the lake shore, and the vent was tentatively placed there.

These vent positions are now clearly inconsistent with the observed grain size distribution. Experience of plinian deposits elsewhere (14,15) is that MP and ML both increase consistently towards source; further, that in lithic-bearing plinian deposits, ballistic lithics are invariably found within 3 to 5 km of the vent, these being blocks having such a size (typically well over 100 mm in diameter) that their fall velocity greatly exceeds the wind velocity and their dispersal is hence virtually unaffected by the wind.

For the Waimihia (Fig. 1) and Taupo pumice (16) deposits, MP and ML increase consistently towards Lake Taupo indicating that

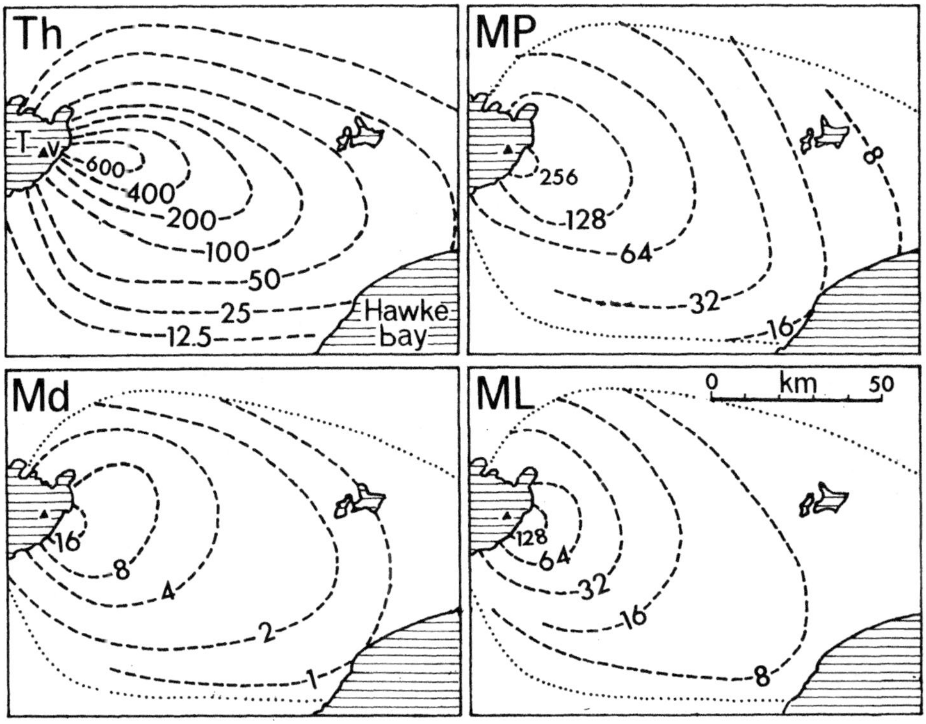

Fig. 1. Maps showing thickness (Th), median diameter (Md), maximum pumice size (MP), and maximum lithic size (ML) for the Waimihia pumice deposit, from unpublished data by the author. Th values are in centimetres; values of Md, MP, and ML are in millimetres. T - Lake Taupo. v - Horomatangi Reefs vent.

the vent lies farther west. The median grain size (Md) from sieved samples likewise increases, and the content of crystals from component analyses decreases (Fig. 2) towards the lake shore. A few lithics just big enough to be regarded as ballistic are found near the lake indicating that the vent is likely to be situated 3 to 5 km offshore. The bathymetry of Lake Taupo (17) reveals the existence of a submerged feature (Horomatangi Reefs) 5 km offshore in approximately the correct position, and this is regarded as a constructional feature at the vent of these two eruptions.

The lesson of this vent-location exercise is that for the most widely dispersed pyroclastic fall deposits the greatest thickness may be developed several tens of kilometres downwind from source, and the isopach map alone is thus not a reliable vent position indicator.

3. VOLUME DETERMINATION OF FALL DEPOSITS

The volume determination of fall deposits has always posed a problem, and various solutions have been tried (18-21). The author has previously used an "area" plot of area enclosed by isopachs versus thickness (as 19 and 21), but there has always been uncertainty regarding the validity of the line that must be extrapolated at the low-thickness end of the area plot, and of what lower limiting isopach to choose.

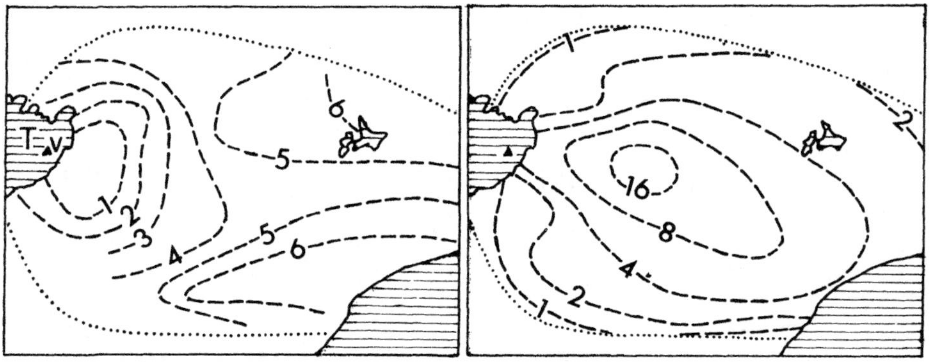

Fig. 2. Maps showing variations in the content of free crystals in the Waimihia plinian pumice deposit. Left - as a multiple of the 3.3 wt.% magmatic crystal content. Right - mass in g per cm^2; integration of this map gives the total mass of free crystals ($0.43\ 10^{15}$g) needed for volume determination.

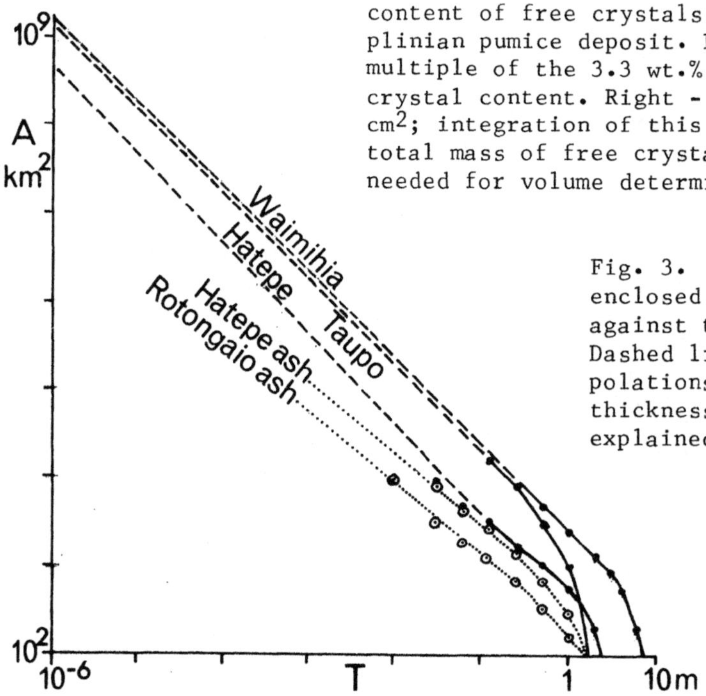

Fig. 3. Plot of area (A) enclosed by isopachs against thickness (T). Dashed lines show extrapolations to the low-thickness end as explained in the text.

A new method has been applied to the Taupo (16), Waimihia (Fig. 2) and Hatepe plinian pumice deposits of Taupo volcano, based on the content of free crystals in the deposit. The mass percentage of crystals in large pumice from the same deposit is taken to be representative of the magmatic content, and it is assumed that crystals are liberated when a portion of the magma is fragmented to an assemblage of particles equal in size to or smaller than the largest crystals (finer than 2 mm for these three deposits).

During eolian fractionation of this erupted assemblage, the free crystals tend to fall closer to vent than the lighter vitric particles; if the total mass of free crystals can be determined (by integration of a map such as Fig.2 right), then the mass of sub-2 mm vitric particles can be estimated.

Application of this method to the Taupo pumice reveals that the total volume is about 24 km^3 (2.5 times greater than was previously estimated), of which 20% have fallen on land within about 200 km of vent.

Taking this crystal-determined volume, an extrapolated straight line on the "area" plot and a limiting thickness can then be selected so as to yield the same volume. When this is done, extrapolated lines extended to the same limiting thickness of 1μm are found for the three plinian deposits to have virtually identical slopes (Fig. 3), suggesting the possibility of a ready applicability to all plinian deposits.

4. MEASURES OF "BIGNESS" APPLIED TO EXPLOSIVE ERUPTIONS

At this stage in the study of pyroclastic deposits it is appropriate to consider carefully what is meant by "bigness" or "violence" applied to explosive eruptions. Reference to published works reveals that measures and terms are not used in any systematic way, and a need clearly exists to systematise the usage.

It is accordingly proposed that "magnitude" should refer specifically to the mass or volume of ejecta or other materials produced in the eruption, "intensity" to the mass or volume discharge rate from the vent, "dispersive power" to the eruptive plume size expressed as the areal extent over which the fall deposits are dispersed, and "violence" to the extent that the momentum or translational kinetic energy with which the pyroclastic assemblage leaves the vent area controls its dispersal. Note that dispersive power and intensity are often closely related, since the plume height depends on the energy discharge rate (22,23), but only where emission is concentrated from a single vent.

The Taupo plinian pumice with a 6 km^3 dense rock equivalent volume was a magnitude 7.8 eruption on Tsuya's scale of magnitudes (24), which is not exceptional for plinian events, but it shows the extreme of dispersive power for which the appelation "ultraplinian" seems appropriate, in consequence of an extremely high eruption intensity.

5. WHOLE-DEPOSIT GRAIN SIZE POPULATIONS

Remarkably little is known about the nature of whole-deposit grain size populations. Previous work (25,20,21,26) indicated that phreatomagmatic eruptions generate a much finer population than plinian ones. New studies have now been made in which whole-deposit populations have been estimated from multiple sample granulometric analysis of pyroclastic deposits combined with the determination of total volume from the content of free crystals (16, and unpub. work by the author). These new studies reveal that ultraplinian and the more widely dispersed plinian deposits differ rather little from phreatomagmatic ones except in having a bigger coarse "tail" (Fig. 4), and suggest that there may well be a direct relationship between dispersive power and fragmentation degree in plinian events, such that with increasing power the plinian population converges on the phreatomagmatic one. This trend is shown clearly by the D/F plot of Fig. 5.

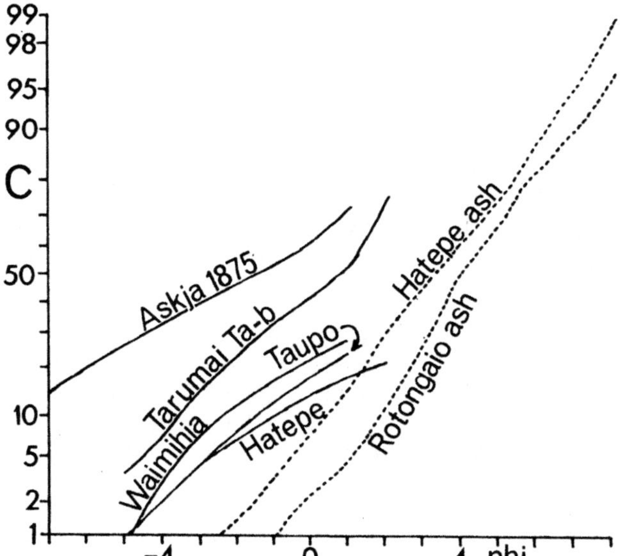

Fig. 4. Whole-deposit populations plotted on probability paper showing the cumulative weight percentage (C) coarser than the given grain size, for 5 plinian deposits (continuous lines) and 2 phreatoplinian ones (dashed lines). Phi $= -\log_2 m$, where m is the grain size in mm (from 16,20,26 and unpub. work by the author).

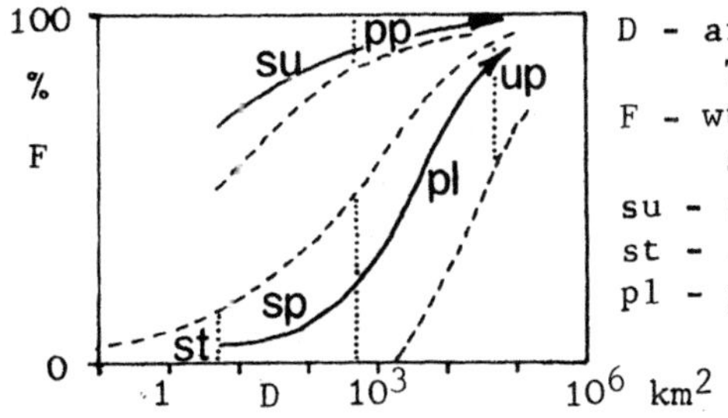

D - area enclosed by isopach for 0.01 Tm (Tm - maximum thickness)
F - wt.% finer than 1 mm where dispersal axis crosses 0.1 Tm isopach
su - surtseyan; pp - phreatoplinian;
st - strombolian; sp - subplinian;
pl - plinian; up - ultraplinian

Fig. 5. D/F plot (25) for fall deposits, showing how with increasing D value the plinian field converges on the phreatoplinian one.

6. FLUSHING OF ASH BY RAIN AND BY ERUPTED WATER.

Pyroclastic fall deposits normally show good sorting and fractionation (using "sorting" to denote the range in particle sizes coexisting in the deposit at any given point or level, and "fractionation" to denote the variation laterally in grain size and constitution). A poorly sorted fall deposit can be formed in several ways. One is when an eruption is strongly pulsatory so that finer material falling from the plume when it is low accumulates together with coarser falling from the plume when it is higher; the resulting deposit is poorly sorted but well fractionated. Another is when pyroclastic material is flushed out of a plume by water to produce a deposit which is poorly fractionated as well.

Two common features of ashes interpreted as being rain-flushed is that they contain accretionary lapilli and show a thickness variation which is related more to the distribution of the rainshower than to the vent position. Thus one ash from Rabaul volcano shows no significant thickness variation over distances 10 to 30 km from vent, and one of Faial has isopachs closed around an area displaced laterally from the vent (Fig. 6).

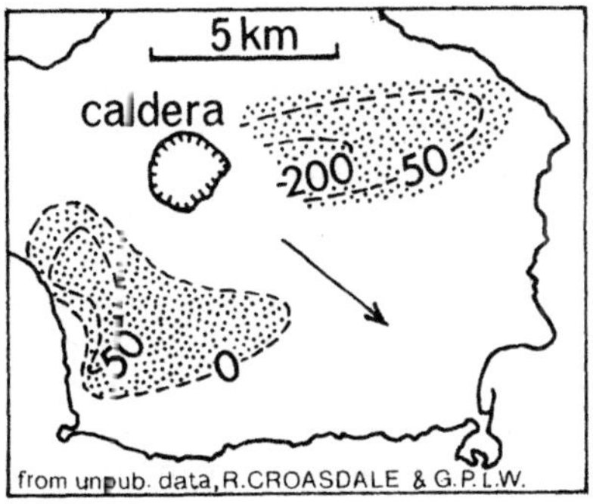

Fig. 6. Sketch map of Faial (Azores) showing distribution of two ashes interpreted as of rain-flushed origin. Thicknesses in cm. Both ashes originated from vents in the caldera, and the eastern one immediately followed a pumice fall dispersed as shown by the arrow; its easterly dispersal may reflect the rainshower shape.

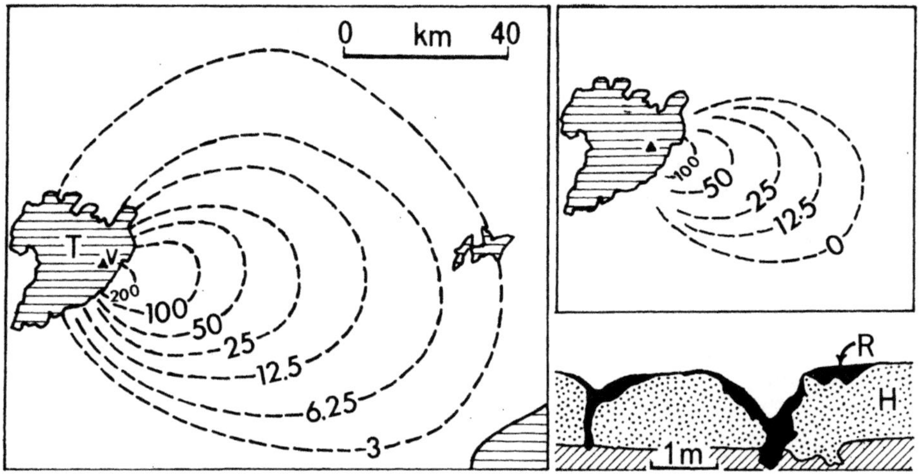

Fig. 7. Isopach map of the Hatepe phreatomagmatic ash. Thickness values are in cm. T - Lake Taupo; v - vent position. The map on the right shows the approximate thickness of Hatepe ash stripped off by water erosion immediately after its accumulation, and below is a drawing of a typical section showing the appearance of the erosion surface. H - Hatepe ash; R - Rotongaio ash resting on the erosion surface. From unpub. work by the author.

Two particular phreatomagmatic ashes - Hatepe and Rotongaio - from Taupo volcano are found to show practically no sorting or fractionation except for the coarsest particles. Although they show abundant signs of accumulating in a wet condition, they lack accretionary lapilli and they show a remarkably good exponential thinning out from source (Figs. 3 and 7). It is therefore believed that both ashes were brought to earth by water which was an integral part of the ash plume, being water from Lake Taupo erupted together with the ash.

A deeply water-gullied surface separates these two ashes (Fig. 7), and features of the gullies indicate very rapid erosion during a single brief pluvial episode. During this episode about 0.3 km^3 of underlying ash and pumice were eroded off, and the quantity eroded per unit area decreased systematically away from Lake Taupo showing that the precipitation followed an exponential decay law similar to that which determined accumulation of the ashes. These features strongly suggest that the gully erosion was volcanogene, being due to runoff from lake water erupted during an interlude when little or no ash was being discharged. It was not due to runoff from a chance rainstorm. The proportion of this lake water which rose in the eruptive column as water droplets rather than vapour which subsequently condensed is uncertain, but is suspected to have been quite high.

7. ORIGIN OF CO-IGNIMBRITE ASHES

In recent years it has become apparent that large-volume ignimbrites are typically associated with large and widespread air-fall ashes of "co-ignimbrite" type which are synchronous with the emplacement of the ignimbrite (27). The marked concentration of free crystals which is commonly found in ignimbrite and is explained by the substantial loss of vitric dust may be correlated with the large amount of vitric dust residing in the co-ignimbrite ash. No thorough study of a large-volume co-ignimbrite ash appears yet to have been made with the specific aim of documenting its origin.

A study is now being made of the 42,000 y old Rotoehu ash, part of which is demonstrably synchronous with the Rotoiti ignimbrite and is interbedded with ignimbrite flow units (7). Certain coarser layers in this ash are found to have an astonishingly wide dispersal shown by the high crystal content out to more than 100 km from source (28). The isopleths for maximum pumice or maximum lithic sizes (for particles of centimetre size or smaller) enclose areas comparable with or greater than those for the most powerful known plinian events. These coarser layers are due to quite exceptionally powerful volcanic events, and have been interpreted as due to great explosions generated where ash flows entered water. Coarser layers in the Oruanui ash (12) have recently been found to have a similarly wide dispersal (Fig. 8).

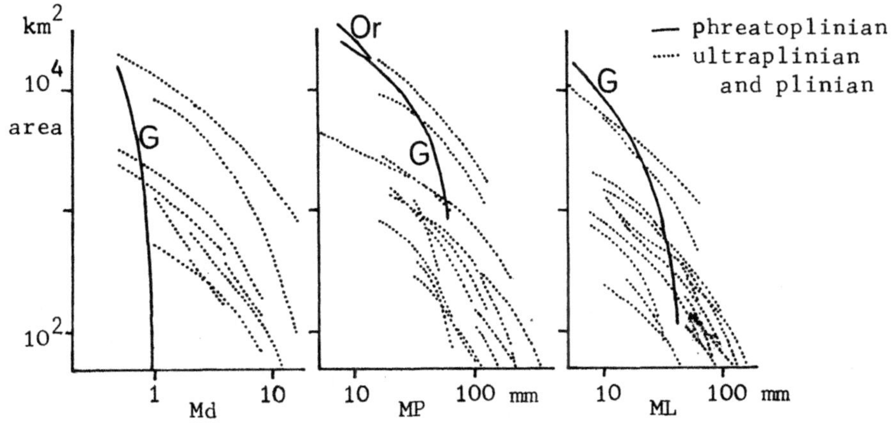

Fig. 8. Areas enclosed by isopleths of Md, MP and ML for the Rotoehu G and Oruanui ashes. The different form of the phreatoplinian curves reflects a different plume expansion mechanism.

8. CHARACTERISTICS OF LOW-ASPECT RATIO IGNIMBRITES

The shape of a tabular rock body such as a lava extrusion or ignimbrite sheet is conveniently expressed by means of the aspect ratio, this being the ratio of average thickness to lateral spread. Ignimbrite sheets commonly have an a.r. near 1 : 1000, but the non-welded 1800-y old Taupo ignimbrite with its average thickness of less than 2 m over a near-circular distribution area 170 km across has an exceptionally low a.r. near 1 : 100,000 (29). The significance of this low a.r. is discussed elsewhere in this volume. Here attention is drawn to the idea that the Taupo ignimbrite marks one end of a spectrum of ignimbrite types and shows a number of features which, though not unique, are better developed in it than in high-aspect ratio ignimbrites.

One feature of the Taupo ignimbrite is that it not only occurs in conventional valley ponds, but also occurs as a thin layer mantling the landscape in between. This mantling "ignimbrite veneer deposit" differs from the pond ignimbrite in showing a crude layering and lacking the coarsest pumice, and is interpreted to have been left behind by the "tail" of the ash flow after the flow travelled across the ground.

Another feature is that the Taupo ignimbrite includes a striking variant consisting largely of coarse and well sorted pumice. This "fines-depleted ignimbrite" is interpreted as being derived from normal ash-flow material by a near-complete loss of fine particles (30), and this origin is supported by the strong resulting enrichment in free crystals. The loss is attributed to a winnowing out of fines resulting in part from the ingestion of forest.

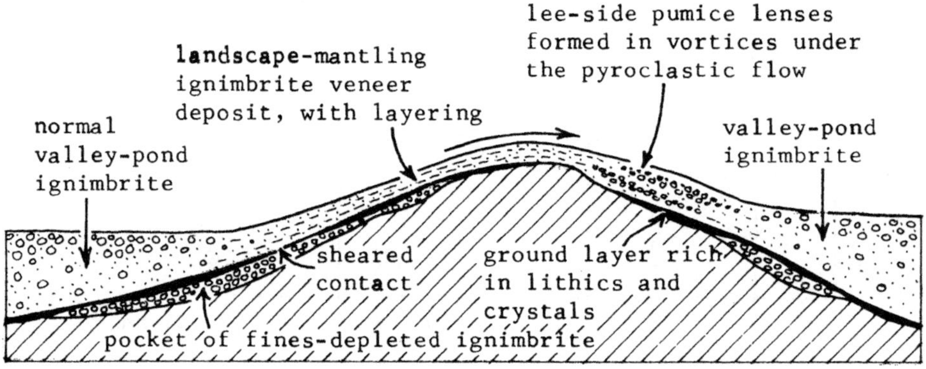

Fig. 9. Schematic section across a small part of the Taupo ignimbrite showing distribution of the ignimbrite variants.

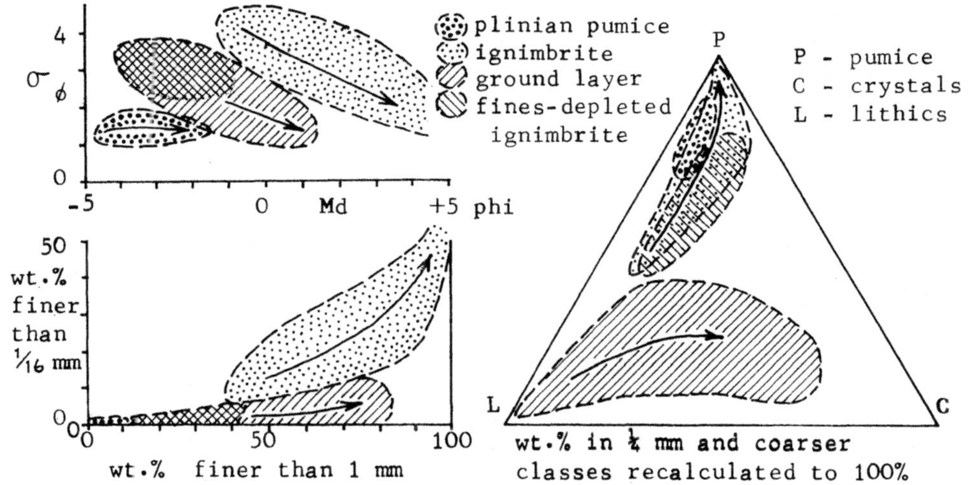

Fig. 10. Characteristics of some variants of the Taupo ignimbrite. Arrows give general trend with increasing distance from source. The diagram top left plots the graphic standard deviation (sigma phi) against median diameter of sieved samples. The pre-ignimbrite Taupo plinian pumice is plotted for comparison.

The Taupo ignimbrite includes another variant consisting largely of lithics, free crystals, and fragments of the densest pumice, forming a "ground layer" underlying normal ignimbrite and attributed to the sedimentation of heavy particles from the head of the ash flow (31).

Fines-depleted ignimbrite and ground layer, both highly variable in thickness, underlie normal ignimbrite and are often separated from it by a sheared contact (Fig. 9). The former is attributed to portions of the ash flow which became detached from and moved in advance of the head. They were so strongly fluidised that they rapidly lost fines and came to rest. They were then over-ridden by the head of the ash-flow (from which the ground layer was forming) and then by the body of the ash flow which, by its passage, developed the sheared and near-planar basal surface.

The relevance of these studies to pyroclastic geology is firstly that deposits having the mantling form or well-sorted nature (Fig. 10) of pyroclastic fall deposits can develop in several ways from ash flows, and secondly that sharp contacts such as might be attributed to normal erosion can develop as a result of shearing within an ignimbrite.

In addition to these various facies the Taupo ignimbrite shows remarkably large systematic lateral variations in grain size and constitution, such that at its distal end the ignimbrite trends towards, and closely approaches, a constitution of 100 % of vitric fine ash and dust.

9. FLAT AND INVERSE VOLCANOES AND THEIR SIGNIFICANCE

What may be claimed as a general principle of volcanology has emerged from these pyroclastic studies, namely that the more powerful or violent a volcanic event, the less impressive-looking is the resulting pyroclastic deposit when viewed on the scale of a limited-extent outcrop (the converse is not necessarily true, since many insignificant-looking deposits result from insignificant eruptions). Thus in the mildest pyroclastic fall activity, practically all of the ejecta fall within 100 m of vent centre and are then redistributed to form a prominent cone having stable 33° slopes, whereas in the most powerful plinian eruptions the material is widely dispersed to form a thin but very extensive layer.

This may be illustrated by the Taupo plinian pumice deposit which has a volume of 24 km^3 but nowhere exceeds 1.8 m thick. The half-volume isopach - the isopach which contains half of the total volume - encloses more than 10^6 km^2. If the deposit were piled up to form a 33° cone, the cone would stand more than 2 km high and the half-volume isopach would enclose only 22 km^2.

The Taupo plinian pumice sheet has an exceedingly low aspect ratio. So have the Taupo ignimbrite and many other pyroclastic deposits of the Taupo volcano. The reason is believed to be that their eruptions were exceedingly high-intensity ones.

When a volcano is constructed from a succession of sheets each of which has a low aspect ratio, there is virtually no possibility of it ever achieving a steeply conical form, and this leads to the idea that the shape of volcanoes may be related in a relatively simple manner to the intensity of its eruptions. Low-intensity eruptions form steep cones, and high-intensity eruptions form "flat" volcanoes having slopes typically of less than 2°.

In the extreme case of the Taupo volcano, dispersal of the pyroclastic products is so wide that within a broad central area the layers are not sufficiently thick to balance the subsidence resulting from the removal of magma, and an "inverse" volcano having the form of a low plateau around a very broad central depression has developed.

ACKNOWLEDGEMENTS

This research was done as Captain James Cook Research Fellow of the Royal Society of New Zealand, while on leave from Imperial College, London.

REFERENCES

1. Grange, L.I.: 1931, N.Z.J. Sci. Technol. 12, pp. 228-240.
2. Baumgart, I.L.: 1954, N.Z.J. Sci. Technol. B35, pp. 456-467.
3. Baumgart, I.L. and Healy, J.: 1956, Proc. 8th Pacif. Sci. Congr. 2, pp. 113-125.
4. Healy, J., Vucetich, C.G. and Pullar, W.A.: 1964, N.Z. Geol. Surv. Bull., n.s. 73.
5. Vucetich, C.G. and Pullar, W.A.: 1969, N.Z.J. Geol. Geophys. 12, pp. 784-837.
6. Pullar, W.A. and Birrell, K.S.: 1973, N.Z. Soil Surv. Rep. 1, 2.
7. Nairn, I.A.: 1972, N.Z.J. Geol. Geophys. 15, pp. 251-261.
8. Vucetich, C.G. and Pullar, W.A.: 1973, N.Z.J. Geol. Geophys. 16, pp. 745-780.
9. Nairn, I.A. and Kohn, B.P.: 1973, N.Z.J. Geol. Geophys. 16, pp. 269-279.
10. Howorth, R.: 1975, N.Z.J. Geol. Geophys. 18, pp. 683-712.
11. Vucetich, C.G. and Howorth, R.: 1976, N.Z.J. Geol. Geophys. 19, pp. 51-70.
12. Self, S. and Sparks, R.S.J.: 1978, Bull. Volcan. 41, pp. 196-212.
13. Nairn, I.A.: 1979, N.Z.J. Geol. Geophys. 22, pp. 363-378.
14. Walker, G.P.L. and Croasdale, R.: 1971, J. Geol. Soc. Lond. 127, pp. 17-55.
15. Booth, B., Croasdale, R. and Walker, G.P.L.: 1978, Phil. Trans. R. Soc. A 288, pp. 271-319.
16. Walker, G.P.L.: 1980, J. Volcan. Geotherm. Res. 8, pp. 69-94.
17. Irwin, J.: 1972, N.Z. Oceanogr. Inst. Lake Chart Ser.
18. Cole, J.W. and Stephenson, T.M.: 1972, Victoria Univ. Wellgtn. Geol. Dept. Pub. 1, pp. 13-15.
19. Booth, B.: 1973, Proc. Geol. Assoc. Lond. 84, pp. 353-370.
20. Suzuki, T., Katsui, Y. and Nakamura, T.: 1973, Bull. Volcan. Soc. Japan 18, pp. 47-64.
21. Rose, W.I., Bonis, S., Stoiber, R.E., Keller, M. and Bickford, T.: 1973, Bull. Volcan. 37, pp. 338-364.
22. Wilson, L., Sparks, R.S.J., Huang, T.C. and Watkins, N.D.: 1978, J. Geophys. Res. 83, pp. 1829-1836.
23. Settle, M.: 1978, J. Volcan. Geotherm. Res. 3, pp. 1727-1739.
24. Tsuya, H.: 1955, Bull. Earthqu. Res. Inst. Univ. Tokoyo 33, pp. 341-384.
25. Walker, G.P.L.: 1973, Geol. Rdsch. 62, pp. 431-446.

26. Sparks, R.S.J., Wilson, L. and Sigurdsson, H.: 1979, Phil. Trans. R. Soc. A 172 (in press).
27. Sparks, R.S.J. and Walker, G.P.L.: 1977, J. Volcan. Geotherm. Res. 2, pp. 329-341.
28. Walker, G.P.L.: 1979, Nature 281, pp. 642-646.
29. Walker, G.P.L., Heming, R.F. and Wilson, C.J.N.: 1980, Nature, 283, pp. 286-287 and 286, p. 912.
30. Walker, G.P.L., Wilson, C.J.N. and Froggatt, P.C.: 1980, Geology 8, pp. 245-249.
31. Walker, G.P.L., Self, S. and Froggatt, P.C.: 1980, J. Volcan. Geotherm. Res. (in press).

POSTSCRIPT

Research done since the NATO meeting has indicated the presence of a young non-welded rhyolitic ignimbrite in the eastern part of Auckland city. If, as seems likely at the time of writing, this ignimbrite originated from a vent in Lake Taupo, the distance travelled was about 220 km measured in a straight line, or about 240 km measured along a plausible pathway, and the overall drop in elevation over this distance is 350 m.

This is probably the farthest-travelled example of an ignimbrite presently known.

TEPHRA LAYER "a"

Jón Benjamínsson

National Energy Authority
Grensásvegur 9, 109 Reykjavik
Iceland

ABSTRACT. A dark tephra layer, named "a", is found in soils in East and Northeast Iceland. Its volume is approximately 2 km^3 and it covers about 40% of Iceland. An isopach map and grain-size study suggests that the tephra deposit forms a distinctly two-lobed layer and was erupted in the year 1477 from the central volcano Kverkfjöll, at the northern margin of Vatnajökull.

An Icelandic chartulary from the year 1477 A.D. describes a tephra-fall in Eyjafjördur in North Iceland. This is thought to be represented by tephra layer "a" which is found in East and Northeast Iceland. The layer was first recognized by Thorarinsson (1) and named by him "a" because it was the youngest dark basaltic layer found in soils in Northeast Iceland. This work is partly based on unpublished data made available by Thorarinsson.

An isopach map (Fig. 1) has been prepared from thickness measurements in 174 soil profiles. In East Iceland the greyish-white tephra from the 1362 A.D. eruption of Oræfajökull (2), was used as a marker horizon but farther to the northeast the layer H1 from the Hekla eruption 1104 A.D. (3) was used. From the isopach map the volume and distribution of the layer can be estimated. Its volume is approximately 2 km^3 and the tephra covers 39,000 km^2 or 40% of Iceland. It is the greatest dark tephra layer of historical time in Iceland. The isopachs clearly indicate two dispersal directions, one to the ENE and another to the NNW.

Grain morphology was studied with a binocular microscope. Twenty-two samples were passed through a set of sieves and the

S. Self and R. S. J. Sparks (eds.), Tephra Studies, 331–335.

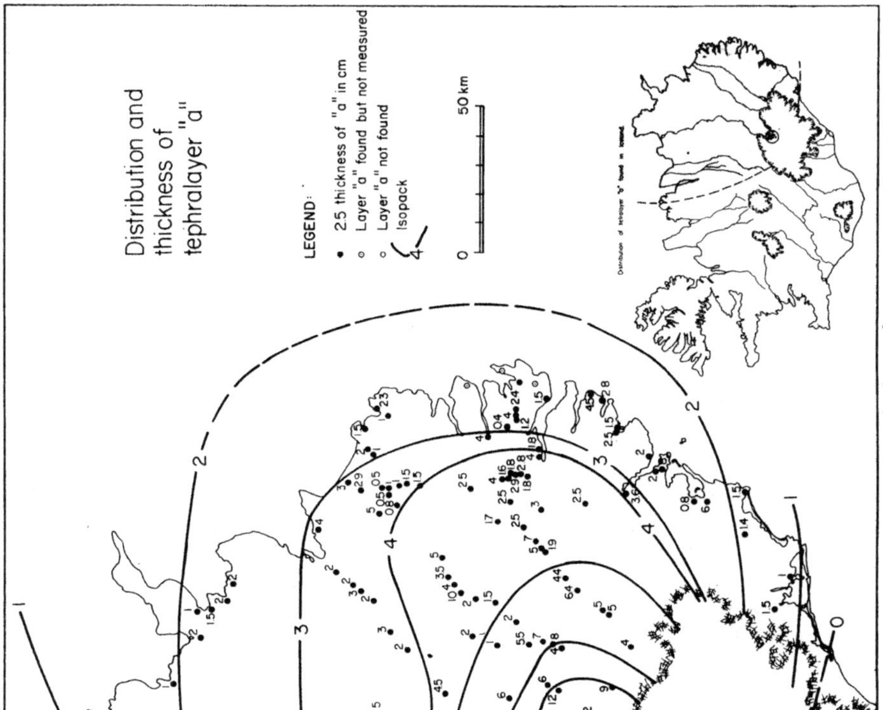

Fig. 1. Isopach map of tephra layer "a"

grain-size parameters calculated (4). The tephra consists mostly of dark glass grains, showing a slight difference in morphology between the two lobes. The shape and surface of grains in the ENE lobe have rather complex angular, rough characters and contain traces of crystals. At the sampling point nearest to the source (∿40 km distance), xenolithic clasts of palagonite tuff are present among the largest grains, as well as highly elongated glass-grains and round grains. The samples at the greatest distance from the source have also platy grains with a smooth surface, but crystals are very sparse or absent. The grains of the

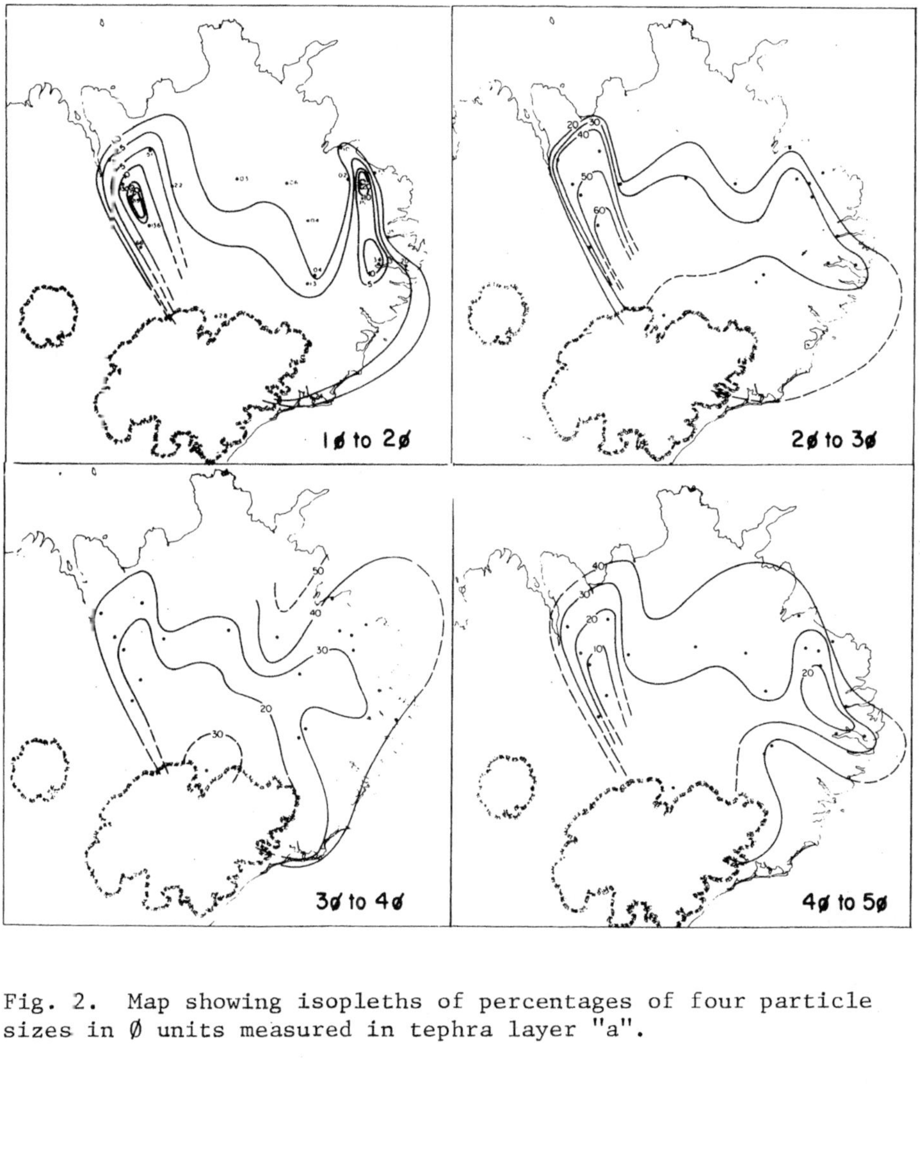

Fig. 2. Map showing isopleths of percentages of four particle sizes in Ø units measured in tephra layer "a".

NNW lobe are more rounded, with a very rough surface and with crystals occurring more frequently. The refractive index of glass fragments from tephra layer "a" lies between 1.55 and 1.59.

The presence of two lobes is well demonstrated by the particle size distribution. Figure 2 shows isopleths of percentages of four particle sizes, a method used by Fisher (5) on the Mt. Mazama tephra. All four groups show the same trend, that is NNW and ENE dispersal. The isopleths further indicate that the particle size distribution does not increase systematically with distance from the source. The two distribution parameters of mean grain-size (M_Z) and sorting So_I) show also the same trend (Fig. 3).

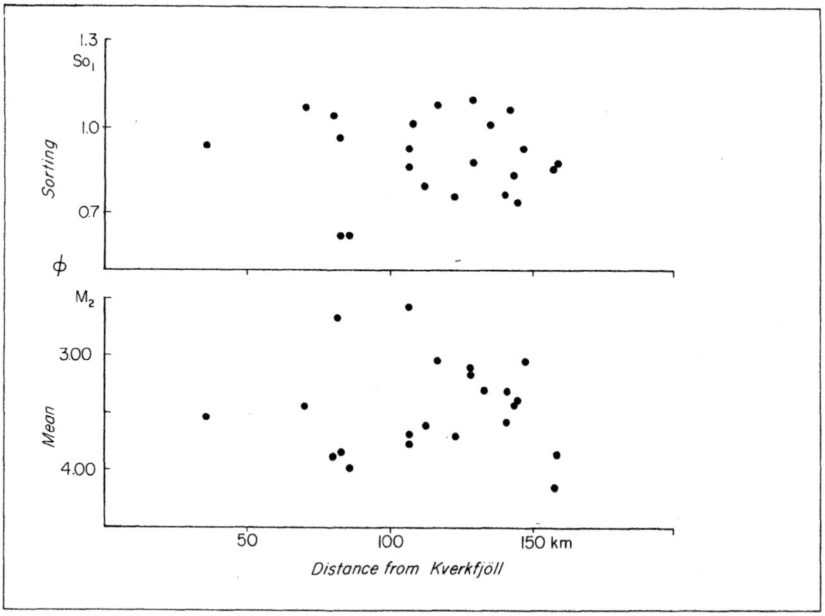

Fig. 3. Plots of grain-size parameters, sorting (So_I) and mean (M_Z) versus against distance from Kverkfjöll.

The two lobes are most likely due to a change in wind direction from SW to SE, a common occurrence when atmospheric depressions pass south of Iceland

The distribution and fine grain size of layer "a" can be explained by a tephra fall from a phreatomagmatic eruption of a subglacial volcano (6). The most likely source for the tephra layer is the central volcano Kverkfjöll at the northern margin of the Vatnajökull ice sheet.

REFERENCES

1. Thorarinsson, S.: 1951, Geog. Annal. 33, pp. 1-89.
2. Thorarinsson, S.: 1958, Acta Naturalia Islandica, Vol. II, No. 2, 99 p.
3. Thorarinsson, S.: 1967, The eruption of Helka 1947-1948, I. Soc. Scien. Islandica, 170 p.
4. Folk, R.L. and Ward, W.C.: 1957, Jour. Sediment. Petrol. 27, pp. 3-26.
5. Fisher, R.V.: 1964, Jour. of Geophysical Res. Vol. 69, pp. 341-355.
6. Benjamínsson, J.: 1975, unpublished B.S. thesis, University of Iceland, 38 p. (in Icelandic).

PERALKALINE IGNIMBRITE SEQUENCES ON MAYOR ISLAND, NEW ZEALAND

M.D. Buck

Macquarie University
North Ryde, NSW, Australia 2113

ABSTRACT

The peralkaline ignimbrite deposits on Mayor Island do not readily conform to the idealised sequence of pyroclastic flows due to their distinctive grading of lithic clasts. Possibly, a unique occurrence is an inverse grading of lithic clasts in a flow unit. Evidence suggests that some pyroclastic flows may be of such high density and poor expansion that they do not separate into the normal ground-hugging flow and overriding ash cloud components. It appears that such flows are a result of a combination of high viscosity and low water content of Mayor Island peralkaline magmas, and abnormal conditions at the vent at the time of discharge, such as a parasitic vent discharging ignimbrites while contemporaneous plinian-type pumice eruptions occur from the main vent.

INTRODUCTION

Mayor Island is an extinct, isolated peralkaline rhyolite volcano lying near the edge of the continental shelf about 26 km offshore in the Western Vay of Plenty, North Island, New Zealand (Fig. 1). The peralkaline character of Mayor Island contrasts markedly with the nearby calc-alkaline Taupo Volcanic Zone and it is thought that its location records the impingement of a tensional rift on the continental crust of the North Island (1).

The island consists of a composite volcanic cone that is built up of pumiceous pyroclastics and thick comendite lava flows. The pyroclastics presently form a thick mantle over much of the

S. Self and R. S. J. Sparks (eds.), Tephra Studies, 337–345.

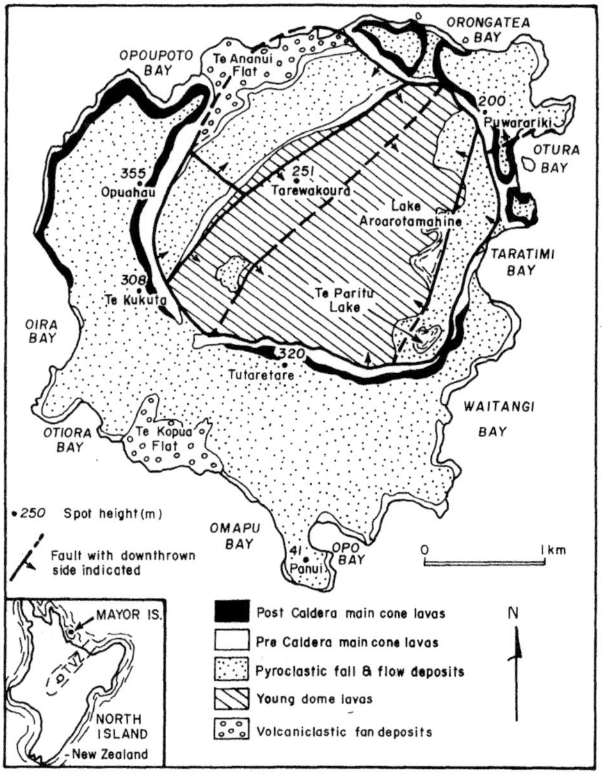

Figure 1. Geology map of Mayor Island. Inset shows regional location of Mayor Island with regard to the North Island, New Zealand and the Taupo Volcanic Zone (TVZ).

island (Fig. 1) as well as being interbedded with the lava flows. Pyroclastic flow deposits form the bulk of the pyroclastic sequences (2) and thin, incipient to non-welded ignimbrites are most noteworthy.

Ignimbrite eruptions commonly pass through a sequence of phases which produce a characteristic sequence of deposits. Sparks et al. (3) first constructed an idealised depositional flow-unit model characterising the main phases of a pyroclastic flow and Fisher (4) has subsequently expanded this model from his observations on the Bandelier Tuff sequence. Thus, the idealised sequence (Fig. 2A) now consists of a ground surge deposit (layer a), a pyroclastic flow unit (layer b), an ash cloud surge unit (layer c) and a thin unit capping the sequence called a tephra fallout unit (4) or a co-ignimbrite ash (5) (layer d).

Most of the ignimbrite sequences on Mayor Island show the

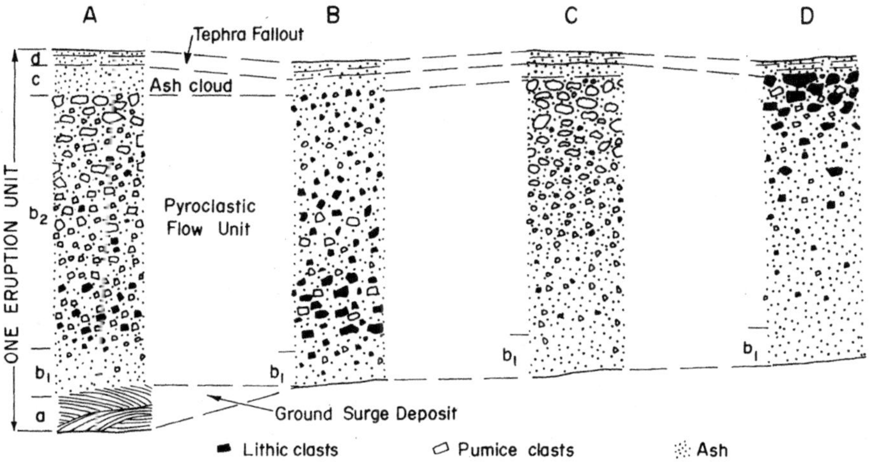

Figure 2. Ignimbrite sequence on Major Island. A. The complete ideal sequence (after Fisher (4)). The pyroclastic flow unit includes layers b_1 and b_2, with an underlying ground surge deposit (layer a), an ash cloud deposit (layer c) and a tephra fallout deposit (layer d). B. Lithic fragment-rich flow unit with normal grading. Ground surge unit absent. C. Pumice fragment-rich flow unit with inverse grading. Boundary between layers b_1 and b_2 becomes obscure. Ground surge unit absent. D. Lithic fragment-rich flow unit with inverse grading. Ground surge and ash cloud units are absent.

same succession of units noted by Fisher (4) in his expanded model but significant differences occur in grading and in the amount and size of clasts of different composition in layer b. Therefore, additional comments are presented to explain the distinctive features seen in these Mayor Island peralkaline ignimbrites.

MAYOR ISLAND IGNIMBRITES

Ignimbrite sequences on Mayor Island that record a single pyroclastic flow eruption are usually less than 5 m thick and the main body of the ignimbrite flow unit (layer b) is normally unwelded, or sometimes incipiently welded. The major types of ignimbrite flow sequences on Mayor Island are schematically summarized in Fig. 2. The complete idealised sequence (Fig. 2A) occurs only in a succession of ignimbrites outcropping in the caldera wall at the northern end of Te Paritu Lake.

Ground surge units in Mayor Island sequences are usually less than 1 m thick, are well stratified with common low-angle cross-bedding, and, even though they show considerable variation in grain size and sorting between beds, are mostly fine grained. When a ground surge unit is present it passes gradationally upwards into the main flow unit into a conspicuously fine grained layer b_1 which coarsens upwards into layer b_2. Layers b_1 and b_2 make up the bulk of the ignimbrite sequences and together they are normally less than 4-5 m thick, with layer b_2 being about one tenth of that thickness.

As in the idealised sequence, ash cloud units (layer c) sometimes immediately overlie the main flow units on Mayor Island. Fisher (4) describes layer c as consisting of 0.5 to 1 m long lenses that are up to 5 cm thick, and are well bedded and sometimes cross-bedded. However, the layer c ash cloud deposit in most of the Mayor Island ignimbrite sequences are mostly massive and occur as very extensive beds that thicken and thin laterally (Fig. 3). They are finer grained and slightly better sorted than layers b_2.

Tephra fallout deposits (layer d), commonly named by others (5) as co-ignimbrite ashes, sometimes overlie and top the ignimbrite sequences (Fig. 3). They are typically very fine vitric ash beds and, of the few samples analysed, 50 per cent is normally finer than 4.75ϕ (0.37 mm). Sometimes these units, which are generally up to 10 cm thick, show a poorly developed shower bedding.

The main body of the flow sequence is layer b_2 and in the Mayor Island ignimbrites it is this layer that shows considerable variation from the idealised model (Fig. 2). Ideally layer b_2 is relatively homogeneous and poorly sorted with ash and blocks coexisting in the rock (3). Pumice clasts generally show an inverse grading where both size and proportion of clasts increases upwards to the top of layer b_2, while lithic fragments show a normal grading with the larger lithic clasts concentrated at the base of layer b_2 (Fig. 2A). The Mayor Island ignimbrites show that if the pyroclastic flow was depleted in either one of the two components in the coarser sized grades, then the resultant layer b_2 shows a normal grading because of a dominance of lithic clasts (Fig. 2B) or an inverse grading because of a dominance of pumice clasts (Fig. 2C).

Ignimbrite units at Omapu Bay have originated from a small parasitic cone in south Opo Bay and, except for a unit exposed near the base of the sequence at north Omapu Bay, they show inverse grading of pumice fragments (Fig. 3). This succession of ignimbrites also shows an up sequence enrichment of pumice clasts with a coincident impoverishment of lithic clasts. The exceptional unit at north Omapu Bay differs in that layer b_2 contains a predominance of accessory lithic fragments which show an inverse

Figure 3. Several flow units of nonwelded inverse graded ignimbrite that were erupted from a parasitic vent to the right of the photograph. Lowermost and second from top units have recessive tephra fallout units at their top. Note the massive, laterally extensive layer c in the subdivided unit. North Omapu Bay.

grading (Fig. 2D), sometimes with blocks up to 75 cm in diameter at the top of the layer. The finer juvenile components of this unit are essentially glass shards and crystals so that its particle size distribution is much the same as the other types of layer b_2 except that about 20 per cent of the particles are coarser than -4ϕ (16 mm). But, unlike the other ignimbrite sequences, the sequence containing inverse graded lithic clasts in layer b_2 does not have an overlying ash cloud layer, and commonly only has a very thin tephra fallout layer.

INTERPRETATION

Many ignimbrites are now considered to result from deposition from the collapsed part of a vertical eruption column (4,6,7) and it appears to be the process involved in the formation of Mayor Island ignimbrites. Firstly, the presence of layers a through d in some ignimbrites on Mayor Island indicates the same column collapse sequence envisaged by Fisher (4). Secondly, the shower bedding and extreme enrichment of fine vitric material in the tephra fallout layers suggests almost complete differentiation of fragment sizes and compositions probably only possible in the upper reaches of a vertical eruption column. Therefore, a vertical explosion occurred initially and subsequent collapse of the eruption column concentrated the larger vitric fragments, crystals and lithic fragments in the collapsing part of the column because of their greater densities.

Ground surge deposits are considered, because of their associated occurrence with ignimbrites, to be the result of an outwardly moving surge cloud or "ash hurricane" (8) that accompanies or precedes the main pyroclastic flow. Sparks et al. (3) suggest that the ground surge deposit spreads laterally more widely and for less distance distally over the terrain than the associated pyroclastic flow which may be confined to valleys. Thus, the absence of ground surge components in most of the Mayor Island ignimbrites (Fig. 2B-D) suggest that either ground surges did not separate from the main flows or the ground surges had a limited extent and the main flows extended beyond that limit. A complete absence of any ground surge units in the many outcrops of these ignimbrites suggests the former is more likely.

The fine grained basal layer b_1 of ignimbrites differs from layer b_2 only in that it lacks any large fragments. Sparks et al. (3) consider it to result from a regime of the pyroclastic flow from which the large fragments are excluded probably due to high shear and grain dispersive forces near the base of the flow.

In the ideal situation, evidence suggests that pyroclastic flows are high-concentration dispersions that move in a laminar flow similar to debris flows (6) and in the main flow body the pumice clasts would have a lower density than the enclosing matrix, and lithic fragments would have a higher density. Thus, if the matrix is only slightly expanded it behaves as a homogeneous fluid phase (3), so that the pumice clasts would literally float to the top while the lithic fragments would sink with the largest grains sinking the furthest. Therefore, the pumice clasts show an inverse grading (Fig. 2A and C) while the lithic clasts show a normal grading (Fig. 2A and B).

Before interpreting the mechanisms involved in the genesis of the inverse graded lithic fragment-rich ignimbrite on Mayor Island it is best to first consider the mechanisms of formation of layers c and d. Pyroclastic flows generally segregate into two main parts (9), a ground-hugging pyroclastic flow and an overriding ash cloud. The ground-hugging flow forms layers a and b, the bulk of any flow sequence, and the overriding ash cloud deposits thinner, finer grained layers c and sometimes d on top. A detailed description of the genesis of layer c from the ash cloud is recorded by Fisher (4) but, briefly, the normally discontinuous, thin lenses form when the body of the pyroclastic flow decelerates and deposition occurs from turbulent vortex cells that continue to sheer across the top.

Layer c in Mayor Island ignimbrites are unlike those noted by Fisher (4), in that they are in more extensive, thicker lenses and are massive. This suggests that grain support and dispersive mechanisms were continually high in a poorly expanded, turbulent cloud so that deposition occurred en masse only after cessation of the cloud movement. It is also envisaged that the ash cloud was not detached but remained intimately associated with the main flow unit (see later).

The final phase of any ignimbrite eruption is the fallout of fine grained ejecta that elutriates from the upper regions of the ash cloud and/or settles from the vertical explosion column (3). Thus, layer d is the residual material left after all coarser and heavier material has been deposited and it is consequently enriched in vitric ash.

An important consideration in interpreting the inverse graded lithic fragment-rich ignimbrite is that its sequence does not contain an ash cloud layer c. This suggests that either an overriding ash cloud was detached from the flow proper and moving on a different flow path, so that it was deposited elsewhere on the island or even out at sea, or that the ash cloud did not separate from the flow. It is proposed that the latter is more applicable, in that other ignimbrite sequences on Mayor Island normally contain an ash cloud layer c and there is no evidence in other locations to suggest the former case. Therefore, it is conceived that this ignimbrite was deposited from a very poorly expanded pyroclastic flow that flowed in a laminar fashion similar to subaqueous debris flows whereby the lithic clasts are "buoyed up" by strong shear forces set up in a highly concentrated matrix and the tendency for smaller particles to fall downward between the larger particles (kinetic sieve mechanism) would enhance the displacement of the larger particles towards the surface (10). The cessation of lateral movement of the flow resulted in deposition en masse probably after the internal shear stress no longer exceeded the yield strength of the flow (11) and the volatiles separated from the solids after deposition by transpiration or percolation. Furthermore, the presumed absence of an overriding ash cloud suggests that when a shower bedded layer d is present in these ignimbrites, it results directly from fallout from the vertical eruption column.

MAGMA AND VENT CONDITIONS

Sparks et al. (7) and Wilson et al. (12) have demonstrated that the formation of ignimbrites from collapsing columns depends on the magmatic gas content, vent velocity and vent radius.

Mayor Island peralkaline magmas had relatively low water content (13) and therefore, low magmatic gas content since water is a major gas component. This low gas content allied with the considerable viscosity of peralkaline magmas (14) may have caused some decrease from the normal fluidisation of ignimbrite flows, both internally and in their interaction with the atmosphere. Similarly, these factors would have ensured continual slow emission of volatiles from the fragments in the flow so that turbulence could be maintained throughout the flow's motion.

Many of the ignimbrite units mentioned here have been discharged from a parasitic cone on the southern edge of Opo Bay. This parasitic vent was emitting ignimbrites while plinian-type pumice eruptions were being contemporaneously discharged from the main central vent on the island (2). The effect this had on the ignimbrite eruptions is incalculable but it may have caused a slight reduction in their mass output rate and escape velocity at the vent.

The greater number of lithic fragments in the lower ignimbrite units at Omapu Bay suggests that the parasitic vent at south Opo Bay was enlarging during its early phases of activity. The increasing vent size probably caused decreased velocities in the later ignimbrite eruptions.

Thus, the combination of all these factors appears to have caused a marked decrease in the explosivity of some eruptions on Mayor Island so that eruption column heights were comparatively low and ground surge components did not always form. Thus, resultant pyroclastic flows would have been poorly inflated so that overriding ash clouds did not separate but remained attached to the main ignimbrite flow. Continual turbulence caused by the volatiles emitted from the fragments in the underlying flow appears to have induced the massive ash cloud (layer c) deposits.

The paucity of pumice in the inverse graded lithic fragment-rich ignimbrite suggests that its magma had a very low gas content and that there were few magmatic fragments from which volatiles could be emitted. Furthermore, the abundance of accessory lithic fragments that would have originally been cool would have further lowered the temperature of this flow (15) and the result was a very poorly inflated flow from which an ash cloud did not evolve.

Therefore, it appears that the difference in the Mayor Island ignimbrite sequences from the idealised model is a combined result of the peralkalinity and low gas content of magmas, and some abnormal conditions of the vent at the time of discharge.

REFERENCES

1. Cole, J.W.: 1978, N.Z. Jour. Geol. Geophys. 21, pp. 645-647.
2. Buck, M.D., Briggs, R.M. and Nelson, C.S., The pyroclastic deposits and volcanic history of Mayor Island, Bay of Plenty, New Zealand, in prep.
3. Sparks, R.S.J., Self, S. and Walker, G.P.L.: 1973, Geology 1, pp. 115-118.
4. Fisher, R.V.: 1979, J. Volcanol. Geotherm. Res. 6, pp. 305-318.
5. Sparks, R.S.J. and Walker, G.P.L.: 1977, J. Volcanol. Geotherm. Res. 2, pp. 329-341.
6. Sparks, R.S.J. and Wilson, L.: 1976, J. Geol. Soc. London 132, pp. 441-451.
7. Sparks, R.S.J., Wilson, L. and Hulme, G.: 1978, J. Geophys. Res. 83, pp. 1727-1739.
8. Sparks, R.S.J. and Walker, G.P.L.: 1973, Nature, Phys. Sci. 241, pp. 62-64.
9. Smith, R.L.: 1960, Geol. Soc. Amer. Bull. 71, pp. 795-842.
10. Middleton, G.V.: 1970, Geol. Assoc. Canada Spec. Paper 7, pp. 162-170.
11. Fisher, R.V.: 1971, J. Sediment. Petrol. 41, pp. 916-927.
12. Wilson, L., Sparks, R.S.J. and Walker, G.P.L.: 1980, Geophys. J. Roy. Astr. Soc., in press.
13. Nicholls, J. and Carmichael, I.S.E.: 1969, Contrib. Mineral. Petrol. 20, pp. 268-294.
14. Schmincke, H.-U.: 1974, Bull. Volcanologique 38, pp. 594-636.
15. Eichelberger, J.C. and Koch, F.G.: 1979, J. Volcanol. Geotherm. Res. 5, pp. 115-134.

THE THORSMÖRK IGNIMBRITE: A REVIEW

K.A. Jørgensen

Institute of Petrology, University of Copenhagen, Østervoldgade 10
DK-1350 Copenhagen K, Denmark

ABSTRACT. The Thorsmörk ignimbrite of southern Iceland outcrops over 80 km^2 in profiles up to 200 m thick. It is partly welded and shows secondary crystallized wedging out towards the SE. The crystallisation shows similarities to both calcalkaline and peralkaline ignimbrites. The ignimbrite was deposited as a series of pyroclastic flows emanating from the caldera region of the Tindfjallajökull volcano and is associated with a ground-surge deposit. The main magma type was comenditic, and the phenocrysts reveal slight variations in physical conditions of the magma-chamber. The xenolith assemblage indicates a minimum depth of 3 km. The ignimbrite is composite, containing a suite of basic and intermediate glasses in addition to the comendite.

1. INTRODUCTION

The Thorsmörk ignimbrite was originally described by Thorarinsson (1). Later investigations (2-6) provided much of the information in this review. The Thorsmörk ignimbrite (THI) is mainly exposed in the lowland area of Thorsmörk (Fig. 1), where present outcrops cover 30 km^2. Stratigraphic sections into the neighbouring acidic centers indicate that it dates from the penultimate interglacial stage, approximately 200,000 years ago. The THI has been strongly eroded by later fluvio-glacial action, which has removed the major part of the ignimbrite, especially in the northern and central parts of the original outcrop area. This has left a number of excellent sections cutting all the way through the ignimbrite.

S. Self and R. S. J. Sparks (eds.), Tephra Studies, 347–354.

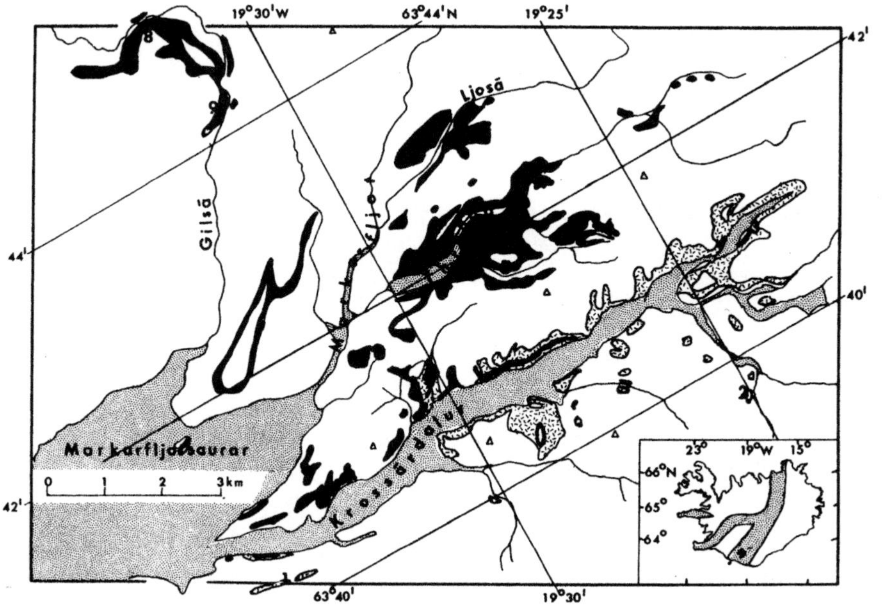

Fig. 1. Outcrop map of the Thorsmörk ignimbrite. Black: welded and crystallized facies dominant. Dashed: incipiently welded to non-welded facies. Stipple: alluvium. Δ: prominent heights. Insert map shows location relative to active volcanic zones of Iceland. Modified from (4).

Outcrop distribution, as well as the welding and crystallization pattern (Fig. 1), points towards a northwesterly source for the THI, most probably in the caldera region of the Tindfjallajökull acid center, a late Pleistocene volcano dominated by mildly alkaline FeTi-basalts and comenditic rhyolites with minor intermediate rocks (7).

2. PALAEOTOPOGRAPHY

The reconstructed pre-eruption surface shows features remarkably similar to the present topography, with the most notable difference being the apparent lack of the Eyjafjallalökull acidic center as an important topographic high to the south, implying that the main bulk of this volcano must have been erupted in post-THI times.

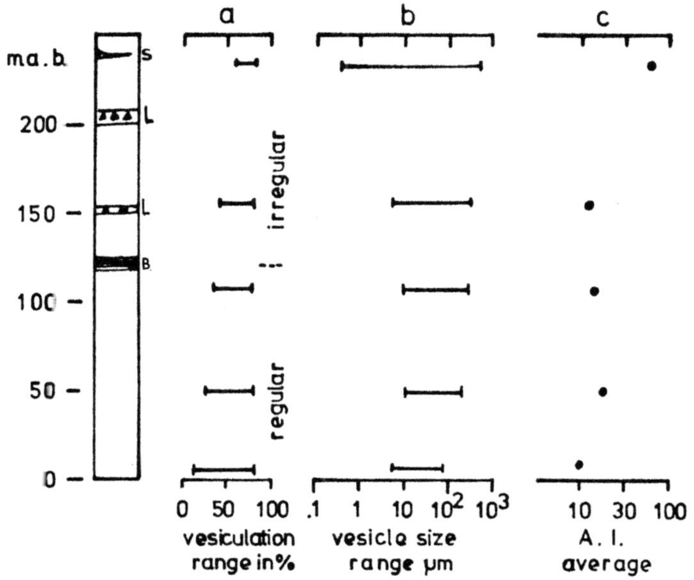

Fig. 2. Vesiculation pattern in relation to stratigraphic position (m.a.b. is meters above base): a: vesiculation of pumices as % voids of surface area (general shape of vesicles); b: vesicle size range in µm (log scale); c: abrasion index (A.I.) of Meyer (8) on feldspars (log scale): B = bedded tuffs; L = prominent xenolith rich horizons; s = intercalated sediments.

3. THE ERUPTION SEQUENCE

Morphological studies of the tephra permit a tentative reconstruction of the eruption sequence. The common occurrence of glassy rhyolite fragments in the lowermost flow-units of the THI, indicates fragmentation of a dome or coulee. The abundance of equant rhyolite shards in the same units may indicate formation in a phreatomagmatic phase of eruption, soon followed by the emission of pyroclastic flows with ever increasing explosivity, as evidenced by an increase in the frequency of lithic bands, increasing abrasion index (8) and possibly by an overall increase in pumice vesicularity, which may be expressed as a change in vesicle shape (Fig. 2). The eruption culminated in a paroxysmal explosion that produced a 15 m. thick deposit consisting of 75 weight % xenoliths. After that the eruption declined with a few highly explosive eruptions at increasing repose periods, as shown by thin intercalated sediments, possibly deposited with drifting snow (niveo-eolian (9)). Minor periods of relative quiesence are revealed in the main phase of pyroclastic flow formation by intercalated sequences of thin, normally graded tephra layers of probable air-fall origin, which may have been disturbed by later pyroclastic flows. These periods of quiesence are further confirmed by a decrease in abrasion index, and by the occurrence of basaltic spheres typical of lava fountaining (10), rather than the otherwise strongly vesiculated basaltic scoria.

4. DEPOSITIONAL FEATURES

The THI consists of a number of flow-units (at least 10), which can be distinguished on the basis of topographic breaks, internal stratification, normal grading of lithics, and reverse grading of pumice, as well as mineralogical changes.

The THI is not underlain by a plinian air-fall deposit, but a pumiceous ground-surge deposit, 40 cm thick, is ubiquitous at the base. This surge deposit is perhaps unusual, in having grain size data compatible with a pyroclastic flow origin (11) in connection with a marked normal grading of phenocrysts, and a slight reverse grading of pumice (Fig. 3).

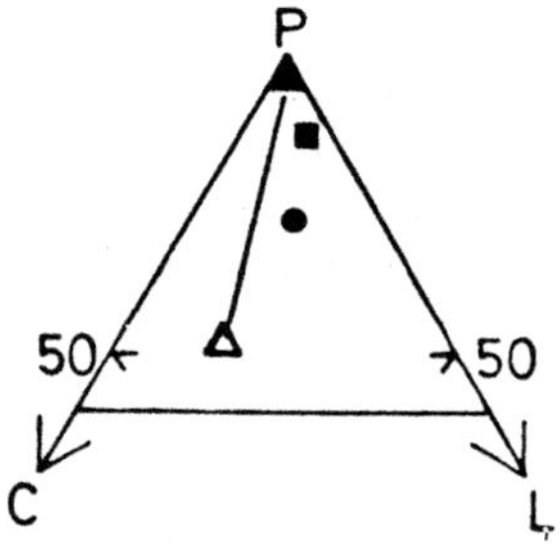

Fig. 3. Crystal-pumice-lithics triangle (12), showing ground-surge △ bottom and ▲ top, as well as ■ basal layer (2a of 13) and ● average ignimbrite. (In weight %).

Ground-surge deposits are also found within the ignimbrite in the distal parts, separating different flow-units. Above the surge deposit is normally found a fine-grained thin layer (2a of (13)), which grades into the bulk of the flow-unit (2b of (13)), which shows slight normal grading of xenoliths and reverse grading of the larger pumice clasts.

5. WELDING, CRYSTALLIZATION AND OTHER FEATURES

Welding in the THI increases significantly towards the NW, where the zone of welding exceeds 70 m, with flattening ratios (14) up to 1:60, while it almost wedges out towards the south. At the base of the welded zone a black vitrophyre is found. It increases in thickness from 0 to 15 m towards the SE, while at the same time it decreases in degree of welding. Secondary crystallization follows welding closely, and passes through a number of stages from vapor phase crystallization of tridymite and alkali feldspar in secondary gas cavities in fiamme, to coarser crystallization of mafic minerals such as arfvedsonite, fayalite, ilmenite and aenigmatite. The crystallization pattern is similar to patterns described from both calc-alkaline (15) and peralkaline (16) ignimbrites.

Both primary and secondary fumaroles (11) have been recognised with the latter forming prominent resistant features. The THI is generally well jointed in the welded part but lacks jointing in the incipiently welded to non-welded parts.

6. PETROLOGY AND MINERALOGY

The juvenile part of the THI consists largely (95%) of slightly peralkaline rhyolite, showing a very limited compositional range throughout. In the lower part of the ignimbrite two other distinctly different compositions are found sporadically, one higher in Si and K than the main rhyolite, the other lower in Si and higher in Ca. Phenocrysts of anorthoclase, Fe-hedenbergite, fayalite, ilmenite, and magnetite, with minor zircon, chevkinite, apatite and pyrrhotite, coexist with the first two compositions, while no coexisting phases have been found in the lower Si, higher Ca composition.

Feldspars show minor stratigraphic evolution, with the most albitic feldspars at the bottom of the THI in the high Si rhyolite and the most K-rich compositions in the uppermost flow units of the ignimbrite (Fig. 4). This probably reflects a minor positive temperature gradient in the original magma-chamber (17). The Fe-hedenbergites show normal zoning compatible with fractional crystallization. The Al-Ti relationships may indicate decreasing fO during crystallization (18), bringing fayalite to the liquidus at a late stage.

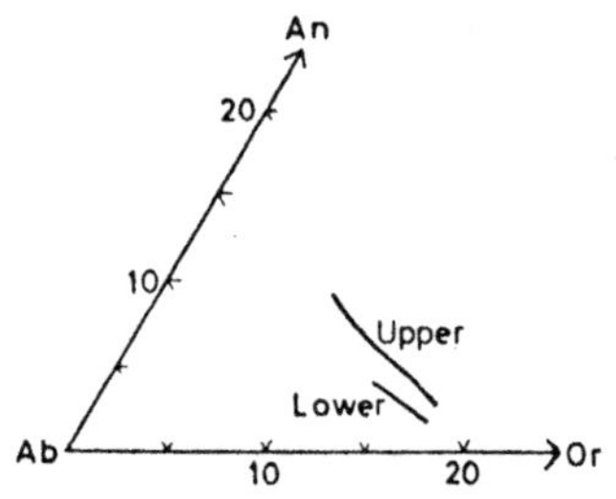

Fig. 4. Range of feldspar phenocryst compositions from THI, showing trends from lowermost high silica sample, and uppermost 'normal' rhyolite sample.

The easily re-equilibrated FeTi-oxides indicate disequilibrium relationships, such as oxyexsolution in ilmenites, spotted magnetites, and a large range in hematite and ulvöspinel contents. Disequilibrium may have been caused by heating of the magma-chamber by some external source during eruption.

Least-squares fractionation calculations show that it is possible to derive the 'normal' comeditic rhyolite through fractional crystallization of the existing phases from the low silica acidic composition, while the origin of the high silica component remains enigmatic.

7. THE MAFIC TEPHRA

Approximately 4% of the juvenile products are a suite of mafic glasses ranging from basalt through intermediate hybrids to dacite. The glasses are extremely hetereogeneous, and show mixing on a microscale, with patches of silicic glass being dissolved in more basic material. The composition of the resulting hybrids change in composition with time, being richer in Fe, Na, Mn, and Zr while at the same time being depleted in Mg and Ti relative to the first appearing hybrids with similar SiO_2 content. At the same time there is an increase in the amount of 50:50 hybrids being formed.

In the last flow-units formed, basic material is absent and mugearitic material scarce, while benmoreitic material abounds. The relationships are highly complex, but are clearly associated with rapid crystallization of pyroxene and plagioclase, with associated assimilation of disequilibrated compositions of the same minerals, especially well developed in the later hybrid compositions.

Banded pumices are present throughout the ignimbrite, but are much more abundant in the upper part and show a change in the mafic phase from basalt to benmoreite from bottom to top.

A special problem of the mixing process is the re-equilibration of the volatile species coexisting in the parental magmas. The appearance of highly titanian magnetite and pyrrhotite globules in the most evolved hybrids, may result from this re-equilibration process.

8. XENOLITHS AND LITHICS

Some 10% of the ignimbrite consists of rock fragments originating from different parts of the hypabyssal environment as well as the volcanic superstructure. The fragments include plutonic rocks, altered hypabyssal and volcanic rocks. Also a minor amount of unaltered volcanics including rhyolite fragments, plagioclase, pyroxene and olivine-phyric and aphyric basalts occur. They were probably picked up by the pyroclastic flows from the outer slopes of the volcano.

The metamorphosed rocks show all stages from zeolite facies to higher greenschist facies. The plutonic rocks are mainly granophyres which are clearly cognate with the comenditic rhyolite. They include peralkaline types bearing aegirine-hedenbergite, arfvedsonite and aenigmatite, sodic Fe-hedenbergite bearing types and types showing extensive replacement of Fe-Mg silicates with haematite rich magnetite, manganoan salite, biotite, hornblende and sphene, indicating progressive cooling of the granophyric envelope (Fig. 5).

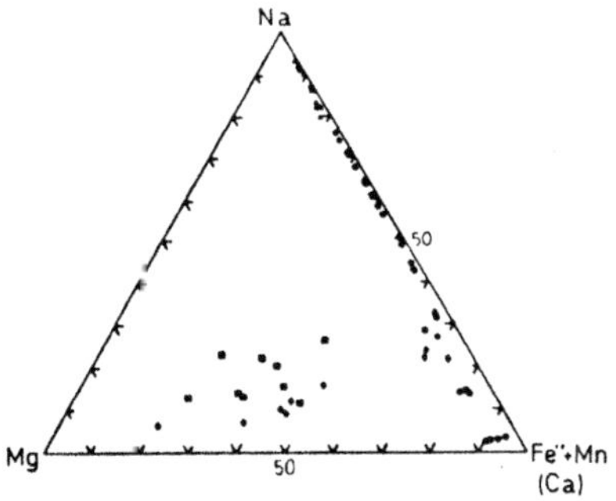

Fig. 5. Compositional range of pyroxenes from cognate granophyres in the system Mg-Fe''+Mn-Na. Δ = arfvedsonites in the system Mg-Ca-Na coexisting with the most sodic pyroxenes.

Comparison with known Icelandic intrusions of comparable type (19), indicates that the granophyres probably were formed somewhat below 3 km depth.

9. CONCLUSIONS

The Thorsmörk Ignimbrite is the largest late Pleistocene extrusive peralkaline body yet recorded from Iceland. It was emplaced as a series of pyroclastic flows, and its depositional and post-emplacement features are very similar to those of other ignimbrites, though it possesses an unusual surge deposit. The eruption may have been triggered by the intrusion of basic magma into an acid magma-chamber, in connection with a rifting episode, but there is little evidence of pre-eruptive convection in the chamber. The hybrids and strongly heterogeneous pumices in the later products however, indicates convection at this time of the eruptions as proposed by Sparks et al. (20).

REFERENCES

1. Thorarinsson, S.: 1969, Natturufrædingurin 10, pp. 139-155.
2. Sigurdsson, H.: 1970, Ph.D. thesis, Durham.
3. Wetzel, R., Wenk, E., Stern, W. and Schwander, H.: 1978, Publ. Stift. Vulk. Inst. Imman. Friedl., 10.
4. Jørgensen, K.A.: 1980, Journ. Volc. Geoth. Res. 8.
5. Jørgensen, K.A.: 1980, Nord. Volc. Rpt.
6. Jørgensen, K.A.: in prep.
7. Larsen, J.G. and Jørgensen, K.A.: Pers. Com.
8. Meyer, J.D.: 1972, Bull. Volc. 35, pp. 358-368.
9. Cailleux, A.: 1978, in: Fairbridge, R.W. and Bourgeois, J., eds. Dowden, Hutchinson and Ross, PA, pp. 501-503.
10. Heiken, G. and Lofgren, G.: 1971, G.S.A. Bull. 82, pp. 1045-1050.
11. Walker, G.P.L.: 1971, J. Geol. 79, pp. 696-714.
12. Walker, G.P.L.: 1972, Cont. Min. Pet. 36, pp. 135-146.
13. Sparks, R.S.J., Self, S. and Walker, G.P.L.: 1973, Geology 1, pp. 115-118.
14. Peterson, D.W.: 1961, U.S.G.S. Prof. Pap. 424-D, pp. D82-D84.
15. Smith, R.L.: 1960, U.S.G.S. Prof. Pap., 354-F, pp. 149-159.
16. Schmincke, H.-U.: 1974, Bull. Volc. 38, pp. 596-636.
17. Seck, H.A.: 1971, N. Jbh. Min. Abh. 115, pp. 315-395.
18. Rønsbo, J., Pedersen, A.K. and Engell, J.: 1977, Lithos 10, pp. 193-204.
19. Blake, D.H.: 1966, J. Geol. 74, pp. 891-907.
20. Sparks, R.S.J., Sigurdsson, H. and Wilson, L.: 1977, Nature 267, pp. 315-318.

TEPHROCHRONOLOGY AND ITS APPLICATION TO PROBLEMS IN NEW-WORLD ARCHAEOLOGY

Virginia Steen-McIntyre

Department of Anthropology
Colorado State University
Fort Collins, Colorado 80523 USA

ABSTRACT. Quaternary tephra layers abound in the Americas, especially in the western and central part. Few of them have been examined in detail. At present, New-World archaeologists use tephra horizons mainly as marker beds for correlation purposes and, where the local tephra sequence has been dated, as rough age indicators for their sites. They are beginning to realize the importance of tephra deposits as: (1) thick caps of sediment that can protect artifacts from disturbance and plunder; (2) indicators of ancient natural disasters; (3) potential sources for radiometric dates; and (4) material suitable for the application of approximate dating methods. Case histories from El Salvador, Central Mexico, and the Pacific Northwest are presented to demonstrate how useful tephra layers are to archaeologists.

1. INTRODUCTION

Archaeology today bears little resemblance to that practiced only a few decades ago. The advent of radiometric dating techniques and the growing realization that Man is not an isolated entity but rather a dynamic part of his natural environment has led the archaeologist far from his traditional field of study in search of essential information. Few modern archaeologists would argue the fact that sediments exposed in site excavations can be as important for the reconstruction of ancient history as the artifacts they contain. Just as few, however, realize the full potential value of one very specific type of sedimentary deposit--the tephra layer.

S. Self and R. S. J. Sparks (eds.), Tephra Studies, 355–372.

In the following sections, I will present examples from El Salvador, Central Mexico, and the Pacific Northwest (Fig. 1), to show how New-World archaeologists are utilizing tephra deposits in their research. While primarily used as rough age indicators of a site and as time-stratigraphic marker horizons for correlation purposes, tephra layers have also been recognized as evidence of ancient natural disasters, as protective coverings for undisturbed living horizons, and as sources for radiometric dates and materials for approximate dating methods.

Figure 1. New-World archaeologic sites and areas discussed in this paper.

2. THE TBJ ERUPTION, EL SALVADOR, CENTRAL AMERICA

At approximately 300 AD, what is now El Salvador experienced a natural disaster of major proportion: a massive explosive volcanic eruption from vents within the Ilopango volcano-tectonic depression, site of the present Lake Ilopango. Within a short period of time, perhaps only days or weeks, the whole countryside was buried under a thick blanket of choking, fine-grained dacitic ash (Fig. 2). The tephra blanket, and the artifacts associated with it, are still being studied. Preliminary findings are in print (1,2,3,4,5), and more detailed accounts will be published shortly (6,7,8).

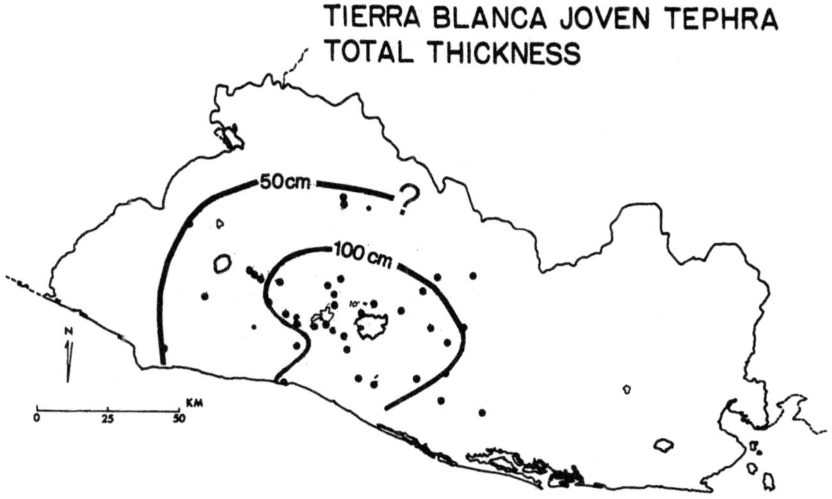

Figure 2. Thickness of the tierra blanca joven (tbj) tephra in El Salvador, as determined by field work in 1978. Source vents for the tephra lie near Lake Ilopango, near the center of the map. The irregular area to the west of the lake is the city of San Salvador (7).

The ash blanket, which we have informally named the tierra blanca joven or tbj tephra to distinguish it from older white ash layers that occur in the area, is composed of a series of mapable units that make it easy to recognize in the field and that enables us to deduce much regarding the nature of the eruption that produced the material (Fig. 3). At the base of the tephra, often found resting on an organic soil that contains Mayan artifacts of Late Preclassic and Protoclassic age (200 BC - ca AD 300), is a local, thin fine ash. This is overlain by a relatively coarse, poorly sorted airfall layer, the basal coarse ash of Figure 3. Recent work shows that there is also a surge unit associated with this deposit (8). Conformably upon this is a fine-grained, airfall ash, the T-2 airfall. It is much thicker than the underlying units and contains numerous accretionary lapilli. Above the T-2, with little or no sign of erosion at the contact, is the surge unit, actually three distinct beds: a coarse-grained, poorly sorted lower unit; a fine-grained ash; and an upper unit composed of both massive and cross-bedded members (8). The surge unit is found in most exposures of tbj tephra within 30 km of the vent. Resting on this, with at most, only slight erosion, is the T-1 ashflow, a fine-grained, pinkish-tan ash with little or no accretionary lapilli, occasional rounded pumice lumps and charcoal fragments,

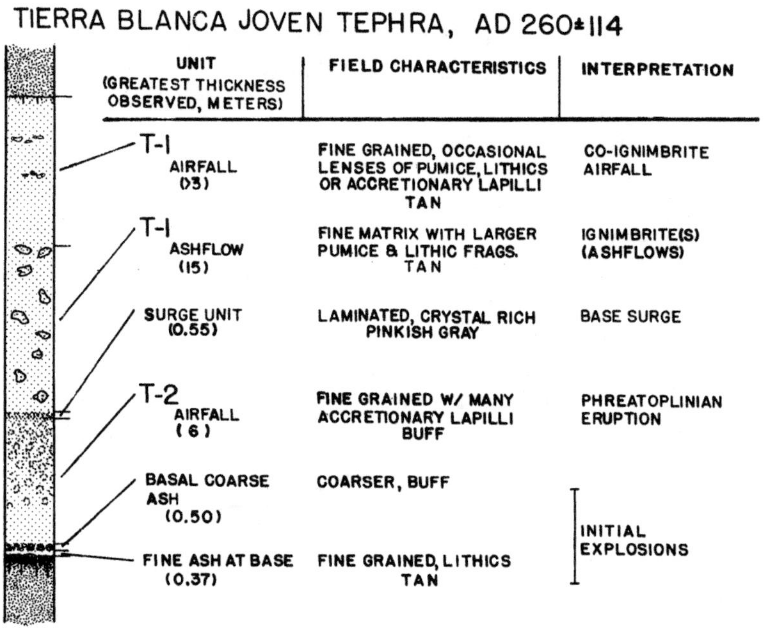

Figure 3. Stratigraphic units of the tbj tephra. Not all units will appear in the same outcrop (7).

and scattered sedimentary structures indicative of base surge deposits. Capping the deposit in several exposures is the T-1 airfall a co-ignimbrite airfall ash.

Joint field work by Hart and Steen-McIntyre in 1978 and continuing studies by Hart have enabled us to piece together the history of the tierra blanca joven eruption, at least in broad aspect. It started with a small explosion near the south end of the present lake that resulted in a local deposit of fine ash. The initial blast was followed soon after by a stronger explosion from a vent near the north side of the lake. It produced the basal coarse ash and the related surge deposit. Then, from a series of one or more vents near the west shore, fine-grained tephra was thrown high into the air, producing the T-2 airfall. The anomalously fine-grained nature of the deposit, as well as numerous accretionary lapilli suggest it was erupted through water--a phreatoplinian eruption (9). A pause ensued, long enough at least for the air to clear of T-2 tephra. Then another small burst, and the culmination of the eruption--a series of gigantic explosions that produced a complex array of ground surge, ashflow, and co-temporal airfall tephra (the surge unit and T-1 deposits). Last to settle was the T-1 airfall ash. This was a natural catastrophe of major proportions.

And what of the many Protoclassic Maya living in the area at that time? Evidence gathered by D. Sheets (6) suggests that the majority were either killed outright or were forced to flee, especially northward into the Guatemalan lowlands. As much as 10,000 km^2 may have been rendered uninhabitable by the eruption (Sheets, written communication, 1980). It is a fact that no diagnostic Early Classis ceramics (Xocco ceramic complex, AD 400-650) have been found in the area (10). By contrast, Late Classic and Early Postclassic ceramics (Payu ceramic complex, AD 650-900; Matzin ceramic complex, AD 900-1200) abound. It is quite possible that the area was abandoned for several tens or perhaps hundreds of years.

Sheets and his colleagues recognized that the tbj tephra eruption from Lake Ilopango afforded a unique opportunity to study the effects of a natural disaster on a prehistoric people. It is believed that evidence for the response of the Protoclassic Maya to the tbj eruption will be found in excavations throughout Mesoamerica. This may take the form of a sudden influx of people in Early Classic time, as evidenced by the increase in complexity of community settlement patterns; or perhaps by the appearance of distinctive pre-eruption pottery styles in post-eruption sediments.

At this stage of the investigation, we do not know how far downwind the tbj tephra extends. However, by comparing what we know of the distribution pattern to what is known about the distribution patterns from historic explosive eruptions (Table 1), it seems very likely that the tbj tephra fall covers most of Mesoamerica and perhaps much of Mexico and the western Gulf of Mexico (Fig. 4). We do know that fresh-looking glass-coated phenocrysts and volcanic glass shards of coarse-silt size are abundant in Early Classic lake-core sediments from the Peten, Guatemala (13), in the "Maya clay" (see 14).

TABLE 1. RELATIONSHIP OF THICKNESS OF TEPHRA MANTLE TO DISTANCE FROM VENT

TEPHRA MANTLE	100 CM ISOPACH	DISTANCE DOWNWIND 1 CM ISOPACH	TRACE AMOUNT
LA SOUFRIÈRE 1902 (11)	∿ 10 KM	200 KM	> 1,300 KM
TBJ	35 KM	?	?
KATMAI 1912 (12)	50 KM	400 KM	> 2,000 KM
[A] ST. HELENS, 1980	?	?	> 1,600 KM

[A] DATA AVAILABLE MAY 21, 1980.

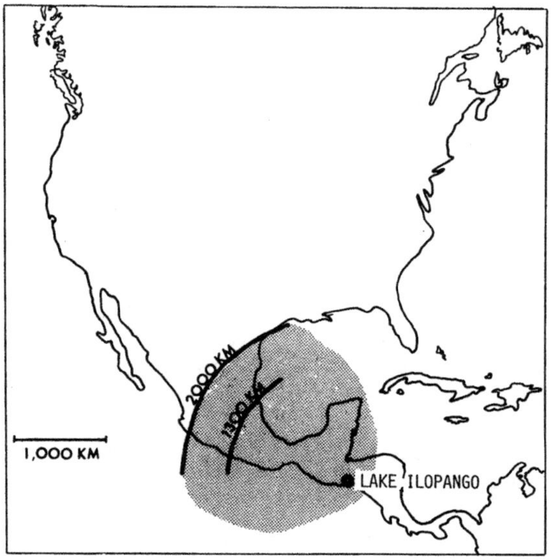

Figure 4. Possible extent of the tbj tephra. Arcs 1,300 and 2,000 km from Ilopango represent the known extent of tephra fall for two eruptions listed in Table 1 (7).

Much work remains to be done on the tbj eruption. Many of the data we need to delineate the pattern of tephra fall are available now from archaeologic sites currently under excavation throughout Middle America. Unfortunately, few archaeologists are aware of the fact. To retrieve this information, we must acquaint these people with our problem. In many cases, this will mean first educating them as to the significance of tephra layers in their own research (15,16).

3. CERÉN ERUPTION, EL SALVADOR, CENTRAL AMERICA

The tbj eruption, described above, was not the only volcanism of archaeology importance that occurred in El Salvador during Mayan time (17,18,19). Perhaps of equal interest, as far as students of the Classic Maya are concerned, is the 600 AD eruption of Laguna Caldera, a parasitic scoria cone located on the northwest flank of the San Salvador volcanic complex, approximately 40 km northwest of the center of Lake Ilopango (Fig. 1). This eruption, the Cerén eruption, produced a basaltic scoria fall deposit that buried the surrounding countryside beneath more than four metres of cinders and ash (17, 19).

Although local in extent, the importance of the Cerén tephra blanket cannot be overemphasized. It buries Cerén site--the "Mayan Pompeii" (20)--preserving intact at least one Classic Maya farmstead and perhaps a whole hamlet or small village (21, 22).

The eruption of Cerén tephra apparently began suddenly, with the deposition of thin beds of ash, lapilli, and occasional tephra bombs up to 30 cm in diameter. The temperature of coarse clasts was greater than 550°C at the moment of deposition (19). The heat and weight of airfall tephra caused the thatch roof of one excavated structure, a farm house, to char and collapse. Soon afterwards, the nature of the eruption changed, apparently with the introduction of water into the magma chamber, and the bulk of the unit was deposited. These are interbedded layers of bombs, lapilli, and ash rich in accretionary lapilli, with some evidence for base surge deposits (17,19).

The disaster that overtook the Mayan farmers living near Laguna Caldera at the time of the eruption is similar in some respects to that which befell the residents of Pompeii and Herculaneum in 79 AD. Limited excavation indicates that the treasure of information buried by the tephra may be rich and varied. There was little chance to flee, much less to remove possessions. Modern local residents reported seeing a living floor, apparently part of the house, strewn with human bones and several polychrome vessels. The feature was removed during construction of grain silos before the site could be examined by archaeologists (21).

What has emerged through controlled excavation at Cerén is a detailed picture of the Classic Maya common family. They worked in a number of localized activity areas. They plastered the floors of their house with clay, then fired them in place. They planted maize in ridged rows. They used pre-formed obsidian tools and ate beans (21,23). Contrary to previous belief, they appreciated and were able to afford well made polychrome pottery such as Copador, Gualpopa, Arambala, Campana, and Nicoya tradeware (10).

Excavations at Cerén site have just begun. In June, 1979 and January, 1980, a geophysical survey of the surrounding area indicated the presence of three buried anomalies, two of which turned out to be cultural in origin when tested with a core drill (22). The prospects seem exciting; not only for new discoveries at Ceren, but for other "Pompeiis" to come to light throughout Mesoamerica as archaeologists look more closely at the innocuous piles of scoria that dot the landscape and wonder what they might contain.

4. HUEYATLACO SITE, VALSEQUILLO AREA, CENTRAL MEXICO

On the north shore of the Valsequillo Reservoir in Central Mexico (Fig. 1), lies another valuable, if controversial, archaeologic site - Hueyatlaco (Fig. 5). It is one of four discovered in 1962 by C. Irwin-Williams and J. Armenta Camacho where extinct Pleistocene vertebrate fossils such as horse, camel, and mastodon are found associated with artifacts of flaked chert in situ (24,25). The sedimentary section at Hueyatlaco also contains four tephra layers (Fig. 6).

Uranium series dates on an articulated camel pelvis associated with bifacial tools at Hueyatlaco gave ages of approximately 250,000 years for the site (^{230}Th, 245,000 ±40,000 years; ^{231}Pa >180,000 years; 26). Archaeologists cannot accept so great an age (25,26) but six converging lines of evidence obtained in 1973 by Steen-McIntyre, Fryxell and Malde through direct tracing of beds in continuous exposures support such an early date for the sediments (27,28).

Evidence to support a pre-Wisconsin age for Hueyatlaco includes: (1) the depth of burial of the artifacts (10 m overburden preserved); and (2) extent of dissection of the sediments by the adjacent Rio Atoyac (more than 50 m). The remaining evidence, (lines 3-6) comes from tephra layers found in the stratigraphic section exposed at the site.

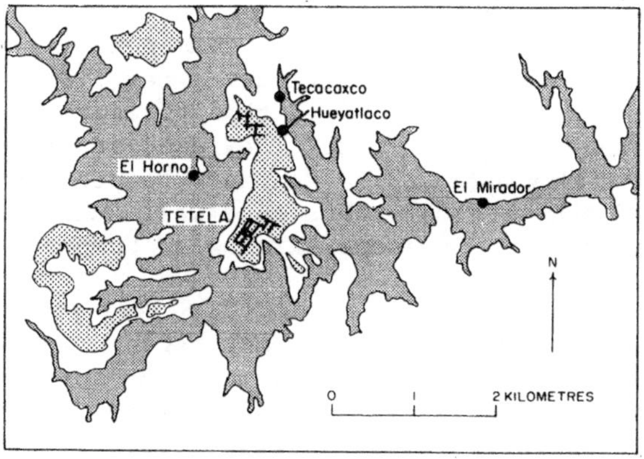

Figure 5. Valsequillo Reservoir area, Central Mexico. The Tetela Peninsula is labeled, as well as four sites where bones of extinct Pleistocene vertebrates and flaked tools were found together in situ. The Tetela brown mud unit, which contains pumice dated by the fission track method, is shown in light grey (28).

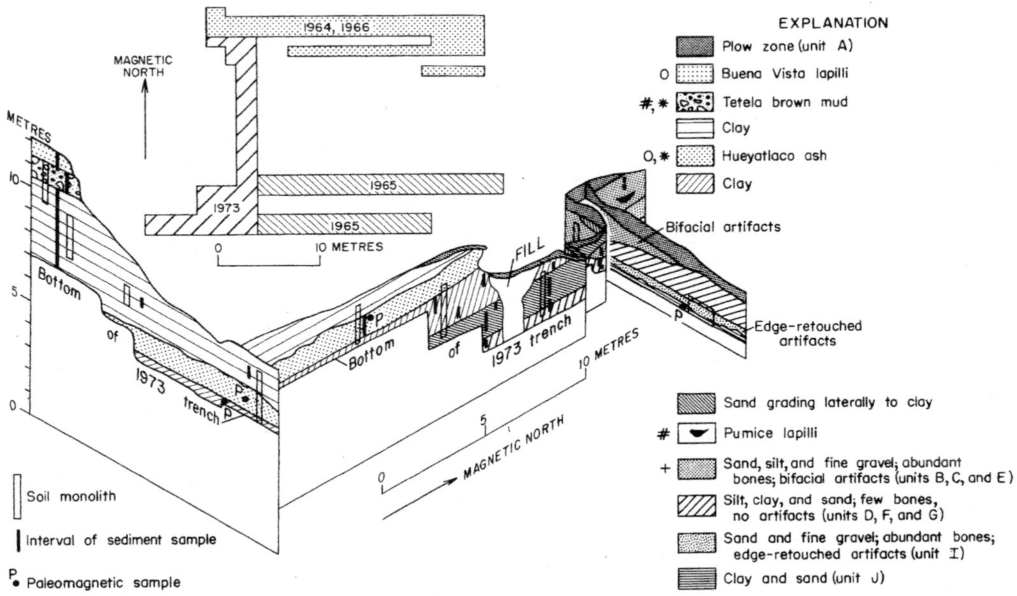

Figure 6. Map and fence diagram of Hueyatlaco site based on mapping done in profile at a scale of 1 to 20 (28).

Four tephra units occur at Hueyatlaco (Fig. 6). Within a zone containing bifacial artifacts is a channel deposit of pumice lapilli. Some four metres higher is the Hueyatlaco ash, a layer of silt-size volcanic ash one metre thick that displays delicate laminar bedding indicative of primary deposition as an airfall into relatively quiet waters. Still higher in the section is the Tetela brown mud, an ancient mud-flow filled with large, rounded blocks of pumice of a single petrographic type. Overlying the Tetela brown mud 250 m northwest of Hueyatlaco is the Buena Vista lapilli, another airfall deposit.

The Hueyatlaco ash and the Tetela brown mud pumice both have been dated by C.W. Naeser, U.S. Geological Survey, using the fission track method on zircon phenocrysts (Hueyatlaco ash, 370,000 ±200,000 years, 2 sigma; Tetela brown mud pumice, 600,000 ±340,000 years, 2 sigma; 28, Table 2). The large plus/minus values result from the fact that only a small number of natural tracks were present. The Buena Vista lapilli contains zircon phenocrysts, but they have not yet been dated. The unnamed lapilli from the channel deposit associated with bifacial tools contains no zircon.

While we could not date the latter tephra directly by the fission track method, we were able, subjectively at least, to estimate an age for it by comparing the extent of weathering of the heavy mineral phenocrysts from within clasts of lapilli and coarse ash to that of similar phenocrysts from the large fragments of Tetela brown mud pumice, a dated tephra unit. Pumice from the brown mud is partially altered, and fresh broken surfaces have a marblelike texture consisting of swirls of waxy brown clay interleaved with fresh pumiceous glass. Hypersthene crystals enclosed in the fresh glass are unaltered, but adjacent crystals protruding into the waxy clay are strongly etched. Hypersthene phenocrysts from the pumice lapilli associated with bifacial tools are even more deeply etched. By contrast, hypersthene from tephra layers 22,000 to 24,000 years old, collected from a similar environment show only rare to incipient etching, and that from younger beds are not etched at all (28).

Tephra hydration dating was used to obtain a rough age estimate for the Buena Vista lapilli and Hueyatlaco ash. The method is described elsewhere

(14,15,29). To estimate age using tephra hydration, the amount of water trapped in closed pumice vesicles of a certain size and shape was noted. This is water of superhydration, a result of the dispersion of water molecules through glass. The water volume was then compared with that found in dated glass samples of similar composition, collected from a similar environment. A comparison of cumulative curves obtained by this method is shown in Fig. 7. It suggests that the Hueyatlaco ash and Buena Vista lapilli (dashed lines) are much older than 40,000 years and closer in age to 251,000 years.

Finally, to present some negative evidence: two of the tephra layers exposed at the site, the Buena Vista lapilli and Hueyatlaco ash, have petrographic characteristics similar to coarse tephra from the flanks of La Malinche, a volcano located 45 km to the northeast. They are dissimilar to the coarse tephra from Popocatépetl and Iztaccíhuatl, two large volcanoes located at roughly the same distance to the west. The tephra sequence on La Malinche is well exposed and has been studied (30,31,32). The oldest exposed beds lie under a buried soil dated at 25,000 years (^{14}C samples W-1911, W-2570, W-2571, 32). After examining this sequence of tephra plus several hundred other samples, I have been unable to correlate the

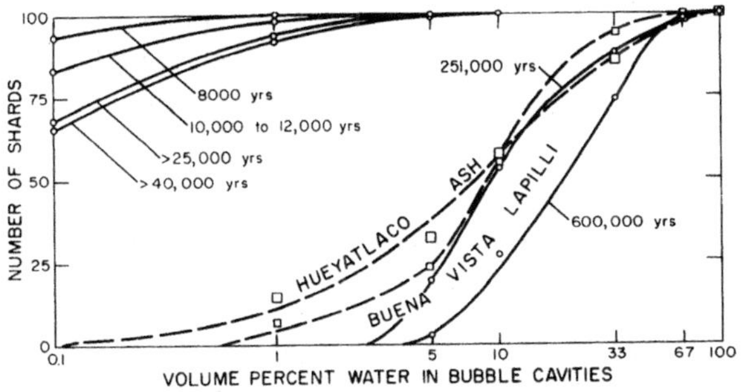

Figure 7. Cumulative curves, each representing the average volume of water observed in spindle-shaped vesicles 10-50 µm long found in 100 glass shards. The dated curves in the upper left of the chart are for samples of tephra collected in the vicinity of Hueyatlaco. The 251,000 year curve is for a sample of yellowstone tephra, collected in the National Park. The 600,000 year curve is for a sample of Pearlette type-O tephra from Saskatchewan (28).

Buena Vista lapilli and Hueyatlaco ash with any of them. I now believe that the correlative deposits on La Malinche are still deeply buried, making the Valsequillo deposits at Hueyatlaco older than any now exposed on La Malinche.

If the artifacts from Hueyatlaco are truly as old as the dates indicate, it would be startling news indeed. Hueyatlaco has yielded sophisticated stone tools such as bifacially-worked knives, scrapers, burins and tanged projectile points. To quote archaeologist Irwin-Williams:

> These tools surely were not in use at Valsequillo more than 200,000 years before the date generally accepted for development of analogous tools in the Old World, nor indeed more than 150,000 years before the appearance of <u>Homo sapiens</u> (26, p. 241).

Limited evidence indicates this may not be as impossible an idea as previously suggested. There exists a uranium-series date for the travertine layer from Sandia Cave, New Mexico. The date is approximately 250,000 years (sample run by H.P. Schwarce, McMaster University, Ontario; C. Vance Haynes, written communication, 1976). The travertine layer buried and sealed off sediments containing bifacial Sandia points, "Folsom" points, and possible fire hearths. At least one artifact, a "Folsom" blade, was found embedded in the lower surface of the travertine crust with no sign of disturbance or recrystallization of the carbonate (33, Plate 7).

A search through a popular book on New World archaeology (34) yields other intriguing data: for example, the resemblance of a leaf-shaped, unfluted, lanceolate point from Sandia Cave to an undated specimen found with the second mammoth near Santa Isabel Iztapan, Mexico (p. 85-86, 97); the Angus, Nebraska find where a fluted point was found in association with an articulated mammoth skeleton in deposits of mid-Pleistocene age (p. 43); crudely-carved stone heads, tens of kilograms in weight, found deeply buried in Pleistocene gravel near Malakoff, Texas (p. 154-155).

The study of Hueyatlaco site is far from complete. Awaiting examination is a ton of crated samples and stabilized columns of sediment, collected directly from the trench walls in 1973. These samples should provide more detailed information on the site, and perhaps suggest additional means for dating it.

5. TEPHRA LAYERS, PACIFIC NORTHWEST

For a final example of the role tephra plays in New-World archaeology, we again move far to the north, this time to the Pacific Northwest (Fig. 1). Here, numerous tephra layers occur in Quaternary sediments. Most of them come from vents in the Cascade Range (Fig. 8). (See appropriate sections in (35) for additional references).

A list of tephra eruptions of special interest to the archaeologist is given in Table 2. Each eruption produced several distinct lobes of tephra, and often extended over a period of hundreds of years. At present, the problem seems to be an overabundance of tephra layers, not a lack of them!

Several studies have been made of Pacific Northwest tephra, with the result that, with time, most of them can be distinguished by using a combination of field, petrographic, and chemical criteria. Lacking up to this moment, however, have been simple and inexpensive methods that would quickly give the archaeologist a rough age estimate for tephra layers exposed at his site, preferably while the field season is still in progress. The tephra-hydration dating method (29,30), described elsewhere in this volume (14), may offer one approach to the problem.

Tephra-hydration dating is similar to obsidian-hydration dating (52,53). Both methods utilize volcanic glass, but in the tephra-hydration method it is in the form of pumiceous shards of fine-sand size. To use the method, one notes the extent of hydration of the glass shards by measuring the apparent thickness of the hydration rinds. Extent of superhydration, the amount of water in sealed glass vesicles that has slowly collected through a process of diffusion (54,55) is visually estimated. This information is then compared to that for dated tephra samples of similar chemical composition, collected from similar weathering environments. Other factors being equal, the older the tephra sample, the more hydrated the glass and the more water in the vesicles.

Table 3 gives apparent rind thickness and water content of vesicles for samples of tephra supplied by

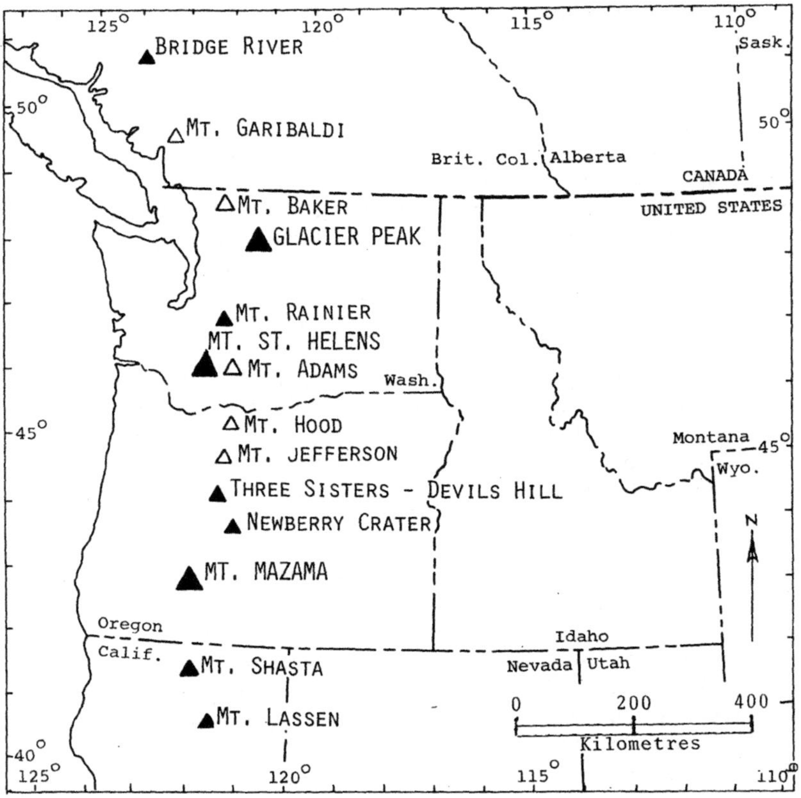

Figure 8. Major volcanoes active in the Cascade Range during Quaternary time. Vents reported to have ejected significant amounts of material are shown in black. Large triangles represent the three major tephra producers: Mount Mazama (Crater Lake), Mount St. Helens, and Glacier Peak.

R. Okazaki, Washington State University. They were collected from various locations in the Pacific Northwest. No attempt was made to match the samples according to weathering environment, although the chemistry of the glass, as suggested by refractive index, is roughly similar.

The tephra-hydration dating method is not yet fully developed. Still, it holds out a promise to archaeologists looking for a rapid means of estimating the age of their sites, and at very little cost.

TABLE 2. TEPHRA ERUPTIONS OF ESPECIAL INTEREST TO ARCHAEOLOGISTS IN THE PACIFIC NORTHWEST

SOURCE VENT	LOCATION (STATE OR PROVINCE)	NAME	APPROXIMATE AGE (YEARS BP)	COMMENTS
UPPER LILLOOET RIVER	BRITISH COLUMBIA	BRIDGE RIVER	1,900; 2,600	(31,32,33)
VARIOUS	BRITISH COLUMBIA	VARIOUS	20,000->37,000	SEVERAL HORIZONS PROBABLY REPRESENT THE DISTAL PORTIONS OF ST. HELENS TEPHRA (34).
GLACIER PEAK	NORTHWEST WASHINGTON	GLACIER PEAK	12,000	THE TEPHRA COMPLEX WAS PROBABLY DEPOSITED BETWEEN ABOUT 11,250 AND 12,760 YEARS AGO (35). SEE ALSO (36).
ST. HELENS	SOUTHWEST WASHINGTON	?	5/18/80 MORNING	
		T	150-200	DATES FROM (37, P. 45), MODIFIED FROM (38), SEE ALSO (39,40,41).
		W	450	
		B	450-2,500	
		P	2,500-3,000	
		Y	3,000-4,000	
		J	8,000-12,000	
		S	12,500-13,500	
		M	18,000-20,000	
		UNNAMED	20,000-35,000	
CRATER LAKE	SOUTHWEST OREGON	MAZAMA	6,600; 7,200	DATES RANGE FROM 6,100 TO 7,600 YEARS BP (33, 42,43,44,45,46).

TABLE 3. HYDRATION DATA FOR [A] SAMPLES OF ST. HELENS AND GLACIER PEAK TEPHRA, PACIFIC NORTHWEST

SOURCE VENT	NAME	APPROXIMATE AGE (YEARS BP)	HYDRATION DATA: EXTENT HYDRATION — APPARENT RIND THICKNESS IN µM (ACTUAL VALUES)	EXTENT SUPERHYDRATION — VOL. % H_2O IN VESICLES OF 100 SHARDS: ≤0.1	≤1	≤5	≤10	≤33	≤67	≤100 %
ST. HELENS	$W_{NORTHEAST}$	450	<1 (<1,<1,<1,<1)	100						
	W_{EAST}	450	1 (1,1,1,1)	97	3					
	$Y_{SOUTHEAST}$	3,400	1.5 (2,1,1,2)	98	2					
	Y_{NORTH}	3,500	[B] 1.5+(1+,1+,2,2) [C] 3.5 (4,3,4,3)	96	4					
	SET $J_{MAIN\ BODY}$	>8,000,<12,000	5 (4,5,4,6)	100						
GLACIER PEAK	UPPER	12,000	5 (4,6,5,5)	64	33	3				
	LOWER	12,000	5.5 (5,6,6,5)	73	24	3				
ST. HELENS	SET S_{UPPER}	13,000	6 (6,5,6,6)	95	5					
	SET M_{BASE}	>18,000 -<20,000	6 (7,5,6,5)	89	11					
	SET C_{BASE}	37,000?	COMPLETELY HYDRATED	87	13					

[A] SAMPLES PROVIDED BY ROSE OKAZAKI, SOILS DEPARTMENT, WASHINGTON STATE UNIVERSITY, PULLMAN.
[B] PUMICEOUS SHARDS, N CA. 1.505.
[C] DENSE SHARDS, N ≤1.501.

REFERENCES

1. Sheets, P.D.: 1976, Research Records No. 9, University Museum Studies, Southern Illinois Univ., Carbondale, 78 p.
2. Steen-McIntyre, V.: 1976, in: Ilopango volcano and the Maya Protoclasic, P.D. Sheets, ed.,

Research Records No. 9, University Museum Studies, Southern Illinois Univ., Carbondale.
3. Steen-McIntyre, V. and Sheets, P.D.: 1978, Geol. Soc. America Abstracts with Programs 10, p. 497.
4. Steen-McIntyre, V. and Hart, W.J., Jr.: 1979, Soc. American Archaeology, 44th Annual Meetings, Vancouver, B.C., Program with Abstracts, p. 75.
5. Sheets, P.D., ed.: 1978, Research of the Protoclassic Project in the Zapotitán Basin, El Salvador: a preliminary report of the 1978 season, Duplicated manuscript, Dept. of Anthropology, U. of Colorado, Boulder, 97 p.
6. Sheets, P.D., ed.: in press, "Volcanic disaster in prehistoric Central America," Univ. of Texas Press, Austin.
7. Hart, W.J., Jr. and Steen-McIntyre, V.: in press, in: "Volcanic disaster in prehistoric Central America," Univ. of Texas Press, Austin.
8. Hart, W.J., Jr.: 1980, "Tephra studies as a tool in Quaternary Research Abstracts," NATO Advanced Science Institute, Laugarvatn, Iceland.
9. Self, S. and Sparks, R.S.J.: 1978, Bull. Volcanologique, 41:3, pp. 1-17.
10. Beaudry, M.: in press, "Volcanic disaster in prehistoric Central America," Univ. of Texas Press, Austin.
11. Anderson, T. and Flett, J.S.: 1902, Proc. Roy. Soc. London 70, pp. 423-445.
12. Wilcox, R.E.: 1959, U.S. Geol. Survey Bull. 1028-N, pp. 409-476.
13. Deevy, E.S., Rice, D.S., Rice, P.M., Vaughan, H.H., Brenner, M. and Flannery, M.S.: 1979, Science 206, pp. 298-306.
14. Steen-McIntyre, V.: this volume, Approximate dating of tephra.
15. Steen-McIntyre, V.: 1977, "A manual for tephrochronology: collection, preparation, petrographic description, and approximate dating of tephra (volcanic ash)." Published by the author, Idaho Springs, Colorado 80452 USA, 167 p.
16. Steen-McIntyre, V.: in preparation, Tephrochronology (volcanic ash chronology) and its application to archaeology, in: Archaeological Geology, G.R. Rapp and J. Gifford, eds.
17. Hart, W.J., Jr.: in press, in: "Volcanic disaster in prehistoric Central America," Univ. of Texas Press, Austin.
18. Hart, W.J., Jr.: in press, El tuff de San Andrés-talpete. La tefra de Cerén. La tefra de El Playón. Las tierras blancas jovenes de Ilopango. All published by El Ministerio de

Educación, San Salvador, El Salvador.
19. Hoblitt, R.: in press, in: "Volcanic disaster in prehistoric Central America," Univ. of Texas Press, Austin.
20. Rensberger, B.: 1978, Archaeologists find a Mayan "Pompeii," New York Times, December 26, 1978.
21. Zier, C.: in press, in: "Volcanic disaster in prehistoric Central America," Univ. of Texas Press, Austin.
22. Sheets, P.D. et al.: in press, "Geophysical exploration for ancient Maya housing at Cerén, El Salvador." National Geographic Soc. Res. Reports.
23. Sheets, P.D.: in press, in: "Volcanic disaster in prehistoric Central America," Univ. of Texas Press, Austin.
24. Irwin-Williams, C.: 1967, in: Pleistocene extinctions, the search for a cause, P.S. Martin and H.E. Wright, Jr., eds. Yale Univ. Press, New Haven and London, pp. 337-347.
25. Irwin-Williams, C.: 1978, in: Cultural continuity in Mesoamerica, D.L. Browman, ed., Aldine, Chicago, pp. 7-22.
26. Szabo, B.J., Malde, H.E. and Irwin-Williams, C.: 1969, Earth and Planetary Sci. Letters 6, pp. 237-244.
27. Steen-McIntyre, V., Fryxell, R. and Malde, H.E.: 1973, Geol. Soc. America Abstracts with Program, 5, p. 820.
28. Steen-McIntyre, V., Fryxell, R. and Malde, H.E.: in press, 1981, Quat. Res.
29. Steen-McIntyre, V.: 1975, in: Quaternary Studies, R.P. Suggate and M.M. Cresswell, eds., Royal Soc. New Zealand, Wellington, pp. 271-278.
30. Steen-McIntyre, V.: 1968, "Petrography of selected Late Quaternary pyroclastic deposits at La Malinche volcano, State of Puebla, Mexico." Duplicated manuscript distributed to participants, Valsequillo Field Conference, November, 1968, 19 p. (see also Geol. Soc. Amer. Abs. with Programs, Mexico City, 1968, p. 289).
31. Heine, K. and Heide-Weise, H.: 1972, Estratigrafía del Pleistoceno reciente y del Holoceno en el volcan de La Malinche y región circunvecina. Comunicaciones 5, Proyecto Puelba-Tlaxcala, Fundación Alemana para la investigación Cientifica, pp. 3-8.
32. Malde, H.E.: 1978, in: U.S. Geol. Survey, Virginia, radiocarbon dates XIV, 20:2, eds. Kelly, L., Spiker, E. and Rubin, M., pp. 299-303.
33. Hibben, F.C.: 1941, "Evidences of early occupation of Sandia Cave, New Mexico and other

sites in the Sandia-Mazano Region," Smithsonian Misc. Collections, 99:23.
34. Wormington, H.M.: 1959, "Ancient Man in North America," Denver Museum of Natural History Popular Series No. 4, 322 p.
35. Westgate, J.A. and Gold, C.M., eds.: 1974, World bibliography and index of Quaternary tephro-chronology, Alberta Printing Services Dept., Univ. Alberta, 528 p.
36. Nasmith, H., Matthews, W.H. and Rouse, G.E.: 1967, Can. Jour. Earth Sci. 4, pp. 163-170.
37. Westgate, J.A. and Dreimanis, A.: 1967, Can. Jour. Earth Sci. 4, pp. 155-161.
38. Westgate, J.A.: 1975, Geol. Soc. America, Abstracts with programs 7, p. 879.
39. Westgate, J.A. and Fulton, R.J.: 1975, Can. Jour. Earth Sci. 12, pp. 489-502.
40. Porter, S.C.: 1978, Quat. Res. 10, pp. 30-41.
41. Westgate, J.A. and Evans, M.E.: 1978, Can. Jour. Earth Sci. 15, pp. 1554-1567.
42. Moody, U.: 1978, "Microstratigraphy, paleo-ecology, and tephrochronology of the Lind Coulee site central Washington." Doctoral dissertation, Dept. of Anthropology, Washington State U., Pullman, 273 p.
43. Crandell, D.R. and Mullineaux, D.R.: 1978, U.S. Geol. Survey Bull. 1382-A, 23 p.
44. Moody, U.: 1977, Geol. Soc. America Abs. with programs 9, pp. 1098-1099.
45. Mullineaux, D.R.: 1974, U.S. Geol. Survey Bull. 1326, 83 p.
46. Mullineaux, D.R., Hyde, J.H. and Rubin, M.: 1975, Jour. Res. U.S. Geol. Survey 3, pp. 329-335.
47. David, P.O.: 1970, Can. Jour. Earth Sci. 7, pp. 1579-1583.
48. Davis, J.O.: 1978, "Quaternary tephrochronology of the Lake Lahontan area, Nevada and California." Nevada Archaeological Survey Res. Paper 7, Uni. Nevada, Reno, 137 p.
49. Fryxell, R. and Daugherty, R.D.: 1962, "Archaeo-logical salvage in the Lower Monumental Reservoir, Washington — interim report. Washington State Univ. Lab. of Archaeology and Geochronology, Inv. 21.
50. Fryxell, R.: 1965, Science 147, pp. 1288-1290.
51. Kittleman, L.R.: 1973, Geol. Soc. America Bull. 84, pp. 2957-2980.
52. Friedman, I. and Smith, R.L.: 1960, Am. Antiquity 25, pp. 476-522.
53. Friedman, I. and Long, W.: 1976, Science 191, pp. 347-352.
54. Roedder, E. and Smith, R.L.: 1965, Geol. Soc. America Sp. paper 82, p. 164.
55. Roedder, E.: 1970, Schwizer mineralog. u Petrog. Mitt. 50, pp. 41-58.

TEPHRA LAYER IN FRANCHTHI CAVE, PELEPONNESOS, GREECE

C.J. Vitaliano[1], S.R. Taylor[2], W.R. Farrand[3] and T.W. Jacobsen[4]

[1]Department of Geology, Indiana University, Bloomington, Indiana; [2]Research School of Earth Science, Australian National University, Canberra, A.C.T.; [3]Department of Geology and Mineralogy, University of Michigan, Ann Arbor, Michigan; [4]Department of Classical Archaeology, Indiana University, Bloomington, Indiana

ABSTRACT. Optical, trace element, and rare earth element data on the pumice-rich layer of volcanic ash discovered near the bottom of the Upper Paleolithic stratigraphic sequence of cultural deposits inside Franchthi Cave at Koilada, on the SW coast of the Argolid Peninsula, have confirmed its correlation with the Y-5 ash layer in the Mediterranean deep-sea cores (Farrand, 1971) (1). Similar data obtained on a sample of Gray Campanian ash from Tufara, near Naples, Italy, confirms the identification of the Y-5 ash and the Gray Campanian ash as suggested by Thunell et al., (1979) (2).

A 5 to 6 cm thick, pale gray pumice-rich tephra layer occurs at about 9.25 m depth in the sediment and rock-fall deposits in Franchthi Cave, on the southwest coast of the Argolid Peninsula of Greece (Fig. 1). This tephra has important bearing on the antiquity of human activity in this region, inasmuch as cultural remains are found beneath it as well as above. It was described by Farrand (1), who correlated it with the Mediterranean deep-sea ash now designated Y-5 (3). This ash layer, originally attributed to the Lower Tephra of Santorini by Ninkovich and Heezen (5), and dated by them as at least 25,000 years old, was subsequently attributed to the Citara-Serrara tuff from Ischia in Italy by Keller (6).

S. Self and R. S. J. Sparks (eds.), Tephra Studies, 373–379.

Thunell and others (2) restudied the Y-5 ash in a series of deep-sea cores. They correlate it with the Gray Campanian ash of Italy, and conclude that it was erupted at least 30,000 and perhaps as much as 38,000 ± 2,000 years ago. Other ages that have been obtained for this ash are 25,000 yrs (7); 26-27,000 years (1); 28-33,000 years (7 and 9); 30,000 years (10); and 40,000 years (11).

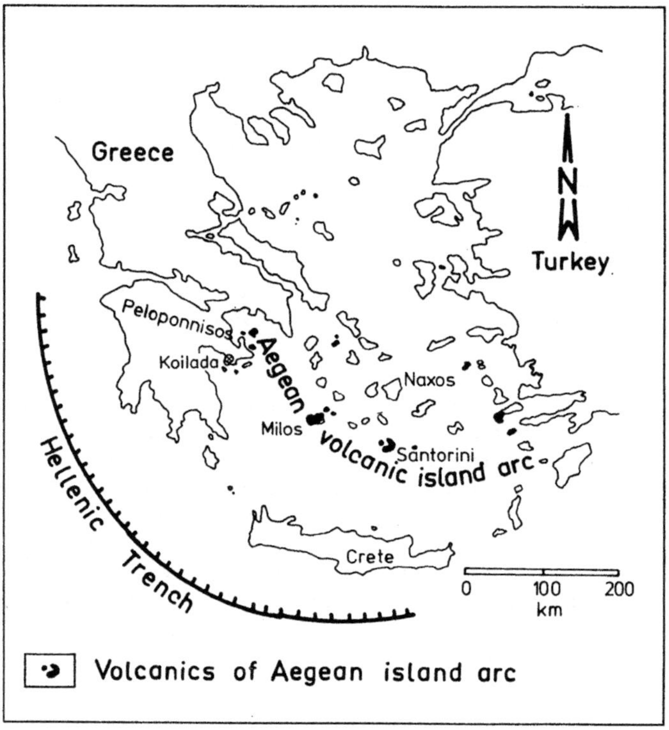

Figure 1. The Aegean Region showing Koilada the locale of the Franchthi Cave. Modified after Hoffman and Keller 1979 (4).

In order to shed further light on the correlation of the ash found in Franchthi Cave, samples of that ash, of the Y-5 ash from deep-sea Mediterranean core TR 172-22, of the "Lower Tephra" from core V10-58, and of the Gray Campanian Tuff from Tufara, near Naples (Fig. 2), were studied optically and chemically for major, trace, and rare-earth elements. The optical data were obtained in the Petrology Laboratory at Indiana University and the chemical data were obtained by XRF and spark source mass spectrometry at the Research School of Earth Sciences,

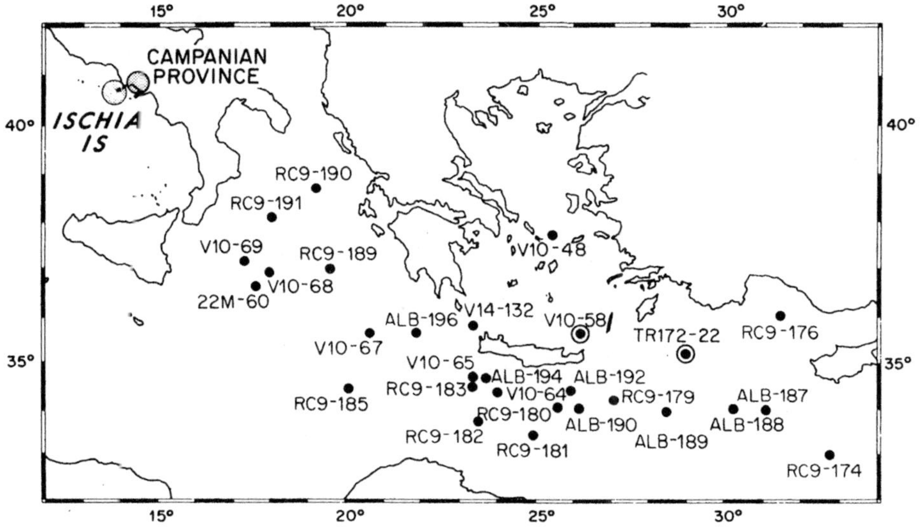

Figure 2. The Aegean Region showing the location of the Campanian Region and the sites of cores V10-58 and TR172-22. Modified after Keller, et al., 1978 (3).

Australian National University, Canberra. A sample of the yellow Citara-Serrara tuff from Ischia was found to be thoroughly indurated and hydrothermally altered and therefore was not analysed.

The refractive indices--determined by means of sodium monochromatic light and adjusted for temperature variation (-dn/dt = 0.0004)--are shown in Table 1.

Table 1. Refractive indices of the ashes in this study

Franchthi, trench a	1.5259±0.0002
Franchthi, trench B	1.5262±0.0002
Y-5(TR172-22)	1.5255±0.0002
Y-5(V10-58)	1.5266±0.0002
Gray Campanian Tuff	1.5257±0.0002

All are virtually identical and could have been derived from the same or closely related source vents.

The chemical composition of the tephras (Table 2) shows that all can be classified as trachytic, with $K_2O/Na_2O > 1$. The major element content, the distribution pattern of the

Table 2. Selected Major Elements

	1	2	3	4	5	6
SiO_2	61.9	62.74	59.0	62.0	61.88	61.8
MgO	0.73	0.49	0.7	0.60	0.62	0.64
Na_2O	3.82	3.83	5.8	4.30	5.12	4.11
K_2O	7.50	7.52	6.1	7.97	7.24	7.58

1. Franchthi Cave, S.R. Taylor, Analyst: (XRF)
2. Y-5 (Core TR 172-22): Thunell, et al., 1979: (EMP)
3. Citara-Serrara Tuff (basal part) Spiaggia, Citara: Keller, et al., 1978
4. Gray Campanian Tuff; Tufara: S.R. Taylor, Analyst: (EMP)
5. Average Campanian Ignimbrite (Gray Facies): Barberi, et al., 1978
6. Y-5 (Core V10 58): S.R. Taylor, Analyst: (XRF)

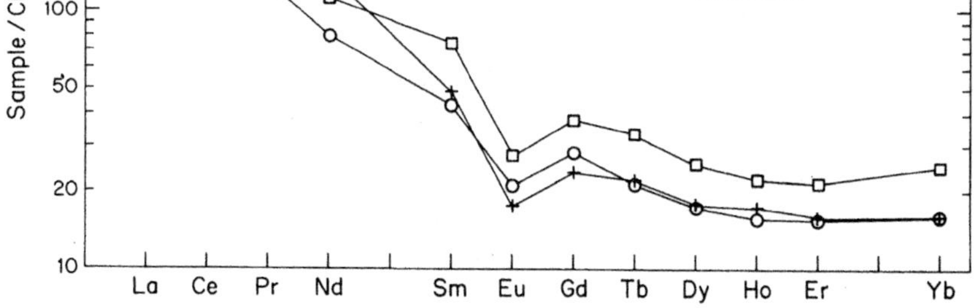

Figure 3. Showing the distribution pattern of the rare-earth elements in the Y-5 ash from the deep-sea cores V10-58 and TR 172-22 and the ash from Franchthi Cave.

rare-earth elements (Figure 3), and the relationship of the trace elements (Figure 4) support their close relationship suggested by the refractive indices.

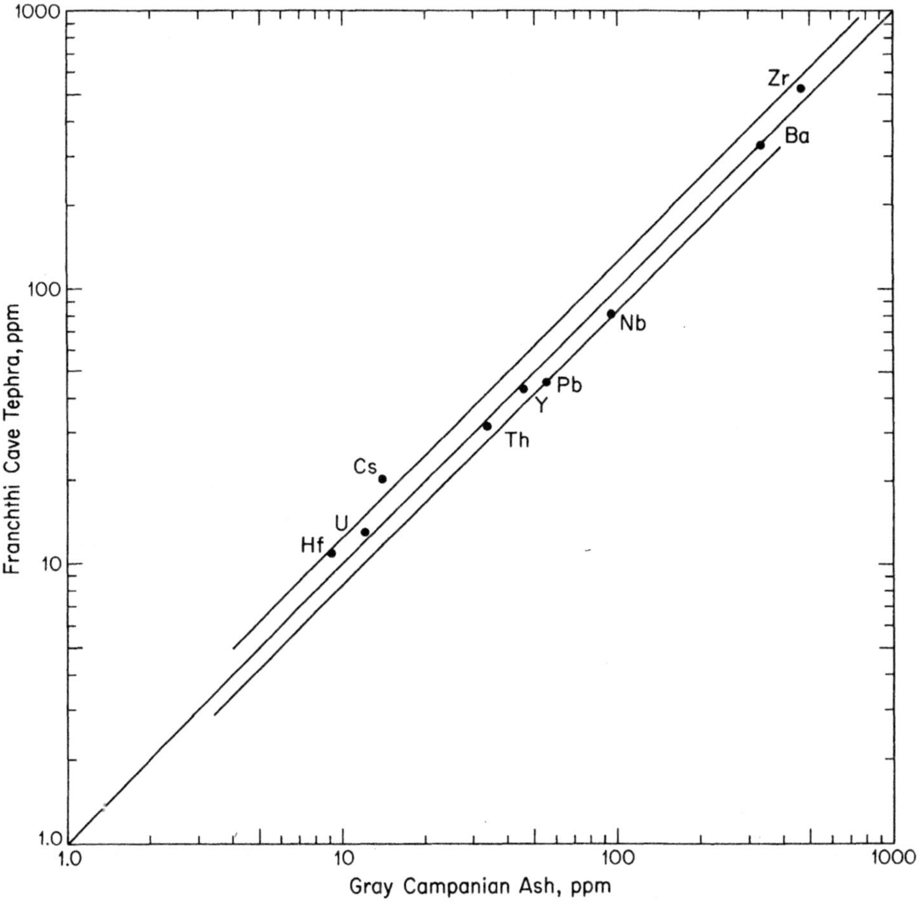

Figure 4. Trace elements relationship of the Franchthi Cave and Gray Campanian ashes.

DISCUSSION AND CONCLUSIONS

Of the 20 or so tephra layers so far discovered in deep-sea cores from the Mediterranean Sea (3), layers Y-3 and Y-7 along with Y-5 have been classified as trachytic (3) and have refractive indices similar although not quite equal (i.e. 1.520±002 and 1.519±0.002 respectively) to those of the tephras in this study. Furthermore ash Y-3 contains olivine and Y-7 contains yellow acmite, minerals not found in Y-5 or the ash from Franchthi Cave. The remainder, which are derived in part from volcanoes of the Hellenic Volcanic Arc (i.e. Santorini, Milos, Kos, Nisyros, and Yali (6), and in part from volcanoes of the Roman Province of Italy (i.e. Sabatini, Vulsini, and Vico, Southern Italy, and Pantelleria), are distinguished from those of this study by notable differences in chemical composition and refractive index of their glasses. Tephras from the volcanoes of the Hellenic Arc are rhyolitic to dacitic in composition, and the refractive indices of the glass shards are much lower (1.510 or less) than those of the glass of the tephras of Table 1. The mineralogy of the Hellenic Arc tephras (augite, hypersthene, plagioclase, ± hornblende) differs from that of the Campanian, Franchthi Cave, and Y-5 ashes which contain aegerine-augite, biotite, apatite, plagioclase, sanidine, hornblende, and an occasional feldspathoid mineral.

A major source of vast quantities of alkali-trachyte tephra in post-Pliocene time was volcanoes located in western Italy (3). These included the volcanoes of the Roman Province, the Phlegrean Fields, Ischia Island, Somma-Vesuvius, and Roccamonfina, the tephras from which all have some characteristics in common with the Franchthi Cave deposit. However, all but one of these tephras, the "Gray Campanian Tuff" can be eliminated as the possible source of the Franchthi tephra. The tephra derived from the volcanoes of the Roman Province all contain leucite and the tephras from Somma-Vesuvius contain melanite in addition to leucite neither of these minerals are found in the ash from the cave. The trachytic eruptive products from Roccamonfina yield an age of 1.2 to 0.368 m.y. (12) which is too old for the Franchthi Cave tephra.

The nearest possible source for a tephra of the composition of the Franchthi Cave deposit is the Campanian region of Italy, specifically the Gray Campanian tephra which is the most voluminous deposit in the Campanian and the Neapolitan region in general.

The optical and chemical similarity of the Franchthi Cave tephra to the Y-5 ash leads us to conclude that both can be correlated with the Gray Campanian tuff. The thickness of the Franchthi Cave deposit (5 to 6 cm) nearly 1,000 km away, testifies to the tremendous magnitude of the eruption, already suggested by the extensive distribution of the Y-5 deep-sea ash.

It is of interest to speculate on how the ash found its way into the cave originally. At present there are two openings in the ceiling some distance back from the northwest-facing mouth of the cave. One of these is probably the result of a relatively recent roof collapse. The presence of the ash inside the cave suggest that the other, located in the rear of the cave, must already have been open at the time of the ash fall, creating a draft which helped draw more ash into the cave for preservation than would ordinarily have been the case.

ACKNOWLEDGEMENTS

We thank D. Ninkovich for the sample of Gray Campanian ash from Tufara, J. Keller for the sample of the Citara-Serrara tuff from Ischia, F.W. McCoy for the sample of the lower ash from core V10-58, and T.C. Huang for the sample of the Y-5 ash from the core TR172-22.

REFERENCES

1. Farrand, W.R.: 1977, Geol. Soc. of Amer. 9, p. 971.
2. Thunell, R., Federman, A., Sparks, S. and Douglas, W.: 1979, Quat. Res. 12, pp. 241-253.
3. Keller, J., Ryan, W.B.F., Ninkovich, D. and Altherr, R.: 1978, Geol. Soc. of Amer. Bull 89, pp. 591-604.
4. Hoffman, C. and Keller, J.: 1979, Lithos 12, pp. 209-219.
5. Ninkovich, D. and Heezen, B.C.: 1965, Colston Papers 17, Butterworths, London, pp. 413-452.
6. Keller, J.: 1971, Acta 1st International Scientific Congress on the volcano Thera, Greece, Greek Archaeological Service, Athens, pp. 152-167.
7. Vergnaud-Grazzini, C. and Rosenberg, H.Y.: 1969, Revue de Géographie Physique et de Géologie Dynamique 11, pp. 279-292.
8. Di Girolamo, P. and Keller, J.: 1972, Berichte der Naturforschenden Gesellschaft zu Freiberg im Breisgau 61-62, pp. 85-92.
9. Stanley, D.J., Knight, R.J., Stuckenrath, R. and Catani, G.: 1978, Nature, 273, pp. 110-113.
10. Barberi, F., Innocenti, F., Lirer, L., Munno, R., Pescatore, T. and Santacroce, R.: 1978, Bull. Volcanologique 41, pp. 10-31.
11. Cita, M.B., Vergnaud-Grazzini, C., Roberts, C., Chamley, H., Chiaranfi, N. and d'Onofrio, S.: 1977, Bull. Volcanologique 40, pp. 11-12.
12. DiGorolamo, P., Lirer, L., Porcelli, C. and Stanzione, D: 1972, Soc. It. Min. Petr. 28, pp. 77-123.

TEPHROCHRONOLOGY AND PALAEOECOLOGY: THE VALUE OF ISOCHRONES

P.C. Buckland, P. Foster, D.W. Perry and D. Savory

Department of Geography, University of Birmingham,
Birmingham, B15 2TT, England

ABSTRACT. The applications of tephrochronology to the correlation of temporally similar yet spatially disjunct flora and fauna are discussed, using as examples recent work in Iceland and Kenya. The paucity of such studies is noted and its relevance to studies in palaeoecology and island biogeography commented upon.

If ecology often seems like a jigsaw puzzle in fog with all pieces at least potentially recoverable, palaeoecology resembles a similar game with most parts irretrievably lost and bedevilled by Time's constant changing of the picture thereon. In any reconstruction of a palaeoenvironment, which employs data from more than a single locality, time becomes a serious problem. Contemporaneity of deposition is frequently impossible to prove and reconstructions which rely upon similarity of facies for correlation may be markedly diachronous (1). In comparison with other problems in the palaeoecology of most of the Phanerozoic, the need for absolute temporal correlation may seem slight, although Raab (2) has recently provided a relevant caveat in relation to the application of island biogeographic theory to the fossil record. In the Quaternary and, more particularly, in the Holocene, however, the level of resolution demanded becomes more critical and isochronous horizons, most readily available in those areas of contemporary vulcanism, become a necessary adjunct to the effective synthesis of palaeoecological data. The nature of the problem was realised when Quaternary palaeoecology was in its infancy and it may be noted that Thorarinsson began tephrochronological studies in the 1930's in association with the pioneer palynologist von Post, using tephra

S. Self and R. S. J. Sparks (eds.), Tephra Studies, 381–389.

to provide correlative horizons in bog stratigraphy (3). Despite this early cooperation and the use of tephra in simple stratigraphic correlation, these two branches of geological research have tended to travel separate paths. Riddiman and Glover (4) used the dispersion of ash bands in North Atlantic deep sea cores as indices of faunal mixing and their importance in the study of vertebrate assemblages from the Plio-Pleistocene of the Lake Turkana, Kenya, has been realised, if not exploited to the full (5). Only rarely have palaeoecologists noted the value of tephra layers in providing not only isochronous correlation horizons but also one of the most effective links between life (biocoenosis) and death (thanatocoenosis) assemblage that is available in the geological record.

Death of an organism usually results in the decay of the soft tissue, dissolution of any hard parts and the transport of the remaining fragments, often to environments of deposition far removed from the animal or plant's normal habitat. The thanatocoenosis recovered by the palaeontologist may therefore contain elements from several biocoenoses and the separation of autochthonous (in situ) components from allochthonous ones is often a difficult step. A catastrophic event may lead to the preservation of autochthonous thanatocoenoses and such assemblages provide ideal starting points for the interpretation of other deposits. The more instantaneous death and burial, the more useful the assemblage. The destruction of St. Pierre and its biota by the superheated gas and ash cloud from Mont Pelee in 1902 might therefore appear an excellent example, yet the extent of destruction and paucity of sealing deposit provide a poor fossil record and one has to turn to the destruction and burial of Herculaneum and Pompeii in A.D. 79 by Vesuvius for a more effective case. Whilst the eruption allowed the escape of at least part of the megafauna, preservation is such that not only do casts of men and other vertebrates remain in the deposit but also insects (6), and plants (7) are well preserved. In many cases, however, death of the whole community does not result directly from the ash fall. Many plants are capable of renewed growth through several centimetres of tephra, some of the more robust insects, particularly the Coleoptera (beetles), may burrow out and death may come to the vertebrate community at varying times after the event, perhaps by gas poisoning or fluorosis (8). Despite these limitations, the tephra horizons, where frequent, do provide indices of rates of sedimentation and sufficient chronostratigraphic control for the hazardous pursuit of applying modern quantitative ecological techniques to chronologically similar but spatially disjunct floral and faunal assemblages from the past.

There remain other problems in the analysis of a palaeolandscape effectively sealed by a pyroclastic mantle. Whilst the chronostratigraphic position of the base of the primary tephra horizon is secure, the nature of the underlying sediment clearly varies as to the former depositional or erosional regime and an element of uncertainty may remain. Preservation will not be uniform over the whole landscape and a variety of taxa may be found fossil in different palaeoenvironments. Root casts and rodent burrows in a loess, for example, cannot be directly compared with plant and arthropod macrofossils in a bog. A restraint in terms of suitable sampling localities is therefore imposed and the isochronous horizons are clearly of most value where the level of preservation is highest, most frequently in bog and lake successions.

Sampling and methods of analysis also impose constraints on the recovered data. Whilst the tephra horizons mean that extensive areas are sealed and it would theoretically be possible to excavate and reconstruct entire plant communities, an effective sampling strategy, employing line transects, quadrats or point quadrats as in modern vegetation studies (9), would not only need skills more appropriate to the archaeologist but also require an investment of time out of all proportion to the returns in information. A technique which concentrates identifiable plant macrofossils, principally seeds, has to be employed and the relationships between macrofossil assemblage and the original plant cover are frequently obscure and there are few studies of modern plant thanatocoenoses for comparison (10). Work upon samples of the surface sealed by the 1357 eruption of Katla in the bog at Ketilsstadir in southern Iceland shows massive variation in macrofossil content between apparently similar samples (Table 1). Carex nutlets dominate two samples but another is swamped by the megaspores of Selaginella selaginoides, presumably due to the inclusion of a sporaginating plant in the samples. Variation is also evident in the character of the insect assemblages from the same samples, a point emphasised by the widely disparate positions of the four faunas in the Clustan dendrogram (Fig. 1). The eruption of 1357 infilled a series of peat cuttings in the bog and thereby fossilizes at a moment in time a mosaic of flora and anthropogenic disturbance of the bog. Ash falls are sufficiently frequent at Ketilsstadir to allow study of lateral variation in the hydrosere through time, although it must be noted that the ash falls themselves may, at least temporarily, modify the succession.

Lake sediments provide sequences of plant macrofossils, largely in the form of the siliceous skeletons of diatoms, which may be wholly autochthonous and therefore not influenced by the disadvantages of partial recovery evident with plant macrofossils. Diatoms are also sensitive indicators of changing water conditions and are often abundant in lakes in volcanic areas (11). In the Naivasha and Nakuru Basins in Central Rift

TABLE 1. Macrofossils from isochronous samples (Katla 1357) at Ketilsstadir, Iceland.

TAXA	SAMPLE			
Planta	KE1/5	BR1/2	KE2/2	KE3/2
	[no. of macrofossils/100g.]			
Selaginella selaginoides (L.) Link.	24.0	24.1	2.6	190.8
Rumex sp.	0	0.2	0	0
Montia fontana L.	0	0.2	0	0
Stellaria media L.	0	0.5	0	0
Cerastium sp.	0	2.0	0	0
Ranunculus c.f. acris L.	0	0.1	0	0
Alchemilla sp.	1.5	1.2	0	0.2
Empetrum nigrum L.	0	1.0	0	0
Gramineae indet.	0	0.9	0	0
Carex spp.	49.8	66.9	0	2.0
Juncus spp.	0.3	8.1	0	0.1
Luzula c.f. multiflora (Retz.) Lej.	1.5	3.8	0	0.1
Insecta - Coleoptera	[no. of individuals/5kg. sample]			
Carabidae				
Patrobus septentrionis (Dej.)	-	1	-	2
Pterostichus diligens (Sturm)	-	1	-	1
P. nigrita (Payk.)	1	1	-	-
Dytiscidae				
Hydroporus nigrita (F.)	-	1	-	-
Hydraenidae				
Hydraena britteni Joy	-	2	-	-
Staphylinidae				
Lesteva longoelytrata (Goez.)	-	2	-	-
Stenus spp.	1	1	-	1
Othius angustus Steph.	-	-	1	-
Gabrius sp.	-	1	-	-
Quedius umbrinus Erich.	1	6	-	1
Aleocharinae indet.	-	9	-	-
Scarabaeidae				
Aphodius lapponum Gyll.	-	1	3	-
Elateridae				
Hypnoidus riparius (F.)	-	1	-	-
Curculionidae				
Otiorhynchus nodosus (Mull.)	-	1	4	-
Hypera suspiciosa (Hbst.)	-	-	4	-

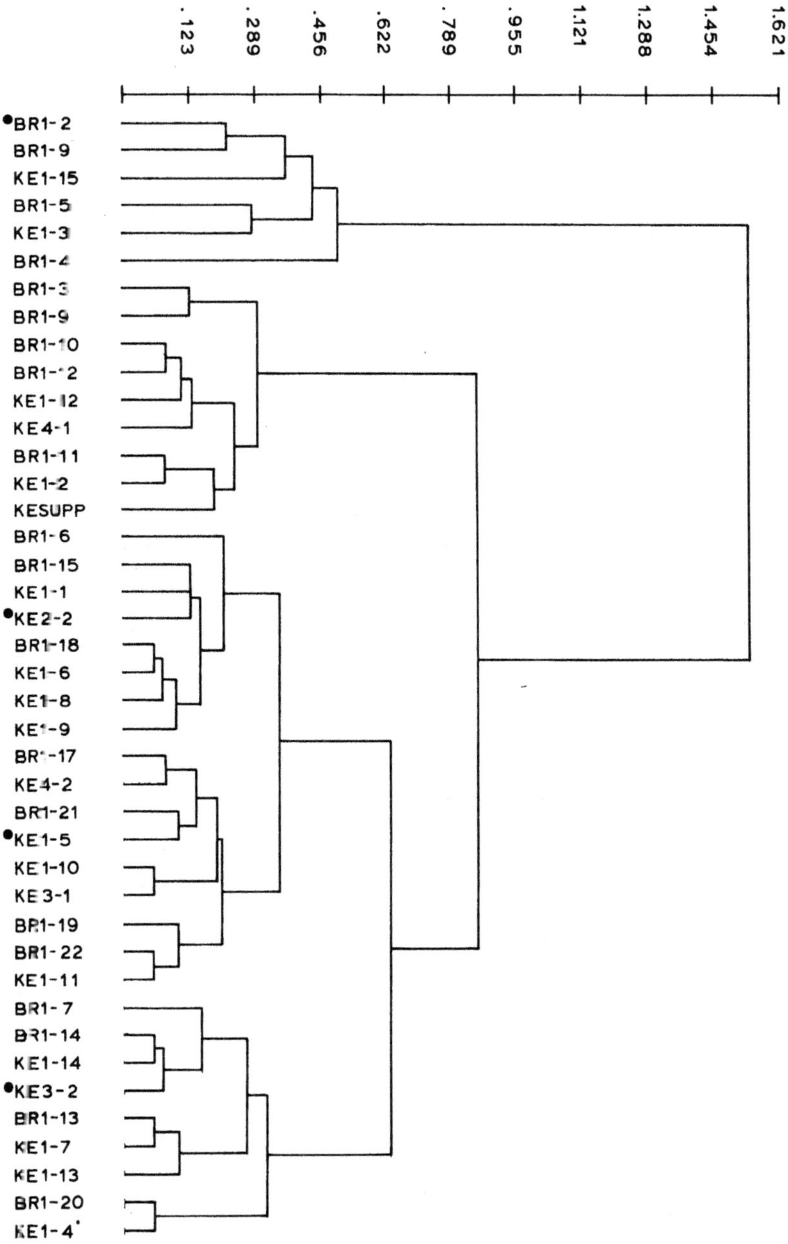

DENDROGRAM OF CLUSTAN PROCEDURE HIERARCHY (WARD'S METHOD) FOR THE KETILSTADIR DATA

• Isochronous samples

Figure 1.

Valley Province, Kenya, a group of recently active volcanoes, from Suswa and Longonot in the south to Menengai in the north, provides interrelated sequences of pyroclastics, which allow the correlation of cores from the lake sediments and exposed stratigraphy in an area where lake levels have fluctuated considerably during the Late Quaternary (12). The succession in the Nakuru and Elmenteita area has been studied extensively by Isaac and his co-workers (13), and the sequence of tephra layers has been extended onto Eburru (14), lying between the Nakuru and Naivasha Basins; it will eventually be possible, using the tephra isochrones, to correlate diatom successions between the lakes. Such a correlation will allow a check on whether lake level fluctuations were synchronous in the two independent basins, providing greater control on the model of climatic change proposed (12).

Fossil vertebrate assemblages do not suffer from the disadvantages that plant macrofossils do in terms of the estimate of numbers of individuals and there is a considerable amount of taphonomic data (15). Recovery of adequate synchronous samples, however, requires considerable investment and it is only in projects such as those around the Turkana Basin in Kenya and Ethiopia that collections with sufficient stratigraphic control (15) to allow consideration of palaeo-communities have been made. In an area of acid pyroclastics in Kenya, with no preservation of bone, Isaac (16) was able to use the distribution of stone artifacts, sealed by the tephra, in a discussion of the ecology of Late Palaeolithic hunting groups.

Direct application of quantitative techniques from ecology to fossil invertebrate assemblages have recently been reviewed by Pielou (17), using vertical variation in Foraminifera assemblages in deep sea cores, but there has been little study of isochronous spatial variation in fossil terrestrial assemblages. As Coope (18), Osborne (19) and others have shown, insects, particularly Coleoptera are sensitive indicators of environmental change and Kenward (20) has recently attempted to apply quantitative ecological methods to the analysis of at least the archaeological part of the data. The samples from Ketilsstadir, Iceland, processed for plant macrofossils, were also examined for insects, using the paraffin (kerosene) flotation technique developed by Coope and Osborne (21). The faunal assemblages recovered were compared mathematically using CLUSTAN, a readily available computer-based cluster analysis package, which has been employed to examine other isochronous assemblages (22). The resultant dendrogram is shown in Figure 1. Although interpretation must take account of ecological factors difficult to introduce into a purely mathematical technique, the objective numerical groupings achieved allow an alternative approach to data frequently difficult to comprehend by simple inspection.

The position of the four isochronous samples in the matrix provide an essential control against which other variation can be assessed. The sequence of faunas, closely dated by the intervening ash layers, presents a detailed picture of the processes of local environmental modification and change in the insect fauna brought about by man since the Landnam, itself, fortuitously, marked by an ash layer (23). Thus, it is apparent, from the work on several sites in southern Iceland, that, shortly after the settlement, the dung beetle Aphodius lapponum, requiring the dung of introduced herbivores to survive, becomes widespread, and, at Holt, 35 km. west of Ketilsstadir, the strongly synanthropic species Typhaea stercorea and Lathridius minutus (group), associated with stored hay, appear before the marker horizon provided by the eruption of Katla in 1000. When the minutiae of faunal and floral change over large, temporally correlated areas are more fully known, that part of the biota which predates human interference can be quantified and its origins discussed. Although the deep sea core data (24) is perhaps insufficiently detailed, the suggestion of Fridriksson (25) and Coope (26) that the greater part of the non-anthropochorous biota of both Iceland and Greenland arrived by ice-rafting from Eurasia during the Late Weichselian (Wisconsinian) to Holocene transition, may eventually be resolved by study of the finer tephra horizons in cores. The presence of Icelandic pumice in Holocene shorelines in Greenland (27) adds some support to this view and Herman's (28) comment upon the extent of low salinity waters, less damaging to terrestrial biota, in the Norwegian Sea during glacial melting is pertinent.

The closely correlated tephrochronological sequences of other remote islands, from the Azores to the altitudinal islands of the Ruwenzori, offer unique opportunities for detailed studies of flora and faunal change, evolution and extinction in small communities through time.

ACKNOWLEDGEMENTS

Research in Kenya was carried out in association with Glynn Isaac of the University of California in 1969-70 with the aid of grants from the Boyce Fund, Oxford and from Wenner-Grenn. Work in Iceland is funded by the Leverhulme Trust and is carried out in association with the Vulcanological Institute, University of Iceland, and the National Museum of Iceland. Tephrochronological aspects are the work of Gudrún Larsen and archaeological ones of Gudrún Sveinbjarnardóttir.

REFERENCES

1. Shaw, A.B.: 1964, in: Time in Stratigraphy, New York.
2. Raab, P.V.: 1980, Lethaia 13, pp. 175-181.
3. Thorarinsson, S.: 1974, in: World Bibliography and Index of Quaternary Tephrochronology, eds. J.A. Westgate and C.M. Gold, pp. xvii-xviii.
4. Ruddiman, W.F. and Glover, L.K.: 1972, Geol. Soc. Amer. Bull. 83, pp. 2817-2836.
5. Findlater, I.C.: 1976, in: Earliest Man and Environments in the Lake Rudolf-Basin, eds. Y. Coppens, F. Clark Howell, G.Ll. Isaac and R.E.F. Leakey, pp. 94-104.
6. Monte, G.dal.: 1956, Redia 41, pp. 23-28.
7. Jashemski, W.F.: 1979, in: Volcanic Activity and Human Ecology, eds. P.D. Sheets and D.K. Grayson, pp. 587-622.
8. Thorarinsson, S.: 1979, in: Volcanic Activity and Human Ecology, eds. P.D. Sheets and D.K. Grayson, pp. 125-159.
9. Kershaw, K.A.: 1973, in: Quantitative and Dynamic Plant Ecology.
10. Birks, H.H. and Birks, H.J.B.: 1980, in: Quaternary Palaeoecology.
11. Richardson, J.L. and Richardson, A.E.: 1972, Ecol. Monographs 42, pp. 499-534.
12. Butzer, K.W., Isaac, G.Ll., Richardson, J.L. and Washbourn-Kamau, C.: 1972, Science 175, pp. 1069-1076.
13. Isaac, G.Ll., Merrick, H.V. and Nelson, C.M.: 1972, in: Palaeoecology of Africa, the Surrounding Islands and Antarctica, ed. E.M. van Zinderen Bakker Sr., 6, pp. 225-232.
14. Buckland, P.C. and Foster, P., in prep. The Late Pleistocene Tephrochronology of Eburru.
15. Behrensmeyer, A.K. and Hill, A.P. (eds.): 1980, in: Fossils in the Making--Vertebrate Taphonomy and Palaeoecology.
16. Isaac, G.Ll.: 1972, in: Man, settlement and urbanism, eds. P.J. Ucko, R. Tringham and G.W. Dimbleby, pp. 165-176.
17. Pielou, E.C.: 1979, Palaeobiology 5, pp. 435-443.
18. Coope, G.R.: 1979, Ann. Rev. Ecol. Syst. 10, pp. 247-267.
19. Osborne, P.J.: 1980, Boreas 9, pp. 139-148.
20. Kenward, H.K.: 1978, in: The Analysis of Archaeological Insect Assemblages: A New Approach.
21. Coope, G.R. and Osborne, P.J.: 1968, Trans. Bristol and Gloucs. Archaeol. Soc. 86, pp. 84-87.
22. Perry, D.W.: 1981, in: The Brigg Boat, ed. S. McGrail, pp. 150-153.
23. Larsen, G., Perry, D.W., Sveinbjarnardóttir, G., Savory, D. and Buckland, P.C., in prep. Tephrochronological, Historical and Palaeoecological Studies of a late Holocene bog succession at Ketilsstadir in Mýrdalur, Iceland.
24. Ruddiman, W.F., Sancetta, C.D. and McIntyre, A.: 1977, Phil. Trans. R. Soc. Lond. B 280, pp. 119-142.

25. Fridriksson, S.: 1969, Jökull 19, pp. 146-157.
26. Coope, G.R.: 1979, in: Carabid Beetles their evolution, natural history and classification, eds. T.L. Erwin, G.E. Ball and D.R. Whitehead, pp. 407-424.
27. Binns, R.E.: 1972, Geol. Soc. Amer. Bull. 83, pp. 2303-2324.
28. Herman, Y.: 1977, in: X Inqua Congress Birmingham 1977 Abstracts, ed. K. Clayton, p. 203.

VOLCANOLOGICAL APPLICATIONS OF PYROCLASTIC STUDIES

George P.L. Walker

Department of Geology, University of Auckland,
Private Bag, Auckland, New Zealand

ABSTRACT. Our understanding of explosive volcanism depends heavily on studies of the products of past eruptions, particularly for eruptions of the type which are so large or violent as to preclude close observation, or so scarce as to provide little or no opportunity for it. Identification of the transport mechanism is a necessary first step before eruptive mechanisms can be determined; identifying criteria are summarised. Pyroclastic studies can help determine the internal plumbing system and magma release mechanisms of volcanoes, for example where they yield the magma output rate and provide evidence for magma mixing, and fall deposits supply the best means of studying compositionally-zoned magma chambers. They are also important in volcanic hazard studies, where the past pattern of explosive activity is used as a guide to the probable future behaviour of a volcano.

1. INTRODUCTION

There are many reasons for studying the pyroclastic products of past explosive eruptions. The deposits can for example be used as stratigraphic marker horizons in archaeology or for measuring sedimentation rates in sedimentary basins. The three reasons discussed below however are specifically volcanic ones: to find out about the explosive eruptions themselves and seek to understand the physical processes involved, to find out about conditions within and underneath the volcano, and to assess volcanic hazard. This list is not intended to be comprehensive, but covers those applications of special interest to the author, to indicate their scope and contributions they can make to volcanology.

S. Self and R. S. J. Sparks (eds.), Tephra Studies, 391–403.

2. FINDING OUT ABOUT EXPLOSIVE ERUPTIONS

Consider explosive eruptions of hawaiian and strombolian styles; being common, many volcanologists have ample opportunities to observe them, and being small and mild, observations can safely be made from a close range and can yield good field measurements of such variables as vent size, exit velocity, and magma temperature, viscosity or gas content. Direct observation of the event therefore yields enough information to elucidate mechanisms, and in consequence these mechanisms are well understood.

Consider now plinian and ignimbrite-forming eruptions and other major explosive events. Being uncommon, very few volcanologists can have or probably ever will have an opportunity to observe one. (Note that only one relatively small-volume plinian outburst, and one rhyolitic eruption of any kind, has yet been observed by volcanologists, and the largest eruption that has been closely observed erupted much less than 5 km^3 of material). Being large or violent, the prospects of being able to make useful visual observations are poor, and the volcanologist may have to choose between pursuing science or ensuring his personal survival.

Understanding of the larger and rarer explosive events must therefore rely very heavily on the study of their pyroclastic products after the event; the larger and rarer the event is, the more complete must be the reliance on this indirect approach.

The first task when studying pyroclastic rocks is to identify the way in which the pyroclastic material was transported from the vent and deposited at the place where it is now seen. Attention is here confined to this aspect since the nature of physical processes which operate at the vent is discussed elsewhere.

Three main transportation mechanisms are now recognised, namely pyroclastic fall, flow, and surge. In the first, pyroclasts fall through the air (and sometimes also settle through water) from the eruptive plume to accumulate as fall deposits, movement laterally from the vent being by some combination of wind drifting, lateral expansion in the ash plume, and initial lateral velocity imparted at the vent. In the second, pyroclasts move over the ground as a hot and concentrated particulate flow (called an ash flow, or pyroclastic flow sensu stricto) in which the continuous phase between particles is gas, and there is a high particle: gas ratio. In the third, pyroclasts are carried laterally entrained in turbulent gas as a ground-hugging dilute particulate flow having a low particle: gas ratio.

Each of these types possesses certain broad characteristics which enable it to be distinguished from the others (Table 1). Experience shows however that it is rarely safe to rely upon a

TABLE 1. Some characteristics of the main pyroclastic types.

1. Pyroclastic fall deposits show:
   (a) mantle bedding
   (b) good to moderate sorting
   (c) more or less exponential decrease in thickness and grain size with distance from vent
   (d) block impact structures

   Exception - water-flushed ash may show (a) only, but gives independent evidence for water flushing (e.g. accretionary lapilli, vesicles, water-splash microbedding)
   Fall deposits can be sufficiently hot when they accumulate to show primary welding near vent.

2. Pyroclastic flow deposits show:
   (a) ponding in depressions, with a nearly level top surface
   (b) irregular thickness variation with distance from vent
   (c) minimal sorting or internal stratification
   (d) evidence for being hot (e.g. welding, pervasive thermal colouration, carbonisation of contained plant remains, uniform direction of thermo-remanent magnetisation of contained clasts)

   Exception - low-aspect ratio ignimbrites include a mantling layer which passes laterally into the valley-pond ignimbrite
   Note 1 - ignimbrite can be defined as a pyroclastic flow deposit made mostly from pumiceous material (pumice, shards)
   Note 2 - primary mudflows (lahars) resemble pyroclastic flow deposits but lack (d): see Fig. 1 and text.

3. Pyroclastic surge deposits show:
   (a) draping of topography
   (b) rapid and irregular or periodic thickness fluctuations
   (c) general decrease in thickness and grain size with distance from source
   (d) commonly erosional base

   Two main types of pyroclastic surges occur:
   A - cold, damp or wet . . . base surges; deposits show:
   (a) good internal stratification or cross stratification
   (b) great grain size variations between contiguous beds
   (c) evidence for dampness (e.g. accretionary lapilli, vesicles, plastering of up-vent side of obstacles)
   (d) association with vents in low-lying or aqueous situations, or vents containing water (crater lakes)

   B - hot, dry . . . surges of nuée ardente type; deposits show:
   (a) little or no internal stratification
   (b) good sorting, depletion in fine or light-weight particles (but these may occur in an overlying fall deposit)
   (c) evidence for being hot

   Exception - very similar deposits underlying ignimbrite can be produced by sedimentation from the pyroclastic flow.

single character, and the overall geometry of the deposit is the surest guide to origin. Mapping of thickness and grain size should ideally be done on eacy pyroclastic sheet when striving for a reasonably unambiguous deduction on the origin. Such a comprehensive study then forms a sound basis from which to construct physical models on the eruptive and transport mechanisms.

Regarding grain size, when making a field study the author consistently measures the maximum pumice (or scoria) and maximum lithic sizes, and commonly also collects samples for granulometric and component analyses. These field measurements are very easily and quickly made, and have the immediate purpose of helping to correlate deposits between different exposures. The measurements are then available for volcanic purposes: to help determine the origin of the deposit, locate the vent position, and study the physical processes involved in eruption, transport, and deposition.

For long, pyroclastic deposits were shunned by volcanologists and sedimentologists alike, and most of the progress in their study dates from the last two decades. The present rate of conceptual advance is high, and there is no doubt much still is to be discovered. The validity of diagnostic criteria must therefore continually be questioned, and the criteria updated to accomodate the new ideas. Thus the concept of base surges and their deposits dates from the Taal eruption of only 15 years ago (1). The recognition of "ignimbrite veneer deposits", which form from flows but combine features of fall, flow and surge deposits, is a more recent though less fundamental example (2).

Pyroclastic types are not discrete and distinct entities but occupy a continuum, and boundaries between types are arbitrary. Thus ash flows and mudflows both travel as concentrated flows and show a similar ponding in depressions and lack of internal stratification and sorting. Two quite different bases can be used to define the boundary between them, one being the deduced emplacement temperature, and the other the nature of the continuous phase during flowage (Fig. 1). When the former is taken as

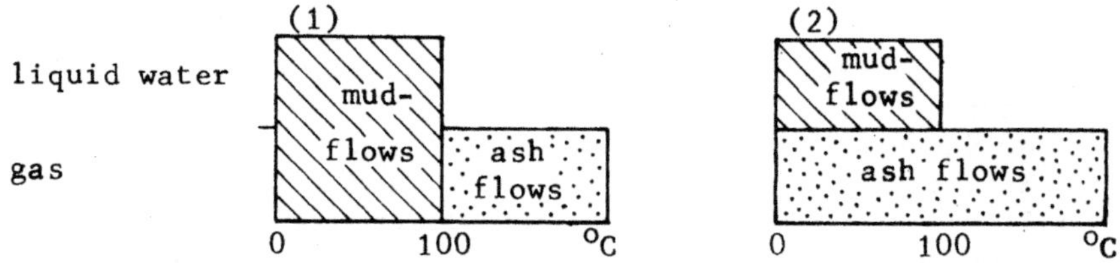

Fig. 1. The different fields of ash flows and mudflows according to whether (1) the temperature, or (2) the nature of the continuous phase (liquid water or gas) is regarded as more important.

the more significant base, a particulate flow which shows no evidence of having been hot will be called a mudflow. When the latter is taken, reliance is placed on features such as the more limited ability of flows having gas as the continuous phase to transport blocks of dense rock, the common development in them of gas jet pipes, and the absence of air bubbles (vesicles) such as are commonly trapped in the muddy matrix of mudflows.

3. FINDING OUT ABOUT THE INTERNAL PLUMBING SYSTEM OF VOLCANOES

There are many different approaches, for example geophysical probing, and mapping of the roots of dissected volcanoes. Pyroclastic studies however also play a significant role, either where tephrochronology provides the means to determine the rates at which volcanic processes operate, or where the pyroclastic rocks themselves record evidence of conditions in magmatic systems.

Consider first the use of tephrochronology in determining the output rate of volcanoes. In a classic study, Nakamura (3) plotted the cumulative output against time over the past 1500 years for the basaltic Oshima volcano, and found that it approximates closely to a straight line (Fig. 2). This is consistent with a mechanism whereby magma leaks upwards at a steady volumetric rate (about 1.5 10^{-2} m^3 s^{-1}) from its deep source into a high level crustal holding chamber. One can go on to speculate that when sufficient magma has accumulated in this chamber it becomes capable of forcing its way to the surface, and an eruption then ensues. The quantity of magma released in each outburst is fairly uniform which suggests that the release mechanism is a very simple one.

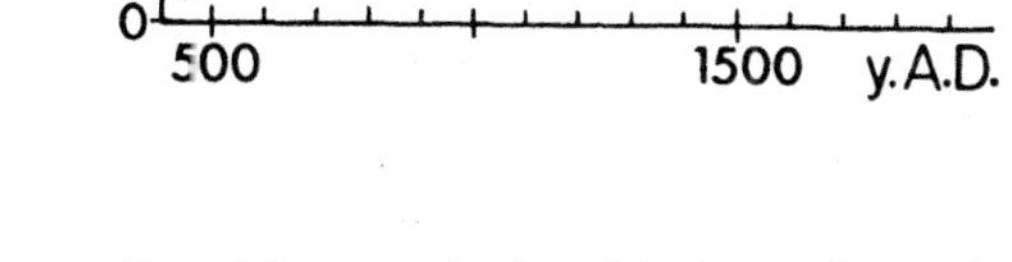

Fig. 2. Cumulative output (C) of Oshima volcano since about 500 A.D., after Nakamura (3). Each dot represents one volcanic eruption.

Eruptions of rhyolitic volcanoes have a generally larger magnitude and lower frequency, and the magma volume released is more variable (Fig. 3). This suggests the operation of different storage and release mechanisms. The activity pattern combined with what is known about the viscosity of rhyolitic magmas is

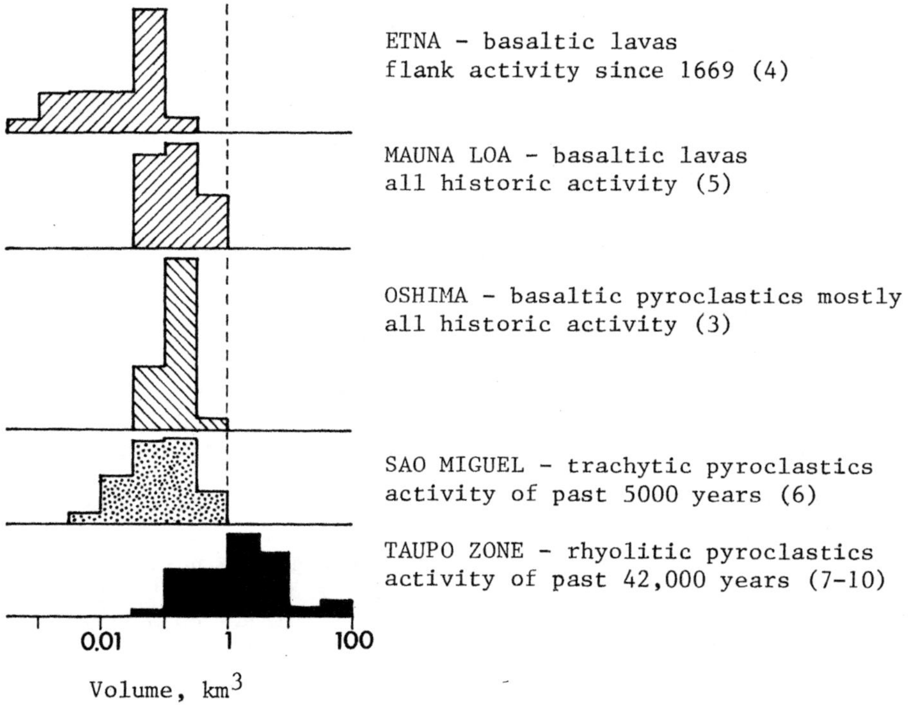

Fig. 3. Histograms showing the percentage of the number of eruptions of given magnitude for a selection of volcanoes. Volumes are dense rock equivalent volumes. The two lava volcanoes Etna and Mauna Loa are included for comparison.

consistent with the occasional uprise to a high crustal level of large-volume magma diapirs. Sometimes a diapir may break through to the surface and generate a large ignimbrite; or it may lodge at a high crustal level and aliquots of magma be later released periodically as a result of basaltic triggering.

Another aspect of pyroclastic studies is that in eruptions which produce fall deposits, the first-erupted fraction appears at the base of the deposit, the last-erupted fraction appears at the top, and the sequence with which the products come out is faithfully preserved in the deposit. This kind of simple relationship does not apply to lava flows. The deposits of pyroclastic flows may show some departures from this simple relationship though they preserve the broad pattern.

In many eruptions involving salic magmas, compositionally-zoned pyroclastic deposits form in which the upper part is more mafic than the lower (Fig. 4a). Sometimes an abrupt boundary occurs between the contrasted types, and sometimes a gradual transition. Streaky mix-pumice commonly also occurs.

Few examples of such deposits have been documented, yet they are very common: examples have been noted by the author on Faial and Sao Miguel in the Azores, Vesuvius (11) and Vulsini (12) in Italy, Apoyeque in Nicaragua, and El Misti in Peru. Compositionally zoned ignimbrites are better known and have been described from the U.S.A. (13-15), Japan (16), Italy (12) and Mexico (17).

A compositionally-zoned pyroclastic deposit must come from a compositionally-zoned magma chamber, and the stratigraphic sequence in the deposit is the reverse of that in the chamber: the first-erupted fraction comes from the top of the chamber, and the last-erupted fraction comes from deepest down (Fig. 4a). There are grounds for believing that an injection of more mafic magma may by an exchange of thermal energy often aid the eruption of the more salic magma (18).

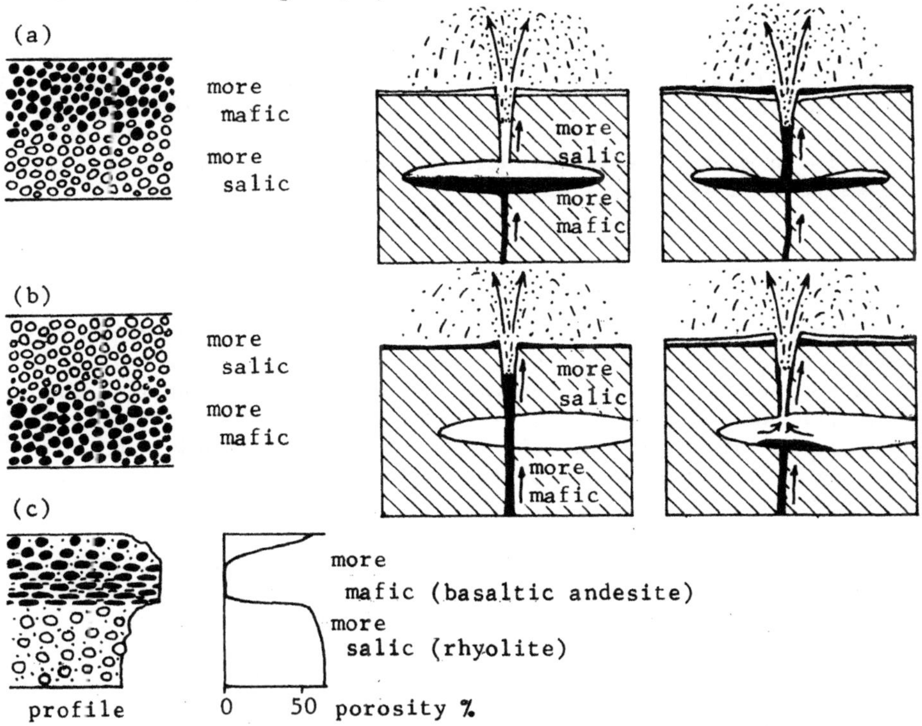

Fig 4 (a) and (b) Two kinds of compositionally-zoned fall deposits, with diagrams showing how they could have formed. (c) Compositionally-zoned ignimbrite, as seen at Acatlan (17) and Acambay in Mexico, in which the more mafic upper part is densely welded and the rhyolitic lower part is largely non-welded.

In a few explosive eruptions, the appearance of a salic magma at the surface was preceded by a basaltic outburst. One example is that which produced the Rotoiti ignimbrite and Rotoehu ash fall in New Zealand (19), and another is the penultimate ignimbrite-producing eruption of Rabaul in New Britan (20). In both, the basaltic event produced a small scoria fall deposit of strombolian /plinian type, and the lack of any soil or erosion break between the basaltic and overlying rhyolitic material indicates that the basaltic outburst immediately preceded the rhyolitic one (Fig. 4b).

This relation is reminiscent of the basalt/rhyolite composite dykes and the composite lavas they fed in eastern Iceland (21). It is evident that uprise of basaltic magma there immediately preceded the uprise of the rhyolitic magma along the same dyke, and the role of basaltic magma was to create a pathway to the surface for the initially more viscous rhyolitic magma. A very similar mechanism may well have operated in the explosive eruptions cited from New Zealand and Rabaul.

Many of the pumice fall deposits of Hekla are compositionally-zoned (22,23), becoming more mafic upwards, and they demonstrate another relationship, namely that the silica content of the first-erupted fraction is a function of the preceding repose period. The longer the repose period, the more silicic is this fraction. Here is an example where pyroclastic studies have a bearing on the rate of magma genesis.

3. ASSESSING VOLCANIC HAZARD

A third application of pyroclastic studies is in compiling dossiers on the eruptive record and behaviour pattern of explosive volcanoes for use in hazard studies, the philosophy being that the past record is likely to be a valid guide to future behaviour. Provided that the dossier is a reasonably complete one, it is unlikely that notable departures will take place in the nature or magnitude of events.

The dossier must obviously include any historical records of explosive eruptions, but generally this record is far too short and incomplete to reveal the full capability of the volcano; even when the historic record spans several centuries, there are very few scientific accounts of volcanic eruptions which took place a century or more ago.

Geological studies must therefore be made on the eruptive products of the volcano, using standard stratigraphical methods to determine the sequence, ^{14}C, fission track, thermoluminescence or other dating techniques to determine the chronology, and soundly-based volcanological methods to interpret the volcanic layers in terms of eruptive styles and the nature of the hazards they would have posed. It is particularly important that the period covered by this study should be long enough to embrace the low-frequency events (e.g. potentially dangerous ignimbrite eruptions or major volcano collapse episodes) as well as the high-frequency ones.

The proper interpretation of the volcanic layers is of crucial importance, yet it is this aspect to which the least attention has hitherto been devoted. It is only in the past 5 to 10 years that pyroclastic products of such well-known catastrophes as Vesuvius 79 and Mont Pelée 1902 have been studied in detail and it has become possible to recognise the products of past pyroclastic surges of the nuée ardente type.

Elsewhere in this book the principle is enunciated that the more powerful or violent a volcanic event, the less impressive-looking is the resulting deposit when viewed on an outcrop scale. Thus the nuée ardente of Mt. Pelée 1902 is recorded now by an innocuous-looking ash layer only about 20 cm thick in the town of St. Pièrre which it totally destroyed. It might be objected that the above principle is negated by ignimbrites, which are often very conspicuous and their emplacement would have been catastrophic. It is true that large-volume ignimbrites are conspicuous; the point to make however is that the emplacement of a low-aspect ratio ignimbrite of modest volume (which does form a thin and inconspicuous sheet) will devastate an area very much larger than will a more normal ignimbrite having the same volume.

In general the layers which record the most dangerous events thus tend to be the most inconspicuous and easily overlooked, have a low survival potential, and have to be sought for most carefully when assessing the past record of a volcano. The compilation of the dossier is a highly skilled task.

The kind of volcanological study which could be used to determine the volcanic hazard is illustrated from Sao Miguel, Azores (6,24). This island has three large stratovolcanoes, namely Sete Cidades, Agua de Pau, and Furnas, distributed along its length. These volcanoes consist of trachyte and basalt, and trachytic pyroclastic fall deposits are particularly conspicuous amongst their products. Each volcano has a caldera, that of Sete Cidades being particularly fine. The narrow "waist" between Sete Cidades and Agua de Pau on which Ponta Delgada city is built is a complex en-echelon basaltic fissure zone which has erupted many basaltic lava flows.

The eruptive activity of Sao Miguel over the past 5000 years includes 29 explosive trachytic eruptions all from vents situated in or near the calderas, and a like number of basaltic eruptions from vents situated in the "waist" or on the flanks of Sete Cidades.

The explosive trachytic eruptions produced mostly pumice fall deposits, and from the study which has been made of them the areas which have been affected by any selected threshold (such as the accumulation of a 25 cm or more thickness) during the 5000 year period can be delineated, and the average recurrence interval for this threshold to be passed can be determined (Fig. 5).

When preparing volcanic hazard zoning maps, it must be remembered that volcanoes are all the time changing their form, being moulded by successive constructive and destructive volcanic events and by erosion, so that areas subject to a particular type of hazard change with time. An example may be cited from the very large island volcano of Tenerife (B. Booth and G.P.L. Walker, in prepn.). Tephrochronology reveals a record of many great phonolitic explosive eruptions during which extensive plinian pumice deposits formed and ignimbrites accumulated preferentially on the lower slopes.

Tenerife is scalloped by a number of big cirque-like valleys ("laderas") which appear to have been produced by major landslides into the sea. One, that of the Guimar valley, contains ignimbrite presumed to have originated from a vent in central Tenerife, but there seems no pathway at present by which a pyroclastic flow could possibly have entered this ladera without travelling along a knife-edge ridge to do so (Fig. 6). The implication is that when the ignimbrite formed, the topography was appreciably different from today's and that further major landslides have taken place since then and modified it.

It was earlier mentioned that conceptual advances are all the time being made; that the state of understanding of volcanism is not static but is in a state of flux. This is healthy, but it means that our knowledge of the capability of volcanoes and of the nature and extent of the hazard they may pose is also changing. A good illustration is provided by the welded air-fall tuffs of Pantelleria, Italy (25) which were produced by eruptions of a type which has not yet been experienced.

Another illustration is provided by ignimbrites of the low-aspect ratio type (26). These are comparatively modest in volume, yet the emplacement of one like the Taupo ignimbrite would be an event far more violent and far-reaching in its effects than the worst volcanic catastrophe yet experienced by modern man. Moreover it is clear that this ignimbrite was not channeled by the topography but swept over almost the entire landscape. It would not have been safe to take refuge from it on the hilltops.

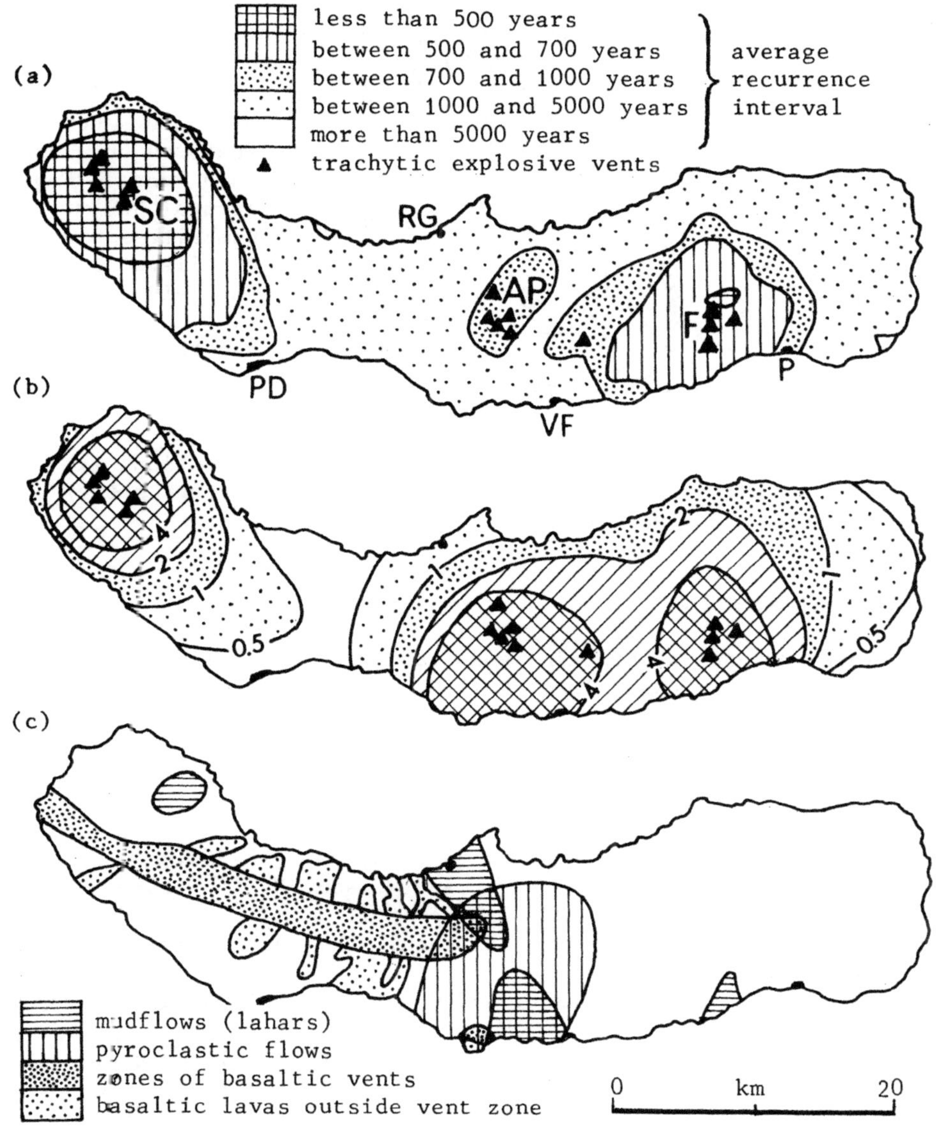

Fig. 5. Maps of Sao Miguel, Azores, showing areas affected by some volcanic happenings in the past 5000 years. These are not volcanic hazard maps since they show past events.
(a) The fall of pumice to a depth of 25 cm or more. Volcanoes: SC - Sete Cidades, AP - Agua de Pau, F - Furnas. Towns: PD - Ponta Delgada, RG - Ribeira Grande, VF - Villa Franca, P - Povoacao.
(b) The maximum depth of pumice (in metres) in any one fall
(c) Other happenings

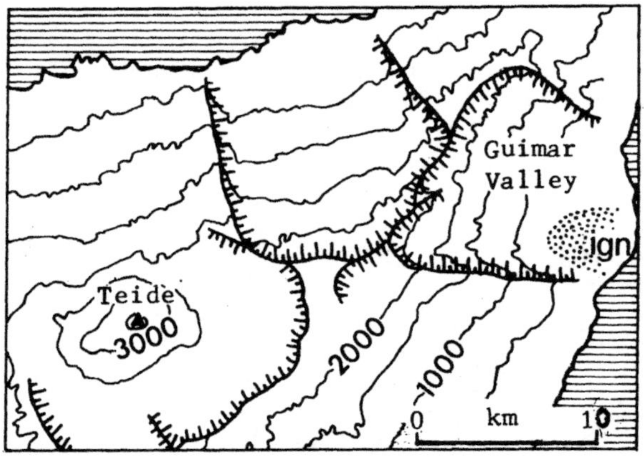

Fig. 6. Sketch map of part of Tenerife showing the location of the Guimar Valley ignimbrite exposure. Collapse Features outlined by hatched lines. Contour interval 500 m. The ignimbrite source area is believed to have been near Teide. From unpublished work by B. Booth and the author.

Perhaps here lies the main justification for continuing research on pyroclastic rocks and seeking to understand better the volcanic processes which generate them.

ACKNOWLEDGEMENTS

The work was done as Captain James Cook Research Fellow of the Royal Society of New Zealand, while on leave from Imperial College, London.

REFERENCES

1. Moore, J.G.: 1967, Bull. Volcan. 30, pp. 337-363.
2. Walker, G.P.L., Heming, R.F. and Wilson, C.J.N.: 1980, Nature 283, pp. 286-287, 286, p. 912.
3. Nakamura, K.: 1964, Bull. Earthqu. Res. Inst. Univ. Tokyo 42, pp. 649-728.
4. Wadge, G., Walker, G.P.L. and Guest, J.E.: 1975, Nature 255, pp. 385-387.
5. Macdonald, G.A.: 1955, Catalogue of active volcanoes of the World, Part III, Hawaiian Islands, I.A.V. Rome.
6. Booth, B., Croasdale, R. and Walker, G.P.L.: 1978, Phil. Trans. R. Soc. A 288, pp. 271-319.
7. Vucetich, C.G. and Pullar, W.A.: 1973, N.Z.J. Geol. Geophys. 16, pp. 745-780.
8. Pullar, W.A. and Birrell, K.S.: 1973, N.Z. Soil Surv. Rep. 1.
9. Howorth, R.: 1975, N.Z.J. Geol. Geophys. 18, pp. 683-712.
10. Vucetich, C.G. and Howorth, R.: 1976, N.Z.J. Geol. Geophys. 19, pp. 51-70.
11. Lirer, L., Pescatore, T., Booth, B. and Walker, G.P.L.: 1973, Bull. Geol. Soc. Am. 84, pp. 759-772.
12. Sparks, R.S.J.: 1975, Geol. Rdsch. 64, pp. 497-523.

13. Williams, H.: 1942, Carnegie Instn. Wash. Pub. 540.
14. Lipman, P.W., Christiansen, R.L. and O'Connor, J.T.: 1966, U.S. Geol. Surv. Prof. Pap., 524-F.
15. Ratte, J.C. and Steven, T.A.: 1967, U.S. Geol. Surv. Prof. Pap., 524-H.
16. Katsui, Y.: 1963, J. Fac. Sci. Hokkaido Univ. IV, 11, pp. 631-650.
17. Wright, J.V. and Walker, G.P.L.: 1977, Geology 5, pp. 729-732.
18. Sparks, R.S.J., Sigurdsson, H. and Wilson, L.: 1977, Nature 267, pp. 315-318.
19. Pullar, W.A. and Nairn, I.A.: 1972, N.Z.J. Geol. Geophys. 15, pp. 446-450.
20. Walker, G.P.L., Heming, R.F., Sprod, T.J. and Walker, H.R.: 1980, Papua New Guin. Geol. Surv. Mem. (in press).
21. Gibson, I.L. and Walker, G.P.L.: 1963, Proc. Geol. Ass. London 74, pp. 301-318.
22. Thorarinsson, S.: 1967, The eruption of Hekla 1947-1947, I, The eruptions of Hekla in historical times, Visindafelag Islendinga, HF Leiftur, Reykjavik.
23. Larsen, G. and Thorarinsson, S.: 1978, Jokull 27, pp. 28-46.
24. Walker, G.P.L. and Croasdale, R.: 1971, J. Geol. Soc. London 127, pp. 17-55.
25. Wright, J.V.: 1980, Geol. Rdsch. 69, pp. 263-291.
26. Walker, G.P.L., Heming, R.F. and Wilson, C.J.N.: 1980, Nature 283, pp. 286-287, 286, p. 912.

SOME EFFECTS OF TEPHRA FALLS ON BUILDINGS

R.J. Blong

Macquarie University, N.S.W., Australia

ABSTRACT. A survey of the volcanological literature reveals that only limited generalisations concerning the effects of volcanic bomb impacts and tephra loads on buildings can be made. By using data from a variety of other sources, however, it is possible to define the impact energies required to penetrate or damage a variety of building materials. These data can then be translated into equivalent bomb sizes and densities. Similarly, snow load effects on building roofs can be converted into data on the likely effects of tephra loads so that the towns, suburbs, buildings, and even portions of buildings, most at risk can be identified. Such data now need to be checked against detailed field surveys of bomb impact and tephra load damage.

INTRODUCTION

Studies of tephra hazard have generally been based on knowledge of probable wind directions and the likely thicknesses of tephra deposition. Although isolated comments outlining the effects of tephra fall and volcanic bomb impact on buildings can be found in the literature it seems that no detailed studies of building performance have been published. In fact it is rare to find together, details of damage, roof material, volcanic bomb diameter and density or roof type, roof slope and tephra load. The present study assembles data on damage to buildings caused by airfall tephra and provides an initial analysis of projectile impacts and tephra loads on buildings.

S. Self and R. S. J. Sparks (eds.), Tephra Studies, 405–420.

DOCUMENTATION OF DAMAGE DURING ERUPTIONS

From consideration of the available data the following comments can be made:

(i) Bombs of 1000 g or less will penetrate weak roofs. At Georgetown, St. Vincent, in 1902 bombs up to 900 g perforated thatched roofs and dented galvanised iron roofs, but left shingle roofs undamaged (1). In 1903 60-80 mm bombs perforated galvanised iron roofs (2). At Manam, Papua New Guinea in 1957 bombs 50 x 75 mm penetrated thatched roofs (3).

(ii) During the 1906 eruption of Vesuvius windows in the towns of Ottaviano and San Giuseppe, including those facing away from the volcano, were often broken by bombs (4). Evidently, the bombs were pumice lapilli, similar to those at Pompeii from the A.D. 79 eruption (5).

(iii) Glacier Hut on Mt. Ruapehu, New Zealand was hit 35 times by ballistic ejecta during the 1975 eruption and perforated with holes up to 300 mm across. Most bombs landed hot and charred wood. Several burnt through the hut floor (6). At Karaizawa 11 km from Asama, Japan in 1783 house roofs were perforated by red hot pumice blocks. Fifty two out of 162 houses were burnt completely (7).

(iv) Flat roofs are more vulnerable than steeply pitched roofs. Perret (8) noted around Vesuvius in 1906, that 100 mm of tephra is usually enough to cause the collapse of flat roofs. He also noted that 60-80 mm in towns around Stromboli in 1912 had no observed effects on flat roofs (9). At Bima 65 km from Tambora in 1815, house roofs collapsed under loads of only 95 mm of tephra (10). During the eruption of Volcan Fuego in 1971 300 mm of tephra caused the collapse of 10-20% of roofs in the town of Yepocapa 8 km distant (11,12). Around Soufriere, St. Vincent, in 1812 150-250 mm caused roofs to fall in. In 1902, 75-125 mm produced the same results (1). Around Sakurazima in 1914 Jagger (14) noted that flat-roofed cottages were crushed whereas those with steeper roofs were less damaged. It has also been noted that tephra saturated by rainfall produces a greater load than dry tephra.

The effects detailed above probably could be generalised to represent the effects on buildings of most falls of tephra and bombs but such comments hardly form the rational basis of an attempt to identify the relative resistances of a variety of buildings and building materials to the impact of volcanic bombs or to varying thicknesses of tephra. However, other sources of information can be used to estimate such effects.

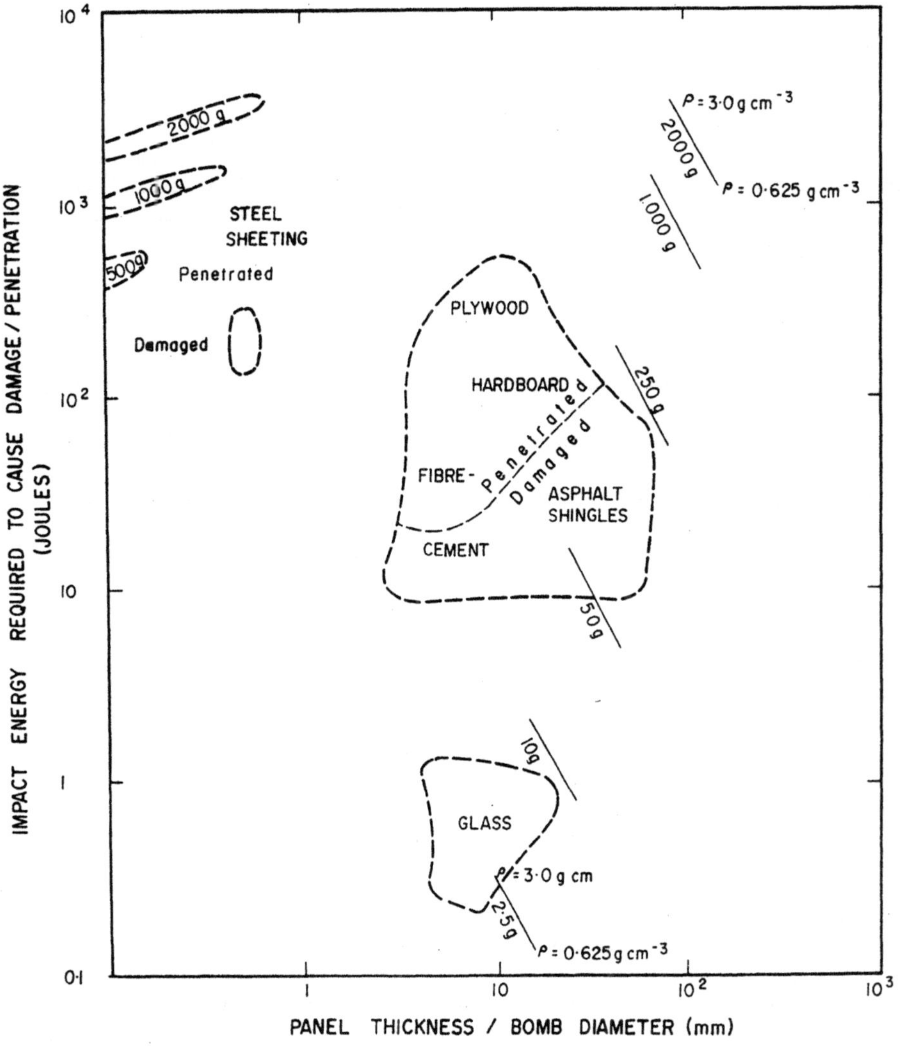

Figure 1. The relationship between volcanic bomb size, density, and impact energy and the damage caused to a range of building materials.

PROJECTILE IMPACTS ON BUILDINGS

The kinetic energy of impact ($=\frac{1}{2}mv^2$) of a volcanic bomb is dependent on both its mass and its terminal velocity. As Walker and others (14) have shown, terminal velocity is a function of bomb diameter and bomb density. They found that experimentally determined terminal velocities were well matched by velocities computed for cylinders and that densities from 0.156 to 10.0 g cm^{-3} covered the complete range of pyroclastic ejecta.

However, most volcanic bombs occur in a much narrower range of densities. Perret (8), for example, exhibited six small bombs from six volcanoes with the densities ranging only from 2.15 to 2.89 g cm^{-3}. It also seems that most pumice lapilli have densities greater than 0.625 g cm^{-3}; most bombs probably occur in the range 0.625 - 3.0 g cm^{-3}.

The terminal velocities presented by Walker and others have been used here to calculate the impact energies (in Joules) of bombs with masses varying from 2.5 to 2000 g. These data are shown as solid sloping lines on the right hand side of Figure 1. For this exercise, the bombs have been assumed cubical in shape; bombs of spherical shape but with the same diameter and density have impact energies equal to about half those of cubical shape.

Three other sets of data from a variety of studies are also shown on Figure 1:

(i) Impact energies required to damage or penetrate a number of building materials are shown in a large field in the centre of the diagram. The location on the diagram of cladding material names indicates the general range of cladding thicknesses and the general range of impact energies recorded in experimental studies. In each study the building materials were fixed according to the local building code.

Experiments, arising from the destruction of Darwin, Australia by Cyclone Tracy in 1974, on plywoods (both 3 and 5 ply), hardboards and fibre-cement boards used 2000 g and 6000 g projectiles. For plywoods 4.5 - 12.0 mm in thickness impact energies at penetration ranged 90 - 500 J. In general, the thicker the plywood the greater the impact energy required for penetration but it was also noted that sharp projectiles penetrated at lower impact energies than did blunt projectiles (15,16).

Fibre - cement cladding 4.5 - 9.5 mm in thickness was penetrated by projectiles with impact energies 20 - 85 J but only damaged in the range 10 - 20 J. Hardboard 9.5 mm thick was penetrated by projectiles with impact energies of 60 - 90 J. Thus, of the three materials in the 8 - 10 mm thickness range, plywoods are the most resistant to projectile penetration.

Asphaltic shingles with various backings were tested in the U.S.A. for resistance to hailstone impact (17). With thicknesses ranging 10 to more than 40 mm impact energies required to damage the shingles ranged 10 - 80 J. Greenfeld (17) also showed that the impact resistance of asphaltic shingles is greater at higher ambient temperatures but that the weathering of shingles decreased resistance.

Greenfeld's study also provided a few results for other roofing materials. Cedar shingles, 10 mm thick, sustained damage with impact energies of only 10 J while 6 mm thick slates were damaged at 10 - 16 J. A test on a 20 mm thick red clay tile produced damage at 20 J.

(ii) The field near the bottom of Figure 1 indicates the range of thicknesses and impact energies required to cause penetration of both annealed and highly tempered window glasses. These data stem from damage caused to window glasses in the U.S.A. by the movement of roof gravels during strong winds. The study by Minor and others (18) determined the velocities required to mobilise 5 g stones and the mean minimum velocities required to penetrate glasses with thicknesses ranging 4.7 - 19.0 mm (Fig. 1). Although these data indicate that windows have a very low penetration resistance Leicester and Reardon (16) have shown that penetration of an 11 mm thick laminated glass required an impact energy of 1500 J.

(iii) The available data for steel cladding and roofing materials are presented in the top left of Figure 1. Experimental studies in Australia indicate that steel sheets 0.42 - 0.6 mm thick with yield strengths of 300 and 550 MPa pulled away from their fastenings at impact energies of 150 - 300 J (15). In another study a 0.5 mm profiled steel sheet was perforated at an impact energy of 450 J (16).

Further data on the penetration resistance of steel sheeting can be derived from Kar's analysis (19) although it should be noted that this was developed for much thicker steel barriers than those used for normal commercial buildings and dwellings. Kar's formula to calculate the thickness of steel penetrated requires data on projectile mass, density, shape and hardness (Brinell Hardness Number - BHN) and knowledge of the ultimate yield strength of the steel. The three ellipses shown in the top left of Figure 1 indicate the impact energies and thicknesses of steel cladding penetrated by projectiles of 500, 1000 and 2000 g mass. The right hand end of each ellipse represents the panel thickness penetrated by a bullet - nosed bomb/projectile of density 3.0 g cm^{-3} with BHN = 200. These parameters provide a reasonable maximum penetration for a volcanic bomb of specified mass. Other points in the ellipses have been calculated for various combinations of density (1.0 - 3.0 g cm^{-3}), shape (flat - nosed to bullet - shaped), BHN (100 - 200) and yield strength (300 - 550 MPa). Further details can be found in Kar (19). The important point for the present study is that the analysis indicates that projectiles of 1000 g or less will not penetrate the commercial range of steel claddings (0.42 - 0.7 mm thickness) and that only the most dense 2000 g missiles will have any effect.

Comparison of the impact energies of the volcanic bombs of specified mass and diameter with the experimental data and analyses for various building materials suggest the following results (Fig. 1):

(i) Even thick sheets of highly tempered glass will be penetrated by bombs of less than 10 g mass. Cubical bombs of only 2.5 g mass but densities > 1.0 g cm^{-3} (10 - 12 mm diameter) can break some window glasses. It seems likely that all bombs > 25 g mass will break most windows (excepting laminated glass) whatever the bomb shape.

(ii) Fibre - cement cladding will be perforated by almost all bombs larger than 500 g mass and damaged by bombs of as little as 125 g mass. Similarly, the diagram indicates that plywood, hardboard, fibre - cement, asphaltic and cedar shingles, slate and clay tiles are all likely to be penetrated or at least severely damaged by volcanic bombs 50 - 60 mm in diameter.

(iii) Bombs with densities > 2.5 g cm^{-3} and masses greater than 2000 g (i.e. diameters > 90 mm) can penetrate commercially available steel cladding materials. Bombs of 250 - 500 g (diameters > 40 - 55 mm) could damage steel sheetings particularly around fastenings.

The results presented in Figure 1 must be treated with some caution as test conditions and assumptions were variable, although fixture of the building materials always complied with the codes extant in the country where the tests were performed. There is some evidence that the impact of a slower heavier mass does more damage than a faster lighter mass with the same impact energy (20). Furthermore, asphaltic shingles, and possibly other materials as well, are more resistant to impact at higher temperatures (17). Similarly, there is a slight increase in the penetration resistance of sheet steel with increasing ambient temperature (19). Projectile shape is of considerable importance: presumably the impact energies recorded on Figure 1 for the penetration of window glass and for damage to asphalt shingles would be lower if pointed ballistics had been used in the tests. Projectile hardness is also of considerable importance as Kar (19) has shown; comparisons with volcanic bombs should assume that the bombs have cooled and solidified. Clearly, this is not always the case. Although the penetration power of hot malleable bombs will be less than that of cold hard projectiles, ignition of ceilings, furnishings or floors can occur causing even more serious damage (21). Two such examples were given in an earlier section.

It should also be remembered that the results presented assume that the angle of impact approaches 90 degrees and that the bombs are cubical in shape; bombs of spherical shape would have to have higher densities or larger diameters to achieve the same penetration.

In general, the analysis only considers the impact of individual bombs. Repeated impacts, even glancing impacts, may weaken walls or roofs. Even without penetration serious maintenance defects could develop in a building, possibly leading to other modes of failure.

Although the analysis presented provides much greater precision than those normally found in the volcanological literature the range of possible effects indicates that the results are but a first approximation. However, knowledge of the local building codes and practice can be used to improve prediction as minimum thicknesses of cladding materials and methods of fixing are usually specified. Thus, observation of local building styles (conforming to local building codes) would allow the identification of the suburbs, houses, and even portions of buildings, most vulnerable to penetration by volcanic bombs.

Finally, the results presented obviously require testing against a body of field data specifying bomb dimensions, density, and mass, and significant details of the damage inflicted. Only then can further refinements be made.

TEPHRA LOADS ON BUILDINGS

In the absence of detailed data on the effects of sand-sized and finer tephra loads on buildings the obvious and appropriate analogue lies in the effects of snow loads.

Most building codes in cold-climate countries allow, in roof design, for the occurrence of a maximum live load equal to the weight of a once in 30 year snow load. Snow accumulation on a flat roof is usually assumed to be about 80% of snow accumulation on the ground. Some allowance is also made for the expectable 24 hr rainfall during the period of maximum snow load as rainfall on snow increases the live load.

Design loads are generally based on snow densities of 0.2 - 0.24 g cm^{-3} (22,23). As tephra densities probably range 0.4 - 0.7 g cm^{-3} increasing to about 1.0 g cm^{-3} with compaction and rainfall, roofs designed to withstand snow loads should remain structurally sound with tephra loads equal to about 25% of the design snow load thickness.

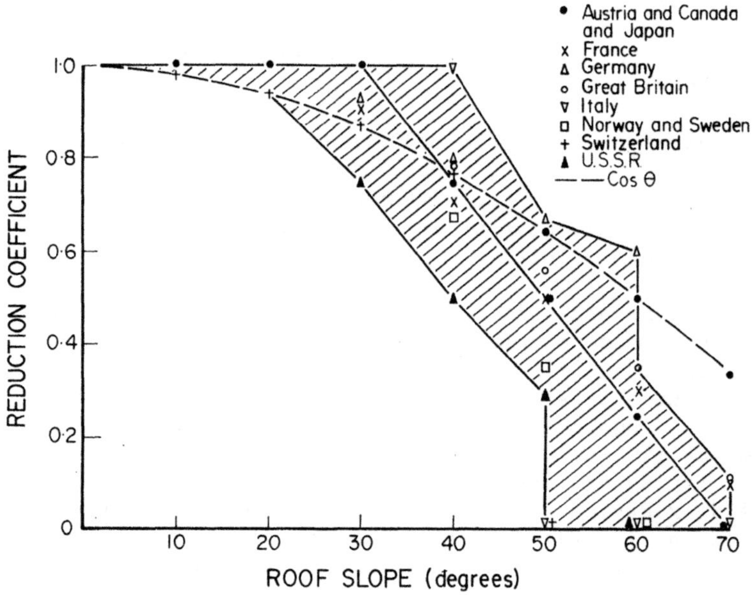

Figure 2. Relationship between roof slope and the reduction coefficient used in (snow) live load design in a number of countries. Data from Taesler (24).

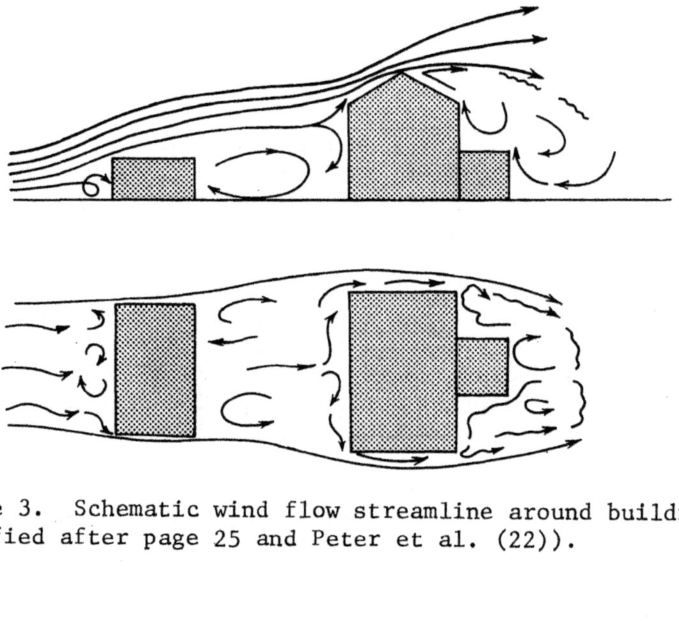

Figure 3. Schematic wind flow streamline around building (modified after page 25 and Peter et al. (22)).

Generally, some reduction in the design load is allowed for roofs pitched at steeper angles (Fig. 2). Despite this common practice, evidence reported by Taesler (34) and Page (25) suggests that snow loads are greater, and design loads are more frequently exceeded, on steeper roofs. This results from the drifting of snow in the wind.

Figure 3 indicates schematic stream lines around a building; where the streamlines are deflected there is increased velocity over and around the building. A flow separation point occurs at the roof peak. In the lee a region of turbulence forms where flow directions are random. Major turbulent regions also form over the lower levels of split level roofs to either leeward or windward of higher level roofs and in areas adjacent to parapets (22).

Thus tephra can be expected to drop out of the wind flow in turbulent wakes where there is no strong flow in any direction. Where streamlines are adjacent to the ground or roof the surface will be swept clear of tephra. Tephra drifts will be at a minimum at the top of a hill whereas the lee sides of hills, depressions and cuts will be places of accumulation. If the analogy with snow deposition is maintained, tephra drifts behind obstructions will have lengths 6 - 9 times the height of the obstruction (26). Similarly, simple rectangular house plans without internal corners or projections will suffer minimum tephra accumulation. Enclosed courtyards will trap tephra (26). As wind speed increases with height above the surface so higher roofs will experience greater wind effects. The wind is more likely to pass over the roof rather than around the ends of a low broad building. For a tall narrow structure a greater part of the wind will pass around the sides (22). In short, the wind profile around a building is affected by its shape, height, orientation and spacing relative to other buildings and objects (25,26).

Wind velocity is obviously also an important consideration. With light winds, tephra will be evenly distributed. With slightly higher velocities, but not fast enough to cause scour of the tephra or saltation, moderate to low drifts will form. With wind speeds above the saltation threshold considerable redistribution of tephra on roofs and the formation of large drifts will occur. Very high winds will cause some areas of the roof to be bared and encourage the formation of large drifts and tephra ramps whereby tephra may be blown up onto roofs (23).

Clearly, the duration of the wind as well as its magnitude is of considerable significance.

With these comments in mind the following general principles can be stated:

(i) With no wind, tephra would blanket ground and buildings to an equal depth; the ratio of the ground load to the roof load will be 1:1 and the roof load will be uniformly distributed. Such conditions are relatively rare and are confined to no wind areas and to windy areas sheltered on all sides by tall trees (27).

(ii) On roofs exposed to the wind, the relatively low density of much of the tephra blanket ensures that little tephra will accumulate. However, where gables or other projections lie transverse to the wind, creating areas of 'aerodynamic shade', accumulation will occur (27).

(iii) Although design loads are reduced for steeper roofs (Fig. 2), contrary to common belief, snow loads on pitched roofs are not significantly less than on flat roofs (24). For example, slide-off of snow from non-metallic roofs of less than 40 - 50 ^{o} seldom occurs (22). The coefficient of static friction between roof material and tephra and between tephra and tephra probably varies from 0.1 to 0.4 (28). Certainly, smooth surfaces such as sheet metal and glass have lower coefficients of friction than asphalt or wood shingles (29). It seems that unbalanced loads on roofs are more likely to be the result of wind scour and deposition than of sliding (23).

(iv) Roof configuration probably has the most important effect on tephra accumulation and roof live loads. Some of the likely accumulations of tephra are shown on Figure 4 based on snow load diagrams in the U.S.S.R. Standard Code of Practice and the National Building Code of Canada (27,29). Where a parapet occurs drift can accumulate to the top of the wall and extend out 3 - 5 m from the wall. Drifts usually occur not only on the windward side of the parapet but also along all other parapets to form a saucer-shaped depression (22). Roof projections such as parapets, chimneys and ventilation shafts have important effects on loading. With chimneys and other projections a depression usually forms against the object with the drift forming about 0.5 m away (22). Roof loads in such areas can be as much as 2 times the ground load. Where low roofs occur below extensive higher roofs (Fig. 4) loads as much as 3 times the ground load are possible (29). Tephra loads in such areas are usually triangular in cross - section with the maximum load against the higher wall. If sliding of tephra from the higher roof does occur impact loads as well as the live load of tephra are possible (22).

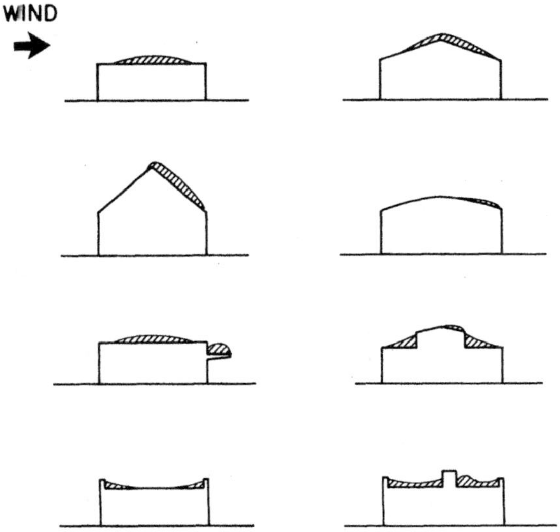

Figure 4. Distribution of tephra loads on a variety of roof styles in situations where drift occurs. (Modified after Schriever (27,29).

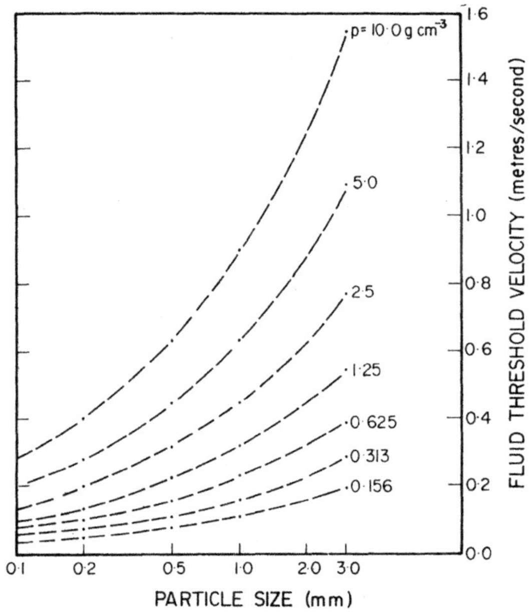

Figure 5. The relationship between tephra particle size, particle density and fluid threshold determined using Bagnold's (31) equation.

(v) Clearly, tephra loads on roofs can be much higher than the ground tephra load, even on steeply pitched roofs. Furthermore, such loads will usually be unbalanced. These assymmetrical loads on roofs are important as loads may not be evenly distributed amongst the various members of the roof support system; some members normally in tension may, under a large unbalanced load, be in compression. Although failure of any one member may be required to cause partial roof collapse, structural failure in buildings is nearly always the result of several factors (22). It is also clear that inadequate bracing may cause wall failure before roof failure occurs.

Schriever and Hansen (30) carried out some interesting tests on conventional timber roof frames built in an area of Canada for the same design load and to satisfy the one building code. They found an extremely wide range of failure loads depending on the size of rafters used, the heel joint details, the type of end support used and the quality of workmanship. Failure loads varied from 88 - 613 kg m^{-2}, equivalent to about 9 - 62 mm of wet tephra. They also found that as roof pitch increased from 14 to 23 ^{o} that the average failure load of trusses increased from 525 to 712 kg m^{-2}. This conclusion is important as it indicates that the steeper pitched roofs were better able to withstand the live load applied (and not necessarily that the live load is less on steeper pitches). Other conclusions reached in the study indicate that increasing the cross-section of truss members and decreasing truss span increases the magnitude of the live load carried at failure. They also noted that a reduction in load-carrying capacity with a reduction in nailing must have been caused by a net increase in total stresses due to greater distortion of the trusses. Deflection of the trusses also increased with increased duration of loading (30).

Although attention here has been focused on the effects of tephra loads on roofs it should be recognised that tephra drifts against walls and/or foundations could lead to failure before roof loads, especially those formed from tephra ramps, become critical. Similarly, loads against chimneys and parapets could lead to failure of these 'obstructions' and, indirectly, to roof failure.

Much of the discussion so far has been based on the premise that the drifting of snow and the resulting load distributions provide suitable models for consideration of the effects of tephra with minor modifications to discount the influence of melting and drainage. Such a comparison seems very reasonable (although for a number of the points made it would not matter if it were not reasonable). However, a comparison of the velocity at which snow and tephra begin to drift is worthy of further consideration. In the case of snow, the fluid threshold at which entrainment begins is presumably a function of crystal size, temperature, compaction and moisture content; in any event, figures of 8 - 12 mph (3.5 - 5.3 m s^{-1}) are mentioned (29).

From Figure 5, based on Bagnold's (31) fluid threshold formula, dry tephra particles 0.1 mm in diameter will begin to move at wind velocities much less than 1 km hr^{-1} (about 0.25 m s^{-1}) and that even 3.0 mm diameter particles with a density of 2.5 g cm^{-3} will move at wind velocities of < 2.9 km hr^{-1} (< 0.8 m s^{-1}). In fact, once movement is initiated, bombardment of the tephra surface by saltating grains initiates further movement. This impact threshold value is considerably lower than the fluid threshold plotted on Figure 5, particularly for the larger grains (31).

As particle size decreases below 0.1 mm the fluid threshold increases. The fluid threshold will also increase if the tephra is wet or sticky or if the surface is crusted. For example, for fine sands the rate of movement varies inversely (approximately) as the square of effective soil surface moisture (32). Numerous other factors (e.g. surface roughness, vegetation cover) also effect the relationship; the values expressed in Figure 5 must be regarded as approximate and close to the lower limit of entrainment/drift-initiation velocities. The diagram suggests, however, that at least some tephras will begin to drift at velocities as low as, and possibly less than, those causing snow drift. The available evidence indicates that the snow drift analogue is a reasonable one.

Some additions to the rather simplistic models of Figure 4 should also be mentioned:

(i) Continued penetration of fine tephra into the ceiling space via eaves ventilation gaps and cladding joints could lead to eventual ceiling collapse or sagging under the accumulated load.

(ii) Similarly, penetration of fine tephra could cause a short-circuit in the electrical system, leading to fire.

(iii) Deposition of acid tephra and associated aerosols could, through corrosion, lead to premature ageing and weakening of cladding and other building materials.

(iv) Deposition and redeposition of tephra following rainfall could lead to overloading and collapse of the roof gutter system, clogging of drains, water damage to ceilings, wall and sub-floor spaces, short-circuiting of the electrical system and so on.

Finally, although fine tephra and volcanic bomb effects have been considered separately here it is reasonable to expect some interaction effects. For example, penetration of cladding materials by bombs would lead to subsequent increased deposition of fine tephra in ceiling spaces, wall cavities and the interior of the house. Similarly, an early fall of fine tephra could decrease the impact energy of later volcanic bombs.

SUMMARY AND CONCLUSIONS

Although occasional comments about the effects of volcanic bombs and tephra falls on buildings can be found in the literature, such comments contribute only a little to a serious study of tephra as a hazard. However, studies of other natural hazards provide a great deal of useful data.

Figure 1 outlines the effects of volcanic bombs on a variety of building materials. Clearly, the resistance of most of the building materials considered is low; it is also clear that increased thicknesses of such materials only marginally improve impact resistance. Nonetheless, the penetration resistances of building materials span 4 orders of magnitude of impact energy. Thus, the diagram can be used as a basis for the recognition of the most vulnerable suburbs, structures, and even portions of buildings.

Similarly, Figure 4 and the related data go some way toward identifying the types of structures and the parts of structures that are most vulnerable to loading by fine tephra. It seems that flat roofs are more likely to collapse than steeply pitched roofs. However the reason for this lies not in the sliding of tephra from steeper roofs, but in the more load resistant construction of pitched roofs and the presence of papapets and similar 'aerodynamic shade-producing' obstacles on flat roofs.

As indicated in the discussion a number of factors complicate the models introduced in Figures 1 and 4. Clearly, the analysis presented here is but a first approximation. Nonetheless, it indicates that a tephra hazard map must consider building styles and building materials as well as probabilistic assessments of bomb trajectories and tephra thicknesses. It is imperative that the analysis presented here be rigorously tested against data from detailed field studies. With the field data currently available, much of it summarised here, such testing is not possible.

ACKNOWLEDGEMENTS

Financial support for the field and library research on which this paper is based was provided by Macquarie University and the Australian Research Grants Committee. I would also like to thank Brian Wood of Pereira Wood, Consulting Architects, Sydney for valuable discussions and comments and the librarians and other officers of the Experimental Building Station, Sydney for their considerable assistance.

REFERENCES

1. Anderson, T. and Flett, J.S.: 1903, Roy. Soc. London Phil. Trans. 200A, pp. 353-553.
2. H.M.S.O.: 1903, in: Further correspondence relating to the volcanic eruptions in St. Vincent and Martinique, in 1902 and 1903, Her Majesty's Stationery Office, p. 193.
3. Palfreyman, W.D. and Cooke, R.J.S.: 1976, in: Volcanism in Australasia, ed. R.W. Johnson, Elsevier, pp. 117-131.
4. Hobbs, W.H.: 1906, J. Geol. 14, pp. 636-655.
5. Lacroix, A.: 1906, Smithsonian Inst. Ann. Rept., pp. 223-248.
6. Nairn, I.A.: 1975, Immediate Report; field investigations of 1975 April 24 and 27 Ruapehu eruptions, New Zealand D.S.I.R. Prelim. Rept., p. 35.
7. Aramaki, S.: 1956, Jap. Jnl. Geol. & Geophys. 27, pp. 189-229.
8. Perret, F.A.: 1950, Carnegie Inst. Washington Publ. 549, p. 162.
9. Perret, F.A.: 1912, Smithsonian Inst. Ann. Rept., pp. 285-289.
10. Anon.: 1816, Quart. Jnl. Sci. & Arts 1, pp. 245-258.
11. Bonis, S. and Salazar, O.: 1973, Bull. Volc. 37 (3), pp. 394-400.
12. S.I.C.S.L.P.: 1971, Smithsonian Inst. Center for Shortlived Phenomena Ann. Rept.
13. Jagger, T.A.: 1956, in: My experiments with volcanoes, Hawaiian Volcano Research Association, p. 198.
14. Walker, G.P.L., Wilson, L. and Bowell, E.L.G.: 1971, Geophys. Jnl. Roy. Astron. Soc. 22, pp. 377-383.
15. Grossman, P.U.A. and Mackenzie, C.E.: 1977, C.S.I.R.O. Australia Div. Building Research Tech. Paper 15, p. 10.
16. Leicester, R.H. and Reardon, G.F.: 1977, Commonwealth of Australia Dept. of Construction, Sydney Workshop on guidelines for cyclone product testing and evaluation, p. 5.
17. Greenfeld, S.H.: 1969, U.S. Dept. Commerce, Building Science Series 23, p. 9.
18. Minor, J.E., Beason, W.L. and Harris, P.L.: 1978, Am. Soc. Civ. Eng. Jnl. Structural Division 104, pp. 1749-1760.

19. Kar, A.K.: 1979, Am. Soc. Civ. Eng. Jnl. Structural Division 104, pp. 1871-1878.
20. Grossman, P.U.A.: 1977, Commonwealth of Australia, Dept. of Construction, Sydney Workshop on guidelines for cyclone product testing and evaluation, p. 5.
21. Anon.: 1975, Commonwealth of Australia, Dept. of Construction, Exptl. Building Station, Notes on the Science of Building 137, p. 4.
22. Peter, B.G.W., Dagliesh, W.A. and Schriever, W.R.: 1963, N.R.C. Canada, Division of Building Research, Res. Paper 189, p. 11.
23. Taylor, D.A.: 1979, Can. Jnl. Civ. Eng. 6(1), pp. 85-96.
24. Taesler, R.: 1970, W.M.O. Tech. Note 109, pp. 129-149.
25. Page, J.K.: 1976, W.M.O. Tech. Note 150, p. 64.
26. Schaerer, P.A.: 1972, N.R.C. Canada, Division of Building Research, Can. Building Digest 146, p. 4.
27. Schriever, W.R. and Otstavnov, V.A.: 1968, N.R.C. Canada, Division of Building Research, Res. Paper 366, pp. 13-33.
28. Lynch, C.T. (ed): 1974, in: Handbook of Materials Science, 1, CRC Press, p. 752.
29. Schriever, W.R.: 1978, N.R.C. Canada, Division of Building Research, Can. Building Digest 193, p. 4.
30. Schriever, W.R. and Hansen, A.T.: 1964, Forest Products Jnl. 14(3), pp. 129-136.
31. Bagnold, R.A.: 1941, in: The physics of blown sand and desert dunes, p. 265.
32. Woodruff, N.P. and Siddoway, F.H.: 1965, Proc. Soil Sci. Am. 29, pp. 602-608.

PYROCLASTIC FLOWS AND SURGES: EXAMPLES FROM THE LESSER ANTILLES

A.L. Smith[1], R.V. Fisher[2], M.J. Roobol[1], and J.V. Wright[1]

[1]Department of Geology, University of Puerto Rico, Mayaguez, Puerto Rico 00708

[2]Department of Geological Sciences, University of California, Santa Barbara, California 93106

ABSTRACT

Four main types of pyroclastic flows are characteristic of Lesser Antillean volcanoes. These are pumice flows, scoria flows, semi-vesicular andesite flows and block and ash flows (nuées ardentes). Associated with these are ground surges produced directly either from the crater or from an eruptive column and ash-cloud surges derived from the turbulent upper part of pyroclastic flows. Ash-cloud surges may become detached and move independently of their parent flow and can resegregate to form secondary pyroclastic flows and surges. Field, granulometric and textural characteristics of the deposits from the different types of flows and surges are presented together with a discussion on their modes of origin and potential hazards.

INTRODUCTION

The Lesser Antillean volcanoes are characterized by the occurrence of four main types of pyroclastic flows--pumice flows, semi-vesicular andesite flows, block and ash flows (nuées ardentes) and scoria flows (1). Two main types of surges can also be further subdivided on the basis of composition and vesicularity of their clasts into pumiceous surges, scoriaceous surges etc. (2).

S. Self and R. S. J. Sparks (eds.), Tephra Studies, 421–425.

MECHANISM OF FORMATION

Two main mechanisms have been proposed for the origin of pyroclastic flows in the Lesser Antilles: 1) Collapse of an actively growing dome or lava flow; 2) Collapse of an eruption column.

Block and ash flows (nuées ardentes) have now been shown to have formed not only by the former mechanism (3,4) but also by the rapid collapse of an eruption column produced by an explosion through a small dome (5).

The other types of pyroclastic flows are all thought to have been formed by eruption column collapse. For most eruptions this involves the collapse of columns produced by discrete explosions. However some of the larger eruptive events e.g. the Roseau pyroclastic flow (6) may have been produced by continuous collapse. The apparent lack of associated airfall beds with some sequences of pyroclastic flows and surges suggests that these might have been formed directly from the crater without the production of an eruptive column, the eruptive material just topping the crater rim and moving down the outer slopes under gravity (e.g. Cotopaxi 1877, 7).

Ground surges appear to be produced directly from the crater, by collapse of the outer part of an eruptive column or as a flow front process in the turbulent "head" of a pyroclastic flow. They need not form at the same time as a pyroclastic flow. In contrast the ash cloud surges are directly associated with pyroclastic flows and are derived by the elutriation of material from the high concentration underflow into the turbulent overriding ash cloud (8,5).

DEPOSITS

Normalized sections through various types of pyroclastic flows and surges are given in Figure 1. These sections have been simplified to show the main features of each type. Actual sections through such deposits are often much more complex and also may not necessarily show all the components illustrated in a vertical sequence.

With the possible exception of the block and ash deposits (5) all the pyroclastic flow deposits are characterized by the presence of underlying ground surge beds (layer a). The lowest unit of the flow itself is generally a fine grained unit (layer b_1) although in the case of the pumice flows, pumice dunes can separate this layer from the underlying beds. The main part of the flow has been designated layer b_2. This layer is often

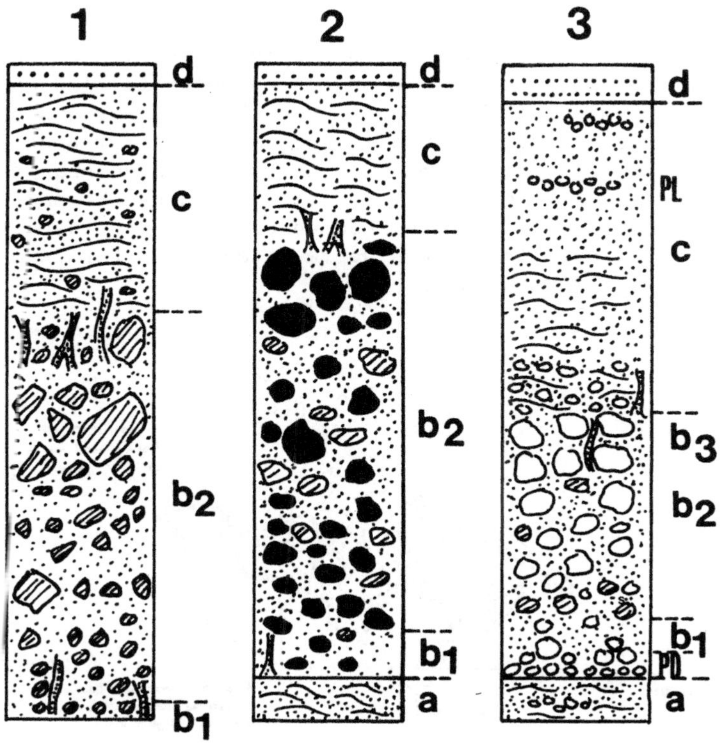

Figure 1. Normalized sections of pyroclastic flow and surge deposits. 1-Block and Ash; 2-Scoria and Ash; 3-Pumice and Ash; Dense Andesite clasts; Basaltic Andesite clasts; Pumice clasts; PL-pumice lenses; PD-pumice dunes.

reversely graded, although both normal and non-graded varieties have been described. In detail layer b_2 can show a complex set of subunits which probably reflect either successive pulses in the actual flow or segregation lenses produced by internal differences in shear stress during laminar flow. The pumice swarms sometimes found at the top of pumice flows (8) have been termed layer b_3.

On top of the main flow unit are shown the ash cloud surge (layer c) and vitric airfall ash (layer d). These sometimes can occur vertically above a pyroclastic flow unit but more commonly are found on the margins of pyroclastic flows where they represent their lateral equivalents (2,5).

ASH CLOUD SURGES

The deposits from ash cloud surges as mentioned above, can occur both directly above the parent pyroclastic flow and as its lateral equivalent. The deposits can show the various unidirectional bed forms associated with surges but planar beds and massive beds are also quite common. Carbonized wood and gas escape structures are also present. In some instances the ash cloud may become detached and move independently of its associated pyroclastic flow. They can also resegregate to form secondary pyroclastic flows and surges (5).

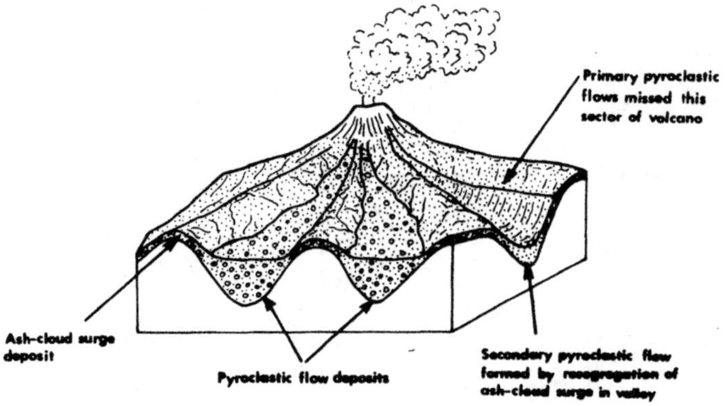

Figure 2. Block diagram showing areal distribution of pyroclastic flow and ash-cloud surge deposits for a typical island arc volcano.

For most Lesser Antillean volcanoes the subaerial distribution of the pyroclastic deposits from an eruption usually follows the pattern shown in Figure 2. The deposits of the high concentration pyroclastic flows are found filling the valleys (for many eruptions these are found in limited sectors of the volcano), while the deposits of the ash cloud surges occur on the intervening areas. As a consequence of the ability of an ash cloud to become detached from its associated flows its deposits invariably extend over a much wider area (5,9). On resegregation secondary pyroclastic flows may fill in topographic depressions not affected by the primary pyroclastic flow (Figure 2).

VOLCANIC HAZARD

On the basis of their high temperature, wide aerial extent and past experience (e.g. eruptions of Mt. Pelée and Soufriere, St. Vincent, 1902) surges, especially ash cloud surges, must be regarded as representing the greatest volcanic hazard present

ACKNOWLEDGEMENTS

Work supported by National Science Foundation Grants EAR 73-00194 and EAR 77-17064 (Smith and Roobol) and EAR 77-14489 (Fisher).

REFERENCES

1. Wright, J.V., Smith, A.L. and Self, S.: 1980, J. Volcanol. Geotherm Res. 8, pp. 315-366.
2. Smith, A.L. and Roobol, M.J.: 1980, in: Thorpe (ed.) Orogenic andesites and related rocks, John Wiley and Sons, Inc., in press.
3. MacDonald, G.A.: 1972, Volcanoes, New Jersey, Prentice-Hall, Inc., 509 p.
4. Williams, H. and McBirney, A.R.: 1979, Volcanology, San Francisco, Freeman, Cooper and Co., 397 p.
5. Fisher, R.V., Smith, A.L. and Roobol, M.J.: 1980, Geology.
6. Carey, S. and Sigurdsson, H.: 1980, J. Volcanol. Geotherm. Res. 7, pp. 67-86.
7. Wolf, Th.: 1878, Neues Jahrb. Min., pp. 113-167.
8. Fisher, R.V.: 1979, J. Volcanol. Geotherm. Res. 6, pp. 305-318.
9. Roobol, M.J. and Smith, A.L.: 1980, Bull. Volcanol., in press.

BASE SURGE DEPOSITS IN JAPAN

Shozo Yokoyama

Faculty of Education, Kumamoto University,
Kumamoto 860, Japan

ABSTRACT. Base surge deposits in Japan are described, with particular reference to the rhyolitic deposits of Mukaiyama volcano, Nii-jima Island.

1. INTRODUCTION

This paper briefly summarises currently recognised base surge (1) deposits in Japan (Table 1). Most are insufficiently investigated. Recognition of base surge deposits is based on the existence of cross-bedded, particularly antidune-like structures, which are here defined as asymmetrical and dune-like, with steeper stoss-side than lee side lamination, and other sedimentary characteristics given by previous workers (e.g. 2,3,4).

It is evident from Table 1 and previous reports that the predominant magma type is basaltic, but that base surges are generated regardless of magma composition. Because all the given examples are interpreted to be related to environments where water was readily available, the contact of magma with water is regarded as essential in the generation of base surges.

2. BED FORMS, VELOCITY OF BASE SURGES, AND MAGMA TYPE

One of the most important problems of base surge deposits is the mechanism of formation of the various sedimentary features (bed forms), such as antidune structures, wavy and/or undulating beds, plane beds, and cross beds. Among the factors influencing their formation, the "power" or velocity of the surges will very

S. Self and R. S. J. Sparks (eds.), Tephra Studies, 427–432.

TABLE 1. Recognized base surge deposits in Japan

Name of volcano or place	District	Rock type	Available water at source of eruption
Rishiri	Hokkaido	basalt	sea
Usu	Hokkaido	dacite	crater-lake
Ichinomegata	NE Honshu	basalt	ground
Numazawa	NE Honshu	dacite	caldera-lake
Oshima	Izu	basalt	sea
Wakago (Niijima)	Izu	basalt	sea
Mukaiyama (Niijima)	Izu	rhyolite	sea
Kozushima	Izu	rhyolite	sea
Miyakejima	Izu	basalt	sea
Senrigahama	Kyushu		crater-lake
Miike	Kyushu	dacite	ground
Tarumizu	Kyushu	dacite	sea
Yamakawa	Kyushu	dacite	sea
Ikeda	Kyushu	dacite	caldera-lake

likely play the most important role. The sequential lateral variation of type and scale of bed forms away from source has been postulated (5,6) to be the result of the general decline in velocity of base surges. Excellent examples of this lateral variation are provided by the deposits of the Laacher See area in the Eifel District of Germany (4), and of the Mukaiyama volcano, Nii-jima Island, Japan (5,7).

The dominant bed forms in base surge deposits occurring around the Laacher See are chute and pool structures near source, replaced by antidune structures downstream, and then plane parallel layers further downstream. Although the lateral changes are not clear-cut this represents the general trend. A decrease in both amplitude and wavelength of the bed forms away from source is also reported (4).

3. MUKAIYAMA BASE SURGE DEPOSITS

Mukaiyama is a rhyolitic volcano located at the southern part of Nii-jima (Figure 1), formed by a single continuous eruption which began in shallow water in A.D. 886. The eruption changed from submarine in the early stage to subaerial in the later stage. The volcanic history is divided into three stages on the basis of the different modes of eruption. The sequence of activity from explosive eruptions in the first stage to

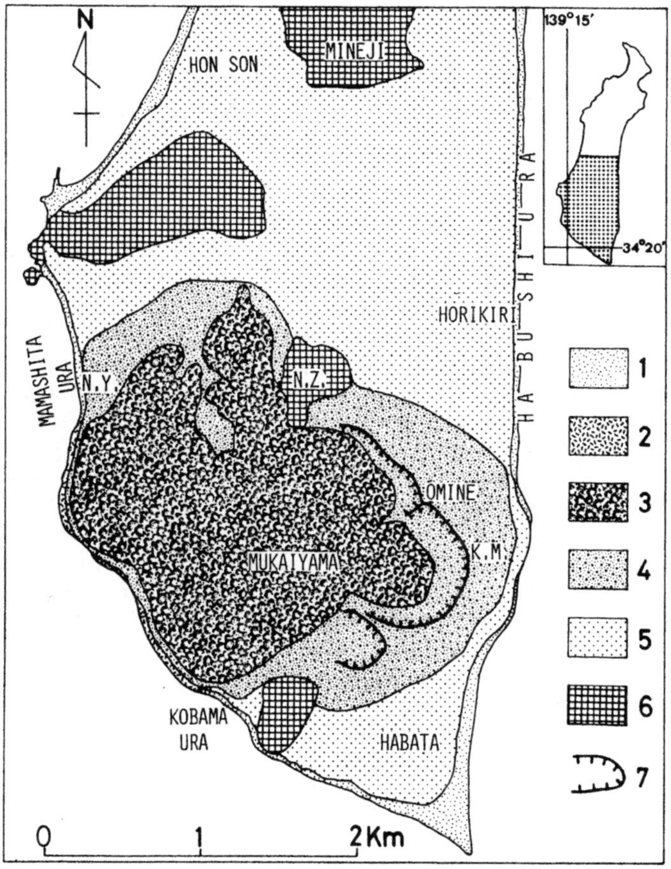

Fig. 1. Geologic map of the southern half of Nii-jima. 1. Beach sand, 2. Talus deposits, 3. Dome lavas (3rd stage), 4. Pyroclastic fall deposits of Omine cone (2nd stage), 5. Mukaiyama base surge deposits (1st stage), 6. Pre-Mukaiyama volcanics (lavas and pyroclastics), 7. Crater.

effusive eruptions in the third stage can be attributed to the change in the degree of contact of rising magma with sea water as the volcanic edifice grew.

In the first stage, a low-lying (less than 100 m above sea level) and flat-topped tuff-ring was formed by repeated base surge eruptions. In the second stage, the eruptions remained explosive. However, no base surges occurred probably because the tuff-ring itself hindered the contact of magma with sea water. The second stage activity resulted in the formation of the Omine pyroclastic cone, with a height of about 200 m and basal diameter of about 2 km, on top of the tuff-ring. In the third stage, the mature volcanic edifice possibly served as an effective interceptor against invasion of sea water into the vent. Thus, the eruption was non explosive and viscous lava effused, forming Mukaiyama lava domes.

Base surges occurred repeatedly in the first stage eruptions. The Mukaiyama base surge deposits are divided into two major groups; the lower coarse grained Mukaiyama-1, and the upper fine grained Mukaiyama-2.

The Mukaiyama-1L, which constitute the lowermost part of the Mukaiyama-1, is characterized by a pile of plane beds. In the Mukaiyama-1U, which constitutes the remaining upper part of the Mukaiyama-1, the lateral transition of sedimentary structures can be seen. This is best observed along the continuous sea cliff in the eastern coast. Here, plane beds, wavy (undulating) beds, large-sized "antidune"structures up to several tens of meters in wave length and middle-sized "antidune" structures appear one after another from south to north, i.e. with increasing distance from source, presumably reflecting the gradual decline of the energy of base surges.

The Mukaiyama-2 is subdivided into a lower part (Mukaiyama-2L) and an upper part (Mukaiyama-2U) by a pyroclastic fall layer. Both are characterized by the superposition of small-sized "antidune" and other cross-stratified structures. However, the Mukaiyama-2L has slightly bigger "antidune" structures and a coarser grain size.

The upward change of type and size of sedimentary structures within the Mukaiyama base surge deposits (Figure 2) may indicate a progressive decline of surge power, suggesting a gradual decrease of contact of magma with sea water, accompanying the growth of the volcano from submarine to subaerial.

The size of large "antidune" structures in Mukaiyama-1U far exceeds any others reported with wave lengths up to 50 m. Two possible explanations for this may be given by 1) the

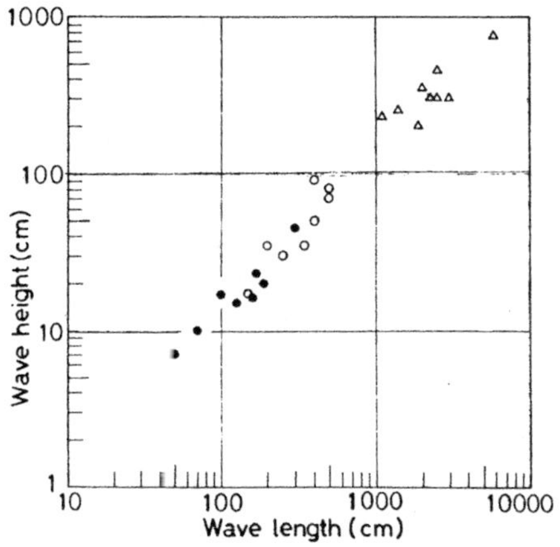

Fig. 2. Relation between wave length and wave height of "antidune" structures in Mukaiyama base surge deposits. Triangle: Mukaiyama-1U base surge deposits, open circle: Mukaiyama-2L base surge deposits, solid circle: Mukaiyama-2U base surge deposits.

existence of cliff exposures which are large enough to expose big sedimentary structures and are nearly parallel to the direction of travel of the base surges. 2) The rhyolitic magma of Mukaiyama volcano may have produced more powerful base surges (and consequently larger bed forms) than most of the hitherto-reported base surge deposits, which are largely of basaltic composition.

In contrast to the Laacher See (4), plane beds prevail near source in Mukaiyama-1U base surge deposits. Downstream, wavy (undulating) beds, large antidune structures and small antidune structures appear one after another (5).

CONCLUSIONS

Previous studies indicate that antidunes in base surge deposits tend to be smaller away from source and this may be interpreted as representing a decrease in velocity (8,1,4,5). The scale (wave length) of antidune structures hitherto reported, which are mainly from basaltic magma, is less than about 20 m. Japanese examples of antidune structures in basaltic base surge deposits are in general smaller (less than several meters in wave length) in comparison with those in dacitic base surge deposits, although the latter contain both large (up to 10-20 m) and small antidune structures. Far exceeding these, very large antidune structures up to several tens of meters in wave length are to be seen in the Mukaiyama-1U base surge deposits indicating exceptionally powerful, high-velocity surges.

One possible explanation for this is the rhyolitic magma of Mukaiyama volcano. Under subaerial conditions, eruptions generally tend to be more explosive with the increase of magma viscosity. If this relation holds in phreatomagmatic eruptions, more acidic magma will generate more powerful base surges, and hence larger bed forms. This seems to be substantiated by the author's field observations in Japan and perhaps by recent witnessed base surges of Taal volcano, Philippines, and Myojinsho volcano, Japan. The estimated maximum velocity of base surges from the basaltic composition magma of Taal volcano is 20-30 m/sec (8). In contrast, the eruption of dacite from Myojinsho produced base surges with an initial velocity exceeding 65 m/sec (1).

REFERENCES

1. Moore, J.G.: 1967, Bull. Volcanol. 30, pp. 337-363.
2. Crowe, B.M. and Fisher, R.V.: 1973, Geol. Soc. America Bull. 84, pp. 663-682.
3. Fisher, R.V. and Waters, A.C.: 1970, Am. Jour. Sci. 268, pp. 157-180.
4. Schmincke, H.-U., Fisher, R.V. and Waters, A.C.: 1973, Sedimentology 20, pp. 553-574.
5. Yokoyama, S. and Tokunaga, T.: 1978, Bull. Volcanol. Soc. Japan 23, pp. 249-262 (in Japanese with English abstract).
6. Wohletz, K.H. and Sheridan, M.F.: 1979, A Model of pyroclastic surge, Geol. Soc. Amer. Spec. Paper 180, pp. 177-194.
7. Tokunaga, T. and Yokoyama, S.: 1979, Geog. Rev. Japan 52, pp. 111-125 (in Japanese with English abstract).
8. Nakamura, K.: 1966, Jour. Geog. 75, pp. 93-104 (in Japanese with English abstract).

TOWARDS A FACIES MODEL FOR IGNIMBRITE-FORMING ERUPTIONS

J.V. Wright[1], S. Self[2], and R.V. Fisher[3]

[1]Department of Geology, University of Puerto Rico, Mayaguez, Puerto Rico 00708

[2]Department of Geology, Arizona State University, Tempe, Arizona 85281

[3]Department of Geological Sciences, University of California, Santa Barbara, California 93106

ABSTRACT

The facies model approach is applied to the products of large explosive salic eruptions using the Lower and Upper Bandelier tuffs as case examples. Both tuffs show a common eruption sequence well established from studies of small-volume ignimbrite eruptions: 1) plinian fall, 2) pyroclastic surge, 3) pumice flow, forming an ignimbrite. A co-ignimbrite lag-fall deposit and "transitional flow units" are interbedded with ignimbrite flow units of the Upper Bandelier tuff near source. Fine ash dispersed as far as Texas from the Lower emption may be a co-ignimbrite ash-fall. Ground surge and ash-cloud surge deposits also occur interbedded with ignimbrite flow units. Pumice dunes are found directly below the base of the lowermost flow unit of the Lower Bandelier ignimbrite.

INTRODUCTION

Sedimentologists have in recent years had considerable success in understanding processes by constructing facies models for lithological associations deposited in particular sedimentary environments. This paper applied this approach to the pyroclastic products of ignimbrite-forming eruptions. In this

S. Self and R. S. J. Sparks (eds.), Tephra Studies, 433–439.

case it is easy to point an analogy with sedimentary rocks by considering models of submarine fans and turbidite facies associations (1,2). In volcanology this approach has only been applied initially to volcaniclastic rocks (3) although in some detail to pyroclastic surges (4). Sheridan (5) recognized a need for this type of study to be carried out on pyroclastic flow deposits.

For our purpose a facies can be considered as an eruptive unit or part of, having distinct spatial and geometrical relations and internal characteristics (e.g. grain size and depositional structures), and a facies model as a generalized summary of the organization of the deposits in space and time.

BANDELIER TUFFS

The Lower and Upper Bandelier tuffs are the products of two voluminous, rhyolitic, ignimbrite-forming eruptions (6), dated 1.4 and 1.1 m.y. (7). Associated with the eruptions was the formation of a large caldera complex, the Valles caldera and including the Toledo caldera (Fig. 1) (6). Seminal works on many of the charactertistics of ignimbrite (ash-flow tuff) volcanism have originated by studies of these deposits and the caldera complex (8,9,10,6,11). The Bandelier tuffs have therefore become models for this type of volcanism. We now extend this earlier work to discuss the eruption sequence or facies associations of the pyroclastic deposits. Note our terminology is purely volcanological and we avoid the stratigraphic nomenclature of the Bandelier tuffs (12).

ERUPTION SEQUENCE

Both tuffs show a common eruption sequence well established from studies of smaller volume ignimbrite eruptions (13,5):

1) plinian pumice fall
2) pyroclastic surge
3) pumice flow, forming an ignimbrite

Plinian fall deposits occur at the base of each Bandelier tuff. Preliminary mapping of the Lower Bandelier plinian deposit suggests that it is one of the most voluminous air-fall deposits yet recognized (> 20km^3).

The two ignimbrites can be divided into a number of flow units. Lithic breccias are interbedded with flow units of Upper Bandelier tuff in the caldera wall (Fig. 1). These are a co-ignimbrite lag-fall deposit and "transitional flow units". A

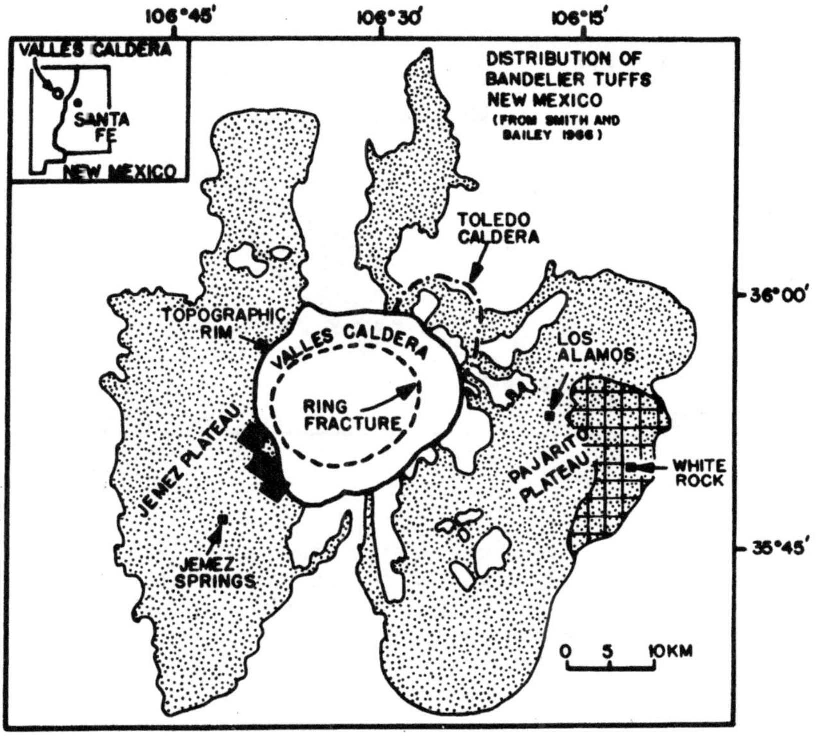

Figure 1. Location map of the Bandelier tuffs. Solid diamonds show the area where the co-ignimbrite lag-fall deposit and transitional flow units have been found exposed; squared rule is the area from which ash-cloud surges have been described (16).

co-ignimbrite lag-fall deposit indicates the ignimbrite formed by collapse of an explosive eruption column (14). Transitional flow units are transitional between lag-fall deposit and ignimbrite flow units (Fig. 2). They are the near vent deposits of pumice flows draining away finer-grained and less dense material from the lag-fall but still capable of carrying some of the largest lithics (15). Towards vent such units will pass laterally into a co-ignimbrite lag-fall deposit, whereas farther away into a true ignimbrite flow unit (Fig. 3). It should be possible with more data to further subdivide true ignimbrite flow units into proximal and distal facies (5). By analogy with turbidite facies associations, transitional flow units can be regarded as a proximal-exotic facies (1).

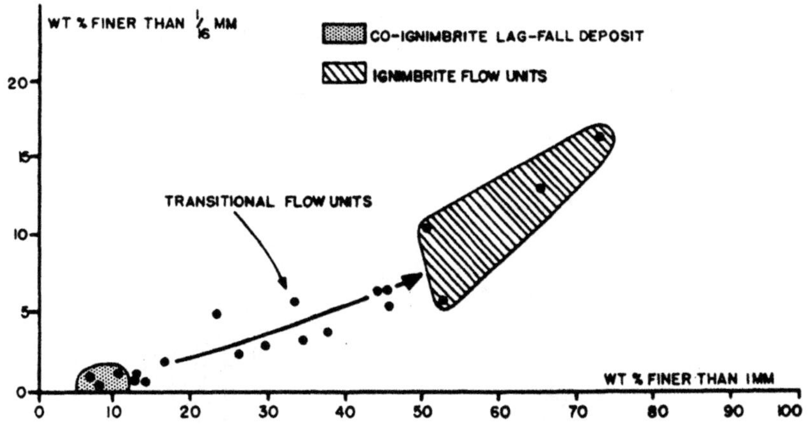

Figure 2. Plot of weight percentage finer than 1/16mm against weight percentage finer than 1mm for the near vent facies and flow units of the Upper Bandelier ignimbrite. Fractionation of lithics from the eruption column and out of the pumice flows is indicated by the trend.

Pyroclastic ground surge deposits occur between each plinian fall deposit and ignimbrite and are the first evidence of column collapse during the eruptions (16). Also directly below the base of the lowermost flow unit of the Lower Bandelier ignimbrite large pumice clasts occur in dune bedforms (height: 0.5-1.5m; wavelength: 5-10m) and discontinous lenses (pumice swarms (17); these pumice swarms have also been found at the top of thin flow units at the base of the Upper Bandelier ignimbrite. The dynamic significance of these features is at present not understood, even whether the dunes belong to the surge or ignimbrite. They may represent fine depleted ignimbrite flow units (18) or be flow foot/head deposits (19,20). We have observed similar pumice dunes in the Bishop Tuff below the lowest flow unit of the ignimbrite.

Ground and ash cloud surge deposits also occur interbedded with ignimbrite flow units. Ash cloud surge deposits of the Upper Bandelier ignimbrite have been described (16).

Fine grained ash from the eruption of the Lower Bandelier tuff is known to be dispersed as far as Texas (21). This may be a co-ignimbrite ash-fall deposit (2).

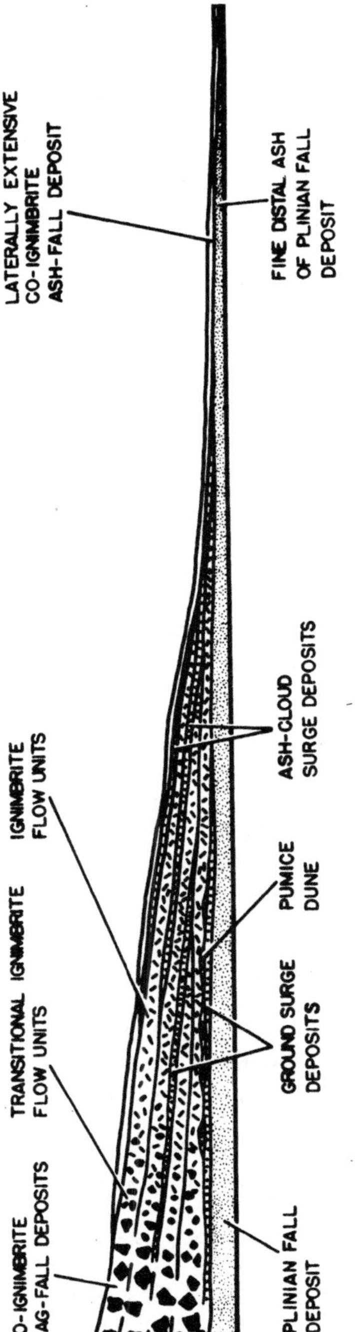

Figure 3. Facies associations of the Bandelier tuffs and model for the products of an ignimbrite-forming eruption.

FACIES MODEL

Our observations of the eruptive sequence or facies associations of the Bandelier tuffs are presented as a model for the products of ignimbrite-forming eruptions in Fig. 3. We specify in our title "Towards a facies model", because only when this data is compared and contrasted with similar studies of other ignimbrites can our model assume the generality to be termed a facies model. Any facies model must fulfill the following requirements (23):

1) be a norm, for purposes of comparison
2) be a framework and guide for future observations
3) be a predictor in new geological situations; they may be important for example in mineral exploration
4) be a basis for interpretation of the dynamics and energetics of the system it represents.

We consider that the model derived from the Bandelier tuffs does hold well for small to medium volume ignimbrites and will fulfill the above four requirements, although an exception here appears to be the Taupo ignimbrite (24,18,19). Our knowledge of the large volume ignimbrites (>1000 km) is limited and thus prevents testing the model on these at the moment.

ACKNOWLEDGEMENTS

J.V.W. thanks the Lindemann Trust Fellowship for funding, S.S. acknowledges support by NASA Grant NSG 5145, and R.V.F. thanks Associated Western Universities, Inc. for partial field support.

REFERENCES

1. Walker, R.G. and Mutti, E.: 1973, Soc. Econ. Palaeont. Miner. Short Course 1, Anaheim, pp. 119-157.
2. Mutti, E. and Ricci Lucchi, F.: 1978, Int. Geol. Rev. 20, pp. 125-166.
3. Lajoie, J.: 1979, GeoSci. Canada, Rep. Ser. 1, pp. 191-200.
4. Wohletz, K.H. and Sheridan, M.F.: 1979, Geol. Soc. America Spec. Pap. 180, pp. 177-194.
5. Sheridan, M.F.: 1979, Geol. Soc. America Spec. Pap. 180, pp. 125-136.
6. Smith, R.L. and Bailey, R.A.: 1966, Bull. Volcanol. 29, pp. 83-104.
7. Doell, R.R., Dalrymple, G.B., Smith, R.L. and Bailey, R.A.: 1968, Geol. Soc. America Mem. 116, pp. 211-248.
8. Smith, R.L.: 1960, Geol. Soc. America Bull. 71, pp. 795-842.

9. Smith, R.L.: 1960, U.S. Geol. Surv. Prof. Pap. 354-F, pp. 149-159.
10. Ross, C.S. and Smith, R.L.: 1961, U.S. Geol. Surv. Prof. Pap. 366, p. 81.
11. Smith, R.L. and Bailey, R.A.: 1968, Geol. Soc. America Mem. 116, pp. 153-210.
12. Bailey, R.A., Smith, R.L. and Ross, C.S.: 1969, U.S. Geol. Surv. Bull. 1274-P, pp. 1-19.
13. Sparks, R.S.J., Self, S. and Walker, G.P.L.: 1973, Geology 1, pp. 115-118.
14. Wright, J.V. and Walker, G.P.L.: 1977, Geology 5, pp. 729-732.
15. Wright, J.V.: 1979, Ph.D. Thesis, Imperial Coll., Univ. London.
16. Fisher, R.V.: 1979, J. Volcanol. Geotherm. Res. 6, pp. 305-318.
17. Crowe, B.M., Linn, G.W., Heiken, G. and Bevier, M.L.: 1978, Los Alamos Sci. Lab., N.W., Informal Rep. LA-7225-MS, pp. 1-57.
18. Walker, G.P.L., Wilson, C.J.N. and Froggatt, P.C.: 1980, Geology 8, pp. 245-249.
19. Walker, G.P.L., Self, S. and Froggatt, P.C.: 1980, J. Volcanol. and Geotherm. Res., in press.
20. Wilson, C.J.N.: 1980, J. Volcanol. Geotherm. Res., in press.
21. Izett, G.A., Wilcox, R.E. and Borchardt, G.A.: 1972, Quat. Res. 2, pp. 554-578.
22. Sparks, R.S.J. and Walker, G.P.L.: 1977, J. Volcanol. Geotherm. Res. 2, pp. 329-341.
23. Walker, R.G.: 1979, GeoSci. Canada, Rep. Ser. 1, pp. 1-7.
24. Walker, G.P.L., Heming, R.F. and Wilson, C.J.N.: 1980, Nature 283, pp. 286-287.

VIOLENCE IN PYROCLASTIC FLOW ERUPTIONS

C.J.N. Wilson[1] and G.P.L. Walker[2]

[1]Geology Department, Imperial College
London SW7 2BP, U.K.

[2]Geology Department, University of Auckland
Private Bag, Auckland, New Zealand

ABSTRACT. Three parameters, magnitude, intensity, and violence, can be used to characterise pyroclastic flow eruptions. Violence reflects the vigour with which a pyroclastic flow is emplaced. It is described quantitatively by the height of hills climbed by the flow (yielding flow-velocity estimates), the overall morphology of the deposit and by the proportion of the flow which forms an ignimbrite veneer deposit. Using the Taupo ignimbrite (New Zealand) as an example of an extremely violent flow, we relate violence to the eruption intensity. We consider that eruption magnitude and intensity control the gross distribution of a pyroclastic flow deposit.

1. INTRODUCTION

Several parameters are needed to describe explosive eruptions (1), including the magnitude (total erupted volume), intensity (discharge rate) and an estimate of violence. The last parameter, as applied to pyroclastic flow eruptions, is considered here. We define "violence" as the vigour with which a given volume of pyroclastic flow material is emplaced, that is, the momentum or translational kinetic energy possessed by the flow, which reflects the flow velocity. We consider here the effects which may be attributed to violence, the means of describing them and their interpretation, with especial reference to the c. 1820 yr BP (2) Taupo ignimbrite in New Zealand.

S. Self and R. S. J. Sparks (eds.), Tephra Studies, 441–448.

We use the term "pyroclastic flow" to denote a hot particulate flow composed mainly of juvenile volcanic material and having a high solids-concentration (that is, a low gas: particle ratio). Although studies on pyroclastic flow deposits have concentrated on ignimbrites (that is, flow deposits in which the juvenile component is pumiceous), the principles described here should be applicable to all pyroclastic flow deposits and also, with caution, to the deposits of rock avalanches and debris flows.

2. EVIDENCE FOR VIOLENCE

Violence, reflecting a high kinetic energy and hence a high velocity, may be assessed either quantitatively or qualitatively from the study of the resulting deposits.

Quantitatively, a high flow velocity is expressed by three aspects. The first is the height climbed by the flow up topographic obstacles. Using the simple potential energy: kinetic energy relationship

$$gh = \tfrac{1}{2}v^2$$

where h is the height climbed, v is the velocity and g the gravitational acceleration, heights climbed by the flow can be converted to minimum-velocity estimates. One refinement is to allow for the frictional losses (3), but here we merely wish to consider the minimum flow velocity. Height-climbed data from the Taupo ignimbrite, Campanian ignimbrite, Italy (4), two Alaskan ignimbrites (5) and the Ito ignimbrite, Japan (6), yield estimated velocities between 70 and 200 metres per second.

These velocities may be compared with those of rock avalanches, where no eruption-column collapse was available to give the flows additional energy (7,8). The average velocity of the 1970 Huascaran avalanche (Peru) was estimated as 75 to 95 metres per second (9) and a velocity of at least 100 metres per second estimated from height-climbed data for the Saidmarreh avalanche in Iran (10). Thus, a high flow velocity is by no means unique to pyroclastic flows.

The second measure of a high velocity is the overall morphology of the flow deposit, which can be expressed in terms of its aspect ratio (defined as the ratio of the average thickness of the deposit to the diameter of the circle which has the same area as the deposit). We have distinguished (11) between ignimbrites having a high aspect ratio (less than 1:1000) such as the Valley of Ten Thousand Smokes ignimbrite (12,13) and the Bishop Tuff (14) (1:400 and 1:250 respectively), and those having a low aspect ratio such as the Taupo and Rabaul (New Britain) ignimbrites (1:70,000 and 1:7000 respectively). The lower the aspect ratio, the higher the inferred violence of emplacement of the deposit.

The third measure of a high velocity is the proportion of the flow which forms an ignimbrite veneer deposit (IVD). The IVD (15,16) is a layer that mantles the landscape and passes laterally into, and connects outcrops of, the "conventional" valley-ponded ignimbrite. This IVD is interpreted (16) as representing the basal and trailing parts of the flow, left behind as the flow travelled across the landscape. The areal and volumetric percentage of the deposit made up by the IVD is thought to increase with the flow velocity. In the Taupo ignimbrite the IVD constitutes probably about 80% of the area and 50% of the volume, and these proportions are higher in the Rabaul ignimbrite (15). With small-volume, high-velocity flows it is possible that the entire flow deposit could consist of IVD, whereas a flow emplaced at a low velocity may well produce no IVD at all (thus apparently no substantial IVD is associated with the Valley of Ten Thousand Smokes ignimbrite).

Qualitatively, a high flow velocity is expressed by five features. The first is the erosion of the surface over which the flow passes. There are many places where the Taupo flow has demonstrably eroded more than one metre of its floor, and in some examples like the 1783 Kambara flow in Japan (17) and the Taupo ignimbrite the amount of erosion at the base of the flow sometimes exceeds the thickness of the flow deposit. Closely related to this is the extent to which lithic fragments contained in the flow have been derived from the underlying floor.

The second feature is the occurrence of certain bedforms within an ignimbrite, which appear to result when the pyroclastic flow passed over an irregular topography so fast that the flow locally left the ground. The distinctive lee-side pumice lenses (16) are interpreted to have formed in the resulting vortices.

The third feature is the occurrence of some ignimbrite facies which can be deduced to have formed indirectly as a result of a high flow velocity. These facies include fines-depleted ignimbrite (18) and the IVD (15,16).

The fourth feature is the evidence that the flow direction was often transverse to the valley systems, showing that momentum was more important than the existing topography in determining the flow direction. In the Taupo ignimbrite the flow direction is indicated by the predominantly vent-radial orientation of carbonised logs and smaller fragments of vegetation contained in it (19). There are many places where the flow is thus deduced to have flowed transversely across quite steep-sided valleys and ridges having more than 500 metres of relief.

The fifth feature is the uniformity of distribution of the flow deposit in all directions about the vent irrespective of the pre-existing topography. High velocity flows are less controlled by topography and thus are able to spread more evenly about the vent. The Taupo ignimbrite is spread over a near-circular area having a radius of 80 ± 10 kilometres and extends no further in directions where the ground is flat than in directions where it is mountainous. In sharp contrast, the Valley of Ten Thousand Smokes ignimbrite shows a very irregularly shaped distribution, being confined almost entirely to the lowest valleys around the vent (see map in reference 12).

It might be thought that the ability of the flow to carry large lithic fragments would be another valid indicator of violence. Three factors, however, complicate matters:

First, the ability of a flow to transport dense clasts depends on its fluidization state, which may not be correlated with the violence. Poor fluidized flows, because of their high yield strength (20), may be capable of transporting very large blocks, whereas strongly fluidized flows have a low yield strength and large dense blocks can sediment out rapidly from them.

Second, an ignimbrite may comprise several facies (21), each of which may have a different lithic size at any given distance from source.

Third, the erosive capability of a flow may be such that lithics are derived locally from the floor (e.g. from stream gravels) and the lithic size may represent more the availability of loose surface debris than the carrying capacity (i.e. violence) of the flow.

For these reasons, we consider that whilst measurements of the largest lithic fragments in a deposit are useful to indicate the source vent, they do not provide a valid basis for the comparison of degrees of violence during the emplacement of pyroclastic flows.

3. CAUSES OF VIOLENCE

It is recognised (22,23) that in many rock avalanches and debris flows the distance travelled by the flows is related to their volumes; the greater the volume, the further the distance travelled for a given height dropped. In many pyroclastic flows the same appears to be true (24), though the uncertain factor of the eruption-column collapse height in ignimbrites (7,8), makes interpretation of the data difficult (25-27). We consider violence, as defined herein, to be a separate parameter of a pyroclastic flow eruption. Thus whilst the volume (or mass) of material will exert a strong control on the distance travelled from vent by a flow, the violence of the flow will determine how it distributes itself across the landscape and, to a lesser extent, how far it will go.

We consider that the violence of a flow is dependent on the discharge rate (intensity). A high discharge rate is important in promoting a high kinetic energy for two reasons:

First, it makes a larger mass of material available within a given time period.

Second, it increases the height of the gas-thrust part of the eruption column (7,8): the greater the height from which column-collapse occurs, the higher will be the initial velocity of the flow.

We can estimate the discharge rate during the Taupo ignimbrite eruption as follows. The ignimbrite consists of a single flow-unit which is divided (15,16) into the IVD and valley-ponded ignimbrite facies. The IVD commonly shows a kind of layering, which we consider to be due to the passage of a series of waves or pulses of material moving out from vent at high speed. The number of layers decreases away from source until, beyond about 40 kilometres, only one is seen. As the front of the flow propagates outward from vent, we envisage these waves of material to leave the vent area in rapid succession, move outwards and catch up and coalesce with the flow front to produce, by the time the flow front has reached about 40 kilometres from vent, a single wave of material. This wave then continues outwards until all its material is used up. The implication of this model is that the entire volume of the ignimbrite is erupted in the time taken for the flow front to reach 40 kilometres from vent. Knowing from height-climbed data that parts of the flow were travelling at velocities well in excess of 150 metres per second from 20 to at least 50 kilometres from vent, we make the conservative assumption that the flow-front itself was moving out from vent at an average of 100 metres per second. Thus the time taken for the flow-front to reach 40 kilometres from vent was about 400 seconds. The ignimbrite has a total estimated volume of about 30 cubic kilometres (about 12 cubic kilometres dense-rock equivalent (DRE)), giving an average eruption rate of about 3×10^7 cubic metres per second DRE, a rate about 30 times greater than that estimated for the preceding Taupo ultraplinian phase (1), or for the Valley of Ten Thousand Smokes ignimbrite (28).

We consider that for the Taupo ignimbrite, its great violence was primarily due to this extremely high discharge rate. Thus whilst violence as defined and described in this paper can be separately documented as a parameter of a pyroclastic flow eruption, we consider that when an eruption takes place on more or less level ground it primarily reflects the eruption intensity. For pyroclastic flows which originate on a high and steep cone, a high velocity may be generated by the topography, but where (as at Taupo) the vent site is in a topographic depression, the velocity of the flow has to be generated entirely by column collapse. Theoretical considerations (8) show that high initial flow velocities are feasible with column collapse.

Thus we conclude that two main eruption parameters control the gross morphology and structure of pyroclastic flow deposits:

First, the eruption magnitude, which controls the distance travelled by the flow for a given degree of violence.

Second, the eruption intensity, which controls the way in which a given volume of material will be spread across the landscape and also, to a lesser extent, the distance travelled by the flow.

These two factors are by no means simply related. Some deposits have a large volume, but were probably erupted at a low intensity; for example, the Fish Creek Mountains Tuff (29) and the Rio Caliente ignimbrite, Mexico (21), form multiple flow-unit deposits having a high aspect ratio. Others, such as the Taupo ignimbrite, have only a modest volume but were erupted at a high intensity to form single flow-unit deposits having a low aspect ratio. In a high intensity eruption individual flow-units are larger in volume and hence will travel further (22-24) than the smaller flow-units generated by a low intensity eruption (21).

4. CONTRASTING STYLES OF PYROCLASTIC FLOW BEHAVIOUR

A spectrum of ignimbrite types can be recognised according to the violence of their emplacement, the end-members of which reflect two fundamentally different styles of pyroclastic flow behaviour.

At one end of the spectrum are relatively low-velocity flows, the emplacement of which is determined more by their mass than by their velocity. Expressed simply, they stop because they run out of velocity. Such flows will have a clearly definable termination which, unless modified by fluidization processes (20), will form a steep flow-front to the deposit. One example of this type appears to be the Valley of Ten Thousand Smokes ignimbrite, which has an extremely asymmetric distribution about the vent, flowed in a way that was strongly controlled by the topography (12) and towards its distal end had insufficient velocity to climb a low moraine or to knock over trees (30).

At the other end of the spectrum are those high-velocity flows, the emplacement of which is determined more by their velocity than by their mass. Expressed simply, they end where they do because they run out of material. The Taupo ignimbrite is an excellent example of this type. It has travelled radially outwards from vent for a similar distance in all directions, more or less regardless of the topography, whilst flow-direction indicators (19) show that it has often flowed transversely to valleys and has not been channelled by them. Height-climbed data indicate that in many areas just short of its distal limit the flow was still moving at more than 80 metres per second. The ignimbrite gradually thins to zero at this distal limit and lacks a definable flow-front. A probable non-volcanic example is the Huascaran rock avalanche (9), which passed without stopping into a muddy-water flood, leaving behind a deposit analogous to the IVD.

We note that these two end-member styles of behaviour correspond with the end members of two other groupings of pyroclastic flow deposits. The low-velocity and high-velocity end-members described above coincide with the end-member types of flow as defined by the aspect-ratio concept (11); the low-velocity flows have high aspect-ratios, whereas the high-velocity flows have low aspect-ratios. Also, the low-velocity, high aspect-ratio deposits tend to be made up of large numbers of flow units, forming compound ignimbrites (21), whereas the high-velocity, low aspect-ratio deposits tend to be composed of a single flow unit. These differences are all caused by differences in the eruption intensity, with its controlling influence on the violence of the flows.

ACKNOWLEDGEMENTS

We thank the U.K. Natural Environment Research Council for a studentship (C.J.N.W.) and the Royal Society of New Zealand for a Captain James Cook Research Fellowship (G.P.L.W.). Drs. S. Self and R.S.J. Sparks offered helpful criticisms of earlier versions of this paper.

REFERENCES

1. Walker, G.P.L.: 1980, J. Volcan. Geotherm. Res. 8, pp. 69-94.
2. Healy, J., Vucetich, C.G. and Pullar, W.A.: 1964, N.Z. Geol. Surv. Bull., n.s. 73.

3. Francis, P.W. and Baker, M.C.W.: 1977, Nature 270, pp. 164-165.
4. Barberi, F., Innocenti, F., Lirer, L., Munno, R., Pescatore, T. and Santacroce, R.: 1978, Bull. Volcan. 41, pp. 10-31.
5. Miller, T.P. and Smith, R.L.: 1977, Geology 5, pp. 173-176.
6. Yokoyama, S.: 1974, Sci. Rep. Tokyo Kyoiku Daigaku, Sec. C, 12, pp. 17-62.
7. Sparks, R.S.J. and Wilson, L.: 1976, J. Geol. Soc. Lond. 132, pp. 441-451.
8. Sparks, R.S.J., Wilson, L. and Hulme, G.: 1978, J. Geophys. Res. 83, pp. 1727-1739.
9. Plafker, G., Ericksen, G.E. and Fernandez Concha, J.: 1971, Bull. Seism. Soc. Am. 61, pp. 543-578.
10. Watson, R.A. and Wright, H.E.: 1969, Geol. Soc. Am. Spec. Pap. 123, pp. 115-139.
11. Walker, G.P.L., Heming, R.F. and Wilson, C.J.N.: 1980, Nature 283, pp. 286-287.
12. Fenner, C.N.: 1923, Natl. Geog. Soc. Contributed Tech. Pap., Katmai Ser., No. 1.
13. Sbar, M.L. and Matumoto, T.: 1972, Bull. Volcan. 35, pp. 335-349.
14. Gilbert, C.M.: 1938, Geol. Soc. Am. Bull. 49, pp. 1829-1862.
15. Walker, G.P.L., Heming, R.F. and Wilson, C.J.N.: 1980, Nature 286, P. 912.
16. Walker, G.P.L., Wilson, C.J.N. and Froggatt, P.C., in press, J. Volcan. Geotherm. Res.
17. Aramaki, S.: 1956, Japan J. Geol. Geogr. 27, pp. 189-229.
18. Walker, G.P.L., Wilson, C.J.N. and Froggatt, P.C.: 1980, Geology 8, pp. 245-249.
19. Wilson, C.J.N. and Walker, G.P.L., in preparation.
20. Wilson, C.J.N., in press, J. Volcan. Geotherm. Res.
21. Wright, J.V.: 1979, Formation, transport and deposition of ignimbrites and welded tuffs, Ph.D. Thesis, Univ. London.
22. Scheidegger, A.E.: 1973, Rock Mech. 5, pp. 231-236.
23. Hsu, K.J.: 1975, Geol. Soc. Am. Bull. 86, pp. 129-140.
24. Ui, T.: 1973, Bull. Volcan. Soc. Japan 18, pp. 153-168.
25. Francis, P.W., Roobol, M.J., Walker, G.P.L., Cobbold, P.R. and Coward, M.: 1974, Geol. Rundsch. 63, pp. 357-388.
26. Sparks, R.S.J.: 1976, Sedimentology 23, pp. 147-188.
27. Sheridan, M.F.: 1979, Geol. Soc. Am. Spec. Pap. 180, pp. 125-136.
28. Curtis, G.H.: 1968, Geol. Soc. Am. Mem. 116, pp. 153-210.
29. McKee, E.H.: 1970, U.S. Geol. Surv. Prof. Pap., p. 681.
30. Johnston, D.A. and Hildreth, E.W.: 1980, Geol. Soc. Am., Abstr. Prog. 12, p. 113.

GLACIER PEAK VOLCANO: TEPHROCHRONOLOGY, ERUPTION HISTORY AND VOLCANIC HAZARDS

J.E. Beget

University of Washington
Seattle, Washington 98195

ABSTRACT. Glacier Peak volcano in northern Washington state has erupted many times during the last 6000 years. The ages of deposits produced during past volcanic eruptions has been determined by tephrochronology and radiocarbon dating. Eruptions of voluminous assemblages of pyroclastic-flow deposits and lahars are usually separated by dormant intervals, lasting as long as 2500 years. This behavior pattern suggests that the volcano, which probably last erupted only a few hundred years ago, may erupt again. Potential volcanic hazards should be considered in land-use planning decisions in areas down-valley from the volcano.

1. INTRODUCTION

Glacier Peak, a remote volcano in northern Washington state (Fig. 1), has erupted many times during the last several thousand years. Although there is no historic record of activity, the volcano may have erupted as recently as 200-300 years ago. Glacier Peak is currently dormant but is probably not extinct.

The probability is low that Glacier Peak will erupt again in the near future. Nonetheless, the destruction of property and hazard to human life which might result during a future eruption indicates that volcanic hazards should be considered in planning for future development near the volcano. The recent eruptions of Mount St. Helens in southern Washington state may provide a new impetus for the implementation of governmentally-mandated land-use policies which could minimize effects from eruptions at Cascade volcanoes. Future eruptions of Glacier

S. Self and R. S. J. Sparks (eds.), Tephra Studies, 449–455.

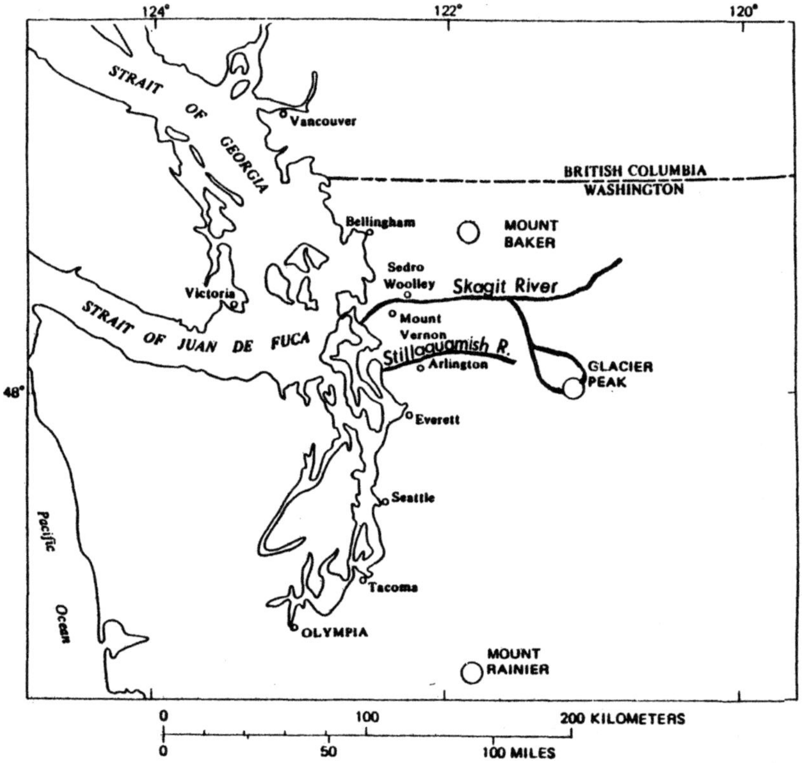

Figure 1. Glacier Peak and vicinity, Washington state. Lahars from Glacier Peak produced during eruptions travelled as far as 100 km down the Stilliguamish and Skagit River valleys. Lahars generated during future eruptions may also travel tens of kilometers down-valley.

Peak, if they were to resemble those which have occurred during the last 6000 years, would affect areas as much as 100 km downstream from the volcano.

Glacier Peak has had little geological documentation and was thought to have been extinct for the last 12,000 years (1,2). Until this study the stratigraphy, extent and volume of volcanic deposits produced during Holocene eruptions was totally unknown. This report presents an inferred chronology of eruptions, based on data generated by detailed studies of the origin and sequence of the volcano's eruptive products, together with tephrochronological studies of the deposits, and 16 radiocarbon dates on lahars and pyroclastic-flow deposits. The radiocarbon dates are on logs and soils buried by and incorporated within volcanic deposits during eruptions of Glacier Peak. Widespread tephra deposits from Glacier Peak (3), Mount Mazama (4) and Mount St. Helens (4) provide well-dated marker beds which can be used to determine the approximate age of some volcanic deposits.

This report summarizes the record of Holocene volcanic activity at Glacier Peak. The data suggest that since about 6000 years ago the volcano has erupted during as many as 9 separate episodes. The volcano has now been dormant for at least 200 years. The last major eruption which involved the production of multiple pyroclastic flows and lahars occurred about 1100 yr. B.P. It is likely that the eruption chronology presented here is incomplete. Eruptions included in this report include only those which produced deposits large enough to be preserved and recognized. Many small eruptions may have occurred for which stratigraphic evidence has not yet been recognized. For example, only a few small deposits were produced during the many historic eruptions of Mount St. Helens, Mount Baker and Mount Rainier during the 19th century (5).

2. ERUPTION CHRONOLOGY

The known eruptions of the last 6000 years at Glacier Peak can be subdivided into two groups on the basis of tephrochronology (Table 1). An older series of eruptions occurred approximately 5500-5100 yr. B.P. Deposits of this ago overlie thick tephra layers erupted by Glacier Peak in late Pleistocene time. Deposits of layer O from Mount Mazama, erupted about 6700 yr. B.P., do not occur on the surface of the mid-Holocene volcanic assemblage, indicating that it is at least less than 6700 years old. Also deposits of tephra layer Yn, erupted from Mount St. Helens about 3400 yr. B.P., do occur in many places on the surfaces of the mid-Holocene age volcanic deposits. The younger group of eruptions produced several sets of deposits which are younger than tephra layer Yn.

A major series of eruptions occurred at the summit of Glacier Peak during mid-Holocene time. Four radiocarbon dates on deposits from this period have ranged between 5500 and 5100 yr. B.P. Eruptions probably occurred intermittently during this interval. During these eruptions block and ash flows were repeatedly generated near the summit of Glacier Peak. Thick and voluminous fans of pyroclastic debris were constructed on the east and west sides of Glacier Peak. Exposures in the eastern fan indicate it was built by at least 50 separate flows. The debris fans near the volcano contain at least 9 km^3 of dacitic debris. As much as 3 km^3 of material may have been reworked in laharic assemblages deposited at this time. These lahars extend as far as 100 km downstream in the Skagit River drainage.

Table 1. Eruptions and dormant intervals at Glacier Peak in postglacial time. Only deposits produced during the last 6000 years are discussed in this report.

(The circles represent specific eruptions that have been dated or closely bracketed by radiocarbon age determinations; the vertical boxes represent dormant intervals).

YR. B.P.	
200	Dormant interval of 200-300 years
	Mudflows, minor tephra eruptions, phreatic eruptions (?)
	Mudflow, large flood
SCALE CHANGE	Apparent dormant interval of unknown duration
1000	Mudflows, pyroclastic flows, dome eruptions (?)
	Apparent dormant interval of 600-700 years
	Pyroclastic flows, mudflows, dome eruptions (?)
2000	
	Apparent dormant interval of 900-1000 years
	Mudflows, dome eruptions (?)
3000	
4000	Apparent dormant interval of 2300 years
5000	
	Repeated eruptions of pyroclastic flows and domes; mudflows
	Tephra eruptions
6000	
BREAK IN SCALE	Apparent dormant interval of 5700-6000 years
11000	Intermittent tephra eruptions of large volume, pyroclastic flows, mudflows, dome eruptions
12000	Possible dormant interval of unknown duration
	Tephra eruption of large volume (?)
	Mudflow
13000	

Pyroclastic-flow deposits extend at least 10 km from the volcano on both its east and west flanks. Ash-cloud deposits mantle upland surfaces and nonvolcanic ridges near Glacier Peak at elevations as much as 500 m above the surface of the pyroclastic-debris fans. These debris fans buried valleys to depths of at least 300 m on the east and 200 m on the west. Temporary lakes were impounded on the east and west flanks of the volcano as the pyroclastic apron produced during successive eruptions backfilled valleys below the volcano. At least 7 peat layers and buried soils have been recognized in lacustrine sections interbedded within the pyroclastic fan on the west flank of the volcano. The soils and peat layers probably record fluctuations of the lake's surface altitude and depth in response to eruptions, in a manner analogous to the recent change in the depth of Spirit Lake following the eruptions of Mount St. Helens.

There is no evidence of eruptions occurring between 5100 and 2800 yr. B.P. A weak soil is present on the mid-Holocene volcanic deposits in places where they have been buried by more recent lahars and pyroclastic-flow deposits. This supports the conclusion that a hiatus as much as several thousand years in length may have occurred after about 5100 yr. B.P.

Eruptions recommenced about 2800 yr. B.P. These eruptions also produced block and ash flows, lahars and ash-cloud deposits. The lahars and pyroclastic-flow deposits are preserved in terraces in valleys near the volcano. These terraces are remnants of formerly more extensive valley fills. At least 3 major periods of volcanic eruptions occurred. These have been dated at 2800, 1800 and 1100 years B.P. These younger eruptions each produced valleys fills which formerly contained approximately 0.5-1.5 km^3 of volcanic material. Lahars produced these eruptions extend today at least as far as 30 km from the volcano. Flooding produced during eruptions may have affected low-lying areas at even greater distances from the volcano. A laharic deposit in the Skagit River valley more than 100 km from the volcano is the same radiometric age as the 1800 year old volcanic assemblage at Glacier Peak, and was probably produced during the same eruptions.

A large flood deposit, containing boulders as much as 1 m in diameter, together with a lahar, occurs as far as 30 km down-valley from Glacier Peak. The volcanic event which produced these deposits occurred between 300 and about 1100 years ago.

The youngest lahars which may have been produced during eruptions at Glacier Peak are found in one valley on the eastern side of the volcano. These deposits consist of as many as 3 lahars which travelled at least 8 km from the volcano. These lahars cannot be much older than a young forest growing on them, in which the oldest tree is about 186 years old.

Glacier Peak has produced only minor airfall tephra deposits during the last 6000 years. A complex unit consisting of several thin tephra layers has been designated layer D. This unit is interpreted as being the product of phreato-magmatic eruptions. Layer D extends only to about 20 km east of Glacier Peak. Layer D is tentatively thought to have formed during an early stage of dome emplacement which accompanied the eruptions at Glacier Peak about 5500 years ago.

A coarse lithic tephra deposit, designated layer A, locally overlies deposits of layer Yn on the east side of Glacier Peak. This lithic tephra may have resulted from phreatic explosions near the summit of Glacier Peak. Layer A is less than 3400 but more than 200 years old.

The youngest tephra deposits, designated layer X, consist of isolated lapilli and fine ash. The deposits do not appear weathered. These tephra deposits occur on old moraines but not on the most recent moraines. Trees growing on ash-mantled moraines are as much as 316 years old. Trees growing on ash-free moraines are as much as 90 years old. It is possible that deposits of layer X were produced contemporaneously with the most recent (ca. 186 year old) lahars. The tephra deposits are tentatively interpreted as resulting from phreatomagmatic explosions which may have been associated with dome growth at the glaciated summit of the volcano. Some local Indians may have witnessed an eruption of Glacier Peak about 1750 (6), close to the time during which the most recent lahars and layer X were deposited.

3. POTENTIAL HAZARDS FROM FUTURE ERUPTIONS

During the last 6000 years voluminous pyroclastic-flow and lahar assemblages have been produced on the average of every 900-1100 years. Volcanic events have occurred about every 300-400 years during the last 1800 years at the volcano. The last large eruptions occurred 1100 yr. B.P. and small eruptions occurred as recently as 200 years ago. This record of volcanism indicates that Glacier Peak should be considered a "live" volcano which could potentially erupt in the near future. During any future eruptions the chief hazards would probably be due to pyroclastic flows near the volcano and to lahars and volcanically-generated flooding at distances of tens of kilometers from Glacier Peak. The distribution of deposits from past eruptions indicates that many thousands of people live in valleys west of Glacier Peak that might be affected by large eruptions at the volcano in the future.

4. ACKNOWLEDGMENTS

I would like to thank Dr. S.C. Porter of the University of Washington and Dr. D.R. Crandell of the U.S. Geological Survey for helpful suggestions made during this study. This work was partially supported by the Volcanic Hazards Project of the U.S. Geological Survey.

REFERENCES

1. Tabor, R.W. and Crowder, D.F.: 1969, U.S. Geol. Surv. Prof. Paper 604.
2. Harris, S.L.: 1976, in: Fire and Ice, Mountaineers-Pacific Search, Seattle.
3. Porter, S.C.: 1978, Quat. Res 10, p. 30.
4. Mullineaux, D.R.: 1974, U.S. Geol. Surv. Bull 1326.
5. Crandell, D.R. and Mullineaux, D.R.: 1975, Sci. 187, p. 438.

A TERMINOLOGY FOR PYROCLASTIC DEPOSITS

J.V. Wright[1], A.L. Smith[1] and S. Self[2]

[1]Department of Geology, University of Puerto Rico, Mayaguez, Puerto Rico 00708

[2]Department of Geology, Arizona State University, Tempe, Arizona 85281

ABSTRACT

No unique classification for pyroclastic rocks can be made and at least two systems are required; 1): a genetic classification to interpret the genesis of a deposit and 2): a lithological classification which may be solely descriptive, but also which may be used to discriminate on a lithological basis the mechanisms which produce a particular pyroclastic deposit. Genetic classification schemes are presented for various types of fall, flow and surge deposits. A lithological classification is based on the grain size limits and distribution, constituent fragments, and degree and type of welding.

PROPOSED CLASSIFICATION

In the past few years work on young pyroclastic deposits has shown that the existing terminology is of limited use for our studies. This paper summarises a terminology we have recently published elsewhere (1).

There is no unique classification system for pyroclastic rocks and at best two quite different systems are needed. One is to interpret the genesis of deposits which can then be related to a volcano's history, behavior pattern and eruptive mechanisms. The other system is lithological. Such a system is primarily descriptive, describing the major characteristics of a deposit such as grain size. However, these features are

S. Self and R. S. J. Sparks (eds.), Tephra Studies, 457–463.

themselves often diagnostic of a particular process and allow conclusions to be made on the mode of origin.

Genetic Classification

The genesis of a pyroclastic deposit is partly deduced from its lithology but also from its overall geometry and field relations. A genetic classification can only be rigorously applied to very young, well exposed Quaternary deposits.

Pyroclastic Falls. The classification scheme of Walker (2) is adopted (Fig. 1). This quantitative scheme relies on accurate mapping of the distribution of a fall deposit and detailed granulometric analysis.

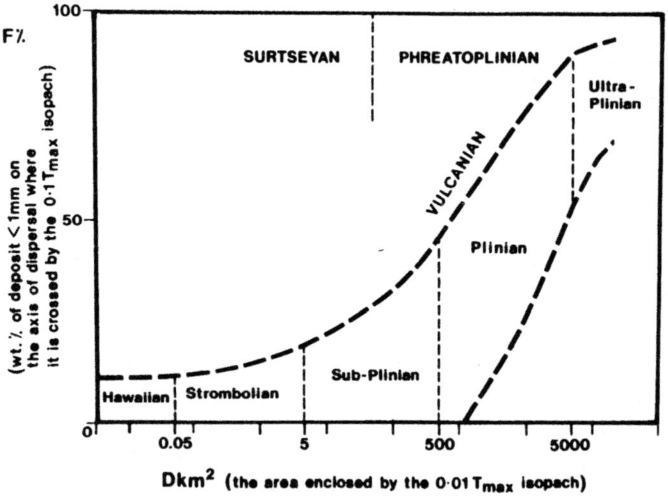

Figure 1. Classification of pyroclastic fall deposits modified from Walker (2).

Pyroclastic Flows. Our proposed classification is presented in Table 1 and a brief description of pyroclastic flow deposits is given in Table 2. With reference to Table 1 two points are worth noting: 1) For the deposits of pumice flows we do not use ash-flow tuff (3,4,5) but prefer the older term ignimbrite (6). This because of the grain size restrictions enforced by the former; it is now known that in many deposits of pumice flows ash-sized particles (> 2mm) are not dominant (7,8). 2) The exact meaning of nuée ardente has become rather ambiguous and the term has been used to include all types of pyroclastic flows. The authors feel that the term should be avoided altogether, the pyroclastic flows produced by "Pelean-type" eruptions being

TABLE 1 GENETIC CLASSIFICATION OF PYROCLASTIC FLOWS

ESSENTIAL FRAGMENT	ERUPTIVE MECHANISM	PYROCLASTIC FLOW	DEPOSIT	COMMENTS
VESICULATED	ERUPTION COLUMN COLLAPSE	PUMICE FLOW	IGNIMBRITE: PUMICE AND ASH DEPOSIT	Large volume deposits envisaged to form by continuous collapse of a plinian eruption column (21). Salic in composition Small volume deposits probably formed by interrupted column collapse as in the case of scoria flows described below. Intermediate to salic in composition.
	ERUPTION COLUMN COLLAPSE	SCORIA FLOW	SCORIA AND ASH DEPOSIT	Small volume deposits probably formed by interrupted eruption column collapse produced by short explosions (see 22). Basalt to andesite in composition.
	ERUPTION COLUMN COLLAPSE	VESICULAR ANDESITE FLOW	VESICULAR ANDESITE AND ASH DEPOSIT	Small volume deposits composed of vesiculated angular andesite clasts.
	ERUPTION COLUMN COLLAPSE; LAVA/DOME COLLAPSE — EXPLOSIVE	BLOCK AND ASH FLOW; NUEE ARDENTE	BLOCK AND ASH DEPOSIT	Small volume deposits, usually andesitic or dacitic in composition. Produced both by explosive collapse of an actively growing dome or lava flow and by the collapse of a vertical eruption column as recognised in the early eruptions (eg. May 8 and 20) of Mt.Pelée 1902 (23).
NON-VESICULATED	LAVA/DOME COLLAPSE — GRAVITATIONAL	BLOCK AND ASH FLOW; NUEE ARDENTE	BLOCK AND ASH DEPOSIT	Small volume deposits usually andesitic or dacitic in composition. These have been termed hot avalance deposits (24).

Table 2. SUMMARY DESCRIPTIONS OF TYPES OF PYROCLASTIC FLOW DEPOSIT

Deposit	Description
Ignimbrite Pumice and Ash	Unsorted ash deposits containing variable amounts of rounded salic pumice lapilli and blocks up to 1m in diameter. In flow units pumice fragments can be reversely graded while the lithic clasts can show normal grading; ungraded flow units are as common. A fine grained basal layer is found at the bottom of flow units. They sometimes contain fossil fumarole pipes and carbonized wood. The coarser smaller volume deposits usually form valley infills while the larger volume deposits may form large ignimbrite sheets. Sometimes they may show one or more zones of welding.
Scoria and Ash	Topographically controlled, unsorted ash deposits containing basalt to andesite vesicular lapilli and scoriaceous ropey surfaced clasts up to 1m in diameter. They may in some circumstances contain large non-vesicular cognate lithic clasts. Fine grained basal layers are found at the bottom of flow units. Fossil fumarole pipes and carbonized wood may also be present. The presence of levées, channels and steep flow fronts indicate a high yield strength during transport of the moving pyroclastic flow.
Vesicular Andesite and Ash	Topographically controlled unsorted ash deposits containing intermediate vesicular (between pumice and non-vesicular juvenile clasts) andesite lapilli, blocks and bombs. Fine grained basal layers, fossil fumarole pipes and carbonized wood all may be present.
Block and Ash	Topographically controlled, unsorted ash deposits containing large, generally non-vesicular, jointed, cognate lithic blocks which can exceed 5m in diameter. The deposits are generally reversely graded. Fine grained basal layers are again present. Again they may contain fossil fumarole pipes and carbonized wood. Surface manifestations include the presence of levées, steep flow fronts and the presence of large surface blocks, all of which again indicate a high yield strength during transport of the flow.

called block and ash-flows (9); or nuée ardente should be restricted to the original definition (10,11) and only be used for these pyroclastic flows produced by the collapse of an actively growing lava or dome (12,8). Both definitions are used in Tables 1 and 2.

Pyroclastic Surges. The proposed classification is presented in Table 3 and a brief description of surge deposits given in Table 4.

TABLE 3 GENETIC CLASSIFICATION OF PYROCLASTIC SURGES

ESSENTIAL FRAGMENT	ERUPTIVE MECHANISM	TYPE OF SURGE	COMMENTS
VESICULATED - NON-VESICULATED	COLLAPSE OF A PHREATOMAGMATIC ERUPTION COLUMN	BASE SURGE	Base surges result from the explosive interaction of magmatic material and water and are consequently cool.
VESICULATED - NON-VESICULATED	ACCOMPANYING PYROCLASTIC FLOWS ERUPTED BY MECHANISMS GIVEN IN TABLE 1	GROUND SURGE	Ground surge, although originally introduced (25) to encompass all pyroclastic surges, is here used to describe those surges found at the base of pyroclastic flow deposits, as well as those produced without any accompanying pyroclastic flow.
VESICULATED - (non-vesiculated)	ALSO ASSOCIATED WITH AIR-FALL DEPOSITS BY COLLAPSE OF AN ERUPTION COLUMN BUT WITHOUT GENERATION OF PYROCLASTIC FLOW	GROUND SURGE	
VESICULATED - NON-VESICULATED		ASH CLOUD SURGE	Ash cloud surges (26) are the turbulent, low density flows derived from the overriding gas-ash cloud of pyroclastic flows. These may in some cases become detached from the parent pyroclastic flow and move independently.

Table 4. SUMMARY DESCRIPTIONS OF TYPES OF PYROCLASTIC SURGE DEPOSIT

Deposit	Description
Base Surge	Stratified and laminated deposits containing juvenile vesiculated fragments ranging from pumice to non-vesiculated cognate lithic clasts, ash and crystals with occassional accessory lithics (larger ballistic ones may show bomb sags near-vent) and deposits produced in some phreatic eruptions which are composed totally of accessory lithics. Juvenile fragments are usually less than 10 cm in diameter due to the high fragmentation caused by the water/magma interaction. Deposits show unidirectional bedforms. Generally they are associated with maar volcanoes and tuff rings. When basaltic in composition they are usually altered to palagonite.
Ground Surge	Generally less than 1m thick, composed of ash, juvenile vesiculated fragments, crystals and lithics in varying proportions depending on constituents in the eruption column. Typically enriched in denser components (less well vesiculated juvenile fragments, crystals and lithics) compared to accompanying pyroclastic flow. Again they show unidirectional bedforms; carbonized wood and small fumarole pipes may be present.
Ash Cloud Surge	Stratified deposits found at the top of and as lateral equivalents flow units of pyroclastic flows. They show unidirectional bedforms, pinch and swell structures and may occur as descrete separated lenses. Grain size and proportions of components depend on the parent pyroclastic flow. Can contain small fumarole pipes.

Lithological Classification

The main basis of lithological classification are:

1) The grain size limits of the pyroclasts and the overall size distribution of the deposits.
2) The constituent fragments of the deposit.
3) The degree and type of welding.

Both 1 and 2 can be used to help discriminate the genesis of a particular pyroclastic deposit in the older geological record.

Grain Size. With regard to grain size limits of pyroclastic fragments the system of Fisher (13) is adopted (Table 5). Concerning overall grain size distribution, granulometric analysis of non-welded and unlithified pyroclastic deposits can be an important discriminant in deducing their mechanism of formation (14,15,16,17). Pyroclastic flow deposits generally show extremely poor sorting, while pyroclastic fall deposits are better sorted; pyroclastic surge deposits tend to overlap the separate fields for flow and fall deposits (18,17).

Table 5. GRAIN SIZE LIMITS FOR PYROCLASTIC FRAGMENTS AFTER FISHER (13)

Grain size (mm)	Pyroclastic Fragments	
	Coarse	Blocks and bombs
256		
	Fine	
64		
	Lapilli	
2		
	Coarse	Ash
1/16		
	Fine	

Constituent Fragments. A summary of the dominant components in a pyroclastic deposit provides a qualitative lithological description as well as providing some information as to the genesis (Table 6).

TABLE 6. SUMMARY OF THE COMPONENTS IN PYROCLASTIC DEPOSITS

A. Pyroclastic Flows and Surges

Type of Flow or Surge	Essential Components		Other Components
	Vesicular	Non-vesicular	
Pumice flow/surge	Pumice	Crystals	Accessory and accidental lithics
Scoria flow/surge	Scoria	Crystals	Cognate, accessory and accidental
Lava debris flow/surge Nuée ardente	Poor-moderate vesicular clasts	Cognate lithics and crystals	Accidental lithics

B. Pyroclastic Falls

Predominant Grain size	Type of Fall	Essential Components*		Other Components
		Vesicular	Non-vesicular	
> 64mm	Agglomerate	Pumice/Scoria		Cognate and accessory lithics
	Breccia		Cognate and/or accessory lithics	
> 2mm	Lapilli deposit	Pumice/Scoria	Cognate and/or accessory lithics	Crystals
< 2mm	Ash deposit	Pumice/Scoria	Crystals and/or cognate and/or accessory lithics	

* Depending on type of deposit

Welding. This process involves the sintering together of hot vesicular fragments and glass shards under a compactional load (3,4,5,19,20). Welding has only been generally described from ignimbrites; no descriptions in the literature are known to the authors of welding occurring in other denser-clast pyroclastic flows, however certain vesicular andesite and ash deposits from Mt. Pelée, Martinique are welded (A.L. Smith and M.J. Roobol, unpub. data). Welded air-fall tuffs have been documented (20) and welded pyroclastic surge deposits are known.

ACKNOWLEDGEMENTS

We thank funding by NERC, NSF (Grants EAR 73-00194 and 77-17064), NASA (Grant NSG 5145), American Philosophical Society, Lindemann Trust Fellowship and the Universities of Wellington and Puerto Rico.

REFERENCES

1. Wright, J.V., Smith, A.L. and Self, S.: 1980, J. Volcanol. Geotherm Res., in press.
2. Walker, G.P.L.: 1973, Geol. Rundsch. 62, pp. 431-446.
3. Smith, R.L.: 1960a, Geol. Soc. America Bull. 71, pp. 795-842.
4. Smith, R.L.: 1960b, U.S. Geol. Surv. Prof. Pap. 354-F, pp. 149-159.
5. Ross, C.S. and Smith, R.L.: 1961, U.S. Geol. Surv. Prof. Pap. 366, 81 p.
6. Marshall, P.: 1935, Trans. R. Soc. New Zealand 64, pp. 1-44.
7. Wright, J.V. and Walker, G.P.L.: 1980, J. Volcanol. Geotherm. Res., in press.
8. Smith, A.L. and Roobol, M.J.: 1980, in: Thorpe (ed.) Orogenic andesite and related rocks, John Wiley and Sons Inc., in press.
9. Perret, F.A.: 1937, Carnegie Inst. Washington Pub. 458, 126 p.
10. LaCroix, A.: 1903, Comptes Rendus 135, pp. 871-876.
11. LaCroix, A.: 1904, Masson et Cie, Paris, 662 p.
12. Rose, W.I., Jr., Pearson, T. and Bonis, S.: 1977, Bull. Volcanol. 40-1, pp. 1-16.
13. Fisher, R.V.: 1961, Geol. Soc. America Bull. 72, pp. 1409-1414.
14. Murai, I.: 1961, Bull. Earthqu. Res. Inst. 39, pp. 133-254.
15. Walker, G.P.L.: 1971, J. Geol. 79, pp. 696-714.
16. Sheridan, M.F.: 1971, J. Geophys. Res. 76, pp. 5627-5634.
17. Sparks, R.S.J.: 1976, Sedimentology 23, pp. 147-188.
18. Roobol, M.J. and Smith, A.L.: 1975, Bull. Volcanol. 39-2, pp. 1-28.
19. Ragan, D.M. and Sheridan, M.F.: 1972, Geol. Soc. America Bull. 83, pp. 95-106.
20. Sparks, R.S.J. and Wright, J.V.: 1979, Geol. Soc. America Spec. Pap. 180, pp. 155-166.
21. Sparks, R.S.J., Wilson, L. and Hulme, G.: 1978, J. Geophys. Res. 83, B4, pp. 1727-1739.
22. Nairn, I.A. and Self, S.: 1978, J. Volcanol. Geotherm. Res. 3, pp. 39-60.
23. Fisher, R.V., Smith, A.L. and Roobol, M.J.: 1980, Geology, in press.
24. Francis, P.W., Roobol, M.J., Walker, G.P.L., Cobbold, P.R., Coward, M.P.: 1974, Geol. Rundsch. 63, pp. 357-388.
25. Sparks, R.S.J. and Walker, G.P.L: 1973, Nature Phys. Sci. 241, pp. 62-64.
26. Fisher, R.V.: 1979, J. Volcanol. Geotherm. Res. 6, pp. 305-318.

SUBJECT INDEX

ADDRESS LIST

Participants at NATO ASI

(d) = directors, (l) = lecturers, (p) = participants

S.O. Agrell (p)	Department of Mineralogy and Petrology, University of Cambridge, Cambridge CB2 3EW, U.K.
J. Beget (p)	Department of Geology, University of Washington, Seattle, Washington 98195, U.S.A.
J. Benjamínsson (p)	National Energy Authority, Grensásvegur 9, 108 Reykjavík, Iceland
R. Blong (p)	School of Earth Sciences, Macquarie University, North Ryde, New South Wales 2113, Australia
H. Bogadóttir (p)	National Energy Authority, Grensásvegur 9, 108 Reykjavík, Iceland
T.J. Bornhorst (p)	Department of Geology, The University of New Mexico, Albuquerque, New Mexico 87131, U.S.A.
S. Brazier (p)	Department of Mineralogy and Petrology, University of Cambridge, Cambridge CB2 3EW, U.K.
M.D. Buck (p)	School of Earth Sciences, Macquarie University, North Ryde, New South Wales 2113, Australia
P. Buckland (p)	Department of Geography, The University of Birmingham, Birmingham B15 2TT, U.K.
A.B. Cormie (p)	Department of Archaeology, Simon Fraser University, Burnaby, B.C. V5A 1S6, Canada
T.H. Druitt (p)	Department of Mineralogy and Petrology, University of Cambridge, Cambridge CB2 3EW, U.K.
S. Fine (p)	Institut for Petrologi, Oster Voldgade 10, 1350 Köbenhavn K., Denmark

P. Francis (p)	Department of Earth Sciences, The Open University, Milton Keynes, Bucks., MK7 6AA, U.K.
M. Gorton (p)	Department of Geology, University of Toronto, Toronto M5S 1A1, Canada
G. Gudbergsson (p)	Agricultural Research Institute, Keldnaholt, V/Vesturlandsveg, 110 Reykjavík, Iceland
H. Haflidason (p)	Grant Institute of Geology, West Mains Road, Edinburgh EH9 3JW, Scotland, U.K.
W.J. Hart, Jr. (p)	Department of Geological Sciences, Rutgers College, The State University of New Jersey, New Brunswick, New Jersey 08903, U.S.A.
G. Heiken (p)	Earth Sciences Group, Los Alamos Scientific Laboratory, Los Alamos, New Mexico 87545, U.S.A.
M. Hermelin (p)	Instituto Nacional de Investigaciones, Geologico-Mineras, Carrera 30, No. 51-59, Bogota, Columbia, South America
K.A. Jörgensen (p)	Institut for Petrologi, Köbenhavns Universitet, Oster Volgade 10, DK-1350, Köbenhavn K., Denmark
I. Kaldal (p)	National Energy Authority, Grensásvegur 9, 108 Reykjavík, Iceland
J. Keller (l)	Mineralogisches Institut, Albert-Ludwigs Universität, D-7800 Freiburg 1, Albertstrasse 23b, W. Germany
P. Kyle (p)	Institute of Polar Studies, The Ohio State University, Columbus, Ohio 43210, U.S.A.
G. Larsen (p)	Nordic Volcanological Institute, University of Iceland, Reykjavík, Iceland
M.T. Ledbetter (p)	Department of Geology, University of Georgia, Athens, Georgia 30602, U.S.A.

L. Lirer (p)	Instituto di Geologia e Geofisica, Lago S. Marcellino, 10, 80138 Napoli, Italy
F.W. McCoy (p)	Lamont-Doherty Geological Observatory, Columbia University, Palisades, New York 10964, U.S.A.
H. Machida (l)	Department of Geography, Tokyo Metropolitan University, Fukazawa 2-1-1, Setagaya 158, Tokyo, Japan
W. Morche (p)	Mineralogisches Institut, Albert-Ludwigs Universitat, D-7800, Freiburg 1, Albertstrasse 23b, West Germany
H. Moriwaki (p)	Department of Geography, Tokyo Metropolitan University, Fukazawa 2-1-1, Setagaya, Tokyo, Japan
C.W. Naeser (l)	U.S. Geological Survey, Branch of Isotope Geology, Denver Federal Center, Denver, Colorado 80225, U.S.A.
A.K. Pedersen (p)	Geologisk Museum, Köbenhavns Universitet, Oster Volgade 5-7, 1350 Köbenhavn K., Denmark
S.C. Porter (l)	Department of Geological Sciences, University of Washington, Seattle, Washington 98195, U.S.A.
P.A. Riezebos (p)	Fysisch geografisch en boden kundig laboratorium, Universiteit van Amsterdam, Dapperstraat 115, Amsterdam-Oost, Holland
W.I. Rose, Jr. (l)	Department of Geology and Geological Engineering, Michigan Technological University, Michigan 49931, U.S.A.
N.W. Rutter (p)	Department of Geology, The University of Alberta, Edmonton, Canada T6G 2E3
H.U. Schmincke (l)	Institut fur Mineralogie, Ruhr-Universität Bochum, Universitatsstrasse 150, Postfach 10 21 48, D-4630 Bochum 1, West Germany
S. Self (d)	Department of Geology, Arizona State University, Tempe, Arizona 85281, U.S.A.

H. Sigurdsson (1)	Graduate School of Oceanography, University of Rhode Island, Kingston, Rhode Island 02881, U.S.A.
Ó. Smárason (p)	Department of Geology, Imperial College of Science and Technology, London SW7 2BP, U.K.
A.L. Smith (p)	Department of Geology, University of Puerto Rico, Mayaguez, Puerto Rico 00708
R.S.J. Sparks (d)	Department of Mineralogy and Petrology, University of Cambridge, Cambridge CB2 3EW, U.K.
V. Steen-McIntyre (1)	University of Colorado, Box 1167, Idaho Springs, Colorado 80452, U.S.A.
E. Thomas (p)	Department of Geology, Arizona State University, Tempe, Arizona 85281, U.S.A.
S. Thorarinsson (d)(1)	Science Institute, University of Iceland, Dunhaga 3, 107 Reykjavík, Iceland
J.C. Varekamp (p)	Department of Geology, Arizona State University, Tempe, Arizona 85281, U.S.A.
K. Vasti (p)	Nordic Volcanological Institute, University of Iceland, 101 Reykjavík, Iceland
K.L. Verosub (p)	Department of Geology, University of California, Davis, California 95616, U.S.A.
E.G. Vilmundardóttir (p)	National Energy Authority, Department of Water Power, Grensásvegur 9, 108 Reykjavík, Iceland
A. Vinci (p)	Instituto di Petrografia e Giacimenti Minerari, Universita di Parma, 43100 Parma, Via Gramsci 9, Italy
C.J. Vitaliano (p)	Department of Geology, Indiana University, Bloomington, Indiana 47405, U.S.A.
D.B. Vitaliano (p)	Department of Geology, Indiana University, Bloomington, Indiana 47405, U.S.A.

G.P.L. Walker (l)	Department of Geology, University of Auckland, Auckland, New Zealand
J. Westgate (l)	Department of Geology, University of Toronto, Toronto M5S 1A1, Canada
C.J.N. Wilson (p)	Department of Geology, Imperial College, London SW7, U.K.
L. Wilson (p)	Department of Environmental Sciences, University of Lancaster, Lancaster LA1 47Q, U.K.
C.A. Wood (p)	NASA, Code 922, Goddard Space Flight Center, Greenbelt, Maryland 20771, U.S.A.
J.V. Wright (p)	Department of Geology, University of Puerto Rico, Mayaguez, Puerto Rico 00708
R.L. Wunderman (p)	Michigan Technological University, Houghton, Michigan 49931, U.S.A.
S. Yokoyama (p)	Faculty of Education, Kumamoto University, Kumamoto 860, Japan

GPSR Compliance

The European Union's (EU) General Product Safety Regulation (GPSR) is a set of rules that requires consumer products to be safe and our obligations to ensure this.

If you have any concerns about our products, you can contact us on ProductSafety@springernature.com

In case Publisher is established outside the EU, the EU authorized representative is:

Springer Nature Customer Service Center GmbH
Europaplatz 3
69115 Heidelberg, Germany

Batch number: 10816188

Printed by Printforce, the Netherlands